Scot

SPRINGBOARD for PASSING THE GED

Mathematics Test

Scott, Foresman and Company
Lifelong Learning Division

1900 East Lake Avenue
Glenview, Illinois 60025
1–800–323–5482
1–800–323–9501 (Illinois)
1–312–729–3000 (Alaska and Hawaii)

Author

Douglas V. Buchanan

Library of Congress Cataloging-in-Publication Data
Scott, Foresman springboard for passing the GED mathematics test.
Rev. ed. of: Scott, Foresman springboard for passing the GED mathematics test/Jane E. Haasch. c1986.

Includes index.
1. Mathematics—Examinations, questions, etc.
2. General educational development tests.
I. Buchanan Douglas V. Scott, Foresman springboard for passing the GED mathematics test. II. Scott, Foresman and Company. Lifelong Learning Division. III. Title: Springboard for passing the GED mathematics test.
QA43.S386 1987 510'.76 87-9678
ISBN 0-673-24316-8

56WEB929190

Acknowledgments

Illustration Acknowledgments

Positions of illustrations are shown in abbreviated form as follows: top *(t)*, bottom *(b)*, left *(l)*, right *(r)*.

115 *(tl)* Opinion Research Corporation. **116** *(r)* L. Carey Bolster, et al., MATHEMATICS AROUND US. Scott, Foresman and Company, 1978, p. 236. **117** *(l)* U.S. Postal Service. **119** *(l)* L. Carey Bolster, et al., CONSUMER AND CAREER MATHEMATICS. Scott, Foresman and Company, 1985, p. 152. **120** *(r)* L. Carey Bolster, et al., SCOTT, FORESMAN MATHEMATICS, BOOK 8. Scott, Foresman and Company, 1980, p. 342. **121** *(l)* Federal Aviation Administration, TOWERED AIRPORT STATISTICS. 1979. 240 (*br*) Erika Hugo. 250 (*tr*) Erika Hugo.

All cartoon art by Cary Cochrane.

Contents

Algebra 299

What You Should Know About Preparing for the GED Test

What Is the GED Test?

The initials GED stand for General Educational Development. You may have heard the GED Test called the High School Equivalency Test. That is because the test measures your ability against that of graduating high school students. If you pass, you will earn a certificate that is the *equivalent* of a high school diploma. You do not have to go back to school to get it. That's quite an opportunity if you think about it!

Why Take the Test?

Every state in the Union, the District of Columbia, six United States territories and possessions, and many Canadian provinces use GED Test results as the basis for giving high school equivalency credentials. A *credential* is something that credits a person with an achievement. GED credentials give credit for achieving the same amount of learning as high school graduates. Those credentials are accepted for employment, promotion, and licensing just as high school diplomas are.

There's another good reason for taking the test. Many colleges and universities now accept satisfactory GED Test scores in place of completed high school grades to admit students.

Who Takes the Test?

In this country about 44 million people over eighteen do not have high school diplomas. This means that about 28 percent of the adult population of the United States has not graduated from high school.

In recent years, more than one-half million people each year have taken the GED Test. A lot of people must think it makes good sense to have a high school equivalency credit.

Getting Ready for the GED Test

If you are not already enrolled in a GED preparation program and would like help preparing for the test, you should make some phone calls. The office of the superintendent of schools or your local vocational school, community college, or adult education center is a good place to contact. Ask if there are GED preparation courses available nearby. A call or two should be enough to get the information.

Many people prefer to study on their own rather than in a class. If you are one of those independent spirits, you can still benefit by calling one of the places just suggested. You may ask for information on the GED Test and about fees, application forms, and test times. If, however, you can't get the information you need, write to the following address:

General Educational Development
GED Testing Service of the American Council on Education
One Dupont Circle
Washington, DC 20036

What's the GED Test Like?

Except for the essay on the GED Writing Skills Test, all the questions on the GED Test are multiple-choice, and there are five answer choices for each. You will not have to write out any answers. *You will need only to mark one of the five answer spaces for each question.*

The GED Test measures if you have mastered the skills and general knowledge that are usually acquired in a four-year high school education. However, you are not expected to remember many details, definitions, and facts. Therefore, being out of school for some time should not be a handicap at all.

The five subtests on the GED Test have been created to test whether an adult has the knowledge and skills that the average high school graduate has in these areas:

Test 1: Writing Skills
Test 2: Social Studies
Test 3: Science
Test 4: Interpreting Literature and the Arts
Test 5: Mathematics

The GED Test does not base your score on your ability to race against the clock, as some tests do. You will have plenty of time to finish each subtest. Even so, you will want to work steadily and not waste any time. It is important that you answer every question, *even if you just guess,* because your score depends on the number of answers you get right.

Now read on to find out more about the GED Mathematics Test.

The GED Mathematics Test

What's on the Test?

The GED Mathematics Test contains fifty-six questions and takes about one and a half hours to complete. The approximate number of questions from each area of math is shown here:

Arithmetic	28 questions
Geometry	11 questions
Algebra	17 questions

The test is arranged with the easiest questions first and the hardest ones last. Arithmetic, algebra, and geometry questions are all mixed together on the test.

Each problem on the GED Mathematics Test will require at least two skills. First, you will have to read and understand the problem. Second, you must decide what the problem is asking before you can do any computation. For most of the problems you will need a third skill—you must be able to compute the answer. This *Springboard for Passing the GED Mathematics Test* will give you plenty of help in all three areas.

Arithmetic To do well on the Arithmetic section of the GED Test, you first will have to review the basic arithmetic skills of adding, subtracting, multiplying, and dividing. You will have to use these skills with whole numbers, decimals, and percents. You will also apply them to measurement (including perimeter, area, volume, rate of motion, rate of interest, time, and money), number relationships (including sequencing fractions and decimals, ordering and comparing data, and working with exponents and scientific notation), and data analysis (calculating means, medians, ratios, and probabilities as well as interpreting graphs and tables).

Geometry and Algebra If you understand all the material related to arithmetic, you are well on your way to passing the GED Mathematics Test. But you must also know some basic geometry and algebra to pass the Test. On the Test you can expect to find some geometry problems that ask you to work with parallel and perpendicular lines, quadrilaterals, triangles, circles, and coordinate geometry. When it comes to algebra, you can expect several problems some of which will ask you to use formulas, set up and solve various types of equations and inequalities, factor algebraic expressions, and work with fractions in algebraic expressions.

The GED Mathematics Test will have about twelve to eighteen questions based on charts, graphs, diagrams, and illustrations. You may be able to answer some of these questions without any computation. You will have to read numbers on a chart or graph, think about what those numbers mean, and then choose an answer.

Remember that, in general, you will not have to memorize complicated formulas or do lengthy computations in order to score well on the GED Mathematics Test. Almost all the questions are word problems—practical, down-to-earth applications of the skills this book will help you master.

Math and Life: Capitalize on Your Experience

If you've been out of school for a while, you may be anxious about preparing for the GED Mathematics Test. You may think that you've forgotten how to study, or you may feel that you're just not good at math.

If that sounds like the kind of thinking you've been doing, you're forgetting one very important thing you've got on your side: all the math *experience* you have.

For example, trips to the supermarket may already have given you experience—and confidence—in working with numbers. You may compare prices of items, estimate how much money you need to pay your bill, and check the amount of change you get back. If you are working on your house or sewing, you are probably using math without even thinking about it. Salaries, weights, budgets, interest rates, mortgage payments, taxes, rents—all use some math. In reading newspapers and magazines you will come across charts and graphs. It is just this kind of practical math experience that gives you a good head start in your preparation for the GED Mathematics Test.

How to Use This Book

What Is a Springboard?

You probably already know what a real springboard can do for athletes like divers and tumblers. A springboard is used to increase the height of leaps. In the same way, this *Springboard* is your take-off point for making the big leap—passing the GED Mathematics Test. This book starts each lesson at a very basic level. Then it brings you right up to the point where you can answer questions like those on the actual GED Test.

Taking the Skills Survey

The Skills Survey can be used as a pretest to show you how much math you already know. The Skills Survey is made up of problems, including word problems, just like those on the actual GED Test.

After you take the Skills Survey, carefully read the answer explanations and check the chart to see which skills you need to work on. When you come to a lesson in the book that teaches one of those skills, you can give more time and attention to it. In that way you can plan your studies to meet your own special needs.

When you work through the Skills Survey, you may do well on certain skills or in a certain area, such as fractions. If that happens, it would still be a good idea to read quickly through those lessons and then answer all "The Real Thing" GED-level questions in them. That will strengthen the skills you already have and give you practice in answering questions like those on the GED Test.

Working the Lessons

The lessons in this book carefully guide you through all the steps you need to understand the material. Each lesson is easy to read and understand.

First, the lesson presents important ideas about a math skill. Next, in *Here's an Example,* you'll find a specific example of how to use the skill that has just been explained. A section called *Try It Yourself* helps you practice the skill by working through a word problem. Then comes a *Warm-up* exercise, which is a set of computation problems to help you practice the math operation you just studied. You can check your answers quickly and easily in the *Warm-up Answers* section nearby. Another section, called *Coming to Terms,* defines important math words used in the lesson. Although you do not have to memorize these terms, you may find it helpful to learn their meaning. You can do this by reading each definition twice, thinking about it, and then reading it again. You'll also find many *Test-Taking Tips.* These are practical suggestions about how to answer certain types of questions that often appear on the GED Mathematics Test.

Each lesson then introduces a very important step, one that helps you leap from a lower level to a higher, GED, one. This step is a special section called *On the Springboard.* It has problems and questions that will bridge the gap between performing simple math operations and working actual GED math word problems. *On the Springboard* is not only a jumping-off point, but also a step that helps you gain confidence in your ability to answer questions like those on the actual test. If you have any problems working the Springboard questions correctly, this book will tell you exactly what to review before going on. After all, you need to be in top shape before taking the final leap.

That leap is the section called *"The Real Thing."* There you will find word problems just like those on the GED Test. Because you've just completed *On the Springboard,* your chances of success with "The Real Thing" will be greatly increased.

Zeroing In on Word Problems

Each major unit in *Arithmetic* contains a special lesson on how to solve word problems. These word-problem lessons will review the material you just studied. They will give you practice in working two-step and three-step problems like many of those on the GED Mathematics Test. They also will give you special ideas and helps for solving word problems.

Using the Math Formulas Page

On the inside front cover of this book you will find a math formulas page. This same page will be given to you when you take the GED Mathematics Test. You don't have to memorize these widely used formulas. Try, however, to solve *"The Real Thing"* problems in each lesson using the formulas page, not the explanations about formulas in the lessons themselves. Do the same when you solve the *Extra Practice* problems and take the *Posttests.* Referring to the formulas page will help you get used to working the way you'll have to when you take the GED Test.

Reading the Answer Explanations

Answers for *On the Springboard* and "The Real Thing" in each section are near the end of the section, after all the lessons. Solid color strips run along the edges of these pages to help you locate them quickly. On these pages, you'll find not only each correct answer but also an explanation of why that answer makes sense. Reading the explanation, even if you got the answer right, will help you check your thinking and strengthen what you have already learned. Finally, it will increase your confidence in your ability to solve problems.

Keeping Track

After the answer explanations for "The Real Thing" problems in each lesson, you'll find a "Keeping Track" box that will help you do just that—keep track of how well you are doing. There you can record how many "Real Thing" answers you got right. Once you've finished all "The Real Thing" questions in a section, you can transfer your scores to the "Keeping Track" chart that follows the explanations. You'll be able to see which skills you've learned and which ones you need to review.

Taking Advantage of the Extra Practice

Following each section is a set of questions called *Extra Practice* for that particular content area of the GED Mathematics Test: Arithmetic, Geometry, and Algebra. These questions measure the same skills that the GED Test measures in each area. Doing the Extra Practice lets you know whether you've learned the skills needed to do well in that particular area on the GED Test.

Using the Progress Chart

The Progress Chart on the inside back cover is for you to keep a record of how many questions you got correct in each of the sets of Extra Practice. Notice how many correct answers you need to get a passing score, a good score, and a very good score. A good or very good score means you are ready to go on to the next section. If you just barely get a passing score or less, take the time to go back over those lessons that gave you trouble.

Taking the Posttests

After you've finished studying all three mathematics skills areas, done the Extra Practice, and reviewed any material you need to, make an appointment with yourself to take the first of the two Posttests. When you finish, check your answers. Then find your score and the explanation of your results.

If you pass, you are ready to take the GED Mathematics Test. If you don't, you can review those sections of the book for which you still need practice and then take the second Posttest.

Using the Index

When you need to review certain skills, the Index at the back of this book can help you. It lists all the important topics in the book and the page numbers on which the topics are discussed. In addition, all the mathematics terms whose definitions appear in *Coming to Terms* throughout the book appear in the Index in **bold** type. That helps you review their meanings *and* gives you practice in using an index.

In preparing you for the GED Test, this *Springboard* takes nothing for granted, leaves nothing to chance. After all, that's how the training of a championship athlete works, and the training of a successful GED candidate is no different.

Math Anxiety: Proving the Math Myths False

Even though people use math every day, they refuse to let their experience help them solve word problems. They have a hard time believing that life experiences can help them study. Even the word *mathematics* strikes terror in their hearts.

Remember that the GED Mathematics Test deals with practical problems, not with cold calculations. Most GED problems are word problems about real-life situations. Many of the things you've been doing all your life relate to the word problems in this book and on the test.

Somewhere along the line, you may have gotten the idea that some people have math ability and some don't. You may think that those who are good at math can work problems almost instantly. Then, of course, you may panic if you have to work hard to get an answer. You may even stop working in the middle of a problem. Having "math in your blood," however, is only a myth. Math is a set of skills that are learned, not inherited. Determined, serious study can put math "in your blood" too.

You may also have the idea that imaginative people have a hard time with a subject as logical and precise as math. This thought causes many people who might have the best chance of succeeding in math to turn away from it. In fact, many problems are best solved with imagination and creativity. When one idea seems to fail, a creative person can come up with another. Imaginative people can find new ways to solve problems. You've had to find ways to solve your own problems in life. You can find the solutions to math problems as well.

Finally, you may have heard that math is easier for men than for women. What *is* true is that women are more likely than men to have math anxiety. When they were girls, women often were not given toys that developed building skills, problem solving, and understanding of shapes and spaces. Later they may have received hints that understanding math is not quite "ladylike." Therefore, many women avoid math whenever they can. Today it's easier to overcome this way of thinking. As women move into all kinds of work, sports, skills, and study that were once reserved for men, they are becoming as successful as men in math.

As you begin to work through this book, you may make many mistakes. If you become frustrated or depressed, you may start to think all the math myths are true. Just remember that many people feel this way and don't let this slow you down. Put aside your fear, pass the GED Mathematics Test, and make math work for you in your life.

Test Anxiety: How to Cope with It

When you think about actually taking the GED Test, you may feel anxious. This is perfectly normal. Plenty of people—no matter how prepared, intelligent, or self-confident they may be—feel test anxiety. Recent studies show that men are just as apt to have test anxiety as women and that younger people are as prone to it as older ones.

Text anxiety can make you so afraid of failing that you put off studying. It can make you panic so that you do not think clearly. You may begin to jump from one area to the next in an impossible attempt to learn everything at once without any plan.

Test anxiety can work against you during the test too. You may feel so anxious that you can't concentrate on the questions. That is the most dangerous result of text anxiety. Just when you need every bit of energy, you waste that energy thinking, "I didn't answer the last question fast enough," or "I'm probably behind everyone else now," or "I never did understand fractions."

Making Anxiety Work for You

Most people do get nervous about taking a test. The successful ones are those who make that nervousness work *for* them. Test anxiety can actually help you if you learn to channel it correctly. For example, anxiety can cause many people to put their noses to the grindstone and spend plenty of time preparing for a test. They try harder to answer the test questions—even difficult ones. Their anxiety makes them alert and careful.

If you use this book properly, you can avoid the build-up of harmful test anxiety. You can make your anxiety work for you by using your energy to prepare thoroughly for the exam.

As you work through this book, be honest in grading yourself on the sections called *On the Springboard* and "The Real Thing." You can fool yourself by saying, "Oh, I marked answer 5, but I thought it might be 3, so I'll give myself credit for that answer." But if you do this, you will begin to feel uneasy. You may doubt that you really have covered the material. Then you have created your own test anxiety. So don't rush your study. The extra time it takes you to review sections is well worth it.

Positive Thinking Can Raise Your Scores

While you've been away from school, you've done many things that show your basic strengths. You may have held down a job; received raises or promotions; had children; supported a family; bought, built, or rented a home; saved some money; traveled; acquired interesting hobbies; or made good friends. However, when facing a new challenge like a test, you may forget that you've done such things. You may feel unsure of yourself because now you're entering unknown territory—and the ground begins to feel shaky!

Look at the Accomplishments Chart below. Take time now to write down five things you've done since you left school that you feel good about. Don't think, "I'm proud of that, but it's too silly to write down." Be honest and put down whatever you want. After all, no one but you has to see your list.

Once you've written your list, ask yourself if any of those things took hard work, courage, patience, or the ability to put your experience to work for you. Then put checks in the boxes where you deserve them.

If you start feeling bad about your abilities later, come back to this chart and take another good look at those checks. Then tell yourself that you have no reason to feel bad.

Feeling depressed or anxious takes a surprising amount of energy. By thinking positively, you will find it easier to prepare for this new challenge in a steady, organized way.

Relax!

Following that order is not always as easy as it might seem. If you are one of the many people who feel anxious about taking a test, you can *learn* to relax.

Exercise can be one of the best ways to relax your mind. You might try jogging or doing simple stretching exercises for fifteen minutes after each study session. Don't think about what you have just studied. For this short period of time, think only of how you will soon pass the GED Test.

Another way to relax is to find a quiet spot where you can be alone for fifteen minutes after every study session. Close your eyes. Think about how you are working steadily to pass the GED Test and that you will soon achieve your goal.

Still another way to relax is to tense and then relax your muscles. For example, you might try tensing your muscles while saying to yourself, "I am working hard to pass the GED Test." Then relax your muscles and say, "And I am going to pass it soon." Each time, appreciate the relaxation for a little while and think about how good it feels to be working toward something you want.

Research shows that people forget less of what they have studied if they relax immediately after studying. That's another good reason to do one of the relaxation activities right after studying for the GED Test.

Accomplishment	Hard work	Courage	Patience	Experience
1.				
2.				
3.				
4.				
5.				

The Endurance Factor

Some people confuse two things: studying in a steady way and studying until they are tired or bored or both. You will be wasting a lot of your time if you study too long or if you study when you are too tired to concentrate.

Decide before you begin to study what hours you will set aside each day to prepare for the GED Test. Probably you should spend between forty-five minutes and two hours per session. You can choose to study once, or, if you have the time, two or even three times a day.

Set up your schedule so that you can use your periods of greatest energy for study. Some people study best in the early morning hours, others in the evening. You will have to decide which times are best for you.

If you have a job, try to make use of your lunch hour and coffee breaks. If you take a bus or train to work, you can use your travel time for study. If you do this regularly, chances are you'll be able to squeeze in a surprising amount of extra study time.

Remember, whenever you feel yourself concentrating poorly or when you feel tired, close your book and do something else. You might do some of the relaxation exercises mentioned earlier.

Tips for Passing the GED Mathematics Test

Long-Range Planning

1. As you prepare for the GED Mathematics Test, keep in mind that what you are learning will benefit you for the rest of your life, not just until you pass the test. Certainly, passing the test is foremost in your mind right now, and it should be. But don't assume that once the test is over, you will no longer use what you are learning. Consider the skills you are now mastering to be a permanent part of your life. This attitude will help you value what you learn. It will also give you a greater sense of purpose as you prepare for the GED Test.

2. Give yourself plenty of time to prepare for the GED Mathematics Test. Don't think you can do all your studying in one weekend. Your brain can't continue to work at its peak if you cram too much into it at once.

3. If you think you might have a vision problem, have your eyes checked before you begin to study for the test. All your knowledge and ability may not help if you have trouble seeing and end up misreading some of the questions.

4. Be sure that your study conditions help you, not hold you back. Check to see that you have good lighting. Use a desk or table that is large enough for all your study materials. Make certain that you are not too warm or too cold. If you are, you may not be able to concentrate as well as you could. Use a comfortable chair that supports your back well.

5. Keep all your study materials in one place and have them ready before you sit down to study. You will probably want to have a couple of pencils for calculations, a pen for notes, some scratch paper, and a notebook.

6. Begin slowly and build your endurance. Don't study more than an hour at a time until you are sure that covering this amount of material doesn't wear you out. Preparing for the GED Mathematics Test is, in this sense, like preparing for running a marathon race.

7. As you study, take notes. If you jot down main ideas and important facts, you reinforce what you are learning. Then later you can use these notes to review. But be careful not to write down too much. Too many notes are just as bad as no notes at all. If you are in a GED class, your notes can also keep track of questions to ask your teacher. Take time to think carefully about his or her answers. Make sure that you really do understand what has been explained. Don't be afraid to ask for further explanations. The first big step in understanding something is being able to ask questions about it.

8. People study best at different speeds. This book will help you decide how long to spend on each section. The Answer explanations, Keeping Track charts, and Progress Chart can help you judge your progress and decide if you are moving at the right pace. Make the most of these self-checks as you work through the lessons.

9. Be sure to review what you have studied each day. It is easy to forget what you have read if you do not go over it once more before closing your book. You can review by going through your notes, by looking at the headings in your book, and by asking yourself to explain what was covered in each section. Try explaining aloud to yourself, as if you were a teacher talking to a class, the important ideas that you just read. This builds your confidence. You won't just *think* you know it—you can actually *hear* that you know it. This is probably the best way to see whether you really understood what was being taught.

10. Put your newly acquired skills to use in your daily life. Practice with other math problems that you have on the job or at home.

Short-Range Planning— The Last 24 Hours

1. If you have prepared for a long time in advance, do something relaxing the night before the GED Test. You might want to go to a movie or to a sports event.

2. Getting a good night's sleep before the exam is one of the best things you can do for yourself. Cramming for any exam is not wise. It doesn't make sense at all for the GED Test. To pass the GED Test, you must read carefully and use common sense. A good night's sleep will help you do just that.

3. It's not smart to drink too much coffee or soft drinks with caffeine or take any other kind of stimulant. After an initial "high," you may begin to feel nervous or tired. This jittery sensation won't help you concentrate.

4. Steer clear of tranquilizers as well. Even though they help you feel less nervous, they will also affect your ability to think quickly and read carefully. The trade-off is just not worth it when you're taking an important test.

5. Don't dress too warmly when you go to take the test. Psychologists have found that if people feel slightly cool, they tend to do better on a test. You may become drowsy if you are too warm.

6. If you are left-handed, get a desk with a left-handed writing board. Ask the examiner about this before the exam, if possible.

During the Test

1. No matter how much or how little people study for a test, they often go into it with the attitude they developed toward test-taking long ago. Some students get too rattled to make good use of what they have learned with all their study. Others pride themselves on not getting nervous and take the test without being serious about it. They trust that their good luck will see them through and don't really give the test their complete attention. What is the best attitude to take during the GED Test? The best attitude is to be very serious about doing your best, but never panic.

2. The people who give the GED Test know that many people are anxious about taking the test. They will make every effort to make you feel at ease. Before you begin the test, they will let you know where the restrooms and smoking areas are located. They will also try to make sure that the testing room is quiet and at a comfortable temperature and that the lighting is

good. A wall clock will probably be in the room. If not, about every fifteen minutes the examiner will announce the time remaining during the test. Even so, you may want to wear a watch so you will know exactly how much time you have to finish.

3. Some GED questions are missed only because test-takers don't mark answer sheets correctly. In some places the answers are corrected by a machine that can recognize only a completely filled-in space as a correct answer. If the machine sees an answer that is marked in some other way—like this (1) , or like this (1), or like this (1) —it cannot count that mark as a correct answer.

Remember that the machine is fair. However, if you confuse it with markings that it does not understand, it will mark your answer wrong even if it is correct. To avoid this, fill in the space so that the answer is definitely clear, but not so hard that you can't erase later if you want to change an answer. Check to make sure you have not made any extra marks near that answer or filled in any other space in that row of choices.

4. Put both the answer sheet and the test booklet on your desk so that they are easy to see and reach. Keep your place by putting one hand near the question and the other hand next to the number of that question on the answer sheet. Otherwise, one of your hands must move back and forth from the answer sheet to the test booklet. This wastes time and increases your chances of making an error.

5. Try to answer every question on the test. Your scores will be based on the number of answers you get right. You do not get any points taken off for marking wrong answers. *So don't be afraid to guess.* Try to answer every question on the exam.

6. Each question counts exactly the same when the final score is being determined. Don't spend too much time on a difficult question and then leave others blank. Try to ration your time.

7. Read the questions carefully. Read each question at least twice. If you don't understand the question, almost any answer can look right. Then compute the answer *before* you read the five possible choices. After you compute the answer, pick the answer from among the choices.

8. You may first want to go through and answer the questions that are easy for you. Then you can go back and answer those that take more time. *Warning:* If you follow this suggestion, be very careful to mark the answer space whose number is the same as that of the question you are answering. Test-takers often skip one question, say number 19, but they forget to skip an answer on the answer sheet. Then, when they mark the answer for question number 20, they put that answer by mistake in the space for number 19. If they don't catch their mistake, they can actually fail the test just because of this mix-up. *Make sure you have the right answer in the right answer space.* When you do skip a question, make a very light *X* beside the number for it on your answer sheet so that you can find it easily when you come back. Remember to erase the *X* when you've answered the question.

9. From time to time breathe deeply and stretch. Did you know that stretching is the most natural exercise to help you feel refreshed and relaxed?

10. On multiple-choice tests that have five answer choices, like the GED Math Test, you can sometimes eliminate one or more of the choices by estimating the answer. Think about the problem and decide which of the answers make sense. Eliminate the others. Then compute the answer. If you have time, go back and check your answers. If you have no idea as to how to work a problem, take a guess. In algebra problems, try substituting each answer to see which one works.

11. Don't choose any answer on the basis of its number. In other words, don't choose answer (3) just because it's been a while since you chose the third answer. On the other hand, if you think answer (1) is correct, choose it even if the answer before was also (1).

12. Try to stick to your first impressions. Don't change your answer once you've marked it unless you are very uncertain about it.

13. Some people work faster, though not necessarily more accurately, than others. If you finish early, don't turn in your test. Use every minute you have to go back and check your answers. Go over each mark to see that you have filled in the correct space. Some students get poor scores just because they are careless. They know that the correct answer is, say, 2, but they accidentally mark the space for 1 or 3. Be sure to check before turning in your exam to see that the answer you have marked is really the number of the answer you have chosen.

Make a date with yourself to reread this section the day before you take the GED Mathematics Test. With these tips and all the study aids this book provides fresh in your mind, you should pass the test with flying colors.

Skills Survey

Directions

On the following pages is a Skills Survey. Here is where you get the chance to show yourself how much you already know.

Don't look at this Skills Survey as a test. That may only make you nervous, and you won't do as well as you otherwise would. Instead, look at this Skills Survey as a personal guide. First, it is a guide to what is on the GED Test. It has word problems, tables, diagrams, and graphs from the three mathematics areas: arithmetic, geometry, and algebra. It has the same kind of multiple-choice questions that the GED Test has. And it asks you to use many of the same skills that the GED Test does. That is the second way in which you can use this survey as a guide. The survey shows you which skills you are already good at, which skills you need to practice, and which skills you need to learn.

To use this guide correctly, you will need to be good to yourself. Get yourself a comfortable chair next to a desk or table in a well-lighted area. Most important, demand some peace and quiet. You will want to be able to concentrate so you can work to the best of your ability.

There is no time limit on this survey. Reading information and answering questions quickly are skills in themselves. They are skills you will get to practice as you work through the lessons, the Extra Practice sections, and the two Posttests in this book. For now, forget about the clock.

Read problems on the Skills Survey carefully. Next read the answer options. Then answer as many as you can. Try to answer every problem.

If this is your own book, you can mark your answers in the answer ovals after the questions. Completely fill in the circle of the number you have chosen as the correct answer. For example, if you think the fourth choice is the correct answer to a question, fill in oval 4 like this:

Filling in the oval completely will give you practice in marking an answer sheet correctly. That will help you when you go to take the actual GED Mathematics Test.

If this is not your book, number a separate sheet of paper from 1 to 28. You can write the number of each answer you have chosen after the number of the question.

Are you ready to find out how much you already know? Then turn the page and start the Skills Survey.

Directions: Choose the one best answer to each problem.

1. Thirty-two people each ate 5 ice cream cones in a day. If there are 16 cups in a gallon and each cone requires 1 cup of ice cream, how many gallons of ice cream were eaten?

   (1) 1 (2) 5 (3) 10 (4) 40
   (5) Insufficient data is given to solve the problem.

   ① ② ③ ④ ⑤

2. According to the electric bill, an apartment dweller used 335 kilowatt-hours of electricity. Approximately how many kilowatt-hours were used each day?

   (1) 0.09 (2) 1.8 (3) 9.3 (4) 10.8
   (5) Insufficient data is given to solve the problem.

   ① ② ③ ④ ⑤

3. In a basketball game, certain players scored the points shown in the table. Find the median number of points scored.

Player	Points
2	6
5	26
6	5
9	0
11	15
12	33
15	2

   (1) 0 (2) 6 (3) 12 (4) 13
   (5) Insufficient data is given to solve the problem.

   ① ② ③ ④ ⑤

4. A carpenter is hired to build a flower box. The labor charge is $25.00. The material costs $10.90. Flower pots to be set inside the box cost 59 cents each. If 11 pots are to be put inside the box, what is the total cost, including labor, material, and pots?

   (1) $17.39 (2) $35.90 (3) $36.49
   (4) $42.39 (5) $401.39

   ① ② ③ ④ ⑤

5. A carpenter needed three two-by-fours, one $4\frac{3}{4}$ feet long and two $3\frac{1}{2}$ feet long. After cutting these from a 20-foot piece, how many feet are left?

   (1) $8\frac{1}{4}$ (2) $11\frac{3}{4}$ (3) $12\frac{1}{4}$
   (4) 13 (5) 40

   ① ② ③ ④ ⑤

6. A choral group raised $960 for a 3-day tour. If this was $\frac{4}{5}$ of the amount needed, how much more did they have to raise?

   (1) $240 (2) $300 (3) $660
   (4) $760 (5) $1,200

   ① ② ③ ④ ⑤

7. If 3 pieces of pegboard are bought for 98 cents each, and sales tax is 5%, how much change will be received from a five-dollar bill?

   (1) $1.91 (2) $1.92 (3) $3.08
   (4) $3.09 (5) $3.97

   ① ② ③ ④ ⑤

8. An amusement park decided to increase its yearly operating days from 180 to 220. What percent increase is this?

   (1) $18\frac{2}{11}\%$ (2) $22\frac{2}{9}\%$ (3) 82%
   (4) $122\frac{2}{9}\%$ (5) 550%

   ① ② ③ ④ ⑤

9. A toaster is on sale at $16\frac{2}{3}\%$ discount. How much is the sale price?

   (1) $1.92 (2) $2.00
   (3) $10.00 (4) $10.08
   (5) Insufficient data is given to solve the problem.

   ① ② ③ ④ ⑤

10. In scientific notation, the number .0056234 is written as

   (1) 0.056234 (2) 56.234×10^{-4}
   (3) 56.234×10^{4} (4) 5.6234
   (5) 5.6234×10^{-3}

   ① ② ③ ④ ⑤

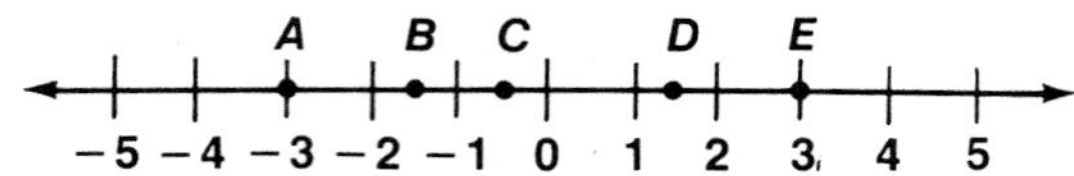

11. Where would you find $\frac{-26}{9}$ on the number line above?

(1) between D and E
(2) at E
(3) Between B and C
(4) between A and B
(5) Insufficient information is given to solve the problem.

(1) (2) (3) (4) (5)

12. Find the measurement of angle E.

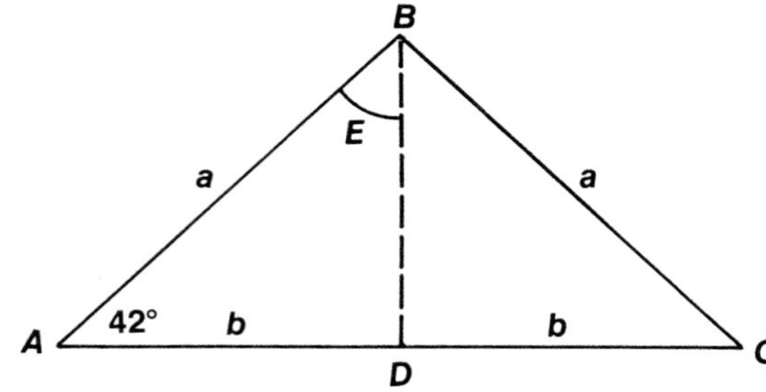

(1) 21° (2) 42° (3) 48°
(4) 90° (5) 96°

(1) (2) (3) (4) (5)

13. A survey used this diagram to find the distance across a river, represented by the length ED.

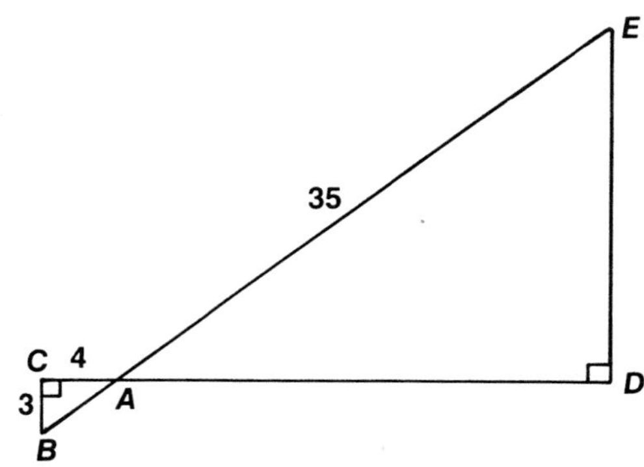

Find the length of side ED.

(1) 0.42 (2) 2.14 (3) 4.2
(4) 5.0 (5) 21.0

(1) (2) (3) (4) (5)

14. Find the approximate volume in cubic feet of a chemical storage tank which is 10 feet tall and has a diameter of 12 feet.

(1) 37.7 (2) 188.4 (3) 376.8
(4) 452.2 (5) 1,130.4

(1) (2) (3) (4) (5)

15. Find the midpoint of line segment PQ.

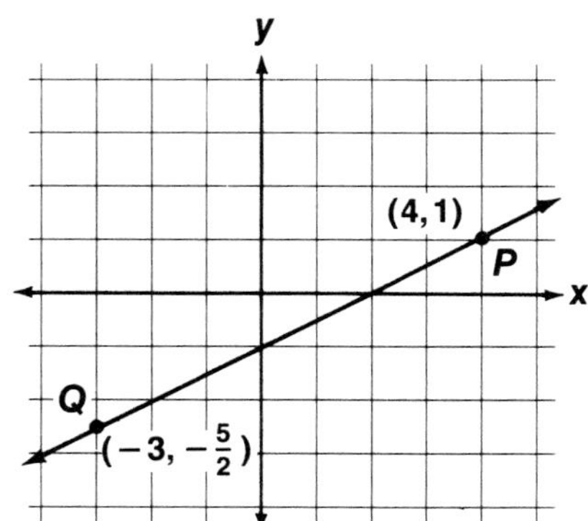

(1) $(\frac{7}{2}, \frac{7}{4})$ (2) $(7, -7)$
(3) $(\frac{-7}{2}, \frac{-7}{4})$ (4) $(\frac{-1}{2}, \frac{3}{4})$
(5) $(\frac{1}{2}, \frac{-3}{4})$

(1) (2) (3) (4) (5)

16. Which line in the figure has zero slope?

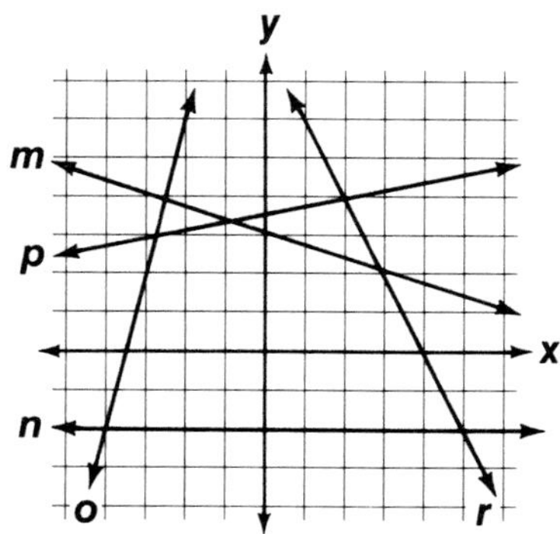

(1) m (2) n (3) p (4) r
(5) A line cannot have zero slope.

(1) (2) (3) (4) (5)

17. In each lot of 50 spark plugs, 3 are expected to be defective. In each lot of 25 mufflers, 2 are expected to be defective. How much higher is the probability of finding a defective muffler?

(1) .06 (2) .08 (3) 1.67
(4) 1.33 (5) 5

(1) (2) (3) (4) (5)

18. Find $x + 4(y - z)$ when $x = -3$, $y = 2$, and $z = 3$.

(1) -23 (2) -7 (3) -1
(4) 0 (5) 1

(1) (2) (3) (4) (5)

Items 19 and 20 refer to the following table.

Telephone Service Plans

Plan	Units per Month	Cost of Units in Plan
A	0	$0.00
B	30	1.10
C	80	2.90
D	110	4.00

Cost of Additional Units
4.5¢ each

19. Which equation describes the cost in dollars for a telephone customer who has Plan D and uses x units in a month?

(1) $(4.00)x$
(2) $2.90 + 04.5x$
(3) $4.00 + 1.10x$ (4) $4.5x$
(5) $4.00 + (x - 110)(4.5)$

(1) (2) (3) (4) (5)

20. The Lau family has Plan C. They used 73 units in September. What was their cost for the month?

(1) $1.10 (2) $2.90 (3) $3.04
(4) 3.29 (5) $19.35

(1) (2) (3) (4) (5)

21. If 10 less than 4 times a number is 14, what is the number?

(1) -4 (2) -1 (3) 1
(4) 4 (5) 6

(1) (2) (3) (4) (5)

22. A father is 41 years old and his son is 9. The father weighs 80 pounds more than the son. In how many years will the father be 3 times as old as the son?

(1) 1.5 (2) 1.6 (3) 3.6
(4) 4 (5) 7

(1) (2) (3) (4) (5)

23. A small business obtained a loan at 8% for 2 years to buy improved equipment. The business repaid $1,450. What was the amount borrowed?

(1) $100 (2) $200 (3) $625
(4) $1,250 (5) $2,500

(1) (2) (3) (4) (5)

24. Find $\frac{3x}{2} \div \frac{6x^2}{4}$

(1) $\frac{9x^3}{4}$ (2) $\frac{9x^2}{4}$ (3) $12x$
(4) $\frac{1}{x}$ (5) $\frac{12}{x}$

(1) (2) (3) (4) (5)

25. Find $\frac{3a}{bc} + \frac{2b}{ac}$

(1) $\frac{3a + 2b}{bc}$ (2) $\frac{6ab}{abc}$ (3) $\frac{3a^2c - 2bc^2}{abc^2}$
(4) $\frac{3a^2 + 2b^2}{abc}$ (5) $\frac{3a + 2b}{abc}$

(1) (2) (3) (4) (5)

26. The sum of two numbers is 28 and their difference is 12. One number is even. Find the smaller number.

(1) 4 (2) 8 (3) 16
(4) 20 (5) 32

(1) (2) (3) (4) (5)

27. Solve $p + 3 < 2(2p + 1)$.

(1) $p > \frac{1}{3}$ (2) $p < \frac{1}{3}$ (3) $p = \frac{2}{3}$
(4) $1 < p$ (5) $\frac{4}{3} < p$

(1) (2) (3) (4) (5)

28. Solve $x^2 + 3x = 28$.

(1) 4 (2) 7 (3) 14 or 2
(4) -7 or 4 (5) 4 or -7

(1) (2) (3) (4) (5)

Answers: Skills Survey

1. (3) 10
Each cone is 1 cup, so find the number of cups used by multiplying 32 by 5 (160). There are 16 cups in a gallon. Divide 160 by 16 to find the number of gallons.

$160 \div 16 = 10$

2. (5) Insufficient data is given to solve the problem.

You do not know the number of days in the period covered by the bill, so you cannot find the kilowatt-hours per day.

3. (2) 6
The median is the number in the middle when all the values are in order from smallest to largest.

0, 2, 5, 6, 15, 26, 33

4. (4) $42.39

labor charge	$25.00
materials	10.90
pots (11 × $0.59)	+ 6.49
	$42.39

5. (1) $8\frac{1}{4}$

Add the lengths of the three pieces.

$$\begin{aligned} 4\tfrac{3}{4} &= 4\tfrac{3}{4} \\ 3\tfrac{1}{2} &= 3\tfrac{2}{4} \\ +3\tfrac{1}{2} &= 3\tfrac{2}{4} \\ \hline 10\tfrac{7}{4} &= 11\tfrac{3}{4} \end{aligned}$$

Then subtract this sum from 20 feet.

$$\begin{aligned} 20 &= 19\tfrac{4}{4} \\ -11\tfrac{3}{4} &= 11\tfrac{3}{4} \\ \hline & \quad 8\tfrac{1}{4} \end{aligned}$$

6. (1) $240
Divide $960 by 4/5 to find the amount needed in all.

$$\frac{960}{1} \div \frac{4}{5} =$$

$$\frac{\overset{240}{\cancel{960}}}{1} \times \frac{5}{\underset{1}{\cancel{4}}} = \frac{1{,}200}{1} = 1{,}200$$

Then subtract the amount already saved from the total needed to find how much more they had to save.

$1,200 − $960 = $240

7. (1) $1.91
Cost of pegboard:
$3 \times \$0.98 = \$2.94$
Find the tax at 5% on $2.94.

amount of tax	?	5	percent tax rate
cost of pegboard	2.94	100	

$5 \times 2.94 = 14.7$

$14.7 \div 100 = \$0.147$ or $0.15

Cost of pegboard with tax is

$$\begin{array}{r} \$2.94 \\ +\ 0.15 \\ \hline \$3.09 \end{array}$$

Change from five-dollar bill:

$$\begin{array}{r} \$5.00 \\ -\ 3.09 \\ \hline \$1.91 \end{array}$$

8. (2) $22\frac{2}{9}\%$

Find the amount of increase by subtracting.

$220 - 180 = 40$ days

amount of increase	40	?	percent increase
original number of operating days	180	100	

$40 \times 100 = 4{,}000$

$$\begin{array}{r} 22 \\ 180\overline{)4{,}000} \\ 360 \\ \hline 400 \\ 360 \\ \hline 40 \end{array} \qquad \frac{40}{180} = \frac{2}{9}\%$$

$22 + \frac{2}{9} = 22\frac{2}{9}$

9. (5) Insufficient data is given to solve the problem.

You do not know the original price of the toaster.

10. (5) 5.6234×10^{-3}
To write a number in scientific notation, first move the decimal point to the right of the first nonzero digit: 5.6234. The point was moved 3 places to the right so the exponent is -3.
$.0056234 = 5.6234 \times 10^{-3}$

11. (4) between *A* and *B*
$\frac{-26}{9} = -(2 + \frac{8}{9})$.

This is between -2 and -3, and very close to -3 because $^{-27}/_{9} = -3$.

12. (3) 48°
Since sides *AB* and *BC* are both *a* units long, triangle *ABC* is isosceles. Since *AD* and *DC* are both *b* units long, *D* is the midpoint of side *AC*. Now *BD* must bisect $\angle B$ because triangle *ABC* is isosceles. Then the measure of $\angle E$ is half the measure of $\angle B$. Now triangle *ABC* is isosceles so $\angle C = \angle A = 42°$, so $\angle B = 180° - 42° - 42° = 96°$. Then $\angle E = {}^{96°}/_{2} = 48°$.

13. (5) 21.0
The sides of a triangle will be in proportion to the sides of a triangle that is similar to it. You are given the length of the hypotenuse (*AE*) of $\triangle ADE$, so first find the hypotenuse of $\triangle ABC$. Then set up a proportion to find *ED*.

$\triangle ABC$:

$$a^2 + b^2 = c^2$$
$$(3)^2 + (4)^2 = c^2$$
$$9 + 16 = c^2$$
$$25 = c^2$$
$$\sqrt{25} = c$$
$$5 = c$$

	sm.△	lg.△	
hypot. AB	5	35	hypot. AE
short leg BC	3	?	short leg ED

$3 \times 35 = 105$
$105 \div 5 = 21$ length of *ED*

14. (5) 1,130.4
The radius *r* is half the diameter, so $r = 12/2 = 6$ feet

Use the formula for volume of a cylinder. The height *h* is 10 feet and the radius is 6 feet.

$$V = \pi r^2 h$$
$$= 3.14(6)^2 10$$
$$= 3.14(36)10$$
$$= 1,130.4$$

15. (5)$(\frac{1}{2}, \frac{-3}{4})$

The *x*-coordinate of the midpoint is the average of the *x*-coordinates of *P* and *Q*. Find the *y*-coordinate the same way.

$$x = \frac{4 + (-3)}{2} = \frac{1}{2}$$

$$y = \frac{1 + -\frac{5}{2}}{2} = \frac{\frac{2}{2} - \frac{5}{2}}{2} =$$

$$\frac{\frac{-3}{2}}{2} = \frac{-3}{4}$$

16. (2) *n*
A zero slope means the line is horizontal.

17. (4) 1.33
Probability of defective spark plug $= 3/50 = 6/100 = 6\%$. Probability of defective muffler $= 2/25 = 8/100 = 8\%$. Now take the ratio of the probabilities.

$$\frac{8/100}{6/100} \text{ or } 8\% \text{ to } 6\% =$$

$$\frac{8}{6} = 1.33$$

18. (2) −7
Substitute the values

$$x = -3, y = 2, z = 3$$
$$x + 4(y - 3) = -3 + 4(2 - 3)$$
$$= -3 + 4(-1)$$
$$= -3 - 4$$
$$= -7$$

19. (5) $4.00 + (x - 110)(4.5)$
Under Plan D, the customer pays $4.00 plus a charge for the units in excess of 110. If *x* units are used, the extra units are $x - 110$ and the charge for the extra units is $(x - 110).5$. Then the total cost is $4.00 + (x - 110)(.5)$.

20. (2) $2.90
Plan C provides 80 units at $2.90. The Lau family used only 73 units.

21. (5) 6
Let *x* be the unknown number "Ten less than" means 10 is subtracted *from* another number. The other number is $4x$.

$$4x - 10 = 14$$
$$4x = 24$$
$$x = 6$$

22. (5) 7
Let *x* be the number of years. The father's age after *x* years will be $x + 41$ and the son's age will be $x + 9$. The father will be 3 times as old as the son.

$$x + 41 = 3(x + 9)$$
$$x + 41 = 3x + 27$$
$$41 = 2x + 27$$
$$14 = 2x$$
$$7 = x$$

23. (4) $1,250
The amount 1,450 equals the amount of the loan plus 2 years interest at 8% so

$$1,450 = x + 2(.08)x$$
$$1,450 = x + .16x$$
$$1,450 = 1.16x$$
$$\$1,250 = x$$

24. (4) $\frac{1}{x}$

$$\frac{3x}{2} \div \frac{6x^2}{4} = \frac{3x}{2} \times \frac{4}{6x^2}$$
$$= \frac{3x\,(2)\,(2)}{2\,(2)\,(3)x^2}$$
$$= \frac{x}{x^2}$$
$$= \frac{1}{x}$$

25. (4) $\frac{3a^2 + 2b^2}{abc}$
Find a common denominator, then add. The least common multiple of the denominator is *abc*.

$$\frac{3a}{bc} \cdot \frac{a}{a} = \frac{3a^2}{abc}$$
$$\frac{2b}{ac} \cdot \frac{b}{b} = \frac{2b^2}{abc}$$

Then

$$\frac{3a}{bc} + \frac{2b}{ac} = \frac{3a^2}{abc} + \frac{2b}{abc} = \frac{3a^2 + 2b^2}{abc}$$

26. (2) 8
Let *b* be the smaller number. Then the larger one is $b + 12$. Their sum is 28.

$$b + (b + 12) = 28$$
$$2b + 12 = 28$$
$$2b = 16$$
$$b = 8$$

27. (1) $p > \frac{1}{3}$

$$p + 3 < 2(2p + 1)$$
$$p + 3 < 4p + 2$$
$$3 < 3p + 2$$
$$1 < 3p$$
$$\frac{1}{3} < p$$

28. (5) 4 or −7
Always try to factor a quadratic equation. Remember to start by getting zero on one side.

$$x^2 + 3x = 28$$
$$x^2 + 3x - 28 = 0$$
$$(x + 7)(x - 4) = 0$$
$$x + 7 = 0 \textbf{ or } x - 4 = 0$$
$$x = -7 \text{ or } x = 4$$

Now circle the number of each item you got wrong on the answer key below.

1. (3)
2. (5)
3. (2)
4. (4)
5. (1)
6. (1)
7. (1)
8. (2)
9. (5)
10. (5)
11. (4)
12. (3)
13. (5)
14. (5)
15. (5)
16. (2)
17. (4)
18. (2)
19. (5)
20. (2)
21. (5)
22. (5)
23. (4)
24. (4)
25. (4)
26. (2)
27. (1)
28. (5)

Using the Survey Results

Question Number	Skill	Lesson
	Arithmetic	
1, 4, 21, 22	*Whole Numbers*	2–7
5, 6, 9	*Fractions*	12–18
2, 7, 8, 23	*Percents*	19–21
3, 17, 20	*Data Analysis*	22, 24, 25
10, 11	*Number Relationships*	26, 28
8, 12, 13, 14	Geometry	33, 36, 37, 40
16, 18, 19, 24, 25, 26, 27, 28	Algebra	43–50

How did you do on the Skills Survey? Above is a chart to show you the skills that were tested by the survey questions. By comparing the answers you got wrong with the chart, you can see which skills you need to concentrate on when you study the lessons in this book. The chart also shows which lessons give you instruction and practice in those areas.

You will still probably want to work through every lesson, of course. That will help you develop skills that weren't tested directly on the Skills Survey as well as strengthen the skills you already have. Those new and strengthened skills will help better your chances of passing the GED Mathematics Test.

Arithmetic

Basic Arithmetic
Data Analysis
Number Relationships
Measurements

When you think of mathematics, you probably think of arithmetic. *Arithmetic* is the area of mathematics that people have to deal with most often. On the GED Mathematics Test, 50 percent of the questions are arithmetic problems. This means that about 28 of the 56 problems on the test are common, everyday arithmetic. You'll be asked to add, subtract, multiply, and divide using whole numbers, fractions, mixed numbers, decimals, and percents. There will be some tables and graphs, positive and negative numbers, and measurements. The GED Mathematics Test will not give you complicated numbers to work with. You will not be asked, for example, to multiply .009512 by $1/14$. The GED Test does not try to trick you with difficult calculations. Instead, it measures how well you can use common sense to solve word problems.

Because the GED Mathematics Test stresses arithmetic, you probably could pass the test by getting all the arithmetic answers right. But you should give your best to the *Geometry* and *Algebra* sections too. You don't have to panic if only the arithmetic sinks in. The rest is like icing on the cake.

Many of the arithmetic problems on the GED Mathematics Test are the kinds of problems you solve every day without even thinking about them. For example, read this sample arithmetic question and see if you can choose the right answer.

1. If a woman earns $6.25 an hour, how much will she earn in a 40-hour week?

(1) $156.25 (2) $240.00 (3) $248.00
(4) $250.00 (5) $312.50

(1) (2) (3) (4) (5)

Was (4) your choice? If it was, you just put your everyday arithmetic experience to work for you. Using such experience is what you'll need to do on the test. If you got it wrong, maybe you didn't see that you had to multiply the hourly wage by the number of hours worked. Or perhaps you made a multiplication error. If so, the *Arithmetic* pages of this book will give you plenty of chances to learn to solve word problems and to do computations quickly and accurately.

Here's another type of arithmetic problem you might find on the GED Mathematics Test. Can you handle it?

2. How many 15-minute interviews can be held in $4\frac{1}{2}$ hours?

(1) $6\frac{3}{4}$ (2) 18 (3) $19\frac{1}{2}$
(4) 33 (5) $67\frac{1}{2}$

(1) (2) (3) (4) (5)

There are several ways you could solve this problem. Like many of the questions on the GED Mathematics Test, there is no *one* right method for arriving at the solution. You could use a direct, common-sense approach and figure out that because there are 4 fifteen-minute sessions in an hour, there would be 16 sessions in 4 hours and 2 in the half hour. This would be 18 sessions altogether, so (2) would be the right answer. You could also solve this problem using fractions or decimals or by converting measurements. Any one of these methods is as right as the others.

You can also answer some arithmetic questions on the GED Mathematics Test just by carefully reading a chart, graph, or table. Study the graph below and then try to answer the question.

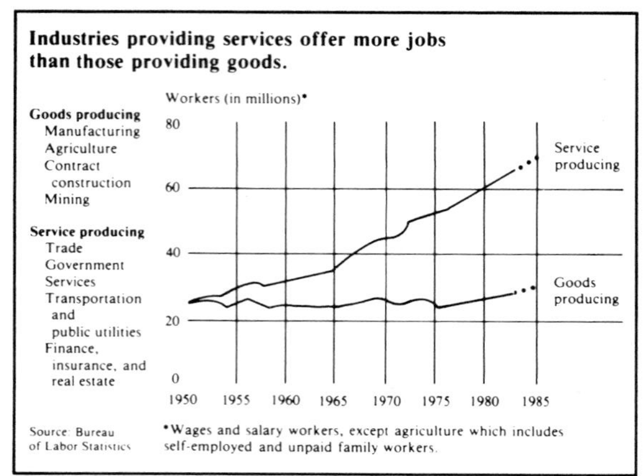

3. How many million workers were employed in service industries in 1980?

   (1) 45 (2) 57 (3) 60
   (4) 62 (5) 70

   ① ② ③ ④ ⑤

Did you see the line at the top of the graph that says "Workers (in millions)"? Did you also notice that the top line on the graph is labeled "Service producing"? Then did you find the spot where the service line crosses the 1980 line? If your answer was (3), you used either logic or your previous experience in reading graphs to find the right answer.

You may already know that the questions on the GED Mathematics Test are *word problems*. One of the most helpful things this *Arithmetic* section will give you is a six-step method for solving word problems. Practice using these steps until they come naturally. When you reach that point, you'll be ready both to take the GED Mathematics Test and to handle math in your daily life with greater confidence and skill.

LESSON 1

Zeroing In on Word Problems

Solving Word Problems

Solving a word problem is like solving a mystery. You have to know *how* to read it—*what clues* to look for on the way. Knowing just these two things will make the GED Mathematics Test much easier for you.

All the *Zeroing in* lessons in this book will give you extra tips and practice with word problems. This lesson starts you off with a system for solving *any* word problem. Whenever you work through a word problem you will be able to use the system outlined here. Try to make using the process automatic and it will help you on the GED Mathematics Test.

A System for Solving Word Problems

A. Read each problem carefully to understand what you are supposed to find. *Read it as many times as necessary.* Sometimes it helps to make a mental picture or a drawing.

B. Figure out what information you need to answer the question and find that information in the problem. It may help you to list and label the information.

C. Decide which operation you need to use. A key word or phrase can be a clue to the operation to use.

You will soon notice that some problems will require more than one operation *or* the same operation more than once. Each time you perform an operation, you take one more step toward solving the word problem. The number of steps in the problem determines the number of times you do the next three things.

D. Set up the computation.

E. Compute.

F. Check your arithmetic.

G. See if your answer makes sense. Read the problem again with your answer in mind and ask yourself if your answer is logical. It may help you to estimate the answer by rounding the numbers and seeing if your answer is in the right ball park.

You can see that this is a three-part system. Before you write anything, you understand what you have to do.

Read, Find, Decide

Next you do some writing.

Set up, Compute, Check

Then you test your answer.

Does the answer make sense?

A Test-Taking Tip

You can round numbers to compute a quick estimate of the answer. For example, in the problem 29 × 198 you could say, "29 is pretty close to 30, and 198 is pretty close to 200, so the answer will be pretty close to 30 × 200, which is 6,000."

Here's an Example

Use the system to solve this problem.

■ At the beginning of the month, the saleswoman's odometer read 27,400 miles. At the end of the month it showed 29,950 miles. How many miles did she travel that month?

Go through the system.

A. Read the problem carefully.

B. Find the information you need. In this case, the number of miles at the beginning and end of the month will be enough.

C. Decide which operation to use. The difference between 29,950 miles (end of month) and 27,400 (beginning of month) will be the answer. To find the difference, you need to subtract.

D. Set up the computation. You need to subtract 27,400 from 29,950, so you would set it up like this:

$$\begin{array}{r} 29{,}950 \\ -\ 27{,}400 \\ \hline \end{array}$$

E. Compute.

$$\begin{array}{r} 29{,}950 \\ -\ 27{,}400 \\ \hline 2{,}550 \end{array}$$

The answer: She traveled 2,550 miles.

F. Check your arithmetic. To check a subtraction problem, add the difference to the lower amount. Your answer should be the larger amount.

$$\begin{array}{r} 2{,}550 \\ +\ 27{,}400 \\ \hline 29{,}950 \end{array}$$

G. Does the answer make sense? Yes, it does. If she traveled 100 miles a day for a month of 30 days, she would have gone 30 × 100 = 3,000 miles. She probably didn't work every weekend, so the answer is reasonable. A simple slip could have given you an answer of 25,550 miles, but step G would alert you that you did something wrong: 25,500 miles in one month is *not* reasonable.

Coming to Terms

Here are some words that can give you a hint about which operation to use to solve a word problem. You don't need to memorize them, but they will be useful as a reference until you can recognize them as clues.

Addition: *sum, plus, add, and, total, increase, raise, both, combined, in all, altogether, additional, as well as, entire*

Subtraction: *less than, more than, difference, decrease, reduce, lost, left, remain, dropped, fell, change, deduct, smaller, greater,* and other *-er* words

Multiplication: *times, each of, as much, twice, multiplied, total, by, every, pair, double, triple*

Division: *split, cut, divided evenly, each, part of, average, equal pieces, every, out of, shared, among, between, distribute, in half, in thirds, in fourths* .

Even with this list you will still have to read the question carefully. Notice that *total* is listed under both *addition* and *multiplication.* You could tell which one was needed *if* you read the question completely and thoroughly. Otherwise, you have a fifty-fifty chance of being wrong.

Try It Yourself

Read the following problem and go through part D of the system to set the problem up.

■ Garment workers at a knitting mill completed 15, 23, 20, 17, and 25 garments during one five-day work week. How many garments altogether were completed that week?

Compare what you did with this.

A. Read.

B. Find. The problem tells you how many garments were completed on each day of the work week.

C. Decide. The word *altogether* is a hint that you need to add.

D. Set up.

$$\begin{array}{r} 15 \\ 23 \\ 20 \\ 17 \\ +\ 25 \\ \hline \end{array}$$

This will solve the problem if you do the addition correctly. Note that you could also show the setup like this:

$15 + 23 + 20 + 17 + 25 =$

Here's another one for you to try. Set up the problem, but don't compute the answer.

■ Two of the busiest airports in California are in Los Angeles and Long Beach. During a recent year, Los Angeles Airport had 523,937 take-offs and landings and Long Beach had 606,323. What was the difference in the number of take-offs and landings at the two airports?

Compare what you did with this.

A. Read.

B. Find. The problem tells how many take-offs and landings there were at each airport.

C. Decide. The word *difference* is a clue that subtraction is needed.

D. Set up.

$$\begin{array}{r} 606{,}323 \\ -\ 523{,}937 \\ \hline \end{array}$$

You could also show the setup like this:

$606{,}323 - 523{,}937 =$

Notice that it doesn't matter where the airports are or in what year all this happened. You have pulled from the problem all that matters, and you have the numbers you need. That is what you do to solve any word problem. Using this system will help you cut through all the words and find out what arithmetic you need to do.

Warm-up

Use the problem-solving system to set up the following problems. You don't have to compute the answers.

1. Ten years ago a hot dog stand sold 6,700 sandwiches. This year it sold 5 times as many. How many did it sell this year?

---

2. If a worker earns $200 a week and works 40 hours a week, what is that worker's hourly wage?

---

3. A hospital parking lot can hold 317 cars. A new parking area that can hold 239 cars was added. How many cars can now be parked in both lots?

---

4. A furniture company planned its deliveries for the week. On Monday it planned 8 deliveries, on Tuesday 6, on Wednesday 9, and on Thursday and Friday 6 deliveries for each day. How many deliveries did it plan to make in that week?

Warm-up Answers

1. $\begin{array}{r} 6{,}700 \\ \times\quad 5 \\ \hline \end{array}$ or $5 \times 6{,}700 =$

2. $40\overline{)200}$ or $200 \div 40 =$

3. $\begin{array}{r} 317 \\ +\ 239 \\ \hline \end{array}$ or $317 + 239 =$

4. $\begin{array}{r} 8 \\ 6 \\ 9 \\ 6 \\ +\ 6 \\ \hline \end{array}$ or $8 + 6 + 9 + 6 + 6 =$

LESSON 2

Adding Whole Numbers

If you drive a car, you've probably kept track of your mileage from time to time. You may need this information to report as an expense, or perhaps you just want to keep track of what kind of mileage your car is getting. To do either of these, you have to add whole numbers.

Place Names in Whole Numbers

Whole numbers are numbers such as 0, 1, 2, 3, 85, 147, and 2,408. They are the basic numbers of arithmetic.

Digits are the single numbers from 0 through 9. There are ten of them: 0, 1, 2, 3, 4, 5, 6, 7, 8, and 9.

In any whole number, each digit is in a certain place, and each place has a name **(place name).** For example, in the whole number 333 each 3 means something different. The right-hand 3 means 3 ones, the middle 3 means 3 tens, and the left-hand 3 means 3 hundreds. Here's a chart of these and other place names for your reference.

Billions	Millions			Thousands			Ones		
billions	hundred millions	ten millions	millions	hundred thousands	ten thousands	thousands	hundreds	tens	ones

Coming to Terms

whole number 0, 1, 2, 3, . . , 24, . . ., 467, and so on

digit any of the single numbers 0, 1, 2, 3, 4, 5, 6, 7, 8, or 9

place name the name that tells in what place (ones, tens, and so on) a digit is located in a whole number

Here's an Example

Look at this number: 4,268,573

3 is in the *ones* place.
7 is in the *tens* place.
5 is in the *hundreds* place.
8 is in the *thousands* place.
6 is in the *ten-thousands* place.
2 is in the *hundred-thousands* place.
4 is in the *millions* place.

Try It Yourself

■ In the number 7,560,934, which digit is in the thousands place? In which place is the digit 6?

If you read the places from the right as ones, tens, hundreds, thousands, ten thousands, and so on, you found that 0 is in the thousands place and 6 is in the ten thousands place.

You need to know all the basic addition facts to add whole numbers. There may be one or two of the facts that slow you down a bit. A quick way to get up to speed is to add two digits on a calculator and say the total just before you hit the equals key. That way you always have the right answer to check what you said.

If you don't have a calculator, you may want to buy addition flash cards or even make your own. Write the possible combinations (0 + 1, 2 + 4, 7 + 3, and so on) on 3 × 5 cards. You will need 100 cards. Write just one combination on each card, and put the answer on the back.

Simple Addition

When you add, you must add similar items. You cannot add gallons and hammers, and you cannot add tens digits to ones digits or hundreds digits to thousands digits. They are different. When you add whole numbers, add ones to ones, tens to tens, hundreds to hundreds, and so on.

Here's an Example

■ Last month 465 people joined an auto club. So far this month 34 more have joined. How many people have joined since the beginning of last month?

To solve, add 465 and 34.

You can set up the problem **horizontally**

465 + 34 =

or **vertically**

$$\begin{array}{r} 465 \\ +\ 34 \\ \hline \end{array}$$

It's easiest to do the addition if you set it up vertically.

Line up the places.

Add the digits in each column.

$$\begin{array}{r} 4\,6\,5 \\ +\ \ \ 3\,4 \\ \hline 4\,9\,9 \end{array}$$

In all, 499 people have joined the auto club.

If you get all the ones places lined up, it's easy to line up the others, even if there are more places than in the example. Look at the following example.

■ A post office handled 8,234 letters, 3 express packages, 12 postcards, and 650 pieces of bulk mail in one day. How many pieces of mail did it handle in all?

To solve, add 8,234 and 3 and 12 and 650. Write the number vertically, lining up from the ones.

Add each column.

$$\begin{array}{r} 8,2\,3\,4 \\ 3 \\ 1\,2 \\ +\ \ \ 6\,5\,0 \\ \hline 8,8\,9\,9 \end{array}$$

The post office handled 8,899 pieces of mail.

Coming to Terms

horizontally written across (in a row)

vertically written up and down (in a column)

Try It Yourself

Read and work through this next problem on a separate piece of paper. Compute the answer before you read the explanation.

■ Last week John added the following number of stamps to his collection of 3,600: 166 stamps, 11 stamps, and 212 stamps. How many stamps does he have in all?

If you lined up the ones digits of the four numbers carefully, you should have 3,989 stamps as your answer.

Warm-up

Add these numbers.

1. 23 + 6 = ______

2. 4 + 12 = ______

3. 2 + 13 + 3 = ______

4. 64 + 34 = ______

5. 27 + 42 = ______

6. 81 + 5 + 12 = ______

7. 206 + 73 = ______

8. 54 + 445 = ______

9. 833 + 152 = ______

10. 2,644 + 335 = ______

11. 101 + 25 + 3,451 = ______

12. 10,305 + 610 = ______

13. 1,034 + 67,305 = ______

14. 6,342,103 + 51,342 + 13 = ______

15. 2,001 + 5,103 + 10,000 = ______

Warm-up Answers
1. 29 **2.** 16 **3.** 18 **4.** 98 **5.** 69 **6.** 98 **7.** 279
8. 499 **9.** 985 **10.** 2,979 **11.** 3,577 **12.** 10,915
13. 68,339 **14.** 6,393,458 **15.** 17,104

✓ A Test-Taking Tip

Words like *sum* and *altogether* on the GED Mathematics Test are usually clues that tell you to add. Some others are *plus, add, and, increase, raise, both, combined, in all, additional, as well as,* and *entire.*

On the Springboard

1. A shipping room filled 50 orders on Monday. On Tuesday it filled 35 more orders. How many orders were filled on those days?

   (1) 15 (2) 85 (3) 535

   ① ② ③

2. On her vacation, a nurse drove 275 miles the first day, 210 miles the second day, and 212 miles on the last day. How many miles did she drive in all?

   (1) 675 (2) 692 (3) 697

   ① ② ③

3. At a ball game 13 box seats, 562 grandstand seats, and 2,304 bleacher seats were sold. How many seats were sold altogether?

   (1) 2,879 (2) 2,869 (3) 2,789

   ① ② ③

Check your answers on page 182. If you answered the Springboard questions correctly, go on to the next section. If you got one or more wrong, reread the material on adding whole numbers and answer the questions again.

Carrying

Sometimes when you add a column of digits, the total is more than 9. When this happens you have to **carry.** First write the right-hand digit in the sum below the column. Carry the left-hand part to the next column. You can write it at the top of that column. When you add that column, you include the number you carried.

You may have to carry more than once. Another word for carrying is *renaming* because that is what you are actually doing; you are renaming the place name of part of the number. For example, suppose you add the right-hand digits and get 13. When you write down the 3 and carry the 1, you are renaming 13 as 1 ten and 3 ones.

Coming to Terms

carrying renaming a sum in addition. For example, if the sum of the ones is 27, you write 7 in the ones column and carry the 2 to the tens column.

Here's an Example

In this problem you have to carry.

■ A copy-machine counter read 5,631 copies at the beginning of the week. In that week 895 more copies were made. Find the new total recorded on the counter.

To solve, add 5,631 and 895.

Step 1 Write the problem vertically. Line up the places as you did before.

$$\begin{array}{r} 5{,}631 \\ +\ \ 895 \\ \hline \end{array}$$

Step 2 Add.

Add the ones.

$$\begin{array}{r} 5{,}631 \\ +\ \ 895 \\ \hline 6 \end{array}$$

Add the tens. You get 12. Write 2 in the tens place and carry the 1 to the hundreds column.

$$\begin{array}{r} \scriptstyle 1\ \ \ \\ 5{,}631 \\ +\ \ 895 \\ \hline 26 \end{array}$$

Add the hundreds. You get 15. Write 5 in the hundreds place and carry the 1 to the thousands column.

$$\begin{array}{r} \scriptstyle 1\ \ 1\ \ \ \\ 5{,}631 \\ +\ \ 895 \\ \hline 526 \end{array}$$

Add the thousands.

$$\begin{array}{r} \scriptstyle 1\ \ 1\ \ \ \\ 5{,}631 \\ +\ \ 895 \\ \hline 6{,}526 \end{array}$$

The new total on the counter was 6,526 copies.

Try It Yourself

Now try working a carrying problem on your own.

■ A family collected campaign buttons. The father had 49, the mother 500, the older daughter 823, and the younger daughter 1,268. How many buttons did the family have altogether?

If you wrote the problem vertically and lined up the places correctly, you should have carried in three columns. The answer is 2,640 buttons.

Warm-up

Add these numbers.

1. 7 + 65 = ______
2. 36 + 28 = ______
3. 75 + 89 = ______
4. 16 + 24 + 5 = ______
5. 49 + 32 + 66 = ______
6. 120 + 78 + 234 = ______
7. 359 + 19 + 4,787 = ______
8. 5,832 + 799 + 67 = ______
9. 899 + 398 + 29 + 471 + 16 = ______
10. 1,429 + 1,863 = ______
11. 34 + 578 + 65 + 1,009 = ______
12. 55,009 + 476,238 = ______
13. 95,489 + 12,566 + 48,605 = ______
14. 2,677,346 + 11,318 = ______
15. 895,480 + 8,134,983 + 976,415 = ______

A Test-Taking Tip

Getting the right answer on the GED Mathematics Test is sometimes as simple as working neatly and accurately. After the test it's too late to tell yourself that you could have done better if you had been more careful in lining up the numbers when you added.

On the Springboard

4. A city police department ticketed 2,690 cars during the week and 450 cars on the weekend. What was the total number of cars ticketed that week?

   (1) 2,140 (2) 3,040 (3) 3,140

   ① ② ③

5. A family needed appliances. They found a refrigerator for $595, a used stove for $210, and a washer for $469. How much money did they need to buy these things?

   (1) $1,774 (2) $1,274 (3) $1,264

   ① ② ③

6. How many calories were consumed by someone who ate three meals containing 560 calories, 345 calories, and 725 calories as well as two snacks containing 100 calories and 210 calories?

   (1) 1,900 (2) 1,930 (3) 1,940

   ① ② ③

7. A band played in a small town. The first night 43 people attended, the second night 67 came, and the third night 89 were there. Find the total attendance for the performances.

   (1) 189 (2) 199 (3) 200

   ① ② ③

You can check your Springboard answers on page 182. If you solved all the problems correctly, you're ready for the next lesson. If not, run through the section on carrying and try again.

By reviewing and practicing all the basic material, as you are doing, you'll find the GED-level questions easy when you get to them in Lesson 4.

Warm-up Answers
1. 72 **2.** 64 **3.** 164 **4.** 45 **5.** 147 **6.** 432 **7.** 5,165
8. 6,698 **9.** 1,813 **10.** 3,292 **11.** 1,686 **12.** 531,247
13. 156,660 **14.** 2,688,664 **15.** 10,006,878

LESSON 3

Subtracting Whole Numbers

If you have a job, you know that what you earn is not the same as your take-home pay. Money is usually deducted to pay for state and federal income taxes and Social Security. Deductions for health or life insurance may also be taken out. Your take-home pay is what is left.

A *deduction* is something that is taken away, or subtracted. To understand how your take-home pay is computed, you must be able to subtract.

To subtract whole numbers you have to know your basic subtraction facts (7 − 2, 4 − 0, 9 − 5, 6 − 6, and so on). Maybe you're a bit rusty on these. You can easily brush up by using a pocket calculator. Or you can buy or make your own flash cards. To make your own, write the possible combinations on 3 × 5 cards. You'll need 55 cards. Write just one combination on each card and put the answer on the back.

Subtracting is a way of finding the **difference** of two numbers. To find the difference, subtract the smaller number from the larger. The answer tells you how much more the larger number is. Or you can look at it the other way around: the answer tells you how much less the smaller number is.

You also compute the difference when you want to do take-away problems to find how much is left.

Coming to Terms

difference the answer you get when you subtract the smaller of two numbers from the larger

Simple Subtraction

When you subtract whole numbers, write the numbers vertically, the larger number on top and the smaller number below it. Be careful to line up the places.

Here's an Example

Look at how to set up a subtraction problem.

■ An office sent out 36 samples. If there were 479 samples at the start, how many were left?

To solve, subtract 36 from 479.

Step 1 Write the problem vertically. Be sure to line up the places.

$$\begin{array}{r} 479 \\ -\ 36 \\ \hline \end{array}$$

Step 2 Subtract. Begin by subtracting the ones.

$$\begin{array}{r} 479 \\ -\ 36 \\ \hline 3 \end{array}$$

Then subtract the tens.

$$\begin{array}{r} 479 \\ -\ 36 \\ \hline 43 \end{array}$$

There is nothing to subtract from the 4 hundreds, so write 4 in the hundreds place of the answer.

$$\begin{array}{r} 479 \\ -\ 36 \\ \hline 443 \end{array}$$

There were 443 samples left.

Now look at another example.

■ Of 9,685 people at a rock concert 423 were adults, the others were teenagers. How many teenagers were at the concert?

To solve, subtract 423 from 9,685.

Step 1 Write the problem vertically.

$$\begin{array}{r} 9{,}685 \\ -\ 423 \\ \hline \end{array}$$

Step 2 Subtract.

$$\begin{array}{r} 9{,}685 \\ -\ 423 \\ \hline 9{,}262 \end{array}$$

There was nothing to subtract from the 9, so you just bring the 9 down.

There were 9,262 teenagers at the concert.

Try It Yourself

Here's one for you to try. Find the answer before you read the solution.

■ If 8,540 stadium seats out of 18,752 were empty, how many seats were filled?

You need to find the difference between all the seats and the empty seats. That will tell you how many seats were filled. Your setup will be

$$\begin{array}{r} 18{,}752 \\ -\ 8{,}540 \\ \hline \end{array}$$

Subtracting should have given you 10,212 as the number of filled seats.

A Test-Taking Tip

It's always a good idea to check your arithmetic on the GED Mathematics Test. You can check any subtraction problem by adding the answer to the number you subtracted. The result should be the top number in the original problem.

Subtract

$$\begin{array}{r} 18{,}752 \\ -\ 8{,}540 \\ \hline 10{,}212 \end{array}$$

Check by adding these two numbers. You get the top number.

Warm-up

Subtract the following.

1. 85 − 14 = ______

2. 349 − 32 = ______

3. 639 − 236 = ______

4. 1,047 − 21 = ______

5. 8,857 − 654 = ______

6. 5,895 − 2,704 = ______

7. 20,481 − 160 = ______

8. 99,999 − 9,999 = ______

9. 38,473 − 8,051 = ______

10. 42,168 − 21,133 = ______

Warm-up Answers
1. 71 **2.** 317 **3.** 403 **4.** 1,026 **5.** 8,203 **6.** 3,191
7. 20,321 **8.** 90,000 **9.** 30,422 **10.** 21,035

A Test-Taking Tip

On the GED Mathematics Test, words like these usually mean you should subtract: *less than, more than, difference, decrease, reduce, lost, left, remain, fell, dropped, change, deduct, smaller, greater,* and other *-er* comparison words.

On the Springboard

1. Paul flew to New York for $148. He returned by train for $32 less. How much did his train trip cost?

   (1) $106 (2) $116 (3) $126

   ① ② ③

2. The home team had 5,792 fans present at the game. There were 431 visiting fans. How many more home fans than visitor fans were there?

   (1) 5,231 (2) 5,361 (3) 6,223

   ① ② ③

3. One company invested $75,698 in a new product. The first month's sales were $3,425. How much more in sales is needed to break even?

   (1) $72,272 (2) $72,273
   (3) $79,123

   ① ② ③

Check your answers on page 182. If you solved all the Springboard problems correctly, you're ready to do subraction with borrowing. If you missed any, review the material and rework the problems you missed to be sure you can get them right. Then you'll be ready to move ahead.

Borrowing Once

Borrowing is tricky for some people. But it won't be for you if you remember that borrowing is simply renaming numbers so you can subtract. For example, 31 means 3 tens and 1 one. You could take one of the tens and call it 10 ones. Then you'd have 2 tens, 10 ones, and 1 one. That makes 2 tens and 11 ones. To rename 31 as 2 tens and 11 ones, you borrowed one of the tens. Subtraction problems often require you to borrow, and renaming makes it easier.

Coming to Terms

borrowing renaming numbers in subtraction

Here's an Example

Watch what happens in this problem.

■ In a recent election, 7,351 people voted on a new library plan. Of these, 4,260 voted for the plan. How many voted against it?

Subtract 4,260 from 7,351.

Step 1 Write the problem vertically.

$$\begin{array}{r} 7{,}351 \\ -\ 4{,}260 \\ \hline \end{array}$$

Step 2 Subtract. Start with the ones.

$$\begin{array}{r} 7{,}351 \\ -\ 4{,}260 \\ \hline 1 \end{array}$$

Move to the tens. You can't subtract 6 tens from 5 tens. But in the next column there are 3 hundreds. Each hundred is worth 10 tens. Borrow 10 tens. With the 5 tens already there, you now have 15 tens. Now you can subtract 6 from 15 in the tens column. When you do that, you get 9, and you write that in the answer.

$$\begin{array}{r} {}^{2\,15} \\ 7{,}351 \\ -\ 4{,}260 \\ \hline 91 \end{array}$$

Now go to the hundreds and subtract 2 from 2. That's 0, so you write 0 in the hundreds place in the answer. Then in the thousands column subtract 4 from 7. You get 3 and you write that in the answer.

$$\begin{array}{r} {}^{2\,15} \\ 7{,}351 \\ -\ 4{,}260 \\ \hline 3{,}091 \end{array}$$

3,091 people voted against the library plan.

Try It Yourself

Try to remember all you've learned about subtraction so far.

■ Of 6,945 raffle tickets, 1,238 were sold. How many tickets were left over?

To get the answer, subtract 1,238 from 6,945. Write the problem vertically.

$$\begin{array}{r} 6{,}945 \\ -\ 1{,}238 \\ \hline \end{array}$$

You had to borrow 10 ones from the tens column. Then you subtracted 8 from 15. The 4 in the tens column became 3. You could subtract everything else without borrowing. There were 5,707 tickets left over.

Warm-up

Subtract the following numbers. You will have to borrow in each problem.

1. 53 − 26 = ______
2. 85 − 9 = ______
3. 442 − 105 = ______
4. 6,379 − 183 = ______
5. 708 − 532 = ______
6. 9,576 − 4,734 = ______
7. 2,155 − 1,435 = ______
8. 67,329 − 22,817 = ______
9. 88,348 − 52,019 = ______
10. 74,075 − 68,054 = ______

Warm-up Answers
1. 27 **2.** 76 **3.** 337 **4.** 6,196 **5.** 176 **6.** 4,842 **7.** 720 **8.** 44,512 **9.** 36,329 **10.** 6,021

On the Springboard

4. A senior flight attendant had flown 97,805 air miles. A junior attendant had flown 6,582 air miles. How many more miles had the senior attendant flown?

   (1) 91,223 (2) 91,323 (3) 91,383

   ① ② ③

5. What is the difference in price between a house selling for $77,695 and one selling for $88,375?

   (1) $720 (2) $10,680 (3) $11,680

   ① ② ③

Check your answers on page 182. If you answered both questions correctly, go on to the next section. If you missed either question, take time out to review the material on borrowing once, then try the questions again. When you're satisfied that you're ready, go on to "Borrowing More Than Once."

Borrowing More Than Once

If Jerry borrows money from Joe to pay his bills, and Joe doesn't have enough to pay his rent, Joe may have to borrow money from Harry. Borrowing more than once in subtraction is similar to this kind of a situation.

Here's an Example

In this problem, you have to borrow more than once.

■ One person paid $5,171 in mortgage interest and another paid $2,222. How much more interest did the first person pay than the second?

Subtract $2,222 from $5,171.

Step 1 Write the problem vertically.

```
  5,171
- 2,222
```

Step 2 Subtract. You will need to borrow to subtract the ones and the hundreds. If you write down everything to show exactly how the borrowing is done, the computation looks like this.

```
 4 11 6 11
  5, 1 7 1
 -2, 2 2 2
  2, 9 4 9
```

The first person paid $2,949 more in interest.

If the subtraction problem has zeros in it, don't let that throw you. You can't borrow from zero but you can borrow from the next digit to the left that is not zero. Just make your way from column to column. Borrow as you need to so that you can subtract in each column. Study the following example step by step. It has a zero in the tens column.

```
                6 10           9
                               6 10 12
  3,702        3,702        3,7 0 2
- 1,567      - 1,567      - 1,5 6 7
                            2,1 3 5
```

Here's a harder example. If you understand it, you'll be able to do just about any subtraction problem that requires borrowing more than once. Study the example carefully.

```
                               9            9 9
             3 10           3 10 10      3 10 10 12
  4,002      4,002          4,0 0 2      4,0 0 2
- 1,456    - 1,456        - 1,4 5 6    - 1,4 5 6
                                         2,5 4 6
```

Just keep borrowing from the next column to the left until you can subtract in each column.

Try It Yourself

Work this problem carefully, subtracting from the right, one column at a time.

■ A car and its passengers weigh 4,004 pounds. The total weight of the passengers is 895 pounds. How much does the car alone weigh?

Try it on your own before looking at the solution. Subtract 895 from 4,004. As soon as you write it vertically, you can see a small difficulty.

```
  4,004
-   895
```

You can't take 5 from 4, so you must borrow. But the first column you can borrow from is the thousands column. If you borrowed from the thousands to the hundreds to the tens to the ones, you did a great job. If not, study the previous example and see if it helps put you on the right track. Then try again. The answer should be 3,109 pounds.

Warm-up

Subtract the following.

1. 9,478 − 8,977 = ______

2. 18,020 − 4,395 = ______

3. 2,350 − 999 = ______

4. 36,000 − 4,888 = ______

5. 99,111 − 8,877 = ______

6. 23,005 − 1,389 = ______

7. 43,945 − 15,876 = ______

8. 329,000 − 2,989 = ______

A Test-Taking Tip

Some problems on the GED Mathematics Test may refer to *gross pay* and *net pay*. *Gross pay* is what you earn before deductions. *Net pay* is your take-home pay—what is left after paying the deductions. In terms of subtraction

Gross − Deductions = Net.

On the Springboard

6. John's gross pay for a week's work is $255. His deductions amount to $89. How much is his weekly net pay?

   (1) $166 (2) $176 (3) $234

   (1) (2) (3)

7. In one year, a country's exports were worth $179 million. The value of its imports in the same year was $217 million. How much more did the country import than it exported in millions of dollars?

   (1) 38 (2) 48 (3) 138

   (1) (2) (3)

Check your answers on page 182. If you got these problems right, give yourself a pat on the back and go on to Lesson 4. If you got either of them wrong, have another look at the material on borrowing. When you can answer both correctly, you will be ready for Lesson 4.

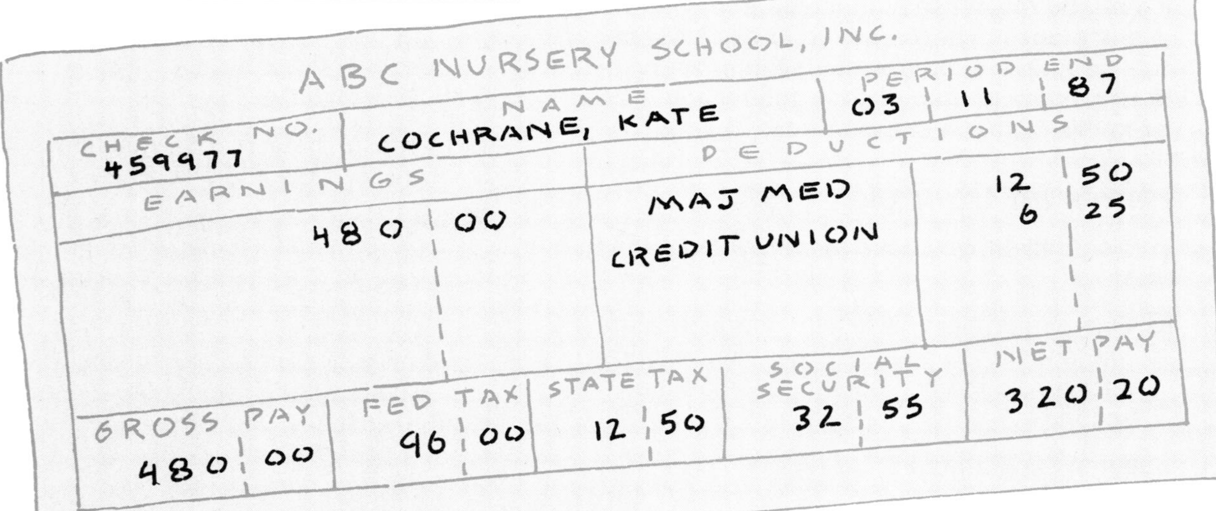

ABC NURSERY SCHOOL, INC.

CHECK NO.	NAME	PERIOD END		
459977	COCHRANE, KATE	03	11	87

EARNINGS		DEDUCTIONS		
480	00	MAJ MED	12	50
		CREDIT UNION	6	25

GROSS PAY	FED TAX	STATE TAX	SOCIAL SECURITY	NET PAY
480.00	96.00	12.50	32.55	320.20

Warm-up Answers
1. 501 **2.** 13,625 **3.** 1,351 **4.** 31,112 **5.** 90,234
6. 21,616 **7.** 28,069 **8.** 326,011

LESSON 4

Zeroing In on Word Problems: Two-Step Problems

The GED Mathematics Test will include practical, real-life problems. Because many practical problems involve more than one step, you'll want to practice solving two-step problems before the real test.

Here's an Example

Remember the system for solving word problems that you used in Lesson 1. Look at this problem.

■ Ben's savings account had a balance of $200. Ben withdrew $45. The interest added the day after the withdrawal was $3. What was Ben's new balance?

There are practical terms in the problem such as *balance* (the amount of money Ben has in his account), *withdrew* (taking money from the account), and *interest* (a payment made to Ben in return for the use of his money).

First, read to find out what the problem is asking for. It's asking for a balance in dollars.

Then decide what operations to use. A *withdrawal* is money subtracted and *interest* is money added. You'll need to subtract, then add.

Step 1	Set up the withdrawal.	$200
	Do the subtraction.	−45
	Check your work.	$155
Step 2	Add the interest.	$155
	Do the addition.	+3
	Check your work.	$158

Ben's new balance was $158.

Try It Yourself

Now try out a two-step problem for yourself. Be careful when you decide which operations to use.

■ A member of a dieting club weighed 135 pounds at the club's first meeting. At the weigh-in for the second meeting she was delighted to have lost 5 pounds. She went on a vacation, and at her next club weigh-in found that she had gained 13 pounds. What was her weight then?

What does the problem ask you to find? If you said it asks you to find the diet club member's weight after her vacation, you're right. The problem tells you how much she weighed when she joined, how much she lost after the first meeting, and how much she gained on vacation. That should be enough to let you figure out her weight after vacation.

You can tell which operations to use from the words *lost* (subtract) and *gained* (added). Just subtract 5 from the 135 and you get 130. Next add 13 and get 143. She weighed 143 pounds after her vacation.

Warm-up

Write down the operations you would use to solve these problems in the order in which you would use them. Don't solve.

1. A home decorator estimated his labor for a job at $635. He then offered the homeowner a $25 discount. The materials cost an additional $150. What was the total cost of the job?

____

2. Three friends went canoeing. They loaded 50 pounds of gear into the canoe. The friends weighed 180 pounds, 145 pounds, and 130 pounds. The canoe could safely hold 500 pounds. How many pounds of gear had to be removed?

____

3. The customer service operators made 60 phone calls on Monday and 55 on Tuesday. Incoming calls totaled 150 for the same two days. How many more incoming calls were there than outgoing calls over the two days?

____

4. A warehouse foreman noted that 6 workers were absent on Wednesday. On Tuesday there were 30 workers on the job. If the total number of workers is 34, how many more workers were absent on Wednesday than on Tuesday?

On the Springboard

Now try solving some two-step problems.

1. A composer wrote 30 songs. Of these she had 11 songs published by one company and 13 by another. How many songs were *not* published?

   (1) 2 (2) 6 (3) 24

   ① ② ③

2. A time-and-motion study showed that the average worker plugged in about 3,000 components a day in the circuit-board section of the factory. Rita installed 1,450 components before lunch and 1,800 after lunch. How many more components did she install than the average worker?

   (1) 250 (2) 350 (3) 3,250

   ① ② ③

3. A restaurant took in a record amount of \$755, \$860, and \$1,115 over one weekend. The previous record for a weekend was \$2,005. How much more money was taken in on the new record weekend?

   (1) \$725 (2) \$735 (3) \$2,730

   ① ② ③

Check your Springboard answers on pages 182–183. If you got these right you are doing well. If you missed any, be sure to go over this lesson again before moving on to "The Real Thing."

Warm-up Answers
1. subtract, then add **2.** add, then subtract **3.** add, then subtract **4.** subtract, then subtract again

The Real Thing 

1. Ron lives with his parents and pays \$185 a month for rent. He could rent his own apartment for \$238 a month. He would have to pay \$23 a month for electricity, \$7 for gas for cooking, and \$35 for telephone service. How much is he saving by living with his parents?

   (1) \$18 (2) \$108 (3) \$118
   (4) \$222 (5) \$488

   ① ② ③ ④ ⑤

2. Lynn is on a 1,200 calorie-a-day diet. For lunch she has a sandwich, which has 268 calories; a cup of skim milk, 88 calories; and an orange, 75 calories. How many more calories is she allowed for the day?

   (1) 431 (2) 769 (3) 779
   (4) 879 (5) 1,631

   ① ② ③ ④ ⑤

3. A business school has an enrollment of 750 students. Of these students, 67 are enrolled in secretarial skills, 54 in word processing, 49 in bookkeeping, and 62 in office management. How many students are enrolled in other courses?

   (1) 222 (2) 232 (3) 518
   (4) 528 (5) 982

   ① ② ③ ④ ⑤

4. Linda lives 208 miles from her mother's house. If she stops at her sister's house on the way to her mother's, it is another 54 miles. If she stops to see her sister on the way to her mother's and on the way back, how many miles must Linda drive?

   (1) 306 (2) 316 (3) 470
   (4) 504 (5) 524

   ① ② ③ ④ ⑤

5. Lewis had 3,500 new telephone directories to deliver last week. He delivered 489 on Monday, 512 on Tuesday, 544 on Wednesday, and 530 on Thursday. How many directories did he have to deliver on Friday?

   (1) 1,415 (2) 1,425 (3) 1,515
   (4) 1,525 (5) 5,585

   ① ② ③ ④ ⑤

Check your answers and record your score on page 195.

LESSON 5

Multiplying Whole Numbers

A parking garage has 7 levels. Each level has parking spaces for 148 cars. How many cars can the garage hold? To solve a problem like this, you have to know how to multiply whole numbers.

First, be sure you know all the basic multiplication facts. These are the multiplication combinations that involve multiplying a one-digit number by a one-digit number: 2×8, 5×3, 0×4, and so on. To check yourself on these or to practice them, you can use a calculator. Can you give the answers quickly? Do you get the same answer as the calculator? If it's not possible to use a calculator, you can make a set of flash cards. You will need 100 cards. Write one combination on each card and put the answer on the back. As you practice, you'll discover that the job is not as big as it seems. Multiplying by 0 always gives 0 as an answer: $2 \times 0 = 0$, $5 \times 0 = 0$, and so on. Multiplying by 1 always gives the same number: $7 \times 1 = 7$, $4 \times 1 = 4$, and so on. Once you're sure of the basic facts, you're on the way to doing any kind of multiplication problem.

Multiplying by a One-Digit Number

To multiply two whole numbers, write the problem vertically. You'll probably find it easiest to write the larger number on top and the smaller number on the bottom. The number you write on the bottom is called the **multiplier.** Be sure to line up the places.

Coming to Terms

multiplier the bottom number in a multiplication problem

A Test-Taking Tip

When you see the word *double* in a question on the GED Mathematics Test, it means to multiply by 2. The word *triple* means to multiply by 3.

Here's an Example

Look at this problem. Remember the meaning of *double.*

■ A furniture factory made 431 tables in June. In July management put in new equipment and doubled the output. How many tables were produced in July?

To solve, multiply 431 by 2.

Step 1 Write the problem vertically, with 2 on the bottom.

$$\begin{array}{r} 431 \\ \times \quad 2 \\ \hline \end{array}$$

Step 2 Begin with the ones. Think 2×1. Write the answer 2 in the ones place.

$$\begin{array}{r} 431 \\ \times \quad 2 \\ \hline 2 \end{array}$$

Step 3 Move to the tens. Think 2×3. Write the answer 6 in the tens place.

$$\begin{array}{r} 431 \\ \times \quad 2 \\ \hline 62 \end{array}$$

Step 4 Move to the hundreds. Think 2×4. Write the answer 8 in the hundreds place.

$$\begin{array}{r} 431 \\ \times \quad 2 \\ \hline 862 \end{array}$$

The answer is 862 tables.

Sometimes when you multiply you have to carry. Just as in addition, you write down the right-hand digit and carry the left-hand digit. Look at this example.

■ If a company cafeteria prepares 485 meals each day, how many meals are prepared over 4 days?

To solve, multiply 485 by 4.

Step 1 Write the problem vertically with the smaller number on the bottom.

$$\begin{array}{r} 485 \\ \times \quad 4 \\ \hline \end{array}$$

Step 2 Multiply. Think $4 \times 5 = 20$. Write down the 0 in the ones place in the answer. Carry 2 to the tens column.

$$\begin{array}{r} {\scriptstyle 2} \\ 485 \\ \times \quad 4 \\ \hline 0 \end{array}$$

Step 3 Multiply the tens. Think 4 × 8 = 32. Add the 2 that you carried, which makes 34. Write 4 in the tens place in the answer. Carry the 3 to the hundreds column.

$$\begin{array}{r} {}^{3\,2}\\ 485 \\ \times\quad 4 \\ \hline 40 \end{array}$$

Step 4 Multiply the hundreds. Think 4 × 4 = 16. Add the 3 that you carried, which makes 19. Write the 9 in the hundreds place. Carry the 1 and write it in the thousands place in the answer (since there is nothing more to multiply).

$$\begin{array}{r} {}^{3\,2}\\ 485 \\ \times\quad 4 \\ \hline 1{,}940 \end{array}$$

Over 4 days, the cafeteria prepared 1,940 meals.

Try It Yourself

■ If 7 people each purchased an airline ticket for $567, how much did they pay altogether for their tickets?

Do this problem on your own. Then check your work with the solution. The 567 should be the top number. Multiply the ones first. 7 × 7 is 49. Write the 9 in the ones place of the answer and carry the 4 to the tens column. Multiply the tens. 7 × 6 is 42 and the 4 that you carried makes 46. Write 6 in the tens place of the answer and carry the 4 to the hundreds column. 7 × 5 is 35 and the 4 that you carried makes 39. In the answer, write 9 in the hundreds place and 3 in the thousands place. They paid $3,969 for the tickets.

Warm-up

Multiply these whole numbers.

1. 42 × 2 = ______ **6.** 5 × 305 = ______

2. 9 × 71 = ______ **7.** 421 × 3 = ______

3. 703 × 7 = ______ **8.** 4 × 149 = ______

4. 3 × 474 = ______ **9.** 796 × 8 = ______

5. 830 × 6 = ______ **10.** 9 × 687 = ______

Warm-up Answers
1. 84 **2.** 639 **3.** 4,921 **4.** 1,422 **5.** 4,980 **6.** 1,525
7. 1,263 **8.** 596 **9.** 6,368 **10.** 6,183

✓ A Test-Taking Tip

On the GED Mathematics Test, words like *times, each of, as much, twice, multiplied, by, every, pair, double,* and *triple* usually mean you should multiply.

On the Springboard

1. If you have 6 parts packed in each of 575 cartons, how many parts do you have in all?

 (1) 3,050 (2) 3,440 (3) 3,450

 (1) (2) (3)

2. If there are 220 tables and each will seat 4 people, how many people can be seated?

 (1) 88 (2) 660 (3) 880

 (1) (2) (3)

Check your Springboard answers on page 183. If you answered both questions correctly, congratulations! You are ready to go on to "The Real Thing." If you missed either question, it would be a good idea to review the material on multiplying by one-digit numbers. Then try the questions again. When you're sure you're ready, go on to "The Real Thing."

"The Real Thing"

1. Eric is paid $105 a week. He has $13 deducted from his weekly pay for taxes and Social Security. How much is Eric's take-home pay for 5 weeks?

 (1) $450 (2) $460 (3) $525
 (4) $560 (5) $590

 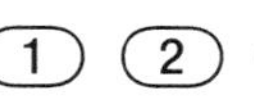  

2. Ms. Morales had $154 in her checking account. She wrote 3 checks for $25 each. How much did she have left in her checking account?

(1) $65 (2) $79 (3) $89
(4) $129 (5) $229

(1) (2) (3) (4) (5)

Check your answers on page 195.

Multiplying by Two-Digit and Three-Digit Numbers

Here's an Example

■ A company paid $78 each for 51 uniforms for its employees. How much did the uniforms cost altogether?

To solve this kind of problem, you need to be able to do the multiplication problem 78 × 51.

Step 1 Write the problem vertically.

```
   78
x  51
```

Step 2 First multiply by the digit in the ones place of the multiplier. Multiply 78 by 1. Be sure to keep the digits lined up as you write the answer.

```
   78
x  51
   78
```

Step 3 Multiply by 5, the digit in the tens place of the multiplier. As you multiply 78 by 5, start with 5 × 8. The answer is 40. Write 0 in the tens place, and carry the 4.

```
    4
   78
x  51
   78
   0
```

Step 4 Next do 7 × 5. The answer is 35. With the 4 you carried, that makes 39. Write down the 39. Notice when you do this that the 3 goes in the thousands column, and the 9 in the hundreds column.

```
    4
   78
x  51
   78
 390
```

Step 5 That's all the multiplying you have to do. Now draw a line under 390 and add.

```
    4
   78
x  51
   78
 390
3,978
```

The uniforms cost $3,978.

Multiplying by a three-digit number takes a little longer but is really no more difficult than the last problem.

■ 517 supermarkets each ordered a sample case of shampoo. Each case held 212 tubes. How many tubes were ordered?

To solve, you must compute 517 × 212.

Step 1 Write the problem vertically.

```
   517
x  212
```

Step 2 Multiply by the digit in the ones place of the multiplier. Try to do the carrying in your head. Be sure to put the right-hand digit (4) of the answer in the ones place.

```
   517
x  212
 1,034
```

Step 3 Multiply by the digit in the tens place of the multiplier. You get 517. Be sure you put the 7 in the tens place (under the 3 in the line above).

```
   517
x  212
 1,034
  5 17
```

Step 4 Multiply by the digits in the hundreds place of the multiplier. As before, do the carrying mentally. The right-hand digit of your answer goes in the hundreds column.

```
   517
x  212
 1,034
  5 17
1,03 4
```

Step 5 Draw a line under this last part of the answer. Add to get the final answer.

```
    517
x   212
  1,034
   5 17
 1,03 4
1 09 604
```

In all, 109,604 bottles were shipped.

Try It Yourself

Try this problem on your own before looking at the answer and explanation.

■ A planning committee decided to seat 12 people at each of 189 tables. How many place settings were needed?

You should have written the problem 189 × 12 vertically. Then you should have multiplied 189 by 2. The answer (378) should have been written with the 8 in the ones column. Then you should have multiplied by the 1 of the multipler. The answer (189) should have been written with the 9 in the tens column. Adding, you get 2,268 as the number of place settings needed.

Try another one.

 There are 144 items in one gross. If you order 25 gross of pencils, how many pencils have you ordered?

You should have written 144 × 25 vertically. Multiply 144 by 5. The answer is 720. The 0 should be in the ones column. Next multiply by 2. You get 288. The 8 on the right should be in the tens column. Then add. The final answer is 3,600 pencils.

A Test-Taking Tip

If one of the numbers you are multiplying ends in zero, you can save time by putting it on the bottom line. For example, to multiply 53 by 70, write the problem vertically.

	53
	x 70
Multiply by 0 ⟶	00
Multiply by 7 ⟶	371
Add ⟶	3,710

Warm-up

Multiply the following.

1. 73 × 31 = ____________
2. 86 × 42 = ____________
3. 555 × 22 = ____________
4. 70 × 236 = ____________
5. 293 × 62 = ____________
6. 636 × 75 = ____________
7. 4,632 × 27 = ____________
8. 46 × 2,843 = ____________
9. 679 × 312 = ____________
10. 3,848 × 690 = ____________
11. 80 × 45 = ____________
12. 321 × 30 = ____________
13. 56 × 65 = ____________
14. 67 × 82 = ____________
15. 120 × 37 = ____________
16. 374 × 900 = ____________
17. 1,324 × 56 = ____________
18. 3,546 × 78 = ____________
19. 354 × 687 = ____________
20. 7,531 × 23 = ____________

On the Springboard

3. How many eggs are there in 53 dozen?

   (1) 156 (2) 635 (3) 636

   ① ② ③

4. If 28 night-school students each took 18 math quizzes in a semester, how many quizzes had to be graded?

   (1) 456 (2) 504 (3) 544

   ① ② ③

Check your answers on page 183. If you got them right, go straight on to "The Real Thing." If you got either of them wrong, why not review the previous section and then try them again?

The Real Thing

3. Last year Rosa paid $380 per month in rent. This year she pays $405 per month. How much more will she pay in rent for the 12 months of this year?

   (1) $75 (2) $180 (3) $200
   (4) $300 (5) $1,500

   ① ② ③ ④ ⑤

4. A company purchased 14 pairs of trousers at $21 each. They spent 3 times as much for jackets. How much did they spend for the jackets?

   (1) $98 (2) $294 (3) $297
   (4) $672 (5) $882

   ① ② ③ ④ ⑤

Warm-up Answers
1. 2,263 **2.** 3,612 **3.** 12,210 **4.** 16,520 **5.** 18,166 **6.** 47,700 **7.** 125,064 **8.** 130,778 **9.** 211,848 **10.** 2,655,120 **11.** 3,600 **12.** 9,630 **13.** 3,640 **14.** 5,494 **15.** 4,440 **16.** 336,600 **17.** 74,144 **18.** 276,588 **19.** 243,198 **20.** 173,213

5. Last year Mr. Young put $45 into a savings account each month. Mrs. Young put $60 a month into the same account. Together how much money did they put into the savings account last year?

(1) $180 (2) $305 (3) $1,260
(4) $1,380 (5) $1,700

① ② ③ ④ ⑤

Check your answers on pages 195–196.

Estimating in Multiplication

The GED Mathematics Test will not ask you to estimate, but estimating can help you save time and check answers. To estimate a multiplication answer, round numbers that have more than one digit to make all digits after the first one zero.

Here's an Example

■ A truck driver has the same route every week. The distance from his starting point to his destination is 516 miles. He drives this distance four times a week. Estimate how many miles he drives each week.

To solve, round 516 to 500. Multiply by 4.

```
  500
x   4
2,000
```

He drives about 2,000 miles each week.

■ A hotel serves about 1,985 room-service meals a month. About how many room-service meals is that per year?

To solve, you must estimate 1,985 × 12. Round 1,985 to 2,000 and round 12 to 10. Then multiply.

```
 2,000
x   10
 0 000
20 00
20,000
```

They serve about 20,000 room-service meals a year.

There is a helpful shortcut for multiplying numbers that end with zeros. Take the zeros from the ends of the numbers. Count how many zeros you have. Multiply the nonzero parts, then, at the right end of the multiplication result, write the number of zeros you took off.

For example, suppose you have to multiply 800 by 60.

Step 1 800 × 60 → Take off the end zeros (three zeros) → 8 × 6 = 48

Step 2 Put three zeros after the 48.

The answer is 48,000.

You can use this short cut in the following example.

■ Estimate the number of hours in one year.

To solve, remember that there are 365 days in a year and 24 hours in each day. You must estimate 365 × 24.

Step 1 Round 365 to 400 and round 24 to 20.

Step 2 Multiply 4 × 2 to get 8, then write three zeros after the 8.

There are about 8,000 hours in a year.

To use estimation to check multiplication answers, first compute the actual answer. Then estimate the answer. If your estimate is nowhere close to the actual answer you computed, double-check your work because you probably made an error.

Try It Yourself

Try this next one on your own before you check the answer.

■ A factory produced 8,902 window cranks each day. After 211 working days, how many window cranks were produced? First multiply to find the answer. Then estimate to check your answer.

Answer ________ Estimate ________

Did you multiply to get 1,878,322? To check did you round 8,902 to 9,000 and 211 to 200 and then multiply 9,000 by 200? Did you get 1,800,000? This estimate is reasonably close to 1,878,322.

Warm-up

Multiply to find the exact answers.

1. 45 × 40 = ______
2. 34 × 22 = ______
3. 13,000 × 620 = ______
4. 695 × 7,821 = ______
5. 840 × 6,004 = ______
6. 2,060 × 410 = ______

Now round and multiply to check your answers.

7. 45 × 40 = ______
8. 34 × 22 = ______
9. 13,000 × 620 = ______
10. 695 × 7,821 = ______
11. 840 × 6,004 = ______
12. 2,060 × 410 = ______

A Test-Taking Tip

Estimating can often help you to eliminate wrong choices and check your answers. The skill of estimating is important to you both in the test itself and in real life.

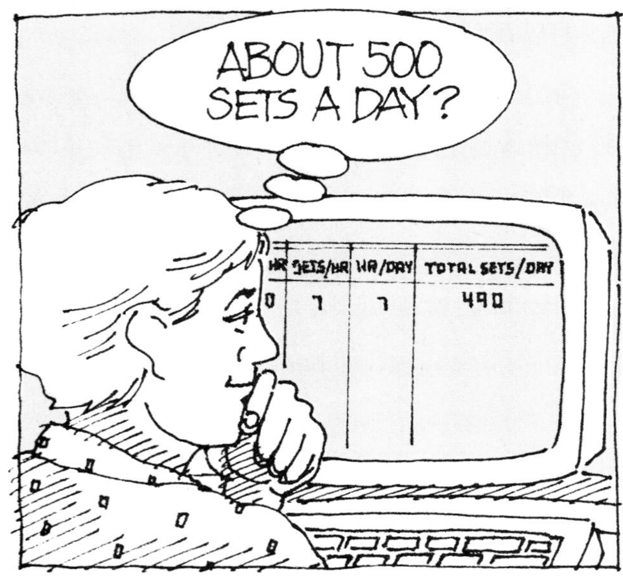

Warm-up Answers
1. 1,800 **2.** 748 **3.** 8,060,000 **4.** 5,435,595 **5.** 5,043,360 **6.** 844,600 **7.** 2,000 **8.** 600 **9.** 6,000,000 **10.** 5,600,000 **11.** 4,800,000 **12.** 800,000

On the Springboard

5. If 675 greenhouses each contain 2,902 plants, estimate how many plants there are in all.

(1) 1,958,850 (2) 2,000,000
(3) 2,100,000

① ② ③

6. Multiply 2,091 cars by $7,020.

(1) $1,505,520 (2) $14,678,820
(3) $15,380,820

① ② ③

Check your answers on pages 183. If you got both Springboard answers right, go on to "The Real Thing." If you did not, reread the material on multiplying by 3-digit multipliers and multiplying numbers that end in zeros. Then answer the questions again.

 The Real Thing ❞

6. A popular rock star's latest record is on sale for $8. If 60,047 copies of the record are sold during the month the record is on sale, what are the total sales for that month?

(1) $48,376 (2) $480,376 (3) $510,076
(4) $4,000,376 (5) $4,800,376

① ② ③ ④ ⑤

7. Mt. McKinley, the highest point in the state of Alaska, is 59 times higher than the highest point in the state of Florida. If the highest point in Florida is 345 feet above sea level, how many feet above sea level is Mt. McKinley?

(1) 18,000 (2) 20,155 (3) 20,355
(4) 20,700 (5) 21,315

① ② ③ ④ ⑤

8. On Monday morning a store had 30 boxes of pencils with 240 pencils in each box. During the day they gave away 3,684 of the pencils. How many pencils were left?

(1) 2,626 (2) 2,964 (3) 3,516
(4) 7,200 (5) 10,884

① ② ③ ④ ⑤

9. The area of Park A is approximately 20 times the combined areas of Parks B and C. The area of Park B is approximately 3,500 acres and the area of Park C is approximately 1,700 acres. Approximately how many acres is the area of Park A?

(1) 3,012 (2) 3,120 (3) 104,000
(4) 301,200 (5) 312,000

① ② ③ ④ ⑤

10. In April, employees of a trucking firm drove an average of 2,957 miles a day over 21 days. In May they drove a total of 71,462 miles. How many miles more did they drive in May than in April?

(1) 9,365 (2) 19,475 (3) 20,365
(4) 53,226 (5) 133,559

① ② ③ ④ ⑤

11. In the 31 days of July, a newspaper sold an average of 1,086 papers a day. They sold 5,374 more papers in June than in July. How many papers did they sell in June?

(1) 9,718 (2) 11,140 (3) 38,830
(4) 38,930 (5) 39,040

① ② ③ ④ ⑤

12. A mail order company sells boxes of fishhooks. Each box contains 144 cards with 12 fishhooks on each card. How many fishhooks are in 18 of these boxes?

(1) 3,888 (2) 7,776 (3) 15,552
(4) 30,004 (5) 31,104

① ② ③ ④ ⑤

13. Margaret spent $537 at a sporting goods store. She bought 24 boxes of golf balls at $11 a box and spent the rest for golf trophies. How much money did she spend for trophies?

(1) $57 (2) $273 (3) $374
(4) $489 (5) $801

① ② ③ ④ ⑤

Check your answers and record all your scores on page 196.

LESSON 6

Dividing Whole Numbers

If you drive a car, you probably check sometimes to see how many miles you've driven between gas stops, and how much gas you've used. You can then figure out your miles per gallon by using division. Knowing how to divide can come in handy for all sorts of things—paying bills, sharing expenses, figuring mileage, and so on.

Writing Division Problems

You can write a division problem in three different ways.

Here's an Example

Look at the three ways to write 96 divided by 3:

$96 \div 3$ $\frac{96}{3}$ $3\overline{)96}$

These three all mean the same thing. As you probably remember, you use the last way of writing the problem when you want to calculate the answer.

Try It Yourself

Now see if you can do it.

■ Write the problem 36 divided by 4 in three different ways.

________ or ________ or ________

You should have written $36 \div 4$, $36/4$, and $4\overline{)36}$. (Do you know what the answer is for each of these? If you said 9, you were right.)

Warm-up

Write each division problem in three ways. You do not have to work the problems.

1. 56 divided by 8

2. 789 divided by 3

3. 903 divided by 7

Dividing by One-Digit Numbers

Before you go on, make sure that you know the division facts by heart. If you need to practice them, you can use a calculator. You will probably find it easier to practice these facts if you buy or make a set of flash cards. To be sure you include *all* the division facts, use a set of 90 cards. On a separate piece of paper, list the multiplication facts that start with 9. Use these to help get all the division facts for dividing by 9.

Multiplication Fact	Division Card (Front)	Division Card (Back)
9 × 0 = 0	0 ÷ 9	0
9 × 1 = 9	9 ÷ 9	1
9 × 2 = 18	18 ÷ 9	2
9 × 3 = 27	27 ÷ 9	3
.	.	.
.	.	.
.	.	.
9 × 9 = 81	81 ÷ 9	9

Repeat this process for the multiplication and division facts that begin with 8, then with those that begin with 7, and so on, down to those that begin with 1. *Do not prepare cards for the multiplication facts that begin with zero.* In arithmetic *you never divide by zero.* Even with a calculator, you'll run into trouble if you try to divide by zero.

When you're confident that you're good enough with the division facts, you're ready to go on.

Warm-up Answers

1. 56 ÷ 8, $\frac{56}{8}$, 8)56 **2.** 789 ÷ 3, $\frac{789}{3}$, 3)789

3. 903 ÷ 7, $\frac{903}{7}$, 7)903

Here's an Example

■ Two landscapers were paid $846 for a job. They split the money evenly. How much did each get?

To solve, divide 846 by 2.

Step 1 Write the problem in this form. 2)846

Step 2 Divide 8 by 2 and put the answer, 4, *directly above* the 8.

```
   4
2)846
```

Step 3 In the same way, divide the 4 by 2, and put the answer, 2, directly above the 4.

```
   42
2)846
```

Step 4 Divide the 6 by 2 and put the answer, 3, directly above the 6.

```
   423
2)846
```

The landscapers each earned $423.

This example was simple because 2 goes into 8, 4, and 6 exactly. But sometimes the number you are dividing by may not go exactly into each digit. Study the next example carefully.

■ A group of 5 performers earned $1,830 for a week's work. How much did each performer earn?

To solve, divide 1,830 by 5.

Step 1 Rewrite the problem in this form. 5)1,830

Step 2 How many 5s in 1? Not even one, so the first division is 18 ÷ 5. The answer is 3 with some left over. Put the 3 above the 8, since the number you divided was 18; the answer goes above the right-hand digit of the number you just divided.

```
     3
5)1,830
```

Step 3 Next multiply the answer 3 by the 5 and write the answer, 15, below the 18.

```
     3
5)1,830
  1 5
```

Step 4 Now subtract 15 from 18. Write the answer, 3.

```
     3
5)1,830
  1 5
    3
```

Step 5 Bring down the next digit of the number you are dividing (in this case it is 3) and write it next to the 3 you got when you subtracted in Step 4.

```
     3
5)1,830
  1 5
    33
```

Step 6 Divide 5 into 33. The answer is 6 and some left over. Put the 6 directly above the 3 in 1,830. Then multiply the 6 by the 5 and put the answer, 30, under the 33 and subtract.

$$\begin{array}{r} 36 \\ 5\overline{)1,830} \\ 15 \\ \hline 33 \\ 30 \\ \hline 3 \end{array}$$

Step 7 Carry on with the same pattern. Bring down the last digit, 0, and divide 5 into 30. Put the answer, 6, right above the 0 in 1,830. Multiply 6 by 5 and put the answer, 30, under the 30 at the bottom. Subtract. The answer is 0. Since there are no more digits left to bring down, you're done.

$$\begin{array}{r} 366 \\ 5\overline{)1,830} \\ 15 \\ \hline 33 \\ 30 \\ \hline 30 \\ 30 \\ \hline 0 \end{array}$$

Each performer earned $366.

Test-Taking Tips

When you solve division problems on the GED Mathematics Test, remember to put a digit in the answer every time you divide. For instance, in $9\overline{)936}$, the 9 doesn't go into 3 so you would put a 0 in the answer and then divide 9 into 36. The answer is 104 *not* 14.

Checking your arithmetic when you take the GED Mathematics Test is very important. Everyone makes mistakes. If you catch your mistakes before you mark your answer, your score will be higher. Division answers are easy to check. Multiply your answer by the number you were dividing by. You should get the number you were dividing. For example, suppose you solve the problem 1,728 ÷ 4 and get 432. To check, multiply 432 by 4. If you do the multiplication you get 1,728. The answer 432 is correct. When you check an answer and know that it's correct, it keeps your confidence up as you work through the test.

Try It Yourself

■ Three prize winners split a $1,200 prize. How much did each person receive?

Did you divide 1,200 by 3? Doing the division 12 ÷ 3 gives you 4. When you divide each 0 in 1,200 by 3, you get 0 each time. The answer is each received $400.

■ There was a dead heat, or tie, for first place among 4 marksmen at the rifle championships. The scores totaled 2,316. How many points had each marksman scored?

You should have divided 2,316 by 4. The first digit of the answer goes above the 3 because you have to divide 23 by 4 to start. The answer is 579. Did you check by multiplying?

Warm-up

Divide these numbers.

1. 963 ÷ 3 = ____________

2. 804 ÷ 4 = ____________

3. 152 ÷ 8 = ____________

4. 144 ÷ 6 = ____________

5. 818 ÷ 2 = ____________

6. 285 ÷ 5 = ____________

7. 2,548 ÷ 4 = ____________

8. 6,055 ÷ 7 = ____________

9. 1,827 ÷ 3 = ____________

10. 2,964 ÷ 3 = ____________

11. 5,272 ÷ 8 = ____________

12. 4,854 ÷ 6 = ____________

A Test-Taking Tip

On the GED Mathematics Test, the following words usually mean that you should divide: *split, cut, divided evenly, each, part of, equal pieces, every, out of, shared, among, between, distribute, in half, in thirds, in fourths.*

Warm-up Answers
1. 321 **2.** 201 **3.** 19 **4.** 24 **5.** 409 **6.** 57 **7.** 637
8. 865 **9.** 609 **10.** 988 **11.** 659 **12.** 809

On the Springboard

1. There were 6 partners who shared equally in a profit of \$51,414 made on a sale. How much did each partner get?

   (1) \$8,568 (2) \$8,569 (3) \$9,569

   (1) (2) (3)

2. A truck removed 5 equal loads of bricks weighing a total of 14,625 pounds. How many pounds did each load weigh?

   (1) 2,925 (3) 3,925 (3) 20,925

   (1) (2) (3)

Check your answers on page 183. If you got both of these right, you are making real progress, so go on to "The Real Thing" and do some GED-level problems. If you got either answer wrong, don't be too bothered. After all, you've made it to Lesson 6! Review dividing by one-digit numbers and try the problems again.

The Real Thing 

1. A shipper put 235 red thermos bottles and 213 blue thermos bottles into cartons. There were 8 bottles in each carton. How many cartons did the shipper use?

   (1) 51 (2) 56 (3) 66
   (4) 501 (5) 506

   (1) (2) (3) (4) (5)

2. An automobile plant produced 98,395 sedans and 56,135 station wagons in 5 months. If the same number of cars were made each month, how many cars were produced monthly?

   (1) 772,650 (2) 77,265 (3) 30,906
   (4) 3,906 (5) 3,096

   (1) (2) (3) (4) (5)

Check your answers on page 196.

Division with Remainders

Not all real-life division problems come out exactly. Sometimes you will have numbers left over; these numbers are called **remainders.** Although the GED Mathematics Test seldom has problems with remainders, some may be included, so you should know how to handle remainders.

Coming to Terms

remainder the amount left over as undivided at the end of a division problem

Here's an Example

■ On Hot Dog Day, three schools were to divide the 709 donated hot dogs as evenly as possible. How many hot dogs should each school receive?

To solve, divide 709 by 3.

Step 1	Write the problem.	3)709
Step 2	Divide 7 by 3. The answer is 2 with some left over. Put the 2 directly above the 7.	2 3)709
Step 3	Start the pattern. Multiply 2 by the 3 and put the answer, 6, under the 7.	2 3)709 6
Step 4	Subtract 6 from 7.	2 3)709 6 1
Step 5	Bring down the next digit, 0.	2 3)709 6 10
Step 6	Now divide the 10 by 3. The answer is 3 and some left over. Put the 3 right above the 0 in 709. Multiply 3 by 3. Write 9 under the 10. Subtract 9 from 10 and bring down the next digit, which is 9.	23 3)709 6 10 9 19

Step 7 Divide 19 by 3. The answer is 6 and some left over. Write the 6 straight above the 9 in 709. Multiply 6 by 3 and write the answer, 18, under the 19. Subtract. There are no more digits to be brought down. After the answer 236, write a small *r* for remainder. Then write the remainder, 1, after the letter *r*.

```
   236 r1
3)709
  6
  10
   9
   19
   18
    1
```

The answer is 236 hot dogs per school, with a remainder of 1.

It's easy to check the answer even though there is a remainder. As usual, multiply the answer by the number you divided by. Then add the remainder. The result should be the number you were dividing. In this case, multiply 236 by 3. You will get 708. Add the remainder, 1, and you have 709. The answer checks.

A Test-Taking Tip

You will not see remainders in the answer choices in the GED Mathematics Test. The question will be phrased so that you will know what to do. The words *as evenly as possible* in the last example show that the remainder can be dropped. Such words as *full, complete, nearest whole number,* and so on will be your clues showing you whether to drop the remainder or round to the next whole number.

Try It Yourself

Try this next one yourself before you check the solution.

■ If 13,000 pounds of produce were distributed to 7 different supermarkets, how many pounds would each superemarket get?

If you lined up the numbers and remembered to bring down the zeros in order, you should have gotten 1,857 *r*1. (If this were an actual question on the GED Mathematics Test, the answer choice to pick would be 1,857.)

Warm-up

Divide these numbers. If there is a remainder, include it with your answer.

1. 885 ÷ 7 = ____________

2. 3,005 ÷ 4 = ____________

3. 29,901 ÷ 3 = ____________

4. 8,427 ÷ 6 = ____________

5. 28,196 ÷ 9 = ____________

6. 11,728 ÷ 5 = ____________

On the Springboard

3. An apartment building used 9 times as much heating oil as a home. If the apartment building used 3,100 gallons, approximately how many gallons were used in the home?

   (1) 315 (2) 344 (3) 27,900

   ① ② ③

4. An assembly line packs 6 items to a carton. If 350 items were packed, how many *full* cartons can be shipped?

   (1) 58 (2) 59 (3) 70

   ① ② ③

5. How many *complete* chicken picnic baskets can be packed using 326 pieces of chicken if a complete basket has 8 pieces?

   (1) 4 (2) 32 (3) 40

   ① ② ③

Check your answers on page 183. If you answered them all correctly, well done. Your next step is to go on to "The Real Thing" and some two-step problems. If you missed any, just review the last section again and try the problems once more.

Warm-up Answers
1. 126 *r*3 **2.** 751 *r*1 **3.** 9,967 **4.** 1,404 *r*3 **5.** 3,132 *r*8
6. 2,345 *r*3

The Real Thing

3. A family went on a 3-day vacation. They spent \$147 on gasoline, \$127 for lodging, \$263 for food, and \$31 for admission fees. They spent approximately the same amount of money each day. About how much did they spend daily?

   (1) \$89 (2) \$122 (3) \$186
   (4) \$189 (5) \$282

   (1) (2) (3) (4) (5)

4. A school secretary had 20 boxes of folders with 48 folders in each box. He gave the same number of folders to each of 9 classrooms. How many folders did each classroom receive?

   (1) 10 (2) 16 (3) 100
   (4) 106 (5) 8,640

   (1) (2) (3) (4) (5)

Check your answers on page 197.

Dividing by Two-Digit Numbers

Dividing by two-digit numbers is almost the same as dividing by one-digit numbers. The division steps are the same. You divide, then multiply, subtract, bring down the next digit, and so on, until there are no more digits left to bring down.

Here's an Example

You may have to do a little guessing to find out how many times a two-digit number goes into a larger number.

■ A toy factory assembled 645 toys in 15 shifts. How many toys were assembled each shift?

To solve, divide 645 by 15.

Step 1 Write the problem.

15)645

Step 2 Decide where to place the first digit in the answer. Since 15 doesn't go into 6 but does go into 64, the first digit will go over the 4.

```
   X
15)645
```

Step 3 Estimate roughly the number of 15s in 64. Maybe two? 2 × 15 is only 30. What about 4 × 15? That's 60, and if you had one more 15 you would get 75 and that's too big. So put down the 4 where the X was and start the pattern.

```
   4
15)645
```

Step 4 Multiply 4 × 15 and put the 60 under the 64. Then subtract.

```
   4
15)645
   60
    4
```

Step 5 Bring down the next digit, 5, and divide 45 by 15. Well, 2 × 15 is 30, so 2 is too small. Try 3. It's just right: 3 × 15 = 45. Put 3 right above the five in 645. Multiply 3 × 15 and put the 45 you get under the 45 now there. Subtract. You get zero, and no digits left to bring down.

```
   43
15)645
   60
    45
    45
     0
```

Each shift assembled 43 toys.

A Test-Taking Tip

Sometimes you see the word **average** in a problem. *Average* means the result you get by dividing the total by the number of items. If a student scored a total of 63 points in 7 tests, the average score would be 63 ÷ 7 or 9 points.

Coming to Terms

average for a group of numbers, the result of adding the numbers and then dividing their sum by how many numbers were in the group

Try It Yourself

Set up the following problem. To begin dividing, round up the number you are dividing by.

■ A cab company had 65,890 passengers in one year. The company had 27 drivers. What was the average number of passengers per driver for the year?

You should have divided 65,890 by 27. To get the first digit, you could have rounded 27 to 30 and said 65 divided by 30 is about 2. From there on, it was the usual pattern, though you had to keep alert at the final zero. The answer is 2,440 *r*10.

Warm-up

Divide these numbers. If there is a remainder, include it in your answer.

1. 8,528 ÷ 82 = ________________

2. 891 ÷ 40 = ________________

3. 8,602 ÷ 17 = ________________

4. 21,840 ÷ 84 = ________________

5. 29,833 ÷ 35 = ________________

6. 31,755 ÷ 31 = ________________

7. 9,584 ÷ 63 = ________________

8. 18,540 ÷ 18 = ________________

On the Springboard

6. A construction company owned 780 feet of land along the lakefront. How many lots 65 feet wide will fit along the lakefront?

   (1) 10 (2) 12 (3) 102

   1 2 3

7. A farmer harvested 6,720 bushels of wheat from 210 acres of land. What was his average number of bushels per acre?

   (1) 30 (2) 31 (3) 32

   1 2 3

Check your answers on page 184. If you answered these correctly, try your hand at "The Real Thing.' If you missed either of them, you may want to reread the material on dividing by two-digit numbers before going on.

Warm-up Answers
1. 104 **2.** 22 *r*11 **3.** 506 **4.** 260 **5.** 852 *r*13 **6.** 1024*r*11 **7.** 152 *r*8 **8.** 1,030

"The Real Thing"

5. A fruit grower sold 4,468 pounds of red apples, 1,342 pounds of yellow apples, and 574 pounds of green apples to 24 stores. Each store bought the same amount. How many pounds of apples did each store buy?

   (1) 218 (2) 224 (3) 242
   (4) 266 (5) 481

   1 2 3 4 5

6. An empty moving van weighed 1,525 pounds. After it was loaded with 27 identical desks, it weighed 4,387 pounds. How many pounds did each desk weigh?

   (1) 12 (2) 16 (3) 102
   (4) 106 (5) 219

   1 2 3 4 5

7. During a sale, a store advertised some chairs at a price of 5 chairs for $520. How much did the library pay for 18 of these chairs?

   (1) $252 (2) $936 (3) $1,152
   (4) $1,188 (5) $1,872

   1 2 3 4 5

8. A manufacturer of canned fruit packs 8 large figs per can and 60 cans to a carton. With 10,590 figs, how many full cartons will they be able to ship?

   (1) 8 (2) 20 (3) 21
   (4) 22 (5) 176

   1 2 3 4 5

9. Linda, Laura, and Joanne share equally in the expenses of an apartment. In January, Linda paid the rent, which was $325. Laura paid the utility bills and bought food. She spent $167. How much in all did Joanne owe her roommates at the end of the month?

   (1) $158 (2) $164 (3) $168
   (4) $241 (5) $246

   1 2 3 4 5

10. On the first day of a deep-sea fishing trip, 7 people caught 468 pounds of fish. On the next day, they caught 380 pounds of fish. If they share the fish equally, about how many pounds of fish will each person get?

(1) 106 (2) 121 (3) 130
(4) 847 (5) 5,929

(1) (2) (3) (4) (5)

11. A highway department plans to install 2,570 feet of snow fence in Sioux County and 3,780 feet of snow fence in Plymouth County. The fencing comes in 50-foot rolls. How many rolls of snow fence will the department have to buy?

(1) 13 (2) 107 (3) 127
(4) 1,270 (5) 1,300

(1) (2) (3) (4) (5)

12. The Andrettis want to buy cookies for the twins' class party. They want 3 cookies for each student. There are 32 other students in the twins' class. If there are 12 cookies in each box, how many boxes will the Andrettis have to buy?

(1) 1 (2) 3 (3) 8 (4) 9 (5) 102

(1) (2) (3) (4) (5)

13. Kiku uses 58 tiles to make a tray. She has 15 boxes of tiles with 120 tiles in each box. How many trays can Kiku make?

(1) 12 (2) 31 (3) 32
(4) 33 (5) 217

(1) (2) (3) (4) (5)

Check your answers and record all your scores on pages 197–198.

LESSON 7

Zeroing In on Multistep Problems

In this lesson, you will be working through some problems that take three or four steps before you get to the final answer. Each step will be simple arithmetic. Careful reading and thinking are the key things in deciding what steps to use. Take care not to jump to conclusions too fast. If you think you know what steps to use, reread the problem. Think about each step and ask yourself if it's getting you closer to the solution.

Don't worry, not every problem on the GED Mathematics Test involves this many steps. But one thing's for sure—if you can solve the more complicated problems, the simpler ones should be really easy.

Here's an Example

Read this problem and notice what kind of thinking goes on to see through the words. Then decide what arithmetic you must do.

■ Mark wanted to earn some extra money by making and selling fishing lures. He spent \$320 on materials and \$159 on tools. He sold 175 of one kind of lure at \$3 each and 250 of another kind at \$2 each. What was his profit?

Read the problem very carefully two or three times. Do that now to be sure you understand the information given in the problem. Remember that *profit* is the result of subtracting the total expenses of doing business from the income of a business: Profit = Income − Expenses. What does the problem ask you to find? Right, it asks you to find how much profit Mark made.

To figure out profit, you first have to know income and expenses. Do you know what Mark's income was? No, but it's easy to figure it out. You know he sold 175 lures for \$3 each, so that's \$3 × 175. He also sold 250 lures for \$2 each, so that's \$2 × 250. If you calculate those two amounts and add them, you get his total income.

Do you know Mark's total expenses? Sure, \$320 + \$159, the amount he spent on materials *plus* the amount he spent on tools.

As soon as you've done the arithmetic for total income and for total expenses, just subtract to find his profit.

Here are the steps and the arithmetic.

Step 1 Calculate the income from each kind of lure.

```
 175     250
x  3    x  2
 525     500
```

He made \$525 on the \$3 lures and \$500 on the \$2 lures.

Step 2 Add to get *total* income.

```
  525
+ 500
1,025
```

His total income was \$1,025.

Step 3 Add his two expenses to get *total* expenses.

```
  320
+ 159
  479
```

His total expenses were \$479.

Step 4 Subtract his total expenses from his total income to get his profit.

```
1,025
- 479
  546
```

Mark's profit was \$546.

Try It Yourself

Here's a problem that you can solve in three steps. Read it carefully to see if you can do it on your own. Then read through the solution given here and compare it with yours.

■ In June, the buyer for the library bought 150 "speaking book" tapes at \$5 each. She also spent \$580 on paperbacks. The monthly amount allowed for such expenses was \$1,500. How much did she have left to spend?

The problem asks how many dollars the buyer had left to spend at the end of the month. To find this amount, you must know how much she had to spend at the start. By subtracting what she spent from what she started with, you find how much she had left.

Step 1 Find how much she spent on tapes.

```
150
x 5
750
```

She spent \$750 on tapes.

Step 2 Add what she spent on tapes and what she spent on books to find the total amount spent.

```
  750
+ 580
1,330
```

The total amount spent was \$1,330.

Step 3 Subtract the total spent from the amount she was allowed.

```
  1,500
- 1,330
    170
```

She had \$170 left to spend.

Not a single part of this problem was difficult if you read the questions carefully and knew where you were going.

Warm-up

Steps for solving the following word problem are given below. Number the steps in the order in which they should be done. One of the steps is not needed at all. Put an X in the blank before the unnecessary step.

■ Three drivers each have a bread route delivering to supermarkets and general stores. On weekdays the first driver has 57 daily stops, the second has 52, and the third 60. On Saturdays they each have 10 fewer stops than on weekdays. How many stops in all did the three have on Saturday?

_______ **a.** Add the number of weekday stops for all of the drivers.

_______ **b.** Find the difference between the total number of stops and the total number of stops they get to skip on Saturday.

_______ **c.** Find the difference between the number of stops the first driver has, and the number the second driver has.

_______ **d.** Multiply 10 by 3 to find the total number of stops the drivers get to skip on Saturday.

On the Springboard

1. An auto plant worker earns $12 per hour. A garment factory worker makes $8 an hour. If both work a 40-hour week for 2 weeks, how much more will the auto worker have earned than the garment worker?

   (1) $80 (2) $160 (3) $320

   (1) (2) (3)

2. The Campbells were away for 3 months. During that time, only 400 kilowatt-hours of electricity per month were used. When they are home, they use approximately 1,100 kilowatt-hours per month. How many kilowatt-hours did they save while they were away?

   (1) 2,070 (2) 2,100 (3) 3,300

   (1) (2) (3)

3. A truck driver drove 5 miles from the truck depot to a farm to pick up produce and then he drove 15 miles to the market. After delivering the produce he returned to the farm, and drove to the depot. If he drove this route 5 days a week, how many miles did he drive in one week?

   (1) 40 (2) 100 (3) 200

   (1) (2) (3)

Now check your Springboard answers on page 184. If you got them all right, you are certainly ready for the multistep problems in "The Real Thing" section. If you missed any, review this section. Try again the ones you missed. The second time around you'll probably get them, and you'll be ready for "The Real Thing."

Warm-up Answers
a. 1 **b.** 3 **c.** X **d.** 2 *Note:* There is another way to do this problem. You can subtract 10 from each of the three numbers and then add the answers together to get the total number of stops for Saturday.

The Real Thing

1. In a particular game, a blue question is worth 4 points, a red one is worth 3, a green question is worth 2, and a yellow one is worth 11 points. How many points does a player have if she answers one yellow question, one red question, and three green questions?

   (1) 18 (2) 19 (3) 20
   (4) 24 (5) 28

   ① ② ③ ④ ⑤

2. The temperature at Sunny Beach on Monday was 84°. On Tuesday it was 89°, on Wednesday and Thursday it was 85°, and on Friday it was 77°. What was the average temperature at the beach during the week?

   (1) 67° (2) 78° (3) 84°
   (4) 85° (5) 105°

   ① ② ③ ④ ⑤

3. Valerie works 8 hours a day on weekdays and 3 hours each day on the weekend. Frida works only 5 days a week, but she has two jobs. Her day job is 7 hours per day and her night job is 3 hours per day. How many more hours per week does Frida work than Valerie?

   (1) 1 (2) 4 (3) 5 (4) 6 (5) 7

   ① ② ③ ④ ⑤

4. A barber works 8 hours per day. He spends 30 minutes per customer when he gives a shampoo and haircut. A haircut without a shampoo takes 20 minutes. How many more customers can he handle on a day when they all get just haircuts than on a day when they all get shampoos and haircuts?

   (1) 6 (2) 8 (3) 10 (4) 16 (5) 24

   ① ② ③ ④ ⑤

5. Linda has $218 in her bank account. She has the following bills to pay: department store account, $52; car payment, $109; electricity, $17; phone, $28; and dentist, $35. She wants to pay as much as possible without being overdrawn on her account. Which bill will she have to pay later?

   (1) department store account
   (2) car payment (3) electricity
   (4) phone (5) dentist

   ① ② ③ ④ ⑤

6. An airline normally has 75 passengers on its flights from New York to Chicago. When it offered a special discount fare, the number of passengers increased to 112. Regular fare was $210 and the discount fare was $150. How much more did it make per flight with the discount fare?

   (1) $1,050 (2) $1,150 (3) $1,620
   (4) $2,220 (5) $2,480

   ① ② ③ ④ ⑤

7. Conrad earns $6 an hour and works a 40-hour week. Dorothy earns $11 an hour and works a 38-hour week. How much more per week does Dorothy earn than Conrad?

   (1) $48 (2) $178 (3) $212
   (4) $658 (5) $1,168

   ① ② ③ ④ ⑤

8. A man opened a savings account with a deposit of $30. Four times per month he deposits $45. In 2 months, how much money will he have deposited into his savings account?

   (1) $120 (2) $210 (3) $360
   (4) $390 (5) $420

   ① ② ③ ④ ⑤

9. The circulation of the *Columbus Ledger* is approximately 26,000 a day on Monday through Friday. The Sunday circulation is approximately 30,000 more than the daily circulation. There is no Saturday paper. Approximately how many copies of this newspaper are sold each week?

(1) 130,000 (2) 156,000 (3) 186,000
(4) 212,000 (5) 216,000

(1) (2) (3) (4) (5)

10. A school secretary is ordering supplies. She needs 48 jars of red paint, 60 jars of yellow, and 72 jars of blue. The paint comes in boxes of 12 jars. How many boxes of each color should she order?

(1) 4 red, 5 yellow, 6 blue
(2) 5 red, 6 yellow, 4 blue
(3) 32 red, 64 yellow, 48 blue
(4) 64 red, 96 yellow, 80 blue
(5) 576 red, 720 yellow, 864 blue

(1) (2) (3) (4) (5)

Check your answers and record your score on pages 198–199.

LESSON 8

Adding and Subtracting Decimals

You may not know it, but decimals are all around you. Every time you go shopping you use decimals. You see prices such as $4.95 and $12.50. These are decimals. When you read the odometer on a car you use decimals. The odometer may show, for example, that the car has been driven 24,688.7 miles. Everywhere you turn you find situations in which it's helpful to understand decimals.

What Are Decimals?

You often use decimals to write amounts that are not exactly equal to a whole number. For example, suppose a can of shaving cream is priced at $1.48. The price is between $1 and $2—more than $1 but less than $2. The number 1.48 is a **decimal.** You know that it's a decimal because of the decimal point between the 1 and the 48. This particular decimal has two **decimal places**—in other words, two digits after the decimal point. (*After* the decimal point always means *to the right* of the decimal point.)

Coming to Terms

decimal a number that has a decimal point and one or more digits after (to the right of) the decimal point

decimal places the number of digits after the decimal point in a decimal number

Here's an Example

■ Look at this number: 9.705.

Is the number a decimal? Yes, because it has a decimal point. How many decimal places does it have? It has three, since there are three digits (7, 0, and 5) after the decimal point.

Try It Yourself

■ Look at this number: 7.0084. Is the number a decimal? If it is, how many decimal places does it have? ____________________

Did you answer that the number is a decimal and that it has four decimal places? If you did, then you've got the idea.

Warm-up

Decide whether each number is a decimal. If it is, write in the blank how many decimal places it has. If it is *not* a decimal, write nothing in the blank.

1. 17.435 ____________________

2. 230 ____________________

3. 459.01 ____________________

4. 0.66 ____________________

5. 2.1 ____________________

6. 45.0378 ____________________

7. 6,984 ____________________

8. 3.14159 ____________________

9. 98.6 ____________________

10. 9.99 ____________________

Place Value in Decimals

In whole numbers, the value of a digit depends on the place or position it occupies in the number. For example, in the number 9,100, the value of the digit 1 is 100 because the 1 is in the hundreds place. In the number 9,010, the value of the 1 is only 10, because it's one place farther to the right, in the tens place.

In decimals too, the value of each digit depends on where it is. Each position that a digit can occupy in a whole number has a place name, and so does each position that a digit can occupy in a decimal. The place names for positions to the left of the decimal point are the same as for whole numbers. The following chart also shows the place names for positions after (to the right of) the decimal point.

Millions	Hundred Thousands	Ten Thousands	Thousands	Hundreds	Tens	Ones	Tenths	Hundredths	Thousandths	Ten-Thousandths	Hundred-Thousandths	Millionths
						3	.5					
					8	4	.5	2				
						0	.9	5	6			
				3	4	7	.0	0	1	4		

Notice that the place names after the decimal point are similar to those to the left, but they all end in the letters *ths.* That *-ths* ending is your clue that you're talking about digits to the right of the decimal point.

In the chart, you can easily tell the value of each digit by noticing the position or column it is in. In the first number, 3.5, the value of the 3 is 3 *ones* and the value of the 5 is 5 *tenths.*

Look at the second number. The value of the 8 on the far left is 8 *tens.* The values of the other digits are 4 *ones,* 5 *tenths,* and 2 *hundredths.*

In decimals, as in whole numbers, you see a pattern of tens. Each time you move from left to right in a decimal, the value of the place is *ten times less.*

Here's an easy way to read a decimal.

Step 1 Read the whole number to the left of the decimal point.

Step 2 Say *and* for the decimal point.

Step 3 Read the number after the decimal point and add the place name for the very last digit.

Warm-up Answers
1. 3 **2.** (not a decimal) **3.** 2 **4.** 2 **5.** 1 **6.** 4 **7.** (not a decimal) **8.** 5 **9.** 1 **10.** 2

Here's an Example

■ Here are the words for the decimals in the chart on page 51.

Write	Say
3.5	three <u>and</u> five <u>tenths</u>
	decimal point — name of the last place after decimal point
84.52	eighty-four <u>and</u> fifty-two <u>hundredths</u>
0.956	nine hundred fifty-six <u>thousandths</u> (You do not write <u>and</u> when there is no whole number to the left of the decimal point. You do not read <u>zero</u> either.)
347.0014	three hundred forty-seven <u>and</u> fourteen <u>ten-thousandths</u>

Another way to read decimals is just to read the digits from left to right. Just be very sure to say *point* or *decimal point* at the right time. This is a very good way to read decimals when you have to read them aloud while someone else writes them down or keys them in on a calculator.

Here's an Example

■ Write "three two point eight three" as a number.

32.83 is correct.

Try It Yourself

■ Write the words for the number 73.049.

____

____

Did you write *and* in place of the decimal point? The last place value is thousandths. If you wrote "seventy-three and forty-nine thousand*ths*," you're doing fine.

■ Now try to write the decimal number for the words two hundred twenty and fourteen thousandths. __________

Did you follow these steps?

Step 1 Find the *and.* Write it as a decimal point.

Step 2 Write numbers in place of the words. Write those numbers before *and* to the left of the decimal point, 220.

Step 3 Find the place name and decide how many places are needed. So far you should have 220. ___ ___ ___ (three places for thousandths)

Step 4 Write the number after *and* so it ends up in the last blank space, 2 2 0. ___ <u>1</u> <u>4</u>

Step 5 Put a zero in the blank space. 2 2 0. <u>0</u> <u>1</u> <u>4</u> or 220.014

Warm-up

Write the words for these decimals. Answer in two ways.

1. 7.2 ____________________

____

2. 0.35 ____________________

____

3. 248.5008 ____________________

____

4. 50.329 ____________________

____

Write the numbers for these words.

5. fifteen and thirty-three thousandths

____

6. four point zero zero three zero five one

____

7. thirty-nine hundredths

____

8. eighty-seven and four tenths

____

9. four and three thousand fifty-one millionths

____

Warm-up Answers

1. seven and two tenths; seven point two **2.** zero and thirty-five hundredths (or simply thirty-five hundredths); zero point three five **3.** two hundred forty-eight and five thousand eight ten-thousandths; two four eight point five zero zero eight **4.** fifty and three hundred twenty nine thousandths; five zero point three two nine **5.** 15.033 **6.** 4.003051 **7.** .39 (or 0.39) **8.** 87.4 **9.** 4.003051 (Notice that this answer is the same as that for question 6. Which was easier to answer?)

Rounding Decimals

It might be very important for an aerospace engineer to know whether the inside of a certain piece of metal tubing measures 1.038 or 1.039 inches. There's only a thousandth of an inch difference, but that thousandth of an inch could be critical. If a plumber was going to use the same piece of tubing, he would probably call it 1.04 inches and leave it at that. The plumber simply rounded to the nearest hundredth because that's good enough for his purpose.

You've probably rounded decimals while shopping. If someone asked you how much you paid for a bottle of shampoo that cost $1.32, you'd probably tell them you paid about a dollar and thirty cents. You rounded $1.32 to the nearest tenth (in other words, to the nearest dime). Rounding decimals is just as easy as rounding whole numbers.

Here's an Example

■ Round 1.038 to the nearest hundredth.

Step 1 Draw a line under the digit that's in the hundredths place. 1.0$\underline{3}$8

Step 2 Look at the next digit to the *right* of the one that you underlined. If it is less than 5, keep the digit that's underlined and drop all that's to the right. If the digit you underlined is 5 or more, *add 1* to the digit you underlined and drop those to the right. In this case, the digit to the right of the 3 is 8. That's more than 5. So add 1 to 3 and drop the 8.

1.0$\underline{3}$8 ⟶ 1.04

↑

More than 5

1.036 rounded to the nearest hundredth is 1.04.

■ Round 23.74 to the nearest tenth.

Step 1 Look at the digit in the tenths place. That digit is 7. 23.$\underline{7}$4

Step 2 Look at the digit to the right of the 7. Is it less than 5? Or is it 5 or greater? It's less than 5. So just drop all the digits to the right of the 7.

23.$\underline{7}$4 ⟶ 13.7

↑

Less than 5

So 23.74 rounded to the nearest tenth is 23.7.

Many times people leave off the word *rounded* and just say *to the nearest tenth* or *to the nearest hundredth.*

Try It Yourself

■ Round 62.507 to the nearest hundredth.

____

Did you add 1 to the hundredths digit and drop the 7? If your answer is 62.51, you're correct.

■ Round 7.6315 to the nearest tenth.

____

Did you notice that 3 (the digit to the right of the 6) is less than 5? Did you keep the digit 6 in the tenths place and drop the 315? If your answer is 7.6, that's correct.

What about 7.295 to the nearest hundredth? Since the digit after the 9 is 5, you want to add 1 to the 9 and drop the 5, correct? But when you add 1 to 9 you get 10. What do you write for the hundredths digit when you write the rounded number? Use what you know about place value and carrying. Write 0 in the hundredths place and carry 1 over to the tenths place.

7.2$\underline{9}$5 → 7.$\overset{1}{\underline{2}}$0 → 7.30

Try It Yourself

■ Round 0.6397 to the nearest thousandth.

____

Did you add 1 to the thousandth digit, 9? When you did that and got 10, did you write 0 in the thousandths place and carry 1 to the hundredths place? If your answer was 0.640, you're correct.

Warm-up

Round these numbers to the indicated place.

1. 795.49 (tenths) ______________________

2. 6.7235 (hundredths) ______________________

3. 0.005 (hundredths) ______________________

4. 0.5691 (thousandths) ______________________

5. 830.2299 (thousandths) ______________________

6. 35.79658 (thousandths) ______________________

Adding Decimals

Adding decimals is a lot like adding whole numbers. You have to be careful, of course, to keep the decimal points lined up. If you're adding decimals that have a different number of places, it also may help to write zeros at the ends of those with fewer places. Doing that does not change the value of the decimal. For example, these decimals all have the same value:

5.4 = 5.40 = 5.400 = 5.4000

Here's an Example

■ A dental assistant worked three days last week: 9.75 hours on Monday, 6.5 hours on Wednesday, and 8.25 hours on Friday. How many hours in all did he work last week?

To solve, add 9.75, 6.5, and 8.25.

Step 1 Write the problem vertically. Keep the decimal points lined up. Put a zero after the 5 in 6.5 so that all of the decimals have two places.

```
  9.75
  6.50
+ 8.25
```

Step 2 Add as usual. Start with the column on the right (in this case, the hundredths column). Be sure the decimal point in the answer is directly below the decimal points in the numbers you are adding.

```
  1 1
  9.75
  6.50
+ 8.25
 24.50
```

He worked 24.50 hours last week. (You can drop the end zero and simply say 24.5 hours.)

Warm-up Answers
1. 795.5 **2.** 6.72 **3.** 0.01 **4.** 0.569 **5.** 830.230 **6.** 35.797

Try It Yourself

■ A company processes waste liquids from photo labs to recover the silver the liquids contain. They recovered 12.2 ounces of silver in January, 9.804 ounces in February, 11.35 ounces in March, and 10.675 ounces in April. What was the total amount of silver recovered?

____

Did you line up the decimals when you wrote the problem vertically? Did you write zeros at the end to make all the numbers three-place decimals? If you did, then you added 12.200, 9.804, 11.350, and 10.675. If you were careful when you added, you found that 44.029 ounces of silver were recovered.

A Test-Taking Tip

If you have to add a whole number to decimals in the GED Mathematics Test, you can make the whole number into a decimal just by putting a decimal point and adding as many zeros as you need at the right-hand end of the number. For example, to compute 23.45 + 46 + 138.379, write 46 as 46.000 and add.

```
   23.450
   46.000
+ 138.379
```

Warm-up

Add the following numbers.

1. 35.632 + 16.5 + 72 + 637.72 =

____

2. 783.2 + 39.555 + 6 = ______________________

3. 92.305 + 101 + 6.2 = ______________________

4. 2.302 + 75.6711 + 19 + 26.1 =

____

5. 1.26951 + 3.01 = ______________________

6. 0.01 + 1.001 + 0.001 + 1 =

____

7. 29.03 + 36.729 + 750.94 = ______________________

8. 2.03 + 0.07 + 2.8 + 2.032 =

____

9. 10.0004 + 5.2 + 16.703 = __________

10. 0.099 + 9.911 = ______________

☑ A Test-Taking Tip

The numbers in word problems have labels. The label may be people, meters, hours, or almost anything else. Don't let a dollar sign scare you. It's just another label. When you solve a GED problem, write down the labels. Then you will know you have found what the question asked for.

On the Springboard

1. A bookkeeper had bills for office supplies. The amounts were $57.57, $404.16, and $6.80. What was the total amount due for office supplies?

   (1) $467.53 (2) $468.43
   (3) $468.53

   ① ② ③

2. Mack delivered loads of steel beams weighing 5 tons, 4.03 tons, and 2.35 tons. What was the total weight in tons of the deliveries?

   (1) 6.43 (2) 11.35 (3) 11.38

   ① ② ③

Check your Springboard answers on page 184. If you got these correct, go on to "Subtracting Decimals." If you missed either one, check your arithmetic. Review if you need to, and, when you've fixed your work, move ahead.

Warm-up Answers
1. 761.852 **2.** 828.755 **3.** 199.505 **4.** 123.0731 **5.** 4.27951
6. 2.012 **7.** 816.699 **8.** 6.932 **9.** 31.9034 **10.** 10.01

Subtracting Decimals

Subtracting decimals is as easy as subtracting whole numbers. Be sure you line up the decimal points and put zeros at the ends of any numbers you need to so that the numbers have the same number of decimal places.

Here's an Example

■ Jeff bought a desk lamp for $14.65. He gave the clerk a $20 bill. What was his change?

To solve, subtract $14.65 from $20.

Step 1 Write the problem vertically. Line up the decimal points and put in zeros as needed.

$$\begin{array}{r} 20.00 \\ -\ 14.65 \\ \hline \end{array}$$

Step 2 Subtract. Be careful with borrowing.

$$\begin{array}{r} {\scriptstyle 1\ \not{1}\!\not{0}\ \ \not{1}\!\not{0}10} \\ {\scriptstyle 9\quad 9} \\ \not{2}\not{0}.\not{0}\not{0} \\ -\ 14.65 \\ \hline 5.35 \end{array}$$

Jeff's change was $5.35.

Try It Yourself

■ A delicatessen salesclerk sliced and sold 2.73 pounds of cheese from a piece that weighed 5.6 pounds. What was the weight of the cheese that was left?

Did you put a zero after the 6 in 5.6 and subtract 2.73 from 5.60? Did you line up the decimal points and were you careful when you borrowed? Then you should have 2.87 pounds as your answer.

Warm-up

Subtract these numbers.

1. 26.784 − 13.053 = __________
2. 9.52 − 4.2 = __________
3. 45.698 − 3.72 = __________
4. 6.6 − 1.75 = __________
5. 348.94 − 72 = __________
6. 70 − 61.39 = __________
7. 35.64 − 12.057 = __________

8. 0.873 − 0.695 = ____________

9. 98.1 − 34.667 = ____________

10. 20 − 0.968 = ____________

On the Springboard

3. Last year, downtown Concord had a record rainfall of 3.5 inches in a day. The Concord airport had 1.68 inches. How many more inches of rain did downtown Concord have?

   (1) 1.82 (2) 1.88 (3) 2.18

   1 2 3

4. An auto mechanic bought a part that cost $9.99. How much change did he receive from a $20 bill?

   (1) $10.01 (2) $11.01 (3) $11.99

   1 2 3

You can check your answers on page 184. If you had no trouble, go on to "The Real Thing." Otherwise, check your arithmetic. Review if you need to. When you've got both problems correct, go on to "The Real Thing."

The Real Thing 

1. Marie deposited checks for $175, $82.51, $125, and $396.28 when she opened a checking account. What was the total amount of her opening deposit?

   (1) $481.79 (2) $497.54 (3) $677.79
   (4) $778.79 (5) $1,521.38

   1 2 3 4 5

2. The Johnsons bought a vacuum cleaner on sale for $49.89. This was $9.98 off the original price. What was the original price of the vacuum cleaner?

   (1) $149.69 (2) $61.87 (3) $59.87
   (4) $49.77 (5) $39.91

   1 2 3 4 5

3. The main span of the Mackinac Bridge is 1,158.245 meters long. Each of the two approaches is 548.84 meters long. The total length of the Mackinac Bridge is how many meters?

   (1) 1,268.013 (2) 1,707.085
   (3) 2,154.825 (4) 2,255.925
   (5) 12.119.925

   1 2 3 4 5

4. Ms. Taylor drove 337.8 miles on Monday and 253.9 miles on Tuesday. How many miles more did she drive on Monday than on Tuesday?

   (1) 83.9 (2) 84.9 (3) 183.9
   (4) 184.9 (5) 591.7

   1 2 3 4 5

5. Mrs. Lovdjieff currently pays $74.63 per month for health insurance. If she insures her son with this policy, she will pay a total of $86.52 per month. What is the monthly cost of insuring Mrs. Lovdjieff's son?

   (1) $8.99 (2) $11.11 (3) $11.89
   (4) $12.11 (5) $12.99

   1 2 3 4 5

6. The winner of a diving competition scored 835.65 points. The runner-up had 697.91 points. How many more points did the winner have than the runner-up?

   (1) 1,533.56 (2) 262.34 (3) 248.74
   (4) 148.96 (5) 137.74

   1 2 3 4 5

Warm-up Answers
1. 13.731 **2.** 5.32 **3.** 41.978 **4.** 4.85 **5.** 276.94
6. 8.61 **7.** 23.583 **8.** 0.178 **9.** 63.433 **10.** 19.032

7. A hobbyist is using matchsticks to build a bridge that will be 40 inches long. He constructed 8.4 inches on Monday, 7.6 inches on Tuesday, 8.25 inches on Wednesday, and 6 inches on Thursday. How many inches of the bridge does he have left to construct?

(1) 9.75 (2) 10 (3) 10.25
(4) 10.75 (5) 15.75

① ② ③ ④ ⑤

8. Juan bought a shirt for $14.79. He was charged $1.04 sales tax. If he gave the clerk $20, how much change did he receive?

(1) $15.83 (2) $5.27 (3) $4.27
(4) $4.17 (5) $3.18

① ② ③ ④ ⑤

9. A cafeteria manager spent $95 for groceries. He spent $37.50 for milk, $18.75 for bread, and the rest for meat. How much did he spend for meat?

(1) $28.75 (2) $38.75 (3) $41.25
(4) $48.75 (5) $72.50

① ② ③ ④ ⑤

10. The average snowfall through April in a large midwestern city usually amounts to 12.4 inches. This year the city received 0.95 inches in December, 1.6 inches in January, 2 inches in February, 0.8 inches in March, and 4.5 inches in April. How many inches less than the average has this year's snowfall been?

(1) 9.85 (2) 4.2 (3) 3.55
(4) 3.45 (5) 2.55

① ② ③ ④ ⑤

Check your answers and record your score on pages 199–200.

LESSON 9

Multiplying Decimals

To multiply decimals, all you need to know is how to multiply whole numbers and where to put the decimal point in the answer.

Simple Multiplication

Begin multiplying decimals by ignoring the decimal points and multiplying as usual. Then count to find the total number of decimal places in the problem. Position the decimal point so that you have that same number of places in the answer.

Here's an Example

■ John's job at a nursing home pays $3.75 per hour. He works 40 hours per week. How much does he earn in a week?

To solve, multiply $3.75 by 40.

Step 1 Write the problem vertically (for now, ignore the dollar sign). Multiply.

```
   3.75
 ×   40
   0 00
  150 0
 150 00
```

Step 2 Count the total number of places after the decimal point in the problem. 3.75 has 2 places, 40 doesn't have any. So the total is 2.

Step 3 Start with the last digit on the right and count back 2 digits to the left. Put the decimal point in front of the second digit. It may help you to draw an arrow.

```
   3.75
 ×   40
   0 00
  150 0
 150.00
```

The answer for the multiplication is 150.00. When you give your final answer, remember that it should be in dollars. John earns $150.00 a week.

Now look at a second problem.

■ One week John worked only 32.75 hours. At his rate of $3.75 an hour, how much did he earn?

To solve, multiply $3.75 by 32.75.

Step 1 Write the multiplication problem vertically, with the longer number at the top. Multiply.

```
    32.75
  × 3.75
   16375
  22925
 9825
1228125
```

Step 2 Count the total number of decimal places in the numbers you multiplied: 2 in 32.75 and 2 in 3.75. The total is 4.

Step 3 Start with the last digit (5) and count 4 places to the left. The place where you stop is the digit in front of which the decimal point goes.

```
    32.75
 ×   3.75
    16375
   22925
  9825
122.8125
```

The result is $122.8125. Because money answers are usually written with just two digits after the decimal point, you may need to round to the nearest hundredth. Your final answer is $122.81. John earned $122.81.

Try It Yourself

It may help you to make a sketch of this problem before solving it.

■ On the frame of a hand loom, the nails are spaced 1.8 centimeters apart. If there are 24 spaces, what is the distance in centimeters from the first to the last nail?

Did you multiply 1.8 by 24? When you did the calculation, you should have gotten 432, and you should have put the decimal point 1 place to the *left* of the last digit (2). When you stated your final answer, did you remember that it was in *centimeters?* The distance is 43.2 centimeters.

A Test-Taking Tip

If a multiplication problem in the GED Mathematics Test has an answer in dollars and cents, you may have to round the answer to the nearest hundredth. For example, suppose imported Swiss cheese sells for $5.28 a pound and a person buys 1.8 pounds. To find the total cost you must multiply $5.28 by 1.8. You'll get $9.504. Since answers in dollars and cents will normally have just 2 digits after the decimal point, you should round this to $9.50.

Warm-up

Do these multiplication problems. Be careful with the placement of the decimal points in your answers.

1. 0.12 × 0.86 = ____________
2. 65.8 × 4 = ____________
3. 3.1 × 8.2 = ____________
4. 0.25 × 0.82 = ____________
5. 5.67 × 0.4 = ____________
6. 0.23 × 17 = ____________
7. 14.24 × 0.5 = ____________
8. 0.46 × 1.13 = ____________
9. 4.5 × 0.213 = ____________
10. 6.05 × 2.2 = ____________

A Test-Taking Tip

You may find the word *cents* in a GED Mathematics Test problem. When you change cents to dollars, remember to move the decimal point two places to the left.

5 cents = $0.05
95 cents = $.95
5.27 cents = $.0527

Warm-up Answers
1. 0.1032 **2.** 263.2 **3.** 25.42 **4.** 0.2050 or 0.205
5. 2.268 **6.** 3.91 **7.** 7.120 or 7.12 **8.** 0.5198
9. 0.9585 **10.** 13.310 or 13.31

On the Springboard

1. The Chens used 251 kilowatt-hours of electricity last month. At $0.062 per kilowatt-hour, how much did they owe?

   (1) $1.56 (2) $15.56 (3) $155.62

   (1) (2) (3)

2. A relay team had 5 runners. If each member ran 0.25 miles, what was the distance in miles they covered altogether?

   (1) 1.05 (2) 1.25 (3) 12.5

   (1) (2) (3)

Check your Springboard answers on page 185. If you answered both correctly, go on to "The Real Thing." If you did not, you may want to take the time to review this section before you go on.

The Real Thing 

1. On her last business trip, Ms. Sanchez averaged 80.75 kilometers per hour. At this rate, how many kilometers did she drive in 9.5 hours?

   (1) 83.125 (2) 113.050 (3) 767.125
   (4) 831.25 (5) 7,671.25

   (1) (2) (3) (4) (5)

2. A man mailed a package that weighed 2.6 pounds. The cost was 63 cents per pound. How much did it cost him to mail the package?

   (1) $0.23 (2) $1.23 (3) $1.64
   (4) $2.34 (5) $16.38

   (1) (2) (3) (4) (5)

Check your answers on page 200.

Adding Zeros in the Answer

Sometimes the multiplication answer does not have enough decimal places. Then you have to add zeros at the left before you place the decimal point.

Here's an Example

Pay special attention to Step 3 in this problem.

■ A variety store had ribbons on sale at 6 cents per yard. A customer needed 0.7 yards. How much should the clerk charge?

If you want to change 6 cents to dollars, it becomes $.06, so you need to multiply $.06 by 0.7.

Step 1 Set up the calculation vertically and multiply.

$$\begin{array}{r} .0\,6 \\ \times\ 0.7 \\ \hline 4\,2 \end{array}$$

Step 2 Count the total number of decimal places in the problem: 2 places in .06 and 1 place in 0.7. The total is 3 places.

Step 3 Count back 3 places from the right-hand digit. Add zeros at the front of the number until you have enough digits for three places. Put in the decimal point 3 places from where you began counting.

$$\begin{array}{r} .06 \\ \times\ 0.7 \\ \hline .042 \end{array}$$

Round to the nearest hundredth since you need an answer in dollars and cents. The clerk should charge $0.04.

☑ A Test-Taking Tip

When working with decimal numbers on the GED Mathematics Test, you may find it helpful to put a zero to the left of the decimal point so that you do not overlook it. In the last problem, we wrote the answer as $0.04 instead of $.04. Leaving off the zero to the left of the decimal point is not wrong, however. Whether you use it or not is up to you.

Try It Yourself

■ The hardware store sells stick-on metric ruler strips at $0.65 per meter. Frank needs 0.13 meters for a rain gauge his son is making for a school project. How much will he be charged?

You should have multiplied 0.65 by 0.13. Ignoring the decimal points, you'll get 845. Since you'll need 4 decimal places, your answer will be .0845. When you wrote the final answer in dollars, did you get $0.08? If you did, you're catching on fine.

Warm-up

Find the answers to the following.

1. 0.064 × 0.81 = ____________

2. 0.7 × 0.12 = ____________

3. 0.05 × 0.09 = ____________

4. 0.219 × 0.04 = ____________

5. 0.007 × 0.04 = ____________

6. 0.056 × 0.14 = ____________

A Test-Taking Tip

There is a way to save some time on the GED Mathematics Test. When you multiply by 0.1, 0.01, 0.001, and so on, you can find the answer very quickly just by moving the decimal point. The decimal point is moved *one place* to the *left* when you multiply by *0.1, two* places to the *left* when you multiply by *0.01, three* places to the *left* when you multiply by *0.001,* and so on. For example, 0.1 × 345.67 = 34.567 and 0.01 × 3.5 = 0.035.

You can also use this shortcut when you multiply by 10, 100, 1,000, and so on. When you multiply by *10,* move the decimal point *one* place to the *right.* When you multiply a number by 100, move the decimal point *two* places to the *right.* When you multiply a number by *1,000,* move the decimal point *three* places to the *right,* and so on.

Warm-up Answers
1. 0.05184 **2.** 0.084 **3.** 0.0045 **4.** 0.00876 **5.** 0.00028
6. 0.00784

On the Springboard

3. If scrap paper is worth $0.005 per pound, how much would you be paid for a 52-pound bundle?

(1) $0.02 (2) $0.26 (3) $1.70

① ② ③

4. A telephone cable can carry 4,000 wires. If each wire is 0.04 inch thick, what is the width in *inches* of 4,000 of the wires laid side by side?

(1) 1.6 (2) 16 (3) 160

① ② ③

Check your Springboard answers on page 185. If you got both of these right, go on to "The Real Thing." If you missed a problem, check your arithmetic. If you need to, review this section before you go on.

 The Real Thing 

3. A tailor bought 5.5 yards of binding tape at $0.015 a yard and a box of thread for $1.36. What was the total cost of these materials?

(1) $1.44 (2) $1.45 (3) $1.69
(4) $1.77 (5) $2.19

① ② ③ ④ ⑤

4. A city ordinance requires 1.5 parking spaces per apartment. Building A has 48 apartments, Building B has 56 apartments, and Building C has 114 apartments. Altogether how many parking spaces are needed for Buildings A, B, and C?

(1) 185.3 (2) 327 (3) 330
(4) 387 (5) 3,270

① ② ③ ④ ⑤

5. Company A offers bulk copying for $0.029 per page. Company B has a price that is only 0.8 that of Company A. How much will an office save per page by giving its large copying jobs to Company B?

(1) $0.00058 (2) $0.00232 (3) $0.0058
(4) $0.0232 (5) $0.232

(1) (2) (3) (4) (5)

6. The cost of a daytime telephone call from Appleton to Sturgeon Bay is 48 cents for the first minute and 29 cents for each additional minute. What would be the charge for a 15-minute call from Appleton to Sturgeon Bay?

(1) $1.93 (2) $3.53 (3) $4.06
(4) $4.54 (5) $5.44

(1) (2) (3) (4) (5)

7. Overtime pay on holidays is figured at 2.5 times a worker's hourly wage. Tony worked a 40-hour week Monday through Friday. He worked 8 hours on Saturday, the Fourth of July holiday. How much did he earn in all that week if his regular rate is $5.65 per hour?

(1) $113 (2) $226 (3) $337
(4) $339 (5) $2,373

(1) (2) (3) (4) (5)

8. A telephone company was able to complete an average international call in 12.5 seconds the first year it was in business. The second year it was able to complete such a call in 0.8 of that time. The third year, the time was down to 0.7 of what it was the second year. What was the time for the average international call the third year?

(1) 0.056 seconds (2) 0.07 seconds
(3) 0.56 seconds (4) 0.7 seconds
(5) 7.0 seconds

(1) (2) (3) (4) (5)

Check your answers and record all your scores on pages 200–201.

LESSON 10

Dividing Decimals

Dividing a Decimal by a Whole Number

Dividing decimals by whole numbers is almost exactly the same as dividing whole numbers. The only difference is that you must decide where the decimal will go in the answer *before* you begin dividing.

Here's an Example

This problem shows you how to divide a decimal by a whole number. Pay special attention to Step 2.

■ A supermarket advertises 6 cans of vegetables for $2.10. Another store has the same vegetables for 33 cents a can. Which is the better buy?

To solve, divide $2.10 by 6 to find the cost per can at the supermarket. Then compare the result with 33 cents.

Step 1 Set up the division problem. 6)2.10

Step 2 Put the decimal point for the answer directly above the decimal point in the number being divided.

```
   .
6)2.10
```

Step 3 Divide just as you would with whole numbers. *Don't* bring down the decimal point when you divide.

```
   .35
6)2.10
  1 8
    30
    30
     0
```

The answer is in dollars. Supermarket cans cost $.35 or 35 cents. The 33-cent cans are the better buy.

Try It Yourself

Be careful where you put the decimal point when you do this calculation.

■ A backpacker took 5 hours to hike 21.5 miles. What was the average distance she covered in an hour?

You should have divided 21.5 by 5. When you set up the division problem, did you put the decimal point for the answer directly above the one in 21.5? If you did and if you were careful with the division, you got 4.3 miles per hour.

Warm-up

Divide the following.

1. 44.8 ÷ 8 ____________

2. 313.8 ÷ 6 = ____________

3. 326.2 ÷ 14 = ____________

4. 269.56 ÷ 46 = ____________

5. 316.2 ÷ 51 = ____________

6. 88.06 ÷ 7 = ____________

7. 208.5 ÷ 5 = ____________

8. 309.33 ÷ 21 = ____________

On the Springboard

1. The cost of a box of 7 plants that all cost the same is $3.57. What does each plant cost?

   (1) $0.05 (2) $0.51 (3) $0.71

   ① ② ③

2. A road crew patched 356.4 miles of road in 40 working days. How many miles did they average each day?

   (1) 8.91 (2) 8.94 (3) 89.1

   ① ② ③

Warm-up Answers
1. 5.6 **2.** 52.3 **3.** 23.3 **4.** 5.86 **5.** 6.2 **6.** 12.58
7. 41.7 **8.** 14.73

Check your Springboard answers on page 185. If you got them both right, go on to "The Real Thing." If you did not, then it may be a good idea to review this section and try the problems again before going on.

The Real Thing 

1. A cafeteria manager ordered 48 ice-cream bars at $5.76. What was the cost of each bar?

   (1) $0.12 (2) $0.14 (3) $0.42
   (4) $1.20 (5) $1.40

   ① ② ③ ④ ⑤

2. The Jacksons rode their bicycles 118.4 miles. They averaged 16 miles per hour. How many hours did the ride take?

   (1) 7.4 (2) 7.9 (3) 8.2
   (4) 8.4 (5) 8.9

   ① ② ③ ④ ⑤

Check your answers on page 201.

Dividing a Decimal by a Decimal

When you divide by a decimal, you must change the decimal you're dividing by into a whole number. The easiest way to do this is to move the decimal point all the way to the right. Be sure you move the decimal point the same number of places to the right in the number you're dividing. Then divide as you would with whole numbers.

When you move the decimal points to the right, what you are really doing is multiplying both numbers by 10 (or by 100; 1,000; and so on). To understand this better, look at these two simple division problems.

$$\begin{array}{r}5\\2\overline{)10}\end{array} \quad \text{and} \quad \begin{array}{r}5\\20\overline{)100}\end{array}$$

Did you notice that the answer in both problems is the same? Both numbers in the first problem were multiplied by 10 to get the numbers in the second problem.

Here's an Example

In this example, pay special attention to how the decimal points are moved in Step 2.

■ A vacationer started with a full tank of gas and traveled 414.4 miles. When he filled the tank again, it took 11.2 gallons of gas. How many miles per gallon did his car get?

To solve, divide 414.4 by 11.2. (Your clue for this is in the words *miles per gallon.*)

Step 1 Set up the division problem. $11.2\overline{)414.4}$

Step 2 Change the decimal you're dividing by into a whole number by moving the decimal point all the way to the right. Then move the decimal point the same number of places to the right in the number you're dividing. In this case, you move both decimal points *one* place to the right. $112.\overline{)4144.}$

Step 3 Put the decimal point for the answer directly above the decimal point in the number you're dividing. Then do the division.

```
        37.
112.)4144.
     336
      784
      784
        0
```

The car got 37 miles per gallon.

Try It Yourself

Remember to move both decimal points when you divide to solve this next problem.

■ A group on a canoe trip paddled 40.5 miles in 4.5 hours. What is their average number of miles per hour? __________

Miles per hour means miles divided by hours. So you should have divided 40.5 by 4.5. Did you first move the decimal point one place to the right in both numbers? The answer is 9 miles per hour.

Warm-up

Divide these. Move the decimal point as needed.

1. 68.88 ÷ 5.6 = __________
2. 205.62 ÷ 2.3 = __________
3. 8.712 ÷ 0.24 = __________
4. 6.46 ÷ 1.7 = __________
5. 4.464 ÷ 0.36 = __________
6. 10.44 ÷ 2.9 = __________
7. 10.44 ÷ .29 = __________
8. 172.8 ÷ 1.2 = __________
9. 16.20 ÷ 3.6 = __________
10. 13.26 ÷ 1.3 = __________

On the Springboard

3. The wholesale price to a grocer for a package of fish fillets is $15.40. The package weighs 5.6 pounds. How much is the grocer paying per pound?

   (1) $2.08 (2) $2.75 (3) $2.78

   ① ② ③

4. A new car traveled 146.20 miles on 6.8 gallons of gas. How many miles to the gallon did the car get?

   (1) 2.15 (2) 20 (3) 21.5

   ① ② ③

Check your Springboard answers on page 185. If you got them both right, well done! Go on to "The Real Thing." If you missed one, check your arithmetic. If you think you need to do so, review this section. When you've got both Springboard answers right, go on to the GED-level problems in "The Real Thing."

Warm-up Answers
1. 12.3 **2.** 89.4 **3.** 36.3 **4.** 3.8 **5.** 12.4 **6.** 3.6
7. 36 **8.** 144 **9.** 4.5 **10.** 10.2

The Real Thing

3. A grocer put 130.5 kilograms of cheese into packages with 1.5 kilograms in each package. How many packages of cheese were there?

   (1) 8.7 (2) 9 (3) 78
   (4) 87 (5) 90

   ① ② ③ ④ ⑤

4. A potato contains 33.12 grams of carbohydrates and a carrot contains 7.2 grams. How many times as many carbohydrates as the carrot does the potato contain?

   (1) 0.46 (2) 4.52 (3) 4.6
   (4) 45.2 (5) 46

   ① ② ③ ④ ⑤

Check your answers on page 201.

Dividing a Whole Number by a Decimal

When you divide a whole number by a decimal, change the decimal to a whole number by moving the decimal point to the right, as you did before. Then add zeros to the end of the whole number being divided so that you have somewhere to move the decimal point.

Here's an Example

This example will show you how the zeros are added to the number being divided.

■ Jesse's son needs meal tickets for the school cafeteria. If meal tickets are $1.25 each, how many tickets can he purchase for $10?

To solve, divide 10 by 1.25.

Step 1 Set up the problem. Put a decimal point after the whole number. $1.25\overline{)10}$

Step 2 Count the number of decimal places in the number you are dividing by. There are 2 decimal places, so put 2 zeros after the decimal point in the number you're dividing. After that, you can move the decimal point 2 places to the right in both numbers.

$1.25\overline{)10.00} \rightarrow 1.25\overline{)10.00.}$

Step 3 Place the decimal point in the answer and divide as usual.

$$\begin{array}{r} 8. \\ 125\overline{)1000.} \\ \underline{1000} \\ 0 \end{array}$$

$10 will buy 8 meal tickets.

Try It Yourself

When you work the next problem, add as many zeros as you need to the right of the decimal point.

■ An architect wants to divide a line 12 inches long on a blueprint into parts that are each 0.75 inches long. How many parts will she get?

You should have divided 12 by 0.75. Did you put a decimal point and two zeros after the 12? After moving both decimal points 2 places, did you divide and get 16 parts as the answer? Then you're right on target!

Warm-up

Divide the following.

1. 105 ÷ 0.035 = ______
2. 96 ÷ 0.24 = ______
3. 165 ÷ 7.5 = ______
4. 615 ÷ 4.1 = ______
5. 2,047 ÷ 8.9 = ______
6. 2,001 ÷ 8.7 = ______
7. 952 ÷ 1.7 = ______
8. 85 ÷ 2.5 = ______
9. 1,767 ÷ 3.1 = ______
10. 336 ÷ .64 = ______

Warm-up Answers
1. 3,000 **2.** 400 **3.** 22 **4.** 150 **5.** 230 **6.** 230
7. 560 **8.** 34 **9.** 570 **10.** 525

✓ A Test-Taking Tip

There's a shortcut for dividing by 10, 100, 1,000, and so on. Just move the decimal point in the number you're dividing as many places to the left as there are zeros in the number you're dividing by. For instance, to divide 456.8 by 10, move the decimal point 1 place to the left and get 45.68. To divide 456.8 by 1,000, move the decimal point 3 places to the left and get 0.4568.

Similarly, if you wanted to *mulitply* by 10, 100, 1,000, and so on, move the decimal point in the number you are multiplying as many places to the *right* as there are zeros in the other number.

On the Springboard

5. How many large floor tiles that are 1.3 feet wide fit across a room 26 feet wide?

   (1) 2 (2) 20 (3) 21

   (1) (2) (3)

6. An automated assembly line uses robots to complete a manufacturing process in 2.5 days. How many times can the process be completed in 365 days?

   (1) 104.6 (2) 140.6 (3) 146

   (1) (2) (3)

Check your Springboard answers on page 185. If you got them both right, you are making real progress. Go on to "The Real Thing." If you missed one, why not review this section and try again? When you're satisfied with your Springboard answers, try the GED-level problems.

The Real Thing

5. The Frame Shop uses 3.5 yards of material for a picture frame. They have 56 yards of material. How many of these frames can they make?

   (1) 196 (2) 160 (3) 16
   (4) 12 (5) 1.6

   (1) (2) (3) (4) (5)

6. A vacuum-cleaner manufacturer spent $420 on parts that cost $.075 each. How many parts did the manufacturer buy?

   (1) 460 (2) 560 (3) 580
   (4) 5600 (5) 6400

   (1) (2) (3) (4) (5)

Check your answers on page 201.

Adding Places

When you're dividing decimals or whole numbers, the answer may not come out exactly even. You may have a remainder at the end. Sometimes you can add zeros to the number you are dividing until the answer does come out exactly even.

Here's an Example

In the next problem, watch how zeros can be added as they are needed.

■ A powerful dye costs $1 for 0.05 ounce. How much would a package containing 2.364 ounces cost?

To solve, divide 2.364 by 0.05.

Step 1 Set up the division problem.

```
0.05)2.364
```

Step 2 Move both decimal points 2 places to the right. Then place the decimal point where it belongs in the answer and divide.

```
   47.2
5)236.4
  20
   36
   35
    1 4
    1 0
      4
```

The answer so far is 47.2 with a remainder of 4.

Step 3 Go on with the division. Simply add zeros after 236.4, bring them down, and divide until you get a zero when you subtract.

```
     47.28
  5)236.40
    20
     36
     35
      1 4
      1 0
        40
        40
         0
```

The answer is $47.28.

In this case, you only needed one more zero to make the answer come out even. In some cases, you may need more zeros.

A Test-Taking Tip

Use your judgment. It might be necessary to add several more zeros to get an exact answer. So, if you're working a GED Mathematics Test problem, look at the answer choices for a clue about how far to carry the division. Usually one more place than you see in the answer choices is far enough.

Try It Yourself

Add zeros to get the answer to the correct number of places.

■ A new car was factory-tested for gear changes. In 3.6 hours, it was driven on the track for a total distance of 250 miles. To the nearest hundredth of a mile, what was the average speed of the car in miles per hour?

You should have divided 250 by 3.6. Did you divide until you had 3 decimal places in the answer? When you rounded that answer to the nearest hundredth, you should have gotten 69.44 miles per hour.

Warm-up

Divide the following.

1. 170 ÷ 0.8 = ________

2. 424 ÷ 5 = ________

3. 288.3 ÷ 1.5 = ________

4. 1.53 ÷ .08 = ________

5. 3.6267 ÷ 0.42 = ________

6. 2.13 ÷ 0.4 = ________

7. 5.16 ÷ .08 = ________

8. 61.2 ÷ 3.2 = ________

9. 8.62 ÷ 50 = ________

10. 86.49 ÷ 0.6 = ________

On the Springboard

7. The cost of delivering a 38-pound package by carrier was $5.65. What was the cost per pound to the nearest cent?

   (1) 15 (2) 14 (3) 13

   ① ② ③

8. A large recreational vehicle traveled 248.5 miles on 28.4 gallons of gas. How many miles per gallon did the vehicle get?

   (1) 8.75 (2) 71.75 (3) 87.5

   ① ② ③

Check your Springboard answers on pages 185–186. If you got both right, go on to "The Real Thing." If you missed either one, it might be a good idea to look through the section again and give it another try.

“The Real Thing”

7. A candy maker has 480 pounds of fudge. He put the same amount of fudge into each of 640 packages. How many pounds did he put into each package?

   (1) 0.75 (2) 0.76 (3) 1.3
   (4) 7.5 (5) 7.6

   ① ② ③ ④ ⑤

Warm-up Answers
1. 212.5 **2.** 84.8 **3.** 192.2 **4.** 19.125 **5.** 8.635
6. 5.325 **7.** 64.5 **8.** 19.125 **9.** 0.1724 **10.** 144.15

8. Mrs. Garcia cut a 47-inch strip into tabs. Each tab was 1.25 inches long. How many tabs did she get?

(1) 386 (2) 376 (3) 38
(4) 37 (5) 3

① ② ③ ④ ⑤

9. A developer bought 6.25 acres of land to sell as 0.25-acre lots. How many lots could he make?

(1) 0.25 (2) 1.56 (3) 2.5
(4) 15.6 (5) 25

① ② ③ ④ ⑤

10. An appliance store bought 12 hand mixers wholesale for $107.40. What was the cost of each mixer?

(1) $8.12 (2) $8.95 (3) $9.12
(4) $10.74 (5) $11.17

① ② ③ ④ ⑤

11. A 34-acre piece of farmland is priced at $27,500. What is the price per acre to the nearest dollar?

(1) $8 (2) $81 (3) $808
(4) $809 (5) $935

① ② ③ ④ ⑤

12. Eduardo has $10.00 for envelopes and stamps. A box of 50 plain envelopes sells for $1.75, tax included. If he buys one box of envelopes and then as many 22-cent stamps as he can with what's left, how many more envelopes will he have than stamps?

(1) 4 (2) 5 (3) 12 (4) 13 (5) 14

① ② ③ ④ ⑤

13. A printing plant charges $8.29 to print 100 copies of a program. At this price, how much will it cost to have 380 programs printed?

(1) $3.15 (2) $30.40 (3) $31.50
(4) $91.19 (5) $315.00

① ② ③ ④ ⑤

Check your answers and record all your scores on page 202.

LESSON 11

Zeroing In on Word Problems: Insufficient Data

Imagine that you are writing a check for groceries. By the check number you can tell that you've written two checks between this one and your last balance, but you can't remember what the amounts were. You know the amount of the new check, and you know the old balance. What you can't tell, however, is whether this check will overdraw your account. Until you receive your canceled checks or remember what the amounts were on the two checks you forgot to record, you don't have enough information to solve your problem.

Problems on the GED Mathematics Test will usually have all the information you need. Once in a while, however, there may not be enough information for you to set up or solve the problem. When this happens, one of your answer options will be "Insufficient data is given to solve the problem." But be careful. This option may also be given when there *is* enough information to solve the problem. Be sure that necessary information is really missing before you choose this answer option.

Coming to Terms

insufficient data information (numbers or other data) that is not enough for setting up or solving a word problem; a possible choice on the GED Mathematics Test

Here's an Example

You can't solve this problem because some necessary information is missing.

■ Howard is preparing to wire a new house. He has 33 outlets to install. Each outlet requires 15 feet of wiring. The wiring is sold only in 25-foot rolls. How much will the wiring cost?

Try to remember the steps in the system for solving word problems. First, read the problem carefully to be sure you know what you're supposed to find out. In this case, the problem asks how much some wiring will cost, so the answer would have to be in dollars and cents. The next step is to decide what information is needed to solve the problem. Here it would be necessary to know how much wiring is needed and how much per roll the wiring costs. But nowhere in the problem is *any* amount of money mentioned. The problem can't be solved without this information. The correct choice on the GED Test would therefore be "Insufficient data is given to solve the problem." You must know or be able to figure out the cost of a 25-foot roll of wire to solve this problem.

Try It Yourself

Identify the information that is missing in this problem.

■ A brickmason laid a walk for a shopping mall. The entire project cost $15,000. The bricks cost 30 cents each. Part of the cost was for labor, part for sand, and part for supplies. How much did the bricks alone cost?

What is missing? ____________________

You can look at this problem in two different ways. If you said the number of bricks is missing, you are correct. If you knew the number of bricks, you could multiply the cost per brick, which is given.

Another way to look at it is to say that, if you knew the costs for labor, sand, and supplies, you could subtract their total to get the cost of all the bricks.

To be sure that a problem has insufficient data, identify what information is missing. Don't choose the "insufficient data" option unless you know what information is needed but not given.

Warm-up

Write *insufficient data* after the problems that don't give you enough information. Then tell what is missing. Do not solve these problems.

1. Cindy works at an ice cream parlor. She is paid $3.50 per hour. If Cindy works 5 days a week, how much does she make a year?

____

____

2. Beckman Brothers will paint any car for $600. A used-car dealer brings in 3 cars to be painted. How long will it take to paint all 3 cars?

____

____

3. Carla and her uncle spent $10.00 to rent a fishing boat. They spent $2.00 for bait, $7.50 for lunch, and $1.75 for gasoline. If Carla's uncle took the cost of the trip out of his social security check, how much did he have left?

____

____

Warm-up Answers
1. insufficient data; need to know how many hours per day she works **2.** insufficient data; need to know how much time it takes to paint one car **3.** insufficient data; need to know the amount of his social security check

On the Springboard

1. A file clerk for an insurance company had 4,000 checks on file. If 350 checks fit in a file drawer, how many drawers are needed?

   (1) 11 (2) 12
   (3) Insufficient data is given to solve the problem.

   ① ② ③

2. On a certain day, a court recorder worked through 5 trials. The average time for the trials was 1 hour. How many words did the recorder take down that day?

   (1) 500 (2) 5,000
   (3) Insufficient data is given to solve the problem.

   ① ② ③

3. Bananas, apples, oranges, and grapefruit were delivered to the produce department of a grocery store. All produce was to be sold in 3-pound packages. If the total amount of fruit delivered was 300 pounds, how many packages of oranges were there?

   (1) 100 (2) 900
   (3) Insufficient data is given to solve the problem.

   (1) (2) (3)

Check your answers on page 186. If you got them right, go on to "The Real Thing." If you didn't, look again at the problems that gave you trouble before going on.

"The Real Thing"

1. A man has two jobs. His monthly salaries before deductions are $975 and $743. Each month a total of $469 is deducted for taxes and benefits. How much is his monthly take-home pay?

   (1) $1,149 (2) $1,249
   (3) $1,259 (4) $1,359
   (5) Insufficient data is given to solve the problem.

   (1) (2) (3) (4) (5)

2. Apples were advertised in one supermarket at 3 pounds for $0.87. In another supermarket apples were $0.68. How much would shoppers save per pound if they bought 3 pounds for $0.87?

   (1) $0.05 (2) $0.06
   (3) $0.19 (4) $0.29
   (5) Insufficient data is given to solve the problem.

   (1) (2) (3) (4) (5)

3. Ms. Nishimura bought 11.5 gallons of gasoline at $1.04 a gallon. How much change did she get back?

   (1) $8.04 (2) $3.90
   (3) $3.04 (4) $0.04
   (5) Insufficient data is given to solve the problem.

   (1) (2) (3) (4) (5)

4. Cedar Point Amusement Park is open 110 days out of the year. When it is open, it has an average daily attendance of 25,975 people. How many people visit Cedar Point Amusement Park during one year?

   (1) 26,085 (2) 285,725
   (3) 2,847,250 (4) 2,857,250
   (5) Insufficient data is given to solve the problem.

   (1) (2) (3) (4) (5)

5. A salesman earns a monthly salary of $650. He also earns a commission of 0.24 of his sales. One month his sales amounted to $3,245. How much did he earn that month?

   (1) $1,428.80 (2) $1,417.60
   (3) $1,318.80 (4) $844.70
   (5) Insufficient data is given to solve the problem.

   (1) (2) (3) (4) (5)

6. During a sale, chicken cost $0.98 a kilogram for whole chickens and $1.30 a kilogram for cut-up chickens. How much does Mr. Gerbosi save if he buys 3 kilograms of cut-up chickens on sale?

   (1) $0.32 (2) $0.68
   (3) $0.96 (4) $1.12
   (5) Insufficient data is given to solve the problem.

   (1) (2) (3) (4) (5)

7. A fast-food chef used 0.5 pound of meat for each Giant Burger. How many pounds of meat were used if 312 Giant Burgers and 428 Midget Burgers were sold?

   (1) 106 (2) 107
   (3) 213 (4) 624
   (5) Insufficient data is given to solve the problem.

   (1) (2) (3) (4) (5)

Check your answers and record your scores on page 202.

LESSON 12

What Are Fractions?

The Meaning of Fractions

Fractions are numbers that let you talk about part of a thing or collection of things. A fraction is written as two numbers separated by a bar. Sometimes the bar is a horizontal line with the numbers above and below it like this:

$$\frac{2}{3} \quad \frac{1}{4} \quad \frac{7}{10}$$

Sometimes the bar is written as a slash (/) with the numbers written on either side of it like this:

2/3 1/4 7/10

The number above the line (or to the left of the slash) is called the **numerator.** The number below the line (or to the right of the slash) is called the **denominator.** Be sure you keep these words straight in your mind, because you'll use them a lot in working with fractions.

Coming to Terms

fraction two numbers separated by a slash or horizontal line; a number used to express part of a thing or collection of things.

numerator the top or first number of a fraction

denominator the bottom or second number of a fraction

Here's an Example

Look carefully at the diagram below to understand how a fraction is expressed.

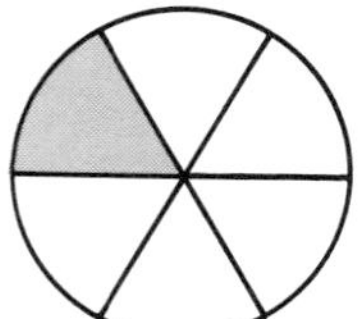

■ A pie was cut into 6 equal parts. One part was eaten. What fraction of the pie was eaten?

To solve, write the number of parts eaten over the total number of parts in the whole pie.

Look at the circle divided into six equal parts. One of the parts is shaded. Think about how you could write the shaded part as a fraction. The numerator tells how many shaded parts. The denominator tells how many equal parts in the whole circle, whether they are shaded or not. For this circle, the fraction that tells how much is shaded is

1 ← shaded part
6 ← total number of equal parts

Try It Yourself

To work the next problem, remember to count the parts in the whole circle and the parts that are shaded.

■ Write the fraction that tells how much of this circle is shaded.

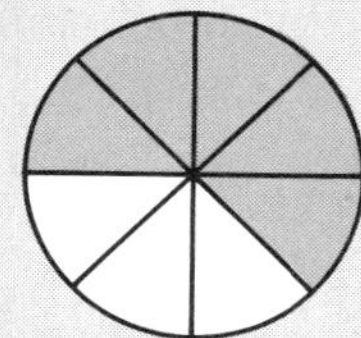

The fraction is ____________________.

The numerator is ____________________.

The denominator is ____________________.

The circle is divided into 8 equal parts and 5 of them are shaded. The fraction that tells how much is shaded is 5/8. The numerator of this fraction is 5 and the denominator is 8. Were these your answers?

Warm-up

Write the fraction represented by each picture. Name the numerator and the denominator of each fraction.

1. 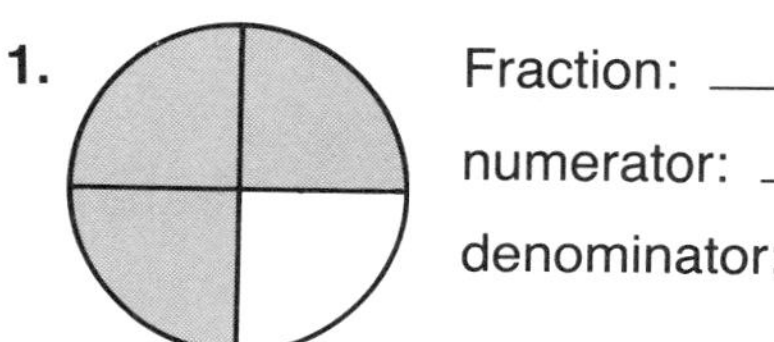

Fraction: ________________

numerator: ________________

denominator: ________________

2. Fraction: ____________

numerator: ____________

denominator: ____________

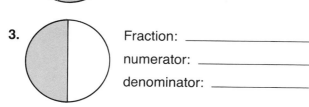

3. Fraction: ____________

numerator: ____________

denominator: ____________

Kinds of Fractions

Fractions fall into two groups:

1. the numerator may be *less* than the denominator, in which case the fraction is called a **proper fraction.**
2. the top number may be greater than or equal to the bottom number, in which case the fraction is called an **improper fraction.**

Here's an Example

■ Write the fraction that tells how much is shaded. Is the fraction proper or improper?

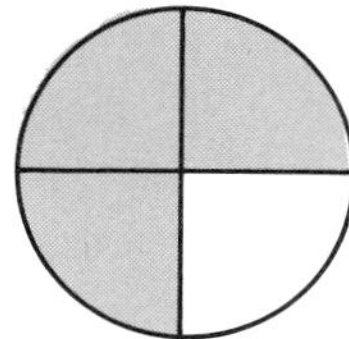

Since 3 out of 4 equal parts are shaded, the fraction is 3/4. The numerator (3) is less than the denominator (4), so 3/4 is a proper fraction.

■ Write the fraction that tells how much is shaded. Is the fraction proper or improper?

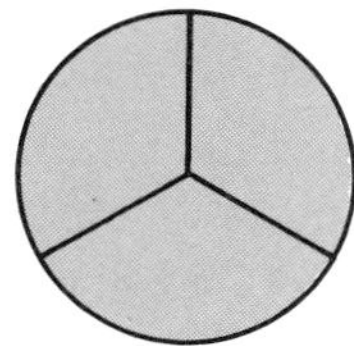

The circle is divided into 3 equal parts and all 3 are shaded. So the fraction is 3/3. Numerator equals denominator, so 3/3 is an improper fraction.

■ Write the fraction that tells how much is shaded. Is the fraction proper or improper?

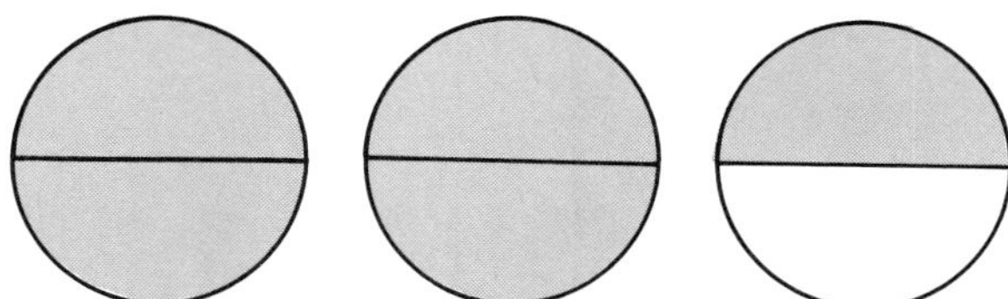

You're right—there's something different going on here. There are three circles, not just one. But each whole circle is divided into 2 equal parts, and 5 of those parts are shaded. So the fraction is 5/2. Since the numerator (5) is greater than the denominator, 5/2 is an improper fraction.

You don't have to use diagrams to decide whether a fraction is proper or improper, but there's something you can observe if you do. Proper fractions represent less than 1 whole circle. Improper fractions represent 1 or more whole circles.

You can also have a whole number and a fraction written together—for example, 2 3/4. This is called a **mixed number.** Here's how you would use circles to picture this number:

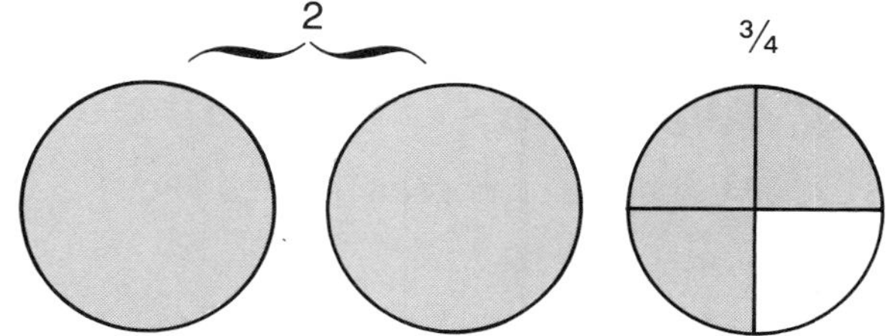

Notice that the first two circles are not divided into parts. Think of them as together representing the whole number 2. The last circle has been divided into 4 equal parts, 3 of which are shaded. It represents the fraction part (3/4) of the mixed number 2 3/4. *Mixed numbers represent more than a whole.*

Coming to Terms

proper fraction a fraction in which the numerator is less than the denominator; represents less than a whole

---

Warm-up Answers

1. Fraction $\frac{3}{4}$, numerator 3, denominator 4
2. Fraction $\frac{4}{5}$, numerator 4, denominator 5
3. Fraction $\frac{1}{2}$, numerator 1, denominator 2

improper fraction a fraction in which the numerator is equal to or greater than the denominator; represents a whole or more than a whole

mixed number a whole number and a fraction written together as one number; represents more than a whole

Try It Yourself

See if you can recognize mixed numbers and the different types of fractions.

■ Tell if each of these is a mixed number, a proper fraction, or an improper fraction.

$6\frac{1}{2}$ ______________________________

$\frac{6}{4}$ ______________________________

$\frac{3}{3}$ ______________________________

$\frac{7}{1}$ ______________________________

$\frac{2}{9}$ ______________________________

Did you say that $6\frac{1}{2}$ is a mixed number, that $\frac{6}{4}$, $\frac{3}{3}$, and $\frac{7}{1}$ are improper fractions, and that $\frac{2}{9}$ is a proper fraction? If you missed any, maybe you should look back over the explanation and examples.

Warm-up

Circle all the proper fractions in each line.

1. $\frac{1}{2}$ $\frac{4}{4}$ $\frac{12}{5}$ $6\frac{1}{4}$ $10\frac{2}{4}$

2. $\frac{5}{3}$ $\frac{8}{9}$ $\frac{4}{4}$ $3\frac{1}{5}$ $\frac{10}{11}$ $\frac{11}{10}$ $\frac{1}{7}$

Circle all the improper fractions in each line.

3. $\frac{7}{8}$ $\frac{3}{3}$ $\frac{15}{1}$ $\frac{11}{6}$ $7\frac{4}{6}$ $14\frac{3}{4}$

4. $\frac{7}{7}$ $5\frac{7}{9}$ $\frac{4}{3}$ $\frac{12}{4}$ $\frac{4}{6}$ $\frac{5}{7}$ $1\frac{3}{4}$

Circle all the mixed numbers in each line.

5. $\frac{5}{6}$ $\frac{7}{7}$ $\frac{11}{4}$ $4\frac{2}{3}$ $5\frac{1}{2}$

6. $4\frac{1}{2}$ $\frac{2}{5}$ $\frac{8}{8}$ $6\frac{7}{8}$ $\frac{3}{4}$ $\frac{2}{10}$ $\frac{16}{5}$

Warm-up Answers

1. $\frac{1}{2}$ **2.** $\frac{8}{9}, \frac{10}{11}, \frac{1}{7}$ **3.** $\frac{3}{3}, \frac{15}{1}, \frac{11}{6}$ **4.** $\frac{7}{7}, \frac{4}{3}, \frac{12}{4}$ **5.** $4\frac{2}{3}, 5\frac{1}{2}$
6. $4\frac{1}{2}, 6\frac{7}{8}$

Renaming Fractions

You know that when you add and subtract whole numbers and decimals you have to use renaming. Adding and subtracting fractions will be easy if you know how to rename fractions.

One way to rename fractions uses multiplication. Look at this diagram.

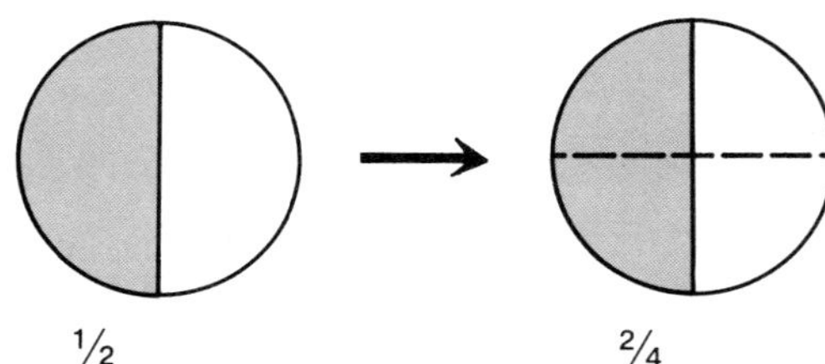

On the left of the arrow, the circle is divided into 2 equal parts; 1 of which is shaded. So on the left the fraction for the shaded part is $\frac{1}{2}$. On the right, a dotted line has been added across the middle. What does this line do to the number of parts? It multiplies the total number of equal parts by 2—instead of two parts, there are now 4. What about the number of *shaded* parts? It multiplies them by 2 also—instead of 1 there are now 2.

$$\frac{1}{2} \rightarrow \frac{1 \times 2}{2 \times 2} \rightarrow \frac{2}{4}$$

If you had drawn two dotted lines, you could have multiplied the number of parts (total and shaded) by 3.

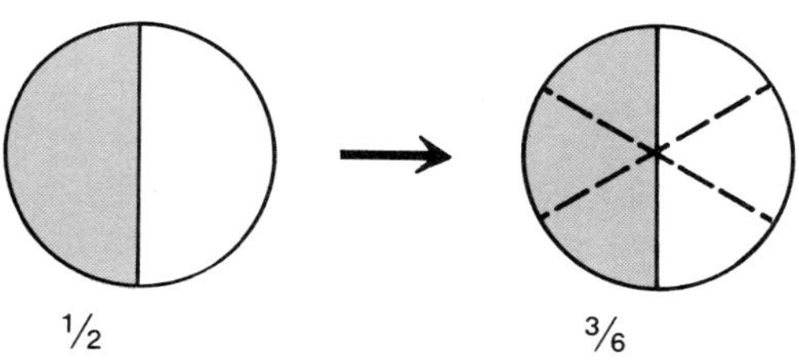

$$\frac{1}{2} \rightarrow \frac{1 \times 3}{2 \times 3} \rightarrow \frac{3}{6}$$

Again, the actual amount shaded has stayed the same, but the number of parts has been multiplied by 3. So $\frac{1}{2}$ can also be renamed $\frac{3}{6}$.

Without drawing more diagrams, you can probably see that you can multiply the number of parts by even larger numbers to rename $\frac{1}{2}$ in still more ways. Look at the next page.

$$\frac{1}{2} \rightarrow \frac{1 \times 4}{2 \times 4} \rightarrow \frac{4}{8}$$
$$\frac{1}{2} \rightarrow \frac{1 \times 5}{2 \times 5} \rightarrow \frac{5}{10}$$
$$\frac{1}{2} \rightarrow \frac{1 \times 7}{2 \times 7} \rightarrow \frac{7}{14}$$

One way to rename a fraction, therefore, is to *multiply both its numerator and denominator by the same number.* (Of course, you shouldn't multiply them by 0. If you did that, you wouldn't rename the fraction, you'd *destroy* it!) Using multiplication of numerator and denominator to rename a fraction is called *renaming the fraction in higher terms.* The word *higher* is used because the numerator and denominator "go up" or get bigger.

Here's an Example

■ Rename $3/8$ in higher terms. Give two names.

You can multiply numerator and denominator by any whole number (greater than 1) that you wish. If you use 4 and 5, you will get $12/32$ and $15/40$.

$$\frac{3}{8} \rightarrow \frac{3 \times 4}{8 \times 4} \rightarrow \frac{12}{32}$$
$$\frac{3}{8} \rightarrow \frac{3 \times 5}{8 \times 5} \rightarrow \frac{15}{40}$$

Many other answers are possible.

■ What is the missing denominator in this problem?

$$\frac{4}{5} = \frac{12}{?}$$

First, notice that 4 can be multiplied by 3 to get 12. So multiply both the numerator and denominator of $4/5$ by 3. You get $4/5 = 12/15$. The missing denominator is 15.

You could put this process in "reverse gear" and *divide* both numerator and denominator by the same number:

$$\frac{7}{14} \rightarrow \frac{7 \div 7}{14 \div 7} \rightarrow \frac{1}{2}$$
$$\frac{3}{6} \rightarrow \frac{3 \div 3}{6 \div 3} \rightarrow \frac{1}{2}$$

Using division to rename a fraction is called *renaming in lower terms.* It's also called **reducing a fraction**, because you reduce the size of numerator and denominator. (You do *not* reduce the amount the fraction represents.)

To rename a fraction in lower terms, there must be a number (other than 1) that divides both the numerator and denominator exactly. When a fraction has been reduced as much as possible, the fraction is said to be in **lowest terms.** The GED Mathematics Test will usually require you to reduce fraction answers to lowest terms.

Here's an Example

■ Reduce $15/20$ to lowest terms.

Ask yourself what number will divide *both* 15 and 20 exactly. (Forget about 1—it divides them exactly, but it won't help you reduce the size of the numerator and denominator.) The number 5 divides both:

$$\frac{15}{20} \rightarrow \frac{15 \div 5}{20 \div 5} \rightarrow \frac{3}{4}$$

There is no number (greater than 1) that divides both 3 and 4 exactly, so $3/4$ is the fraction for $15/20$ that's in lowest terms.

Coming to Terms

reducing a fraction renaming a fraction so that it has a smaller numerator and denominator

lowest terms a fraction that cannot be reduced

Try It Yourself

■ Find the missing number to raise $\frac{2}{5}$ to higher terms:

$$\frac{2}{5} = \frac{8}{?}$$

Did you notice that 2 multiplied by 4 equals 8? If you multiply top and bottom of $\frac{2}{5}$ by 4, you get $\frac{8}{20}$. So the missing number is 20.

■ Find the missing number to reduce $\frac{14}{16}$ to lower terms:

$$\frac{14}{16} = \frac{?}{8}$$

Did you notice that if you divide top and bottom of $\frac{14}{16}$ by 2 you get $\frac{7}{8}$? The missing number is 7.

■ Reduce $\frac{12}{18}$ to lowest terms.

Did you decide that the largest number that divides both 12 and 18 is 6? When you divide top and bottom by 6 you get $\frac{2}{3}$. So $\frac{12}{18} = \frac{2}{3}$.

Warm-up

Find the missing number in order to raise these fractions to higher terms.

1. $\frac{5}{6} = \frac{20}{?}$ ______
2. $\frac{7}{6} = \frac{?}{18}$ ______
3. $\frac{3}{2} = \frac{30}{?}$ ______
4. $\frac{3}{4} = \frac{15}{?}$ ______
5. $\frac{11}{3} = \frac{33}{?}$ ______

Find the missing number in order to reduce these fractions to lower terms.

6. $\frac{4}{8} = \frac{1}{?}$ ______
7. $\frac{10}{12} = \frac{?}{6}$ ______
8. $\frac{15}{20} = \frac{3}{?}$ ______
9. $\frac{12}{32} = \frac{3}{?}$ ______
10. $\frac{36}{48} = \frac{?}{4}$ ______

Reduce these fractions to lowest terms.

11. $\frac{12}{16} =$ ______
12. $\frac{18}{54} =$ ______
13. $\frac{13}{39} =$ ______
14. $\frac{25}{60} =$ ______
15. $\frac{44}{77} =$ ______

Warm-up Answers

1. 24 **2.** 21 **3.** 20 **4.** 20 **5.** 9 **6.** 2 **7.** 5 **8.** 4 **9.** 8 **10.** 3 **11.** $\frac{3}{4}$ **12.** $\frac{1}{3}$ **13.** $\frac{1}{3}$ **14.** $\frac{5}{12}$ **15.** $\frac{4}{7}$

Renaming Improper Fractions and Mixed Numbers

One of the skills you'll need for doing calculations with fractions is renaming improper fractions. When you were working on division of whole numbers, you saw that every division problem can be written in three ways. For example, 16 divided by 3 can be written in these three ways:

$$16 \div 3 \qquad \frac{16}{3} \qquad 3\overline{)16}$$

You can use this idea to change, that is, rename, any improper fraction. Divide the numerator by the denominator. You'll get a whole number or a mixed number.

Here's an Example

■ Change the improper fraction $\frac{16}{3}$ to a mixed number or a whole number.

Step 1 Change $\frac{16}{3}$ to $3\overline{)16}$.

Step 2 Do the division.

$$\begin{array}{r} 5 \\ 3\overline{)16} \\ \underline{15} \\ 1 \end{array}$$

Step 3 Use the remainder, 1, as the numerator and the number you divided by, 3, as the denominator to make the fraction $\frac{1}{3}$. Put it after the 5 that you got when you divided.

$$\begin{array}{r} 5\frac{1}{3} \\ 3\overline{)16} \\ \underline{15} \\ 1 \end{array}$$

So the answer is 5 $\frac{1}{3}$. In other words, $\frac{16}{3} = 5\frac{1}{3}$.

In that first example, there was a remainder. This isn't always so. Look at this example.

■ Change $\frac{26}{13}$ to a mixed number or a whole number.

Step 1 Change $\frac{26}{13}$ to $13\overline{)26}$.

Step 2 Divide. Note that there is no remainder.

$$\begin{array}{r} 2 \\ 13\overline{)26} \\ \underline{26} \end{array}$$

So $\frac{26}{13}$ equals the whole number 2.

Try It Yourself

■ Change $^{32}/_{8}$ to a whole number or mixed number.

____

■ Change $^{50}/_{15}$ to a whole number or mixed number.

____

In the first problem, did you do the division problem $8\overline{)32}$ and get 4 as your answer? In the second problem, did you divide 50 by 15 and get 3 with a remainder of 5? Did you write your answer as the mixed number 3 $^{5}/_{15}$? Did you reduce $^{5}/_{15}$ to lowest terms and write your final answer as 3 $^{1}/_{3}$? Remember, on the GED Mathematics Test answers involving fractions will always show the fraction reduced to lowest terms.

When you change a mixed number to a fraction, you multiply instead of divide.

Here's an Example

■ Change 7 $^{2}/_{3}$ to an improper fraction. Follow these three steps.

Step 1	Multiply the whole-number part of the mixed number (7) by the denominator of the fraction (3).	$7 \times 3 = 21$
Step 2	Add the numerator (2) to what you got when you multiplied.	$21 + 2 = 23$
Step 3	Write the total (23) over the denominator (3).	$\frac{23}{3}$

The answer is $^{23}/_{3}$.

Try It Yourself

■ Change 4 $^{2}/_{5}$ to an improper fraction. _______

You should have multiplied the whole number 4 by the denominator 5 to get 20. Then you should have added the numerator 2 to 20 to get 22. Did you write $^{22}/_{5}$ as your final answer? If you did, then you're doing great.

A Test-Taking Tip

Recognizing various ways of writing fractions can help you. On the GED Mathematics Test you may get an answer that is an improper fraction but find that the answer choices are mixed numbers. Just change your answer to a mixed number to make your choice. Be sure you have reduced all fraction answers to lowest terms.

Warm-up

Write these improper fractions as mixed or whole numbers. If your answer is a mixed number, be sure the fraction part is in lowest terms for your final answer.

1. $\frac{33}{10}$ = _______________
2. $\frac{3}{2}$ = _______________
3. $\frac{10}{5}$ = _______________
4. $\frac{17}{4}$ = _______________
5. $\frac{5}{5}$ = _______________
6. $\frac{24}{6}$ = _______________
7. $\frac{34}{8}$ = _______________
8. $\frac{65}{10}$ = _______________

Write these mixed numbers as improper fractions.

9. $7\frac{5}{6}$ = _______________
10. $12\frac{3}{4}$ = _______________
11. $3\frac{1}{3}$ = _______________
12. $4\frac{2}{3}$ = _______________

Warm-up Answers

1. $3\frac{3}{10}$ **2.** $1\frac{1}{2}$ **3.** 2 **4.** $4\frac{1}{4}$ **5.** 1 **6.** 4 **7.** $4\frac{1}{4}$ **8.** $6\frac{1}{2}$ **9.** $\frac{47}{6}$ **10.** $\frac{51}{4}$ **11.** $\frac{10}{3}$ **12.** $\frac{14}{3}$

LESSON 13

Adding Fractions

The denominator is not just the bottom number of a fraction. It tells what the top number is divided by and also acts as a label saying what the fraction is called. For example, in $2/5$ and $3/5$ the 5 not only tells you that 2 and 3 are divided by 5, it also tells you that the fractions are fifths.

Adding Fractions with the Same Denominator

When fractions have the same denominator, you can add them simply by adding the numerators. For example, think about $3/7 + 2/7$. Just say to yourself, "3 sevenths plus 2 sevenths is 5 sevenths." Saying it like that reminds you that 3 things plus 2 things equal 5 things, whatever they are; in this case, the things happen to be sevenths.

Here's an Example

You can use the same system to solve word problems with fractions that you used for problems that had only whole numbers.

■ A pattern for a child's shorts and top calls for $3/4$ yard of material for each piece. How much material would you need altogether?

To solve, add $3/4$ and $3/4$.

Step 1 Write the addition problem.

$$\frac{3}{4} + \frac{3}{4} =$$

Step 2 Find the denominator that is common to both fractions and write it in the denominator for the answer. Here it is 4.

$$\frac{3}{4} + \frac{3}{4} = \frac{?}{4}$$

Step 3 Now just add the numerators: 3 + 3 = 6. So write 6 as the numerator for the answer.

$$\frac{3}{4} + \frac{3}{4} = \frac{6}{4}$$

The sum is $6/4$, but, as you can see, it is an improper fraction. It's a good idea to change improper fractions to mixed numbers before stating the final answer. It's also a good idea to reduce to lowest terms. You can reduce $6/4$ to $3/2$. Then change $3/2$ to a mixed number. You get $1\ 1/2$.

You would need $1\ 1/2$ yards of material.

Here's an example with mixed numbers.

■ A recipe calls for $2\ 3/4$ cups of white flour and $1\ 1/4$ cups of whole-wheat flour. How much flour will be used altogether?

To solve, add $2\ 3/4$ and $1\ 1/4$.

Step 1 Write the addition problem.

$$2\frac{3}{4} + 1\frac{1}{4} =$$

Step 2 Add the whole numbers first and write that part of the answer.

$$2\frac{3}{4} + 1\frac{1}{4} = 3$$

Step 3 The fractions have the denominator 4 in common. So use it in the answer too. Then add the numerators (3 + 1 = 4) and write 4 for the numerator of the answer.

$$2\frac{3}{4} + 1\frac{1}{4} = 3\frac{4}{4}$$

Step 4 Change $4/4$ to the whole number 1. Since $3\ 4/4$ means $3 + 4/4$, you get 3 + 1, or 4.

Altogether, 4 cups of flour will be used.

Try It Yourself

This problem contains mixed numbers. If the final answer contains a fraction, be sure to reduce it to lowest terms.

■ Felix volunteers to spend $2\ 5/6$ hours every week visiting patients in hospitals. How much time will he spend visiting in the first two weeks?

To solve this problem, you have to add $2\ 5/6$ and $2\ 5/6$. You should have added the whole numbers first to get 4; then the fractions to get $10/6$. The sum is $4\ 10/6$. Change the fraction $10/6$ to the mixed number $1\ 4/6$ and your sum is $4 + 1\ 4/6$ or $5\ 4/6$. Finally, reduce $4/6$ to $2/3$, and you've got it: $5\ 2/3$. Felix will spend $5\ 2/3$ hours visiting patients.

Here's another one to try.

■ Anita ran 2 3/5 miles on Monday, 5 2/5 miles on Tuesday, and 1 1/5 miles on Wednesday. How many miles did she run in all?

Adding three mixed fractions is no different from adding two. Add the whole numbers to get 8 and the fifths to get 6/5. The sum is 8 6/5. Change 6/5 to 1 1/5. When you add 8 and 1 1/5, you get 9 1/5. Anita ran 9 1/5 miles.

Warm-up

Add these fractions and mixed numbers.

1. $\frac{5}{16} + \frac{8}{16} =$ ____ **6.** $1\frac{3}{10} + 5\frac{3}{10} =$ ____

2. $\frac{5}{12} + \frac{5}{12} =$ ____ **7.** $\frac{7}{8} + \frac{5}{8} =$ ____

3. $\frac{3}{8} + \frac{5}{8} =$ ____ **8.** $4\frac{5}{11} + 7\frac{9}{11} =$ ____

4. $2\frac{4}{9} + 6\frac{1}{9} =$ ____ **9.** $2\frac{5}{12} + 8\frac{11}{12} =$ ____

5. $10\frac{3}{5} + 2\frac{2}{5} =$ ____ **10.** $3 + 6\frac{1}{2} =$ ____

On the Springboard

1. Alan's first candy recipe called for $\frac{3}{4}$ cup of sugar and the second needed $\frac{1}{4}$ cup. How much sugar did he need in all?

   (1) $\frac{4}{8}$ cup (2) 1 cup (3) $1\frac{1}{4}$ cup

   ① ② ③

2. Luis is mounting one of his photos for an exhibit. The photo is $8\frac{5}{16}$ inches wide. He wants a border $2\frac{7}{16}$ inches wide on each side. How wide should he cut the board on which he'll mount his photo?

   (1) $4\frac{7}{8}$ inches (2) $10\frac{3}{4}$ inches
   (3) $13\frac{3}{16}$ inches

   ① ② ③

Warm-up Answers

1. $\frac{13}{16}$ **2.** $\frac{5}{6}$ **3.** 1 **4.** $8\frac{5}{9}$ **5.** 13 **6.** $6\frac{3}{5}$ **7.** $1\frac{1}{2}$
8. $12\frac{3}{11}$ **9.** $11\frac{1}{3}$ **10.** $9\frac{1}{2}$

Turn to page 186 to check your answers. If you got both right, go on to "The Real Thing." If you missed one, review this section before going on.

The Real Thing 

1. A customer bought $\frac{5}{8}$ yard of corduroy, $\frac{7}{8}$ yard of shirting, and $\frac{3}{8}$ yard of interfacing. How many yards of material were bought?

   (1) $1\frac{1}{2}$ (2) $1\frac{7}{8}$ (3) $2\frac{1}{2}$
   (4) $2\frac{7}{8}$ (5) 15

   ① ② ③ ④ ⑤

2. June Ellen jogged $2\frac{3}{4}$ miles on Monday and $3\frac{3}{4}$ miles on Friday. How many miles did she jog during the week?

   (1) $5\frac{3}{8}$ (2) $5\frac{1}{2}$ (3) $5\frac{3}{4}$
   (4) $6\frac{1}{2}$ (5) $7\frac{1}{2}$

   ① ② ③ ④ ⑤

3. The weather bureau reported $\frac{9}{10}$ inch of rain on Monday and $\frac{3}{10}$ inch on Wednesday and again on Friday. The other days of the week were without rain. How many inches of rain were reported in all for that week?

   (1) $\frac{15}{20}$ (2) $1\frac{1}{5}$ (3) $1\frac{2}{5}$
   (4) $1\frac{1}{2}$ (5) $2\frac{1}{10}$

   ① ② ③ ④ ⑤

Check your answers on page 202.

Adding Fractions with Different Denominators

You have already added fractions and mixed numbers with the same denominator. To add fractions with different denominators, all you have to do is rename the fractions so they have the same denominator and then add them. To do the renaming, you have to find a **common denominator** for the fractions you're adding.

Coming to Terms

common denominator a number that all the denominators of a group of fractions will divide evenly. For example, $\frac{1}{2}$, $\frac{1}{3}$, $\frac{1}{4}$, and $\frac{1}{6}$ could all use 12 as a common denominator because their denominators all divide 12 exactly. Other common denominators are possible—for instance, 24 or 36.

Here's an Example

Before you can add, you must remember to rename the fractions.

■ It takes Lois $\frac{3}{4}$ hour to drive from home to work. She drives $\frac{1}{2}$ hour longer to get from home to class. How long does it take her to drive from home to class?

To solve this problem, you must add $\frac{3}{4}$ and $\frac{1}{2}$. To do this, you have to find a common denominator and then rename the fractions. One simple way that always works is to multiply the different denominators.

Step 1 Write down the problem.

$$\frac{3}{4} + \frac{1}{2} =$$

Step 2 Multiply the two different denominators and write down the answer.

$$4 \times 2 = 8$$

Step 3 Rename the first fraction to have a denominator of 8. You get $\frac{3}{4} = \frac{6}{8}$. Put $\frac{6}{8}$ on the right-hand side of the equal sign and put a plus sign (+) after it.

$$\frac{3}{4} + \frac{1}{2} = \frac{6}{8} +$$

Step 4 Rename the second fraction to have a denominator of 8. You get $\frac{1}{2} = \frac{4}{8}$. Put $\frac{4}{8}$ after the plus sign.

$$\frac{3}{4} + \frac{1}{2} = \frac{6}{8} + \frac{4}{8}$$

Step 5 Add the new fractions together, and reduce what you get if possible.

$$\frac{6}{8} + \frac{4}{8} = \frac{10}{8}$$

Change $\frac{10}{8}$ to the mixed number $1\frac{2}{8}$. Reduce $\frac{2}{8}$ to $\frac{1}{4}$ and you get $1\frac{1}{4}$. It takes Lois $1\frac{1}{4}$ hours to drive from home to class.

Another Method

Here's a slightly different way of doing the same problem. You choose the one you like best.

Step 1 Write the problem.

$$\frac{3}{4} + \frac{1}{2} =$$

Step 2 Choose the larger denominator, in this case 4, and, in your head or on paper, multiply it by 1, 2, 3, 4, and so on, until you get a number that the other denominator divides evenly. Use that number as the common denominator. In this case, multiply by 1 to get 4 and the job is done; 2 divides 4 evenly.

Step 3 Rename the fractions to have the common denominator 4. Write them down as before. $\frac{3}{4}$ doesn't need renaming. For $\frac{1}{2}$ you get $\frac{1}{2} = \frac{2}{4}$. The problem looks like this:

$$\frac{3}{4} + \frac{1}{2} = \frac{3}{4} + \frac{2}{4}$$

Step 4 Add.

$$\frac{3}{4} + \frac{2}{4} = \frac{5}{4}$$

Now change $\frac{5}{4}$ to a mixed number. You get $1\frac{1}{4}$, the same as before. The second method of finding a common denominator gives you the smallest, or **least common, denominator** right at the beginning. The process of multiplying by 1, 2, 3, 4, and so on, is called finding **multiples.**

Here's one more example in which you must find a common denominator to add the fractions.

■ In a survey conducted for a manufacturer of footwear, $\frac{2}{3}$ of the people said they wore sandals or slippers after they got home from work and $\frac{1}{4}$ said they wore shoes. The rest said they wore only socks or went barefoot. What part of the group wore sandals, slippers, or shoes?

To solve, you need to add $\frac{2}{3}$ and $\frac{1}{4}$. You can find a common denominator using either method.

Step 1 Write down the problem.

$$\frac{2}{3} + \frac{1}{4} =$$

Step 2 Use the denominators 3 and 4 to find a common denominator. You can use 12.

Step 3 Rename the fractions, using 12 as the common denominator.

$$\frac{2}{3} + \frac{1}{4} = \frac{8}{12} + \frac{3}{12}$$

Step 4 Add and reduce as necessary. You get $\frac{11}{12}$. This fraction is already in lowest terms, so you're done.

$\frac{11}{12}$ of the people wore sandals, slippers, or shoes.

Coming to Terms

least common denominator (LCD) the smallest possible denominator you can use when renaming two or more fractions

multiples for any given whole number, the numbers you get by multiplying that number by 1, 2, 3, 4, and so on. For example, the multiples of 3 are 3, 6, 9, 12, 15, 18, and so on.

A Test-Taking Tip

There's an easy way to check if you have renamed a fraction correctly. For example, to check if $\frac{3}{4}$ equals $\frac{6}{8}$, multiply the top number of each by the bottom number of the other. If you get the same answer, you renamed correctly.

$$\frac{3}{4} \times \frac{6}{8}$$

Since 3×8 is 24 and 6×4 is also 24, the fractions are, in fact, equal.

Try It Yourself

Both these problems require renaming.

■ A carpenter built a cabinet of $\frac{1}{2}$-inch plywood and covered it with oak veneer that was $\frac{5}{32}$ of an inch thick. How thick is the finished cabinet? ______

The least common denominator is 32. You could have used 64, but then you would have to reduce the fraction before you were finished. The answer is $\frac{21}{32}$ inch.

■ Two repair jobs took $2\frac{5}{6}$ and $1\frac{1}{4}$ hours. How long did the jobs take altogether? ______

Did you use either 12 or 24 as the denominator? You should end up with $3\frac{13}{12}$. Since $\frac{13}{12}$ is an improper fraction, you should have used division to change it to $1\frac{1}{12}$. This gives $3 + 1\frac{1}{12}$, which is $4\frac{1}{12}$ hours.

Warm-up

Add these fractions and mixed numbers.

1. $\frac{3}{7} + \frac{1}{3} =$ ______
2. $\frac{3}{8} + \frac{10}{16} =$ ______
3. $5\frac{2}{3} + 4\frac{1}{4} =$ ______
4. $3\frac{8}{18} + 5\frac{5}{9} =$ ______
5. $\frac{3}{11} + \frac{19}{22} =$ ______
6. $10\frac{2}{5} + 16\frac{7}{9} =$ ______
7. $5\frac{7}{8} + 3\frac{5}{6} =$ ______
8. $\frac{7}{15} + \frac{3}{20} + \frac{5}{12} =$ ______

On the Springboard

3. The hours logged by three volunteer flag sellers were $9\frac{3}{4}$, $6\frac{5}{6}$, and $2\frac{1}{3}$. How much time altogether was spent selling flags?

(1) $18\frac{11}{12}$ hours (2) $18\frac{5}{6}$ hours
(3) $19\frac{1}{12}$ hours

(1) (2) (3)

4. Wendy worked $3\frac{5}{7}$ weeks at one part-time job and $6\frac{5}{21}$ weeks at another. What was the total time she worked on the two jobs?

(1) $9\frac{20}{21}$ weeks (2) $10\frac{10}{21}$ weeks
(3) $10\frac{20}{21}$ weeks

(1) (2) (3)

Check your Springboard answers on page 186. If you got all of them right, well done! That was a lot to learn. Go on to "The Real Thing." If you missed one or more, it may be a smart move to review this section and do the problems again before you go on.

Warm-up Answers

1. $\frac{16}{21}$ **2.** 1 **3.** $9\frac{11}{12}$ **4.** 9 **5.** $1\frac{3}{22}$ **6.** $27\frac{8}{45}$
7. $9\frac{17}{24}$ **8.** $1\frac{1}{30}$

The Real Thing

4. The basic frame of a house is $3\frac{3}{4}$ inches thick. The outside covering is $\frac{3}{4}$ inch thick and the inside covering is $\frac{3}{8}$ inch thick. The entire wall is how many inches thick?
   (1) $3\frac{9}{16}$ (2) $3\frac{7}{8}$ (3) $4\frac{1}{8}$
   (4) $4\frac{1}{2}$ (5) $4\frac{7}{8}$
   ① ② ③ ④ ⑤

5. A table top is $27\frac{1}{2}$ inches from the floor. There is a shelf $12\frac{3}{4}$ inches above the table top. The shelf is how many inches from the floor?
   (1) $40\frac{1}{4}$ (2) 40 (3) $39\frac{7}{8}$
   (4) $39\frac{1}{4}$ (5) 39
   ① ② ③ ④ ⑤

6. For toppings on a pizza, Bruce used $\frac{1}{2}$ pound of cheese, $\frac{1}{4}$ pound of pepperoni, $\frac{1}{16}$ pound of onions, and $\frac{1}{4}$ pound of bacon. How many pounds of toppings did Bruce use?
   (1) $\frac{1}{4}$ (2) $\frac{5}{16}$ (3) $\frac{13}{16}$
   (4) $1\frac{1}{16}$ (5) $1\frac{1}{2}$
   ① ② ③ ④ ⑤

7. The wingspan of a red admiral butterfly is $2\frac{1}{4}$ inches. The wingspan of a monarch butterfly is $1\frac{1}{2}$ inches wider. What is the wingspan of a monarch butterfly in inches?
   (1) $\frac{3}{4}$ (2) $1\frac{1}{4}$ (3) $3\frac{1}{2}$
   (4) $3\frac{3}{4}$ (5) 4
   ① ② ③ ④ ⑤

8. The di Silva's patio door has 2 panes of glass, each $\frac{3}{16}$ inch thick. The panes enclose an air space $\frac{7}{8}$ inch. How many inches thick is the patio door?
   (1) $1\frac{3}{8}$ (2) $1\frac{1}{4}$ (3) $1\frac{1}{16}$
   (4) $\frac{7}{8}$ (5) $\frac{13}{16}$
   ① ② ③ ④ ⑤

9. Mark loaded trucks for $3\frac{2}{3}$ hours in the morning and $3\frac{3}{4}$ hours in the afternnon. He spent a total of how many hours loading trucks that day?
   (1) $7\frac{11}{12}$ (2) $7\frac{5}{12}$ (3) 7
   (4) $6\frac{11}{12}$ (5) $6\frac{5}{12}$
   ① ② ③ ④ ⑤

10. A band played three pieces of music. The first took $\frac{1}{3}$ of an hour, the next $\frac{1}{6}$ of an hour, and the last $\frac{3}{4}$ of an hour. How many hours did it take to play these three pieces?
    (1) $1\frac{1}{4}$ (2) $1\frac{5}{12}$ (3) $1\frac{2}{3}$
    (4) $1\frac{3}{4}$ (5) $1\frac{11}{12}$
    ① ② ③ ④ ⑤

11. Mrs. Sherry's old boat was $18\frac{3}{4}$ feet long. Her new boat is $4\frac{1}{2}$ feet longer. How long is Mrs. Sherry's new boat?
    (1) $22\frac{1}{4}$ (2) $23\frac{1}{4}$
    (3) $23\frac{1}{2}$ (4) $24\frac{1}{2}$
    (5) Insufficient data is given to solve the problem.
    ① ② ③ ④ ⑤

Check your answers and record all your scores on pages 202–203.

LESSON 14

Subtracting Fractions

If you've ever made things from wood or fabric or if you've done a lot of woodworking, you've probably had to subtract measurements that had fractions in them. Before you can *add* fractions, they have to have the same denominator. The same is true when you subtract.

With the Same Denominator

When fractions have the same denominator, you just subtract the numerators and leave the denominator unchanged.

Here's an Example

Here's a problem for people who like to raid the refrigerator.

■ After a party, $5/6$ of the last pizza was left. In the middle of the night, someone ate another $1/6$. How much pizza was left?

To solve, subtract $1/6$ from $5/6$.

Step 1 Write down the problem.

$$\frac{5}{6} - \frac{1}{6} =$$

Step 2 Subtract the numerators and keep the same denominator.

$$\frac{5}{6} - \frac{1}{6} = \frac{4}{6}$$

Step 3 Reduce the answer if possible. Divide numerator and denominator by 2 to reduce $4/6$ to $2/3$.

So $2/3$ of the pizza was left.

If the numbers are mixed numbers, check to be sure the fractions have the same denominators. If they do, subtract the whole numbers first and then the fractions.

■ When Mary Jo measured the length of the living room on a floor plan, it was $6\,7/8$ inches. She checked and found that the measurement was $1\,5/8$ inches too long. What should the length have been?

To solve, subtract $1\,5/8$ from $6\,7/8$.

Step 1 Write the subtraction problem.

$$6\frac{7}{8} - 1\frac{5}{8} =$$

Step 2 Subtract the whole numbers ($6 - 1 = 5$) and write that part in the answer.

$$6\frac{7}{8} - 1\frac{5}{8} = 5$$

Step 3 Subtract the fractions. When you subtract the numerators, you get $7 - 5 = 2$. Keep the same denominator. Write the fraction in the answer.

$$6\frac{7}{8} - 1\frac{5}{8} = 5\frac{2}{8}$$

Step 4 Reduce the fraction part of the answer if possible. Since $2/8 = 1/4$, you get $5\,1/4$.

The length of the living room on the floor plan should have been $5\,1/4$ inches.

Try It Yourself

■ If $1/4$ yard of fabric was cut from a piece measuring $3/4$ yard, how much material was left over?

If you subtracted and reduced correctly, you got $1/2$ yard.

Now try this problem with mixed numbers.

■ Musicians who had recorded a song that was $5\,9/10$ minutes long had to record the piece again to cut $1\,1/10$ minutes. After they did this, how long was the song?

Did you subtract the whole numbers to get 4 and then the fractions to get $8/10$? Did you reduce $8/10$ and get $4/5$ minutes? If you did, you got $4\,4/5$ minutes, and you're catching on perfectly.

Warm-up

Do these subtractions.

1. $\frac{7}{8} - \frac{3}{8} =$ __________
2. $\frac{6}{13} - \frac{3}{13} =$ __________
3. $8\frac{3}{5} - 4\frac{1}{5} =$ __________
4. $10\frac{13}{16} - 6\frac{5}{16} =$ __________

Warm-up Answers

1. $\frac{1}{2}$ **2.** $\frac{3}{13}$ **3.** $4\frac{2}{5}$ **4.** $4\frac{1}{2}$

On the Springboard

1. A piece of board was $24\frac{3}{8}$ inches long. A carpenter sawed a piece from it to get a board $19\frac{1}{8}$ inches long. How many inches long was the piece he removed?

   (1) $4\frac{1}{4}$ (2) $5\frac{1}{4}$ (3) $5\frac{1}{2}$

   (1) (2) (3)

2. A family used $8\frac{1}{10}$ liters of milk last week and $11\frac{7}{10}$ this week. How many liters more did they use this week than last?

   (1) $3\frac{1}{5}$ (2) $3\frac{3}{5}$ (3) $3\frac{4}{5}$

   (1) (2) (3)

Answers are on page 186.

With Different Denominators

The first thing to do when subtracting fractions with different denominators is to rename them so that they have the same denominator.

Here's an Example

After renaming, you'll see that this subtraction problem is very easy.

■ A car contained 7½ gallons of gas before a trip. After the trip, it had 2¼ gallons left. How much gas was used on the trip?

To solve, subtract 2¼ from 7½.

Step 1 Write the problem.

$$7\frac{1}{2} - 2\frac{1}{4} =$$

Step 2 Find a common denominator for the fractions and rename as needed. The least common denominator is 4.

$$7\frac{2}{4} - 2\frac{1}{4} =$$

Step 3 Subtract and, if possible, reduce.

$$7\frac{2}{4} - 2\frac{1}{4} = 5\frac{1}{4}$$

So 5¼ gallons of gas were used.

Try It Yourself

Use renaming to help you do this one.

■ A piece of turkey weighed 3⅞ pounds. The clerk sliced ½ pound from it. How many pounds did the remaining piece weigh?

Did you find that you could use 8 or 16 as the common denominator? Did you subtract to get either 3⅜ or 3⁶⁄₁₆? If you got 3⁶⁄₁₆, did you reduce to get 3⅜ pounds as your final answer? Then you got it right.

Warm-up

Subtract these fractions and mixed numbers. Reduce answers to lowest terms.

1. $\frac{4}{5} - \frac{1}{10} =$ ______
2. $\frac{2}{3} - \frac{1}{4} =$ ______
3. $6\frac{3}{5} - 2\frac{1}{3} =$ ______
4. $10\frac{1}{6} - 8\frac{1}{10} =$ ______
5. $4\frac{5}{7} - 3\frac{1}{14} =$ ______
6. $21\frac{13}{16} - 15\frac{5}{8} =$ ______

On the Springboard

3. The weight of a load of bricks and concrete was $2\frac{3}{4}$ tons. If the concrete weighed $1\frac{5}{8}$ tons, how many tons did the bricks weigh?

   (1) $\frac{15}{16}$ (2) $1\frac{1}{16}$ (3) $1\frac{1}{8}$

   (1) (2) (3)

4. The cook in the cafeteria had $40\frac{3}{4}$ gallons of chili at the beginning of the lunch period and only $12\frac{1}{2}$ gallons at the end. How many gallons of chili were sold that day?

   (1) $28\frac{1}{4}$ (2) $28\frac{1}{2}$ (3) $29\frac{1}{4}$

   (1) (2) (3)

Check your Springboard answers on page 186.

Warm-up Answers

1. $\frac{7}{10}$ **2.** $\frac{5}{12}$ **3.** $4\frac{4}{15}$ **4.** $2\frac{1}{15}$ **5.** $1\frac{9}{14}$ **6.** $6\frac{3}{16}$

Borrowing

If a fraction has the same number on the top and bottom, the fraction is equal to 1. For example, $2/2 = 1$, $3/3 = 1$, and so on. To see why, remember that fractions can be thought of as division. $2/2 = 2 \div 2 = 1$. The same for $3/3$. Since $3/3 = 3 \div 3$, $3/3 = 1$.

This idea lets you rename whole numbers in a useful way. When you subtract mixed numbers, you may have to rename the whole numbers so that you can borrow.

Here's an Example

■ A piece of wood $5\frac{3}{4}$ feet long is to be cut from a 12-foot board. How much of the board will be left over?

To solve, you need to subtract $5\frac{3}{4}$ from 12.

Step 1 Write the problem.

$$12 - 5\frac{3}{4} =$$

Step 2 You can't just subtract the whole numbers, because then you haven't taken care of the fraction. So rename 12 as $11\frac{4}{4}$ and rewrite the problem.

$$11\frac{4}{4} - 5\frac{3}{4} =$$

Step 3 Now subtract. Subtract the whole numbers first, then the fractions.

$$11\frac{4}{4} - 5\frac{3}{4} = 6\frac{1}{4}$$

$6\frac{1}{4}$ feet of board were left over.

You can use this system even if there are fractions that you can't subtract directly.

■ From a $6\frac{1}{4}$-foot piece of lumber a piece $4\frac{3}{4}$ feet long was cut. How much was left?

To solve, subtract $4\frac{3}{4}$ from $6\frac{1}{4}$.

Step 1 Write down the problem.

$$6\frac{1}{4} - 4\frac{3}{4} =$$

Step 2 Rename the 6 as $5\frac{4}{4}$. Then add on the extra fourth to change $6\frac{1}{4}$ to $5\frac{5}{4}$ and rewrite the problem.

$$5\frac{5}{4} - 4\frac{3}{4} =$$

Step 3 Subtract and reduce if possible.

$$5\frac{5}{4} - 4\frac{3}{4} = 1\frac{2}{4}$$

Reduce $2/4$ to $1/2$.

The piece left is $1\frac{1}{2}$ feet long.

When you rename the mixed number, don't forget to add in the fraction that was there in the first place.

It may be necessary to find a common denominator and borrow in the same problem.

■ In a stock car race, Janice made $4\frac{1}{3}$ laps in the same time it took Nick to make $3\frac{3}{4}$ laps. How far ahead of Nick was Janice?

To solve, subtract $3\frac{3}{4}$ from $4\frac{1}{3}$.

Step 1 Write the problem.

$$4\frac{1}{3} - 3\frac{3}{4} =$$

Step 2 Rename the mixed numbers so the fractions have a common denominator. If you use the least common denominator, 12, the problem changes to

$$4\frac{4}{12} - 3\frac{9}{12} =$$

Step 3 You can't subtract $9/12$ from $4/12$, so rename 4 as $3\frac{12}{12}$. With the $4/12$ already there, $4\frac{4}{12}$ becomes $3\frac{16}{12}$. The subtraction problem now looks like this:

$$3\frac{16}{12} - 3\frac{9}{12} =$$

Step 4 Subtract and reduce if possible. You get $7/12$, which is already in lowest terms.

Janice was $7/12$ laps ahead of Nick.

Try It Yourself

Now try these problems.

■ From an 8-foot piece of pipe, a piece $3\ 5/12$ feet long was cut. How much of the pipe was left? ______________

If you renamed 8 as $7\ 12/12$, you should have ended up with an answer of $4\ 7/12$.

■ A large spool held $11\frac{3}{8}$ yards of cord. If a customer bought $5\ 7/8$ yards, how much cord was left? ______________

You should have renamed 11 as $10\ 8/8$ and then added on the $3/8$ to change the first number to $10\ 11/8$. When you subtracted, did you get $5\ 4/8$? Did you reduce $4/8$ and get $5\ 1/2$ yards as the final answer?

Warm-up

Subtract. Reduce your answers if possible.

1. $8 - 4\frac{3}{5}$ = ______
2. $10\frac{1}{8} - 7\frac{5}{8}$ = ______
3. $5\frac{3}{16} - 1\frac{3}{8}$ = ______
4. $9\frac{2}{3} - 2\frac{4}{5}$ = ______
5. $6\frac{1}{7} - 3\frac{2}{5}$ = ______
6. $13\frac{2}{3} - 9\frac{3}{4}$ = ______
7. $4\frac{1}{10} - \frac{9}{10}$ = ______
8. $5\frac{3}{7} - 2\frac{1}{3}$ = ______
9. $8\frac{9}{20} - 1\frac{4}{5}$ = ______
10. $15\frac{1}{6} - 8\frac{7}{10}$ = ______

A Test-Taking Tip

Don't try to change denominators and borrow at the same time. On the GED Mathematics Test, you may think you're saving time by doing this, but it's easy to make a mistake that way. First, rename the fractions using a common denominator. Then, *decide* whether you really *must* borrow. Sometimes you won't have to borrow at all.

On the Springboard

5. The two-step recipe called for 6 cups of sugar in all. The first step used $4\frac{1}{4}$ cups. How many cups of sugar did the second step use?

   (1) $1\frac{3}{4}$ (2) $2\frac{1}{4}$ (3) $2\frac{3}{4}$

   ① ② ③

6. On a big paint job, the crew used $3\frac{1}{2}$ gallons of paint on the first day. They started with $11\frac{1}{5}$ gallons. How many gallons of paint were left at the end of the day?

   (1) $7\frac{7}{10}$ (2) $8\frac{1}{3}$ (3) $8\frac{3}{10}$

   ① ② ③

Check your Springboard answers on page 186. If you have all of these right, you have made very good progress indeed. Go on to "The Real Thing." If you missed any of the problems, you may want to review this section. When you're satisfied with your answers, go on.

The Real Thing

1. A recipe for one loaf of bread calls for 1 cup of rice flour. A baker wants to make a dozen loaves. If he has only $5\frac{1}{3}$ cups of rice flour, how many more cups of this flour does he need?

   (1) $6\frac{1}{3}$ (2) $6\frac{2}{3}$ (3) $7\frac{1}{3}$
   (4) $7\frac{2}{3}$ (5) $17\frac{1}{3}$

   ① ② ③ ④ ⑤

2. Before hemming, a skirt was $34\frac{1}{4}$ inches long. After hemming, it was $29\frac{3}{4}$ inches long. How many inches was the hem?

   (1) $4\frac{1}{2}$ (2) $4\frac{3}{4}$ (3) 5
   (4) $5\frac{1}{2}$ (5) $5\frac{3}{4}$

   ① ② ③ ④ ⑤

3. A plumber has a pipe that is $13\frac{1}{2}$ feet long. He needs to cut a length of pipe that is $8\frac{5}{8}$ feet long. How many feet of pipe will he have left?

   (1) $5\frac{1}{4}$ (2) 5 (3) $4\frac{7}{8}$
   (4) $4\frac{1}{2}$ (5) $3\frac{7}{8}$

   ① ② ③ ④ ⑤

4. Juanita works a $37\frac{1}{2}$-hour week. By Thursday, she had worked $34\frac{1}{4}$ hours. How many hours did she need to work Friday?

   (1) $2\frac{1}{4}$ (2) $2\frac{3}{4}$ (3) 3
   (4) $3\frac{1}{4}$ (5) $3\frac{3}{4}$

   ① ② ③ ④ ⑤

Warm-up Answers

1. $3\frac{2}{5}$ **2.** $2\frac{1}{2}$ **3.** $3\frac{13}{16}$ **4.** $6\frac{13}{15}$ **5.** $2\frac{26}{35}$ **6.** $3\frac{11}{12}$
7. $3\frac{1}{5}$ **8.** $3\frac{2}{21}$ **9.** $6\frac{13}{20}$ **10.** $6\frac{7}{15}$

5. A storekeeper ordered 3 bolts of material with 25 yards on each bolt. She sold $59\frac{3}{4}$ yards of this material. How many yards of material did she have left?

(1) $15\frac{1}{4}$ (2) $15\frac{3}{4}$ (3) $16\frac{1}{4}$
(4) $25\frac{1}{4}$ (5) $26\frac{1}{4}$

① ② ③ ④ ⑤

6. A wood finish is a mixture of varnish, tung oil, and turpentine. If $\frac{1}{3}$ of the mixture is varnish and $\frac{1}{4}$ turpentine, what fraction is tung oil?

(1) $1\frac{5}{12}$ (2) $\frac{11}{12}$ (3) $\frac{3}{4}$
(4) $\frac{7}{12}$ (5) $\frac{5}{12}$

① ② ③ ④ ⑤

7. The Swensons hiked $39\frac{3}{8}$ miles in three days. They hiked $12\frac{1}{4}$ miles on the first day and $14\frac{5}{8}$ miles on the second day. How many miles did they hike on the third day?

(1) $12\frac{3}{8}$ (2) $12\frac{1}{2}$ (3) $12\frac{5}{8}$
(4) $13\frac{1}{2}$ (5) $13\frac{5}{8}$

① ② ③ ④ ⑤

8. Diane caught two fish that weighed $3\frac{1}{2}$ pounds and $2\frac{3}{4}$ pounds. Her fish weighed $1\frac{3}{4}$ pounds more than those Gwen caught. How many pounds did Gwen's fish weigh?

(1) $3\frac{1}{2}$ (2) $4\frac{1}{4}$ (3) $4\frac{1}{2}$
(4) $5\frac{1}{4}$ (5) $5\frac{1}{2}$

① ② ③ ④ ⑤

9. An empty van weighs $1\frac{7}{8}$ tons. The weight limit of a bridge is $4\frac{1}{2}$ tons. If the van carries a load of $1\frac{3}{4}$ tons, how many tons under the load limit is it?

(1) $\frac{1}{2}$ (2) $\frac{7}{8}$ (3) $4\frac{3}{4}$ (4) $6\frac{1}{4}$
(5) Insufficient data is given to solve the problem.

① ② ③ ④ ⑤

Check your answers and record your score on page 203.

LESSON 15

Multiplying Fractions

Multiplying fractions is easier than adding or subtracting them because you don't have to worry about a common denominator.

Fractions by Fractions

To multiply two fractions together, all you have to do is multiply their numerators and then their denominators.

Here's an Example

This problem will show you how easy multiplying fractions can be.

■ Two napkins are to be made from a piece of fabric $7/8$ yard long. You will need half of the $7/8$-yard piece for each napkin. What fraction of a yard is needed for each napkin?

To find $1/2$ of $7/8$, multiply $1/2$ by $7/8$.

Step 1 Write the problem.

$$\frac{1}{2} \times \frac{7}{8} =$$

Step 2 Multiply the numerators for the numerator of the answer. Multiply the denominators for the denominator of the answer.

$$\frac{1}{2} \times \frac{7}{8} = \frac{7}{16}$$

Each napkin requires $7/16$ yard of fabric.

✓ A Test-Taking Tip

The word *of* is one of the most common words used to tell you that you may have to multiply.

Try It Yourself

You may want to try this one on a piece of scrap paper.

■ A painter began a job with $\frac{3}{4}$ gallon of paint. When the work was finished, half of the paint was left. How much paint was left?

To find half of $\frac{3}{4}$, you should have used $\frac{1}{2} \times \frac{3}{4}$. Did you multiply the numerators and get 3, then the denominators and get 8? If you did, you found that $\frac{3}{8}$ gallon of paint was left.

Warm-up

Multiply these fractions.

1. $\frac{1}{3} \times \frac{2}{7} =$ ________
2. $\frac{1}{4} \times \frac{1}{8} =$ ________
3. $\frac{2}{3} \times \frac{4}{5} =$ ________
4. $\frac{3}{8} \times \frac{2}{5} =$ ________
5. $\frac{5}{6} \times \frac{7}{9} =$ ________

Canceling

You can often **cancel** when you multiply fractions. To cancel, divide the numerator and the denominator by the same number. It's best to cancel before you do the actual multiplication. Canceling before multiplying saves you the work of reducing the final answer.

Coming to Terms

cancel in multiplication of fractions, divide numerator and denominator by the same number

Here's an Example

To use canceling, first look at the numbers you are going to multiply in the numerators and denominators to see if any of them are the same.

■ Three friends shared $\frac{3}{4}$ of a pan of lasagna. Each got $\frac{1}{3}$ of the lasagna. What fraction of the whole pan of lasagna did each get?

To solve, multiply $\frac{1}{3}$ by $\frac{3}{4}$.

Step 1 Write the problem.

$$\frac{1}{3} \times \frac{3}{4}$$

Step 2 Rewrite the problem to show the multiplication of numerators and denominators, but don't do the multiplication yet. Look for any numbers that are the same on the top and bottom, or that can both be divided evenly by some number. Here we have 3 on the top and on the bottom.

$$\frac{1 \times 3}{3 \times 4}$$

Step 3 Divide both 3s by 3. Strike them out and write 1s above or below them.

$$\frac{1 \times \overset{1}{\cancel{3}}}{\underset{1}{\cancel{3}} \times 4}$$

Step 4 Multiply the numbers left at the top and those left at the bottom and write the answer as a fraction.

$$\frac{1 \times 1}{1 \times 4} = \frac{1}{4}$$

Each person will get $\frac{1}{4}$ pan of lasagna.

If you must multiply a whole number by a fraction, first *change the whole number to a fraction by writing it over 1.* For example, $7 = \frac{7}{1}$, $3 = \frac{3}{1}$, $10 = \frac{10}{1}$, and so on. (Remember, $\frac{7}{1}$ means $7 \div 1$ or 7, $\frac{3}{1}$ means $3 \div 1$ or 3, and so on.) The next example uses this idea. It also shows that canceling is just as easy with several fractions as it is with two.

■ A company department has 18 employees, and $\frac{5}{9}$ of the employees are women. Of these women, $\frac{3}{5}$ are working mothers. How many of the women are working mothers?

To solve, find $\frac{5}{9}$ of 18 and multiply by $\frac{3}{5}$. Remember that 18 can be written as $\frac{18}{1}$.

Step 1 Write the problem.

$$\frac{18}{1} \times \frac{5}{9} \times \frac{3}{5}$$

Step 2 Rewrite the problem to show the multiplication of numerators and denominators, but don't do the multiplication yet.

$$\frac{3 \times 18 \times 5}{5 \times 1 \times 9}$$

Step 3 Look for numbers that are the same on top and bottom and for numbers that can both be divided evenly by some number. The number 5 is on the top and the bottom. In addition, the 18 on the top and the 9 on the bottom can be divided evenly by 9. Cancel by 5s.

Warm-up Answers

1. $\frac{2}{21}$ 2. $\frac{1}{32}$ 3. $\frac{8}{15}$ 4. $\frac{3}{20}$ 5. $\frac{35}{54}$

Then divide the 18 on top and the 9 on the bottom by 9.

$$\frac{3 \times \overset{2}{\cancel{18}} \times \overset{1}{\cancel{5}}}{\underset{1}{\cancel{5}} \times 1 \times \underset{1}{\cancel{9}}}$$

Step 4 Multiply the numbers left on the top and those left on the bottom. Think 3 × 2 is 6 and 6 × 1 is 6 for the top. Do the bottom in the same way.

$$\frac{3 \times 2 \times 1}{1 \times 1 \times 1} = \frac{6}{1} \text{ or } 6$$

The department has 6 working mothers.

Try It Yourself

In this problem, you can cancel twice.

■ If 3/4 of your garden was set apart for vegetables, but you only had enough plants for 8/9 of that area, how much of your garden would be planted with vegetables?

Your set up should have been 3/4 × 8/9 and, if you canceled by 3 and by 4, you ended up with 2/3. So 2/3 of the garden was planted with vegetables.

Warm-up

Multiply these fractions. Cancel wherever you can.

1. $\frac{3}{8} \times \frac{1}{4} =$ ____________
2. $\frac{5}{6} \times \frac{1}{3} =$ ____________
3. $\frac{4}{5} \times \frac{9}{10} =$ ____________
4. $\frac{2}{3} \times \frac{5}{8} =$ ____________
5. $\frac{4}{9} \times \frac{3}{16} =$ ____________
6. $\frac{10}{16} \times \frac{4}{15} =$ ____________
7. $\frac{4}{15} \times \frac{3}{8} \times \frac{2}{3} =$ ____________
8. $\frac{1}{3} \times \frac{6}{7} \times \frac{21}{24} =$ ____________
9. $12 \times \frac{5}{8} =$ ____________
10. $\frac{1}{3} \times 15 \times \frac{3}{10} =$ ____________

Warm-up Answers

1. $\frac{3}{32}$ **2.** $\frac{5}{18}$ **3.** $\frac{18}{25}$ **4.** $\frac{5}{12}$ **5.** $\frac{1}{12}$ **6.** $\frac{1}{6}$ **7.** $\frac{1}{15}$ **8.** $\frac{1}{4}$ **9.** $7\frac{1}{2}$ **10.** $1\frac{1}{2}$

On the Springboard

1. One-half of the cakes in the shop window were chocolate cakes. Only $\frac{5}{6}$ of the chocolate cakes had frosting. What fraction of the cakes in the window were chocolate cakes with frosting?

   (1) $\frac{5}{12}$ (2) $\frac{3}{5}$ (3) $\frac{6}{10}$

   ① ② ③

2. On a tree farm, $\frac{5}{6}$ of the trees were evergreen. Of the evergreens, $\frac{2}{5}$ could be used as Christmas trees but only $\frac{2}{5}$ of the Christmas trees would be the right height at sale time. What fraction of all the trees could be sold for Christmas trees?

   (1) $\frac{1}{10}$ (2) $\frac{2}{15}$ (3) $\frac{3}{4}$

   ① ② ③

Check your Springboard answers on page 187. If you got them all right, you have a good grasp of what we've done so far. Go on to "The Real Thing."

"The Real Thing"

1. Tony typed a report in $\frac{3}{4}$ hour. Frank took $\frac{1}{3}$ as long for his report. How long did it take Frank to type his report?

   (1) $2\frac{1}{4}$ (2) $\frac{4}{7}$ (3) $\frac{5}{12}$ (4) $\frac{1}{4}$ (5) $\frac{1}{8}$

   ① ② ③ ④ ⑤

2. Juan's typing speed is 60 words per minute. Betty's typing speed is $\frac{3}{5}$ as fast as Juan's. How many words per minute does Betty type?

   (1) 36 (2) 40 (3) 50
   (4) 60 (5) 80

   ① ② ③ ④ ⑤

Check your answers on page 203.

Multiplying Mixed Numbers by Fractions

When you multiply with mixed numbers, first change them to improper fractions, then multiply as usual.

Here's an Example

Before reading further, be sure you recall how to change mixed numbers to improper fractions.

■ Mary bought $2\frac{1}{4}$ cups of cheese to make blintzes for guests. Some of the guests could not come and she used only half the cheese. How much cheese did she use?

To solve, multiply $2\frac{1}{4}$ by $\frac{1}{2}$.

Step 1 Change $2\frac{1}{4}$ to an improper fraction and write down the problem.

$$\frac{9}{4} \times \frac{1}{2}$$

Step 2 Check for canceling possibilities. Here nothing cancels so go ahead and multiply.

$$\frac{9}{4} \times \frac{1}{2} = \frac{9}{8}$$

Step 3 Since the result is an improper fraction, change it to a mixed number.

$$\frac{9}{8} = 1\frac{1}{8}$$

Mary used $1\frac{1}{8}$ cups of cheese.

Sometimes you may have to multiply one mixed number by another. To do so, just change them all to improper fractions and then multiply.

Try It Yourself

Now try a problem with two mixed numbers.

■ A mason has $3\frac{1}{3}$ bags of cement. If $4\frac{1}{2}$ times this amount is needed for a job, how many bags are needed? ____________

Your set-up should have been $\frac{10}{3} \times \frac{9}{2}$. You can cancel by dividing numerator and denominator by 3 and by 2. Your final answer should be 15 bags of cement.

Warm-up

Multiply. Change all improper fractions in the answers to mixed numbers.

1. $\frac{1}{4} \times 2\frac{1}{3} =$ ____
2. $3\frac{1}{3} \times \frac{3}{10} =$ ____
3. $\frac{2}{3} \times 2\frac{1}{2} =$ ____
4. $\frac{4}{7} \times 6\frac{1}{8} =$ ____
5. $5 \times 7\frac{1}{2} =$ ____
6. $3\frac{1}{2} \times 15 =$ ____
7. $3\frac{1}{12} \times 36 =$ ____
8. $5\frac{5}{8} \times 1\frac{1}{9} =$ ____
9. $10\frac{2}{3} \times 2\frac{3}{4} =$ ____
10. $3\frac{3}{4} \times 5\frac{1}{3} \times 2\frac{1}{2} =$ ____

On the Springboard

3. It takes $6\frac{2}{3}$ yards of fabric to make a sari and a shoulder wrap. How many yards are needed to make 12 such sets?

   (1) $26\frac{2}{3}$ (2) $72\frac{2}{3}$ (3) 80

   (1) (2) (3)

4. There are $8\frac{2}{3}$ pounds of fish food in a bargain pack. How many pounds are there in $5\frac{1}{4}$ bargain packs?

   (1) $47\frac{1}{3}$ (2) $45\frac{1}{2}$ (3) $43\frac{5}{6}$

   (1) (2) (3)

Check your answers on page 187. If you got them right, congratulations. Go on to "The Real Thing." If you missed either, it may be wise to review the section and try them again.

A Test-Taking Tip

Before you begin to multiply, be sure that you change all mixed numbers to improper fractions. To change whole numbers to fractions, write them over 1. If your answer is in the form of an improper fraction, change it to a mixed number. The GED Mathematics Test will give answers as mixed numbers, not as improper fractions.

Warm-up Answers

1 $\frac{7}{12}$ **2.** 1 **3.** $1\frac{2}{3}$ **4.** $3\frac{1}{2}$ **5.** $37\frac{1}{2}$ **6.** $52\frac{1}{2}$
7. 111 **8.** $6\frac{1}{4}$ **9.** $29\frac{1}{3}$ **10.** 50

The Real Thing

3. Ceramic tiles $3\frac{3}{4}$ inches wide were used side by side to form a straight border. If there were 24 tiles, how many inches long was the border?

(1) 18 (2) 21 (3) 72
(4) 90 (5) 120

① ② ③ ④ ⑤

4. A recipe for coffee cake calls for $1\frac{2}{3}$ cups of whole wheat flour. The recipe makes exactly 8 pieces of cake. How many cups of flour are needed to make 30 pieces?

(1) $3\frac{1}{8}$ (2) $6\frac{1}{4}$ (3) $6\frac{1}{3}$
(4) $8\frac{1}{3}$ (5) $10\frac{2}{3}$

① ② ③ ④ ⑤

5. Clifton practices the piano for $1\frac{3}{4}$ hours each weekday and for $2\frac{1}{4}$ hours each day of the weekend. How many hours does he practice in one week?

(1) $14\frac{1}{2}$ (2) $13\frac{1}{4}$ (3) $12\frac{1}{4}$
(4) $8\frac{3}{4}$ (5) $4\frac{1}{2}$

① ② ③ ④ ⑤

6. Mrs. Gregory was paid $1,000. She put $\frac{1}{5}$ in the bank, used $\frac{1}{4}$ to pay various debts, and used the rest for current expenses. How much money did she use for current expenses?

(1) $650 (2) $550 (3) $450
(4) $250 (5) $200

① ② ③ ④ ⑤

Check your answers and record all your scores on page 204.

LESSON 16

Dividing Fractions

Using what you already know you can work out how to divide 6 by a fraction. You already know two ways of finding half of 6. You could divide by 2 or you could multiply it by ½. Now write 2 as $\frac{2}{1}$ and 6 as $\frac{6}{1}$ and look at these two ways of finding half of 6:

$$\frac{6}{1} \times \frac{1}{2} \quad \text{and} \quad \frac{6}{1} \div \frac{2}{1}$$

You can see that dividing by $\frac{2}{1}$ means the same as multiplying by ½. This pattern gives you a simple rule that always works for dividing fractions. *To divide by any fraction, simply turn it upside down and multiply.* In mathematics, you use the term **invert** to mean *turn upside down.*

Here's an Example

■ Write $\frac{5}{8} \div \frac{3}{4}$ as a multiplication problem. Invert the fraction you are dividing by, and instead of dividing, multiply.

$$\frac{5}{8} \times \frac{4}{3}.$$

Even if you're dividing by a mixed number, the rule works fine, because you can write the mixed number as an improper fraction and invert it.

■ Write the division problem $\frac{7}{8} \div 2\frac{1}{3}$ as a multiplication problem.

Since $2\frac{1}{3}$ is the same as $\frac{7}{3}$, just invert $\frac{7}{3}$ and multiply.

$$\frac{7}{8} \times \frac{3}{7}$$

Coming to Terms

invert turn a fraction upside down; that is, put the numerator on the bottom and the denominator on top

Try It Yourself

■ Change these to multiplication problems.

$\frac{3}{7} \div \frac{5}{8}$ $\frac{3}{7} \div 3\frac{1}{2}$

In the first problem, when you invert $5/8$ and multiply, you get $3/7 \times 8/5$. In the second, first change $3\frac{1}{2}$ to $7/2$. When you invert and multiply, you get $3/7 \times 2/7$.

Here's an Example

■ How many pieces of $1/4$-inch thick pegboard are there in a stack 20 inches high?

To solve, divide 20 by $1/4$.

Step 1 Write the problem.

$$20 \div \frac{1}{4}$$

Step 2 Change 20 to $20/1$, invert $1/4$, and multiply. Show this by rewriting the problem.

$$\frac{20}{1} \times \frac{4}{1}$$

Step 3 Multiply.

$$\frac{20}{1} \times \frac{4}{1} = \frac{80}{1} \text{ or } 80$$

There are 80 pieces of pegboard in the stack.

Look at another problem with mixed numbers and canceling. It works the same way.

■ How many books $1\frac{1}{4}$ inches wide will fit on a shelf $17\frac{1}{2}$ inches wide?

To solve, divide $17\frac{1}{2}$ by $1\frac{1}{4}$.

Step 1 Write the problem.

$$17\frac{1}{2} \div 1\frac{1}{4}$$

Step 2 Change the mixed numbers to fractions.

$$\frac{35}{2} \div \frac{5}{4}$$

Step 3 Invert $5/4$, and change division to multiplication.

$$\frac{35}{2} \times \frac{4}{5}$$

Step 4 Cancel where possible and multiply.

$$\frac{\overset{7}{\cancel{35}}}{\underset{1}{\cancel{2}}} \times \frac{\overset{2}{\cancel{4}}}{\underset{1}{\cancel{5}}}$$

Multiplying, you get $14/1$, which equals 14. So 14 books will fit on the shelf.

These four steps are enough for any division problem with fractions or mixed numbers.

Try It Yourself

■ How many $1\frac{1}{2}$-ounce bottles of perfume can be filled using the perfume in a $22\frac{1}{2}$-ounce container? ________________

After you change to mixed numbers, the problem is $45/2 \div 3/2$. When you invert and multiply, it becomes $45/2 \times 2/3$. Canceling and multiplying gives you 15 bottles as the answer.

Warm-up

Divide these fractions. Change any improper fractions in the answers to mixed numbers.

1. $10\frac{1}{2} \div 1\frac{1}{2} =$ ______
2. $\frac{3}{4} \div 6 =$ ______
3. $\frac{5}{8} \div \frac{2}{3} =$ ______
4. $3\frac{2}{3} \div \frac{2}{7} =$ ______
5. $\frac{9}{16} \div \frac{3}{4} =$ ______
6. $\frac{11}{12} \div \frac{3}{4} =$ ______
7. $7\frac{1}{6} \div 5 =$ ______
8. $11 \div 4\frac{5}{7} =$ ______
9. $25 \div \frac{5}{8} =$ ______
10. $4\frac{1}{2} \div 2\frac{5}{8} =$ ______

On the Springboard

1. A total of 15 schoolchildren ran a demonstration relay race, covering a total of $\frac{5}{16}$ of a mile. What fraction of a mile did each runner cover?

   (1) $\frac{1}{48}$ (2) $\frac{1}{16}$ (3) $4\frac{11}{16}$

   (1) (2) (3)

2. There were $16\frac{1}{4}$ pints of water left in a container. How many $1\frac{1}{4}$-pint water bottles could be filled from it?

   (1) 11 (2) 12 (3) 13

   (1) (2) (3)

Warm-up Answers

1. 7 **2.** $\frac{1}{8}$ **3.** $\frac{15}{16}$ **4.** $12\frac{5}{6}$ **5.** $\frac{3}{4}$ **6.** $1\frac{2}{9}$ **7.** $1\frac{13}{30}$ **8.** $2\frac{1}{3}$ **9.** 40 **10.** $1\frac{5}{7}$

Check your answers on page 187. If you got these right, you're ready to move on to "The Real Thing." If you missed either, consider reading the lesson again. When you're satisfied with your answers, go on.

The Real Thing

1. A hospital dietician needs to divide a $22\frac{3}{4}$-ounce box of cereal into 7 equal portions. How many ounces are in each portion?
   (1) $\frac{4}{13}$ (2) $3\frac{1}{4}$ (3) $4\frac{1}{3}$
   (4) 7 (5) $159\frac{1}{4}$
   ① ② ③ ④ ⑤

2. A clerk used $\frac{3}{8}$ yard of yarn to make a bow. How many bows can be made from $\frac{3}{4}$ yard of yarn?
   (1) $\frac{9}{32}$ (2) $\frac{1}{2}$ (3) 1 (4) 2 (5) 3
   ① ② ③ ④ ⑤

3. A tailor can finish a suit in $3\frac{1}{2}$ days. If he works 20 days each month, how many complete suits can he finish in 6 months?
   (1) 70 (2) 35 (3) 34
   (4) 30 (5) 5
   ① ② ③ ④ ⑤

4. A garden is planted in 20-foot rows. Broccoli plants are set out $1\frac{1}{3}$ feet apart. How many plants would be in each row if 2 feet are allowed at each end of a row?
   (1) 11 (2) 13 (3) 15
   (4) 16 (5) 21
   ① ② ③ ④ ⑤

5. Roseanne bought 9 bags of cookies that weighed $1\frac{3}{4}$ pounds each. She separated the cookies into 6 equal gift packages. How many pounds did each gift package weigh?
   (1) $1\frac{1}{6}$ (2) $2\frac{5}{8}$ (3) $3\frac{3}{7}$
   (4) $5\frac{1}{4}$ (5) $15\frac{3}{4}$
   ① ② ③ ④ ⑤

6. In one week, 4 health-club members lost $2\frac{1}{4}$, $1\frac{3}{8}$, $3\frac{1}{2}$, and $2\frac{5}{8}$ pounds. What was the average number of pounds each person lost?
   (1) $2\frac{3}{16}$ (2) $2\frac{1}{4}$ (3) $2\frac{5}{16}$
   (4) $2\frac{7}{16}$ (5) $9\frac{3}{4}$
   ① ② ③ ④ ⑤

7. A woman uses $2\frac{1}{2}$ pounds of apples for each quart of apples that she cans. She canned 45 pounds of Jonathan apples and 36 pounds of Winesap apples. How many quarts of apples did she can?
   (1) 90 (2) $32\frac{2}{5}$ (3) $28\frac{2}{5}$
   (4) 18 (5) $14\frac{2}{5}$
   ① ② ③ ④ ⑤

8. A shipment of books came in 3 boxes. Each box was 15 inches deep. The books were stacked flat and each was $\frac{5}{8}$ inch thick. How many books were in the shipment?
   (1) 9 (2) 24 (3) 27
   (4) 28 (5) 72
   ① ② ③ ④ ⑤

9. Mr. Roth has $16\frac{1}{2}$ yards of fabric and uses $9\frac{1}{4}$ yards to cover a couch. If he uses $\frac{7}{8}$ yard to cover each matching pillow, how many matching pillows can he make?
   (1) 6 (2) 7 (3) 8 (4) 10 (5) 18
   ① ② ③ ④ ⑤

10. Alma had $4\frac{1}{4}$ yards of material. She used $1\frac{7}{12}$ yards to make a table runner and the rest to make aprons. How many aprons was Alma able to make?
    (1) $\frac{2}{3}$ (2) $\frac{3}{4}$ (3) $\frac{11}{12}$ (4) $2\frac{2}{3}$
    (5) Insufficient data is given to solve the problem.
    ① ② ③ ④ ⑤

Check your answers and record your score on pages 204–205.

LESSON 17

Fraction and Decimal Names for Numbers

A man whose name is William could be called Bill or Will or some other such name, but he would still be the same man. In the same way, numbers can have different names without changing their values. Here are various ways of writing the decimal 0.8 and the fraction $4/5$.

Names for 0.8	Names for $4/5$
0.8	$4/5$
0.80	$8/10$
0.800	$12/15$
0.8000	$16/20$
0.80000	$20/25$

Changing Decimals to Fractions

Besides having other decimal names, a decimal can also have fraction names.

Here's an Example

■ Look at the decimal 0.8. You already know that the 8 is in the tenths position.

This decimal could be read as *eight-tenths.* One-tenth is $1/10$, so eight-tenths is $8/10$, which reduces to $4/5$. All the decimals and fractions in the two columns above are equal to $8/10$ or $4/5$.

You just saw how to change a one-place decimal to a fraction. It works the same if the decimal has more than one place. Look at 0.25. The last digit is in the hundredths position. So the fraction equal to 0.25 is $25/100$. There is a simple pattern here. To change any decimal into an equal fraction, drop the decimal point and write the resulting whole number as the numerator. In the denominator, put a 1 followed by as many zeros as there were digits after the decimal point.

Try It Yourself

■ Change 0.75 to a fraction and reduce.

The numerator will be 75 and, because there are two decimal places, the denominator will have two zeros. So 0.75 = $75/100$, which reduces to $3/4$.

■ Change 3.6 to a mixed number.

The part after the decimal point is equal to $6/10$ so the whole decimal equals $3\,6/10$, which reduces to $3\,3/5$.

Warm-up

Change these decimals to fractions or mixed numbers. Reduce where possible.

1. 0.7 = ______	**5.** 0.025 = ______
2. 0.07 = ______	**6.** 2.16 = ______
3. 0.70 = ______	**7.** 21.6 = ______
4. 3.7 = ______	**8.** 4.03 = ______

Changing Fractions to Decimals

To change a fraction to a decimal, you just have to remember that a fraction is a way of writing a division problem. If you actually do the division that the fraction indicates, you will get a decimal for the fraction.

Here's an Example

■ Change $4/5$ to a decimal.

Step 1	Write $4/5$ as a division problem.	5)4
Step 2	Put a decimal point after the 4 and also directly above for the answer.	5)4.
Step 3	Add a couple of zeros after the decimal point in the number you're dividing. Start dividing.	.8 5)4.00 4 0 00

Warm-up Answers

1. $\frac{7}{10}$ **2.** $\frac{7}{100}$ **3.** $\frac{7}{10}$ **4.** $3\frac{7}{10}$ **5.** $\frac{1}{40}$ **6.** $2\frac{4}{25}$ **7.** $21\frac{3}{5}$ **8.** $4\frac{3}{100}$

The answer is .8 (usually written 0.8). You may need to add several zeros before you can finish dividing.

Try It Yourself

Here you will have to add more than two zeros.

■ Change $\frac{5}{8}$ to a decimal.

When you divided 5 by 8, did you find that you had to add 3 zeros? If you divided carefully and placed the decimal point correctly, you should have gotten .625 (or 0.625) as your answer.

Warm-up

Change these fractions to decimals.

1. $\frac{1}{2}$ = ______
2. $\frac{1}{5}$ = ______
3. $\frac{9}{10}$ = ______
4. $\frac{4}{5}$ = ______
5. $\frac{3}{4}$ = ______
6. $\frac{1}{8}$ = ______
7. $\frac{7}{8}$ = ______
8. $\frac{3}{8}$ = ______

Remainders

Sometimes you keep getting a remainder no matter how many zeros you add. Your answer just keeps repeating the same digits, which is why these decimals are often called **recurring,** or *repeating,* **decimals.**

Here's an Example

■ Change $\frac{1}{3}$ to a decimal.

Write it as a division problem as usual and you will soon see that you could go on dividing forever.

```
  0.3333
3)1.000
   9
   10
    9
    10
     9
     1
```

You can write the answer as $0.333\overline{3}$, with a bar over the 3 to show that the 3 keeps repeating. You can also write the last remainder over the number you're dividing by (3) and say that the answer is $.333\frac{1}{3}$. You could also stop and round to the number of decimal places you need. Some recurring decimals have a two-digit part that repeats.

■ Change $\frac{3}{11}$ to a decimal.

You can write the final answer as $0.\overline{27}$ (with a bar over the part that keeps repeating) or as $.2727\frac{3}{11}$. You could also round the decimal to the number of places you need.

```
     .2727
11)3.0000
   2 2
     80
     77
      30
      22
       80
       77
        3
```

Coming to Terms

recurring decimal a decimal with a continuously repeating group of digits, also called a *repeating decimal*

Try It Yourself

■ Find the decimal name for $\frac{1}{6}$. Divide to 4 decimal places then round off to the third place.

Did you divide to 4 decimal places and get 0.1666? If you did, then you should have 0.167 as your rounded answer. (To write the decimal for $\frac{1}{6}$ as a recurring decimal, you can put a bar over the first 6 because that digit repeats forever. The decimal would then be $0.1\overline{6}$. You did not have to do this for the answer, however, since you were asked to round to the third decimal place.)

Warm-up Answers
1. 0.5 **2.** 0.2 **3.** 0.9 **4.** 0.8 **5.** 0.75 **6.** 0.125
7. 0.875 **8.** 0.375

Warm-up

Change these fractions and mixed numbers to decimals. If the answers are recurring decimals, stop after four decimal places and round off to the third place.

1. $\frac{5}{11}$ = ______ **5.** $\frac{1}{9}$ = ______

2. $\frac{5}{6}$ = ______ **6.** $2\frac{3}{4}$ = ______

3. $1\frac{1}{12}$ = ______ **7.** $3\frac{4}{9}$ = ______

4. $2\frac{2}{3}$ = ______ **8.** $\frac{7}{6}$ = ______

Write these fractions as decimals. This time show the answer as a two-place decimal and a fraction that contains the remainder.

9. $\frac{3}{11}$ = ______ **11.** $\frac{1}{6}$ = ______

10. $\frac{2}{9}$ = ______ **12.** $\frac{2}{3}$ = ______

Test-Taking Tips

Generally on the GED Mathematics Test, if the numbers in a problem are written as fractions, the answer will be a fraction. If the numbers in the problem are written as decimals, the answer will be a decimal. If a problem has both fractions and decimals, look at the answers. Change to decimals or fractions based on the answer choices.

Some fractions and decimals are used very often in real life, so they often show up on the GED Mathematics Test. It will help a lot if you learn to recognize the decimals for these fractions by sight. Some of these fractions and decimals have been emphasized in this lesson:

$\frac{1}{2} = 0.5$ $\frac{1}{3} = .33\frac{1}{3}$

$\frac{1}{4} = 0.25$ $\frac{3}{4} = 0.75$

$\frac{2}{3} = .66\frac{2}{3}$

Warm-up Answers
1. 0.455 **2.** 0.833 **3.** 1.083 **4.** 2.667 **5.** 0.111
6. 2.75 **7.** 3.444 **8.** 1.167 **9.** $0.27\frac{3}{11}$ **10.** $0.22\frac{2}{9}$
11. $0.16\frac{2}{3}$ **12.** $0.66\frac{2}{3}$

On the Springboard

1. A kilometer is about 0.625 miles. In lowest terms, a kilometer is about what fraction of a mile?

   (1) $\frac{625}{1,000}$ (2) $\frac{25}{40}$ (3) $\frac{5}{8}$

   ① ② ③

2. The smallest divisions on some rulers are sixteenths of an inch. What is the exact decimal equivalent of $\frac{1}{16}$?

   (1) 0.06 (2) 0.063 (3) 0.0625

   ① ② ③

Check your Springboard answers on page 187. If you got these right, you're ready for "The Real Thing." If you missed either one, you may find it a good idea to review earlier material in this lesson and then try again. When you're satisfied that you're ready, go on to "The Real Thing."

The Real Thing

1. At $12\frac{1}{2}$ cents per tile, how many tiles can be bought for $1.00?

   (1) 8 (2) $12\frac{1}{2}$ (3) 80
   (4) 125 (5) 800

   ① ② ③ ④ ⑤

2. A burger shop used $\frac{1}{4}$ pound of meat for each hamburger. How many pounds of meat do they need for 318 hamburgers?

   (1) 72.5 (2) 78.75
   (3) 79.5 (4) 80.5
   (5) Insufficient data is given to solve the problem.

3. Manuel ran three marathons. His times were $2\frac{1}{10}$ hours, $1\frac{3}{5}$ hours, and $3\frac{1}{2}$ hours. His average time per marathon, expressed as a decimal, was how many hours?

(1) 2.16 (2) 2.4 (3) 6.12
(4) 7.2 (5) 21.6

① ② ③ ④ ⑤

4. A family budgets $\frac{1}{4}$ of its monthly income for housing costs. Of that amount, $\frac{1}{12}$ covers the cost of electricity. If the family's monthly income is \$2,604, how much money is budgeted for electricty?

(1) \$50.08 (2) \$53.25
(3) \$54.00 (4) \$54.25
(5) Insufficient data is given to solve the problem.

① ② ③ ④ ⑤

5. Two swimmers swam timed trials. The first swimmer completed $3\frac{1}{2}$ laps and $2\frac{3}{4}$ laps. The second finished $2\frac{1}{2}$ laps and $4\frac{1}{4}$ laps. How many more laps did the second swimmer complete?

(1) 13 (2) 6.75 (3) 6.25
(4) 0.5 (5) 0.25

① ② ③ ④ ⑤

6. Teresa earns \$4.50 per hour. For all hours over 40, she earns overtime pay at $1\frac{1}{2}$ times her hourly wage. How much did Teresa earn if she worked a 45-hour week?

(1) \$202.50 (2) \$213.75 (3) \$225.00
(4) \$303.75 (5) \$337.50

① ② ③ ④ ⑤

Check your answers and record your scores on pages 205–206.

LESSON 18

Zeroing In on Word Problems: Unnecessary Information

Sometimes more information is given in a problem than is needed to solve it. Such information is unnecessary, or extra. One way to avoid confusion is to use the system for solving word problems that was explained in Lesson 1.

Here's an Example

■ The cost of a large deluxe pizza is \$13.95. This includes a \$1.25 delivery charge. If 5 friends order a large deluxe pizza and give the delivery person a \$2.00 tip, what is each person's share of the total?

The first step is to read the problem carefully to understand what you're supposed to find. In this problem, you are asked to find each person's share of the total cost of a pizza. Ask yourself what you have to know to find the answer. The total amount paid for the pizza and the number of people sharing the cost, right? The problem tells you there were 5 people sharing the cost. So you can get the total cost by adding \$13.95, \$1.25, and \$2.00, right? Wrong! You should only add \$13.95 and \$2.00, because the \$1.25 delivery charge was *already included* in the \$13.95. The amount of the delivery charge is extra information. If you add \$13.95 and \$2.00 and then divide by 5, you'll get the answer.

Try It Yourself

Read this problem carefully. Is there any extra information? Don't solve for the final answer.

■ Alan earned \$384 last week. He worked 6 days of the week and was paid \$6.00 per hour for the first 40 hours he worked. He was paid \$9.00 per hour for overtime. What was Alan's average daily pay?

Did you read carefully? The problem asks you to find Alan's average daily pay. To do that, you need to know his total wages and how many days he worked. The problem gives both of these pieces of information. The rest of the information ($6.00 per hour for the first 40 hours, $9.00 per hour for overtime) is not necessary to solve the problem.

Warm-up

Read each problem carefully. First, list the numbers that are needed to solve the problem. Then, list the numbers that are extra and not necessary. You do not need to solve for the final answers.

1. A factory manufactured 17,560 toasters last month. Of these, 2% did not pass their first inspection and 3% failed the second inspection. What percentage passed both inspections?

Needed: ______________________

Not needed: ______________________

2. Mindy spent $12.50 last month for electricity, $22.50 for telephone, and $250 for rent. If she makes $425 per month after taxes, what fraction of her pay after taxes went to pay utility bills?

Needed: ______________________

Not needed: ______________________

3. It is 10 miles from where Ray lives to where he works. One afternoon he took the bus for ⅗ of the distance. He walked the remaining 4 miles. How many miles did he travel by bus?

Needed: ______________________

Not needed: ______________________

The practice you have just had in *reading* carefully, *finding* what was actually asked for, and *deciding* what information and operations to use will help you spot unnecessary information in the following Springboard questions.

Warm-up Answers
1. *Needed:* 2%, 3%; *Not needed:* 17,560
2. *Needed:* $12.50, $22.50, $425; *Not needed:* $250
3. Here your answer depends on how you plan to solve the problem. If you plan to multiply 10 by ⅗, then you need 10 and ⅗ but do not need 4. If you plan to solve the problem by simple subtraction, you need 10 and 4, but do not need ⅗.

On the Springboard

1. Sue was allowed 12 days of sick leave each year. In March she used $\frac{1}{4}$ of it. In June she spent 2 days in Arizona on a vacation weekend. Then she got sick and stayed away from work for 5 days. How many days of sick leave does she have left?

   (1) 0 (2) 2 (3) 4

   (1) (2) (3)

2. At a track meet in Chicago, a discus thrower threw under his own record by 2.5 meters. Later that week, in a city in Colorado that is 1,520 meters higher above sea level, he threw 65.6 meters. This beat his earlier record by 1.4 meters. How many meters did he throw the discus in Chicago?

   (1) 61.7 (2) 62.4 (3) 64.2

   (1) (2) (3)

Check your Springboard answers on page 187. If you answered these correctly, go on to "The Real Thing." If you missed either, reread the problems that gave you trouble and have another try before you continue.

"The Real Thing"

1. Kevin has $20 to spend for gifts. A gift subscription to a magazine is $8.00 for 8 issues or $12.80 for 16 issues. How much less is the magazine per issue if Kevin chooses the 16-issue subscription for his gift?

   (1) $0.20 (2) $0.80 (3) $0.90
   (4) $4.80 (5) $20.80

   (1) (2) (3) (4) (5)

2. The Gilberts spent $1,200 during a 14-day vacation. They spent $\frac{5}{12}$ of the total amount for motels, $\frac{3}{8}$ for gas and entertainment, and the rest on food. How much did the Gilberts spend on food?

(1) $950 (2) $500 (3) $450
(4) $350 (5) $250

(1) (2) (3) (4) (5)

3. A cab driver works 8 hours a day and covers approximately 150 miles per day. His cab gets 26.7 miles per gallon of gasoline. At this rate, about how many days will the cab run on 12.4 gallons of gasoline?

(1) 1.0 (2) 2.2 (3) 2.5
(4) 3.0 (5) 3.2

(1) (2) (3) (4) (5)

4. There were 48 students in a sailing class that met 3 times a week and cost $124. Of the 48 students, $\frac{3}{4}$ had never had sailing lessons. Of those who had never had lessons, $\frac{1}{4}$ had never sailed. How many of the students had never sailed?

(1) 9 (2) 12 (3) 16
(4) 27 (5) 36

(1) (2) (3) (4) (5)

5. A person takes multivitamins, vitamin E, and vitamin C. The multivitamin capsules cost $6.50 per 100 and the vitamin C capsules $4.95 per 100. If this person takes one capsule of each kind every day, approximately how much does this person spend on vitamins every day?

(1) $0.02 (2) $0.18
(3) $0.20 (4) $1.84
(5) Insufficient data is given to solve the problem.

(1) (2) (3) (4) (5)

Check your answers and record your score on page 206.

LESSON 19

Percents

What Are Percents?

The percent sign (%) is one that you've seen many times. Banks advertise that they pay 5% interest on regular savings accounts. Stores advertise that all merchandise is on sale at 20% off. What does it mean when people use the percent sign? It's easy; it means *out of every 100* or *per hundred.*

If you have had trouble understanding percents before, just remember that you can write any **percent** as a fraction that has 100 as the denominator. The numerator is the number that you see before the % sign. So 5% means 5 out of every 100 or 5/100, and 20 means 20 out of every 100 or 20/100.

You can do the reverse too. If you have a fraction with 100 as the denominator, you can write it as a percent. Just write the numerator and put the % sign after it: 30/100 is 30%, 15/100 is 15%. Just be sure that the fraction does have 100 as its denominator.

Here's an Example

■ Write 25% as a fraction that has a denominator of 100.

You can do it in one step. The number before the % sign is 25, so when you write 25 over 100 you get 25/100.

Try It Yourself

■ Write 50% as a fraction that has a denominator of 100. ________________

Did you write 50 in the numerator? Did you put 100 in the denominator? If you did, you got the right answer, 50/100.

■ Write 7/100 as a percent. ________________

Did you check to be sure that 100 is the denominator? Did you then write the numerator (7) followed by the % sign? If you did, you got 7%.

Coming to Terms

percent out of every hundred. A number written with the % sign after it. *Percent* means the fraction with that number as the numerator and 100 as the denominator (hundredths)

Warm-up

Write each percent as a fraction that has 100 as denominator.

1. 15% = ______
2. 70% = ______
3. 18% = ______
4. 33% = ______
5. 90% = ______
6. 100% = ______

Write each of these fractions as a percent.

7. $^{65}/_{100}$ = ______
8. $^{3}/_{100}$ = ______
9. $^{80}/_{100}$ = ______
10. $^{17}/_{100}$ = ______

Changing Percents to Fractions

You already know how to write a percent as a fraction with a denominator of 100. Often it is possible to reduce the fraction after you've done that. For instance, 50% is $^{50}/_{100}$ or $^{1}/_{2}$, and 20% is $^{20}/_{100}$ or $^{1}/_{5}$.

Sometimes you'll see percents that contain fractions or mixed numbers. They can also be changed to ordinary fractions.

Here's an Example

■ Change $33\frac{1}{3}$% to a fraction.

Step 1 Write the number before the % sign over 100. $\frac{33\frac{1}{3}}{100}$

Step 2 Change the fraction you just wrote to a division problem.

$$33\frac{1}{3} \div 100$$

Step 3 Divide. First, change the numbers to fractions; then, invert and multiply.

$$\frac{100}{3} \div \frac{100}{1} = \frac{\overset{1}{\cancel{100}}}{3} \times \frac{1}{\underset{1}{\cancel{100}}} = \frac{1}{3}$$

So $33\frac{1}{3}$% equals $^{1}/_{3}$.

Try It Yourself

Down payments on homes are often expressed as fractions.

■ If a family makes a down payment of 15% on their new home, what fraction of the total price have they paid?

Did you change 15% to $^{15}/_{100}$? Did you reduce by dividing the top and bottom by 5? If you did, you should have $^{3}/_{20}$ as your answer.

■ City sales tax is $7\frac{1}{2}$%. By what fraction does this tax increase the cost of an item?

Did you write $7\frac{1}{2}$ in the numerator and 100 in the denominator? Did you change that to $7\frac{1}{2} \div 100$ and then to $^{15}/_{2} \div {}^{100}/_{1}$? When you inverted $^{100}/_{1}$, multiplied, and reduced, you should have gotten $^{3}/_{40}$ as your final answer.

Warm-up

Change these percents to fractions. Reduce when possible.

1. 25% = ______
2. 20% = ______
3. 10% = ______
4. $12\frac{1}{2}$% = ______
5. 75% = ______
6. 45% = ______
7. 48% = ______
8. $2\frac{1}{2}$% = ______
9. 150% = ______
10. 60% = ______

Changing Percents to Decimals

You know how to change a percent to a fraction and you know how to change a fraction to a

Warm-up Answers
1. $^{15}/_{100}$ **2.** $^{70}/_{100}$ **3.** $^{18}/_{100}$ **4.** $^{33}/_{100}$ **5.** $^{90}/_{100}$ **6.** $^{100}/_{100}$
7. 65% **8.** 3% **9.** 80% **10.** 17%

Warm-up Answers
1. $^{1}/_{4}$ **2.** $^{1}/_{5}$ **3.** $^{1}/_{10}$ **4.** $^{1}/_{8}$ **5.** $^{3}/_{4}$ **6.** $^{9}/_{20}$ **7.** $^{12}/_{25}$
8. $^{1}/_{40}$ **9.** $1\frac{1}{2}$ **10.** $^{3}/_{5}$

decimal. So you should be able to change a percent to a decimal.

If the number before the percent sign is a whole number or a decimal, changing to a decimal is very easy. Just remember that you need to divide this number by 100 and that you can do that by dropping the % sign and moving the decimal point 2 places to the left. To see why this works, think of 25%. It is equal to $\frac{25}{100}$, which means 25 ÷ 100. If you put a decimal point in the number 25, it comes after the 5. When you move it two places to the left, you'll get .25, or 0.25.

If the number before the % sign is a fraction or mixed number, change it to a decimal. Then move the decimal point two places to the left.

Here's an Example

■ Change 56% to a decimal.

To do this, put the decimal point after the 6 in 56, then move it two places to the left.

.56.

Your answer is .56.

■ Change 7% to a decimal.

When you move the decimal two places to the left, you'll need to put a 0 in front of the 7.

.07.

The answer is .07.

■ Change $37\frac{1}{2}$% to a decimal.

Change $37\frac{1}{2}$ to 37.5 and move the decimal point two places to the left.

.37.5

The answer is .375.

Try It Yourself

■ Change 98% to a decimal.

Did you put the decimal point after the 8 in 98? Did you then move it two places to the left? You should have .98 (or 0.98) as your answer.

■ Change $5\frac{1}{4}$% to a decimal.

Did you change $5\frac{1}{4}$ to the decimal 5.25? If you did and if you then moved the decimal point 2 places to the left, you got the correct answer, .0525 (or 0.0525).

Warm-up

Change these percents to decimals.

1. 35% = ______	**6.** 75% = ______
2. 9% = ______	**7.** $8\frac{1}{4}$% = ______
3. $12\frac{1}{2}$% = ______	**8.** 50% = ______
4. 18.5% = ______	**9.** 125% = ______
5. 30% = ______	**10.** 250% = ______

Changing Decimals to Percents

You've changed percents to decimals. You can also go in the opposite direction and change decimals to percents. To change a percent to a decimal, you dropped the % sign and moved the decimal 2 places to the left. So, to change a decimal to a percent, move the decimal point 2 places to the right and add on the percent sign.

Here's an Example

■ Change 0.45 to a percent.

Move the decimal point two places to the right and add on the percent sign.

0.45.%

Written as a percent, 0.45 is 45%.

■ Change 2.1 to a percent.

Move the decimal point 2 places to the right. Add on the necessary 0 and the percent sign.

2.10.%

The answer is 210%.

Try It Yourself

■ Change 0.39 and 0.4 to percents.

Did you move the decimal point two places to the right for each of these? Did you remember to add on a 0 when you changed 0.4? You're right if you got 39% and 40%.

Warm-up Answers
1. 0.35 **2.** 0.09 **3.** 0.125 **4.** 0.185 **5.** 0.30 or 0.3
6. 0.75 **7.** 0.0825 **8.** 0.50 or 0.5 **9.** 1.25
10. 2.50 or 2.5

Warm-up

Change each decimal to a percent.

1. 0.15 = ______ **5.** 1 = ______

2. 0.5 = ______ **6.** 3.4 = ______

3. 0.125 = ______ **7.** 0.019 = ______

4. .8 = ______ **8.** 0.65 = ______

Changing Fractions to Percents

Remember that the basic meaning of a percent is a fraction with a denominator of 100. So a good way to change a fraction to a percent is to rename the fraction to have a denominator of 100.

Here's an Example

■ Change $3/20$ to a percent.

Step 1 Write the renaming problem. You want to rename $3/20$ to have a denominator of 100, so you want to solve

$$\frac{3}{20} = \frac{?}{100}$$

Step 2 What could you multiply 20 by to get 100? If you don't see the answer right away, divide 100 by 20. You get 5; does 20 × 5 give 100? Yes.

Step 3 Multiply both numerator and denominator of $3/20$ by 5.

$$\frac{3 \times 5}{20 \times 5} = \frac{15}{100}$$

Step 4 Use the numerator of 15/100, which is 15. Put a % sign after it; 15%

So $3/20$ is equal to 15%.

Here's an example where it's not easy to see right away how to rename the fraction to get a denominator of 100.

■ Change $1/3$ to a percent.

Step 1 Write the renaming problem.

$$\frac{1}{3} = \frac{?}{100}$$

Step 2 To find out what you would have to multiply 3 by to get 100, divide 100 by 3.

$$\begin{array}{r} 33\tfrac{1}{3} \\ 3\overline{)100} \\ \underline{9} \\ 10 \\ \underline{9} \\ 1 \end{array}$$

You get $33\frac{1}{3}$. (You can check on a piece of scratch paper that if you multiply 3 by $33\frac{1}{3}$ you really do get 100.)

Step 3 Multiply both the numerator and denominator of $1/3$ by $33\frac{1}{3}$.

$$\frac{1 \times 33\frac{1}{3}}{3 \times 33\frac{1}{3}} = \frac{33\frac{1}{3}}{100}$$

Step 4 Use the numerator of $\frac{33\frac{1}{3}}{100}$ and put a percent sign after it. You get $33\frac{1}{3}\%$

So $1/3$ is equal to $33\frac{1}{3}\%$.

Try It Yourself

■ Change $1/50$ to a percent.

Did you see that if you multiply both numerator and denominator by 2 you can get $\frac{1 \times 2}{50 \times 2}$ or $2/100$? Did you use the numerator of $2/100$ and get 2% as the final answer?

■ Change $2/9$ to a percent.

Did you say that you have to solve $\frac{2}{9} = \frac{?}{100}$ to change to a percent? To find out what to multiply 9 by to get 100, did you divide 100 by 9 and get $11\frac{1}{9}$? Next you should have renamed $2/9$ by multiplying the top and the bottom of $2/9$ by $11\frac{1}{9}$.

You should have gotten $\frac{22\frac{2}{9}}{100}$, so that the final answer is $22\frac{2}{9}\%$.

Warm-up Answers
1. 15% **2.** 50% **3.** 12.5% **4.** 80% **5.** 100% **6.** 340%
7. 1.9% **8.** 65%

Warm-up

Change each fraction or mixed number to a percent. (For the mixed numbers, first change to an improper fraction.)

1. $\frac{1}{25}$ = ______
2. $\frac{1}{4}$ = ______
3. $\frac{7}{10}$ = ______
4. $\frac{1}{2}$ = ______
5. $1\frac{1}{2}$ = ______
6. $\frac{2}{3}$ = ______
7. $\frac{5}{8}$ = ______
8. $2\frac{3}{10}$ = ______

A Test-Taking Tip

Here is a list of some common fractions and their percent equivalents. If you memorize this list, you can save time when you have to change percents and fractions on the GED Mathematics Test.

$\frac{1}{2} = 50\%$	$\frac{1}{5} = 20\%$	$\frac{1}{8} = 12\frac{1}{2}\%$
$\frac{1}{3} = 33\frac{1}{3}\%$	$\frac{2}{5} = 40\%$	$\frac{3}{8} = 37\frac{1}{2}\%$
$\frac{2}{3} = 66\frac{2}{3}\%$	$\frac{3}{5} = 60\%$	$\frac{5}{8} = 62\frac{1}{2}\%$
$\frac{1}{4} = 25\%$	$\frac{4}{5} = 80\%$	$\frac{7}{8} = 87\frac{1}{2}\%$
$\frac{3}{4} = 75\%$	$\frac{1}{6} = 16\frac{2}{3}\%$	$\frac{1}{10} = 10\%$
	$\frac{5}{6} = 83\frac{1}{3}\%$	$\frac{3}{10} = 30\%$
		$\frac{7}{10} = 70\%$
		$\frac{9}{10} = 90\%$

Warm-up Answers

1. 4% **2.** 25% **3.** 70% **4.** 50% **5.** 150% **6.** $66\frac{2}{3}\%$ **7.** $62\frac{1}{2}\%$ or 62.5% **8.** 230%

LESSON 20

Working with Percents

Setting Up a Percent Problem

All percent problems are made up of four things: a *whole,* a *part,* a *percent,* and *100.* Once you can find these things in a problem, you can set up and solve the problem with the help of a special grid. This section is about setting up percent problems using the grid. The grid is simply a square divided into 4 boxes like the one below.

part	percent
whole	**100**

Notice that three of the boxes do not have numbers. You fill in these boxes of the grid with information from the problem. The problem will probably have enough information to let you fill in two of the empty boxes. You put a *?* in the box that stands for what you're trying to find. Here is how to pick out the right numbers for each box.

1. **100.** The lower right-hand box *always* has the number 100 in it.

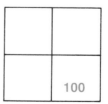

	100

2. **Percent.** The upper right-hand box always is for the percent, *without the percent sign.*

	percent
	100

3. **Whole.** The lower left-hand box is for the whole.

	percent
whole (of)	100

For instance, suppose a problem asks how many days an amusement park is open, and suppose it's open 80% of the year. Then the number of days in a year is the whole, and 365 would go in that box. Be sure to read the problem carefully to find what the *whole* is for the problem. The whole often comes after the word *of* as it did in this example (80% *of* the year).

4. **Part.** The part goes in the upper left-hand box. Usually the part is less than the whole, but it is possible to have a problem where the part is more than the whole. Just be sure you've read the problem carefully.

part	percent
whole (of)	100

Here's an Example

The following problem is not going to be solved. This is to show you how to fill in the grid after you've read the problem with care.

■ Mr. Thao pays $350 for rent each month. This is 25% of his income. How much is his monthly income?

Step 1 Draw the grid and put 100 in the lower right-hand box.

part	percent
whole	100

Step 2 There is a percent given in the problem, 25%. So 25 goes in the percent box, without the % sign.

part	percent 25
whole	100

Step 3 Find the *part* the problem is talking about. The $350 he pays for rent is a part of his monthly salary. So 350 goes in the box labeled *part.*

part 350	percent 25
whole	100

Step 4 $350 is 25% of his monthly income. The problem is asking for his full monthly income. So that must be the *whole* for the problem. Since this whole is the number the problem asks you to find, put a ? in the box labeled "whole."

part 350	percent 25
whole ?	100

Try It Yourself

Just set this problem up using the grid. You do not have to solve it.

■ In one city, 0.6% of the car owners had cars stolen last year. If the city has 79,835 car owners, how many people had cars stolen?

First, you should have drawn the grid. Did you remember to put 100 in the lower right-hand box?

part	percent
whole	100

You were told that 0.6% of the car owners had cars stolen. So 0.6 goes in the percent box. Do you know how many people had cars stolen? No, that's what the problem asks you to find. But that number is *part* of the total number of car owners (79,835). So 79,835 goes in the *whole* box, and ? goes in the *part* box. The grid should look like this:

part	percent
?	0.6
whole	
79,835	100

Warm-up

Write *part, percent,* and *whole* in the correct boxes. Then fill in the numbers and ? to set up the problems.

1. 25% of 48 is what number?

2. 20 is 2% of what number?

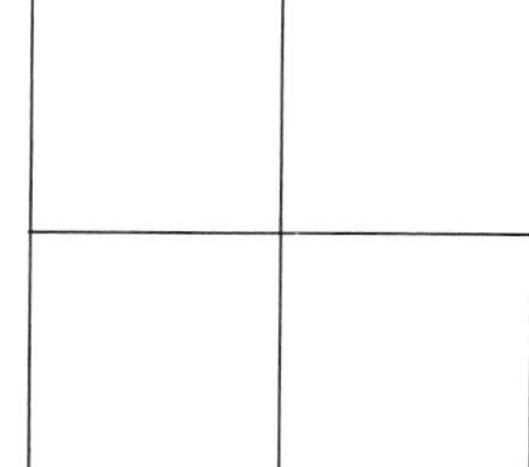

3. What percent of 750 is 150?

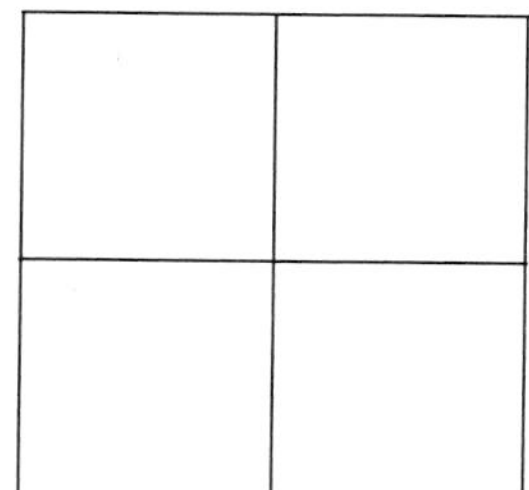

4. What percent of 480 is 12?

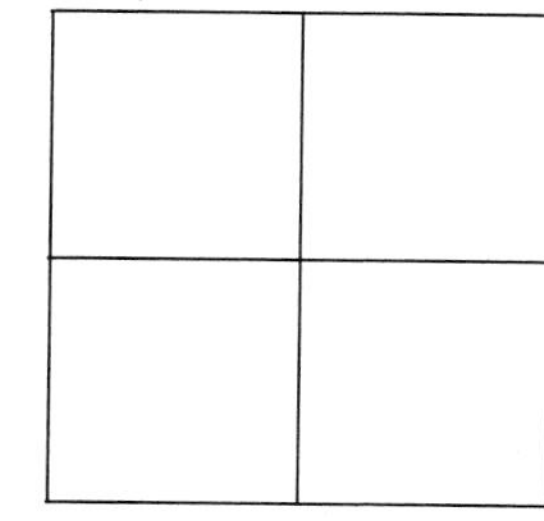

5. 200% of 30 is what number?

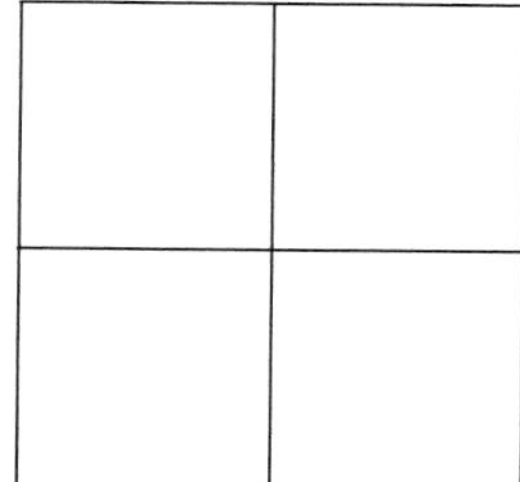

6. 6% of what number is 24?

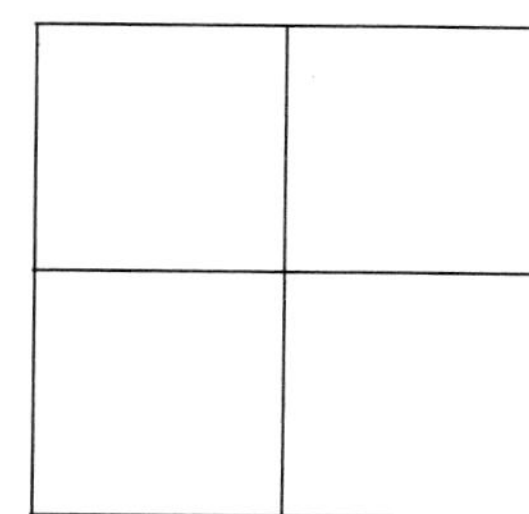

Solving Percent Problems

Now that you can use the grid, solving percent problems can be easy as 1, 2, 3.

1. Put the problem on the grid.
2. Multiply diagonal numbers.
3. Divide by the unused number.

Here's an Example

Now it's time to solve a problem using the grid.

■ Of 350 dentists polled, 28% recommended sugarless gum. How many recommended sugarless gum?

Warm-up Answers

1.

part	percent
?	25
whole	
48	100

2.

part	percent
20	2
whole	
?	100

3.

part	percent
150	?
whole	
750	100

4.

part	percent
12	?
whole	
480	100

5.

part	percent
?	200
whole	
30	100

6.

part	percent
24	6
whole	
?	100

Step 1 Fill in the grid with what you know. You know that the whole group of dentists is 350, the percent is 28. The 100 always goes in the lower right-hand box. What you are looking for is *part* of the whole group of dentists. So *?* goes in the *part* box,

part	percent
?	28
whole	
350	100

Step 2 Multiply the diagonals that have two numbers.

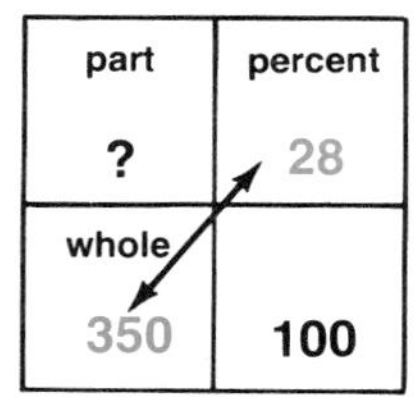

$$350 \times 28 = 9{,}800$$

Step 3 Divide by the unused number, 100.

$$9{,}800 \div 100 = 98$$

Now you've got the answer; 98 dentists recommended the sugarless gum.

Even though the grid system makes things almost automatic, it's good to know why it works. Knowing this can help with filling in the grid. If two fractions are equal, the results of multiplying the diagonal numbers are also equal. For example, $\frac{2}{5} = \frac{40}{100}$ and $2 \times 100 = 200$ and $5 \times 40 = 200$. If you put a *?* in place of any of the four numbers, you could find the *?* number by multiplying the diagonal numbers and dividing by the unused number. For example, $\frac{?}{5} = \frac{40}{100}$, $40 \times 5 = 200$, and $200 \div 100 = 2$. Every percent problem involves two equal fractions with one number not known. The grid is just a neat way to put those fractions down and keep track of what you are doing.

Try It Yourself

Put this on the grid and solve it.

■ A stereo is on sale for 60% of the original price. The sale price is $300. What was the original price? ______________

The whole is what you are looking for, so when you filled in the grid, the *?* should have gone in the box labeled *whole.* The percent is 60, and the part is 300. You should have multiplied 100 by 300 to get 30,000 and then divided 30,000 by 60. Did you do that and get 500? The original price was $500.

Warm-up

Find the missing numbers in these percent problems. They are the same problems you had for the first Warm-up in this lesson on page 103 where you set up the grids. Go back to the grids now and solve them.

1. 25% of 48 is what number?
2. 20 is 2% of what number?
3. What percent of 750 is 150?
4. What percent of 480 is 12?
5. 200% of 30 is what number?
6. 6% of what number is 24?

On the Springboard

1. What percent of the school year of 180 days is 45 days?

   (1) 20% (2) 25% (3) 35%

2. In a glass marble collection of 900 marbles, 60% have blue in them. How many marbles in all have blue in them?

   (1) 540 (2) 450 (3) 360

   ① ② ③

Warm-up Answers
1. 12 **2.** 1,000 **3.** 20% **4.** $2\frac{1}{2}$% or 2.5% **5.** 60 **6.** 400

Check your Springboard answers on pages 187–188. If you got both right, go on to "The Real Thing." If you missed either, why not check to see if you filled in the grid correctly. Was your arithmetic correct? When you're ready, go on to "The Real Thing."

“The Real Thing”

1. A salesperson receives a 5% commission on gross sales each week. If the salesperson received a commission of $218.40 one week, how much were gross sales for that week?

   (1) $10.92 (2) $1,092.00 (3) $2,074.80
   (4) $4,368.00 (5) $87,360.00

   ① ② ③ ④ ⑤

2. A bicyclist spends 10 hours each day of the weekend biking. If a weekend is 48 hours long, what percent is spent biking?

   (1) $41\frac{2}{3}\%$ (2) 240% (3) $20\frac{4}{5}\%$

   (4) $9\frac{3}{5}\%$ (5) $4\frac{1}{6}\%$

   ① ② ③ ④ ⑤

3. JoAnne gets 15% commission on sales as an appliance salesperson. What must her sales be in order to earn a commission of $600?

   (1) $2,000 (2) $3,000
   (3) $4,000 (4) $5,000
   (5) Insufficient data is given to solve the problem.

   ① ② ③ ④ ⑤

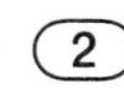

Check your answers on page 206.

Two-Step Problems

Sometimes you can't go straight to the final grid. Some other step comes first. It usually isn't a hard step—often you can do it in your head. But you must read the problem carefully to be sure you do not miss that step before you fill in the grid.

Here's an Example

■ A stereo is for sale at 40% off the original price. The sale price is $300. What was the original price of the stereo?

Notice the word *off.* The sale price is 40% off the original, which means that 40% was taken from the original price. So the sale price is 60% *of* the original. Now the problem and the grid are identical to the last "Try It Yourself," and the answer will be the same, $500.

Some problems need two grids. The answer for the first grid goes into the second. Here's a sample of that kind of problem.

■ Betty got 10% commission on all boat sales she made. She offered her friend Janet 40% of her commission as a finder's fee for any leads who bought a boat. Janet's first lead bought a boat for $80,000. What was Janet's finder's fee?

Step 1 Use the grid to find Betty's commission.

80,000 × 10 = 800,000

800,000 ÷ 100 = 8,000

part	percent
?	10
whole	
80,000	100

Step 2 Use another grid to find 40% of $8,000.

8,000 × 40 = 320,000

320,000 ÷ 100 = 3,200

part	percent
?	40
whole	
8,000	100

If two percents are mentioned, and the second depends on the first, you may have a two-grid problem.

Try It Yourself

■ Forty VCRs were on sale at 65% of the original price of $300. One customer bought more than one VCR and was given an extra discount of 5%. What did this customer pay for each VCR?

The *forty* was a bit of unnecessary information. Since the customer was given an extra 5% discount, the price to him was 70% of $300. The grid gives you (70 × 300) ÷ 100. The answer is $210 for each VCR.

Warm-up

Just set up the grids for these problems. You do not have to find the final answers.

1. Alan's bank paid 5¼% interest per year. He moved the $1,000 in the account to a credit union account that paid 8¼%. What amount in extra interest did he get in the first year?
2. A new inspection system dropped the reject rate of machine parts in a factory from 3% to 2% per week. This made a difference of 120 pieces per week in the number of acceptable parts. How many parts were produced each week?

On the Springboard

3. Of 120 candidates, 70% passed the exam. Of the candidates who passed, 25% did so with honors. How many honor students were there?

(1) 50 (2) 42 (3) 21

① ② ③

4. A storekeeper added 30% of his cost to every article in his store. City tax on the selling price was 5%. What would a customer pay in dollars for an article that cost the storekeeper $200?

(1) 285 (2) 273 (3) 265

① ② ③

Check the Springboard Answers on page 188. If you got all these right, you did very well and you can go on to "The Real Thing." If you missed one, why not review and check your work before going on?

Warm-up Answers

1.

part	percent
?	3
whole	
1,000	100

2.

part	percent
120	1
whole	
?	100

"The Real Thing"

4. The Andersons bought a sofa that was marked 30% off. The original price was $750. They had it delivered for an additional 4% of the sale price. What was the total cost of the sofa with delivery?

(1) $234.00 (2) $504.00 (3) $546.00
(4) $772.50 (5) $803.40

① ② ③ ④ ⑤

5. Jamie earns $1,400 a month. Of this amount, 25% goes for rent and 20% for food. How much does she spend for rent and food in one year?

(1) $840 (2) $1,860 (3) $5,040
(4) $6,300 (5) $7,560

① ② ③ ④ ⑤

Check your answers on pages 206–207.

Increases and Decreases

Sometimes a problem asks by what percent something increased or decreased. These problems too can be solved on the grid with just a small change in the grid.

Here's an Example

This is about increase, but if the numbers were the other way around it would be about decrease. The same method applies to both kinds of problems.

■ A city's population increased from about 400,000 to about 500,000 in ten years. What was the percent increase in population?

By reading the question carefully, you can see that you are dealing with a *change* from an *original* value. So the original value is put in the *whole* box and the change is put in the *part* box.

Step 1 Fill in the grid. Write 100 in the usual box. Write the original population (400,000) in the *whole* box, but label the box *original.* Write the change in population (100,000) in the *part* box, but label the box *change.*

change 100,000	percent ?
original 400,000	100

Step 2 Multiply diagonals as usual and divide by the unused number.

$$100 \times 100{,}000 = 10{,}000{,}000$$
$$10{,}000{,}000 \div 400{,}000 = 25$$

The population increased by 25%.

If the population had *decreased* from 500,000 to 400,000, the *original* box would have 500,000 in it. The change is the same, 100,000. Try it on scratch paper and see if you get 20% as the decrease.

Try It Yourself

Label your grids for this problem with *change* and *original* the way you saw in the example.

■ A computer that sold for $10,000 a few years ago can now be bought for $1,000. What is the percent decrease in price?

Here the change in price is $9,000 and the original price is $10,000. You should have gotten 9,000 × 100 = 900,000. When you divided by 10,000, you should have gotten 90% as your final answer.

These two grids can help you solve most kinds of percent problems.

ordinary percent

part	percent
whole (of)	100

increase/decrease

change	percent
original	100

A Test-Taking Tip

When you're using the grids to solve percent problems, there's a way you can often save work. When you are ready to multiply diagonals and divide by the unused number, set up the calculation as a fraction. (It's all right to do this, since a fraction is a way of showing division.) If you can see an easy way to cancel, you've saved a lot of arithmetic. For example, if you have to do the multiplication 100 × 200,000 and then divide by 400,000, set it up like this:

$$\frac{100 \times 200{,}000}{400{,}000}$$

Before you do the calculation, you can see right away that you can cancel and get this:

$$\frac{100 \times \cancel{200{,}000}^{1}}{\cancel{400{,}000}_{2}} = \frac{100}{2}$$

You can probably do 100 ÷ 2 in your head. You get 50 as the final answer.

Warm-up

In each of these problems, the first number is the original value and the second number a later value. Find the change, if any, and compute the percent change. In your answer, tell whether it is percent increase or percent decrease.

1. From 400 to 300 ____________
2. From 400 to 700 ____________
3. From 400 to 400 ____________
4. From 400 to 480 ____________
5. From 12 to 15 ____________
6. From 12 to 9 ____________

Warm-up Answers
1. 25%, decrease **2.** 75%, increase **3.** 0%, neither increase nor decrease **4.** 20%, increase **5.** 25%, increase **6.** 25%, decrease

On the Springboard

5. Shirley started her trip with 30 pounds of air pressure in her tires. By the time she got to Colorado, the pressure had gone up to 36 pounds. What was the percent increase in pressure?

   (1) 20 (2) 30 (3) 66

   1 2 3

6. The 100% cotton waistband shrank with the first washing from 40 to 38 inches. Find the percent decrease due to shrinkage.

   (1) $2\frac{1}{2}$ (2) 5 (3) 8

   1 2 3

Check your Springboard answers on page 188. If you got these right, well done. Go on to "The Real Thing." If you missed either, it may be a good idea to review both kinds of grid.

The Real Thing 

6. In 1973, gasoline cost $0.40 a gallon. In 1985, it cost $1.76 a gallon. The price of a gallon of gasoline increased by what percent?

   (1) 340% (2) 315% (3) 34%
   (4) 31.5% (5) 3.4%

   1 2 3 4 5

7. Last quarter, 60 students were enrolled in a science course. This quarter, 40 students are enrolled. The number of students enrolled decreased by what percent?

   (1) 60% (2) 50% (3) 40%
   (4) $33\frac{1}{3}$% (5) 20%

   1 2 3 4 5

8. About 75% of a person's body weight is made up of water. Suppose a person weighs 138 pounds. How many pounds of this weight is water?

   (1) 18.4 (2) 34.5 (3) 102.8
   (4) 103.5 (5) 184.0

   1 2 3 4 5

9. After a pay raise, an employee's monthly salary was $972. Before the raise, his monthly salary was $900. By what percent did his salary increase?

   (1) 0.08% (2) 7% (3) 8%
   (4) 93% (5) 108%

   1 2 3 4 5

10. There are 20 trees lining Arbor Road. In the fall, 6 turn gold, 5 turn red, and the rest turn orange. What percent of the trees turn orange in the fall?

    (1) 45% (2) 81.8% (3) $83\frac{1}{3}$%
    (4) 120% (5) 222%

    1 2 3 4 5

11. An employee makes 7% commission on his sales. If his commission one week was $245, what was the amount of his sales?

    (1) $3,500.00 (2) $3,255.00
    (3) $1,715.00 (4) $350.00
    (5) $171.50

    1 2 3 4 5

12. A lumber yard charges $1.90 for a 10-foot board plus 7% for preparation. For a 5-foot board the charge is $1.25 plus 7% for preparation. What will be the total cost of 60 boards?

    (1) $7.98 (2) $111.98
    (3) $114.00 (4) $121.98
    (5) Insufficient data is given to solve the problem.

    1 2 3 4 5

Check your answers and record all your scores on page 207.

LESSON 21

Zeroing In on Word Problems: Item Sets

Now you will look at a special kind of problem on the GED Mathematics Test. The problem will begin with a short paragraph followed by from 2 to 5 questions. You can answer the questions by picking out the information you need from the paragraph. Read the paragraph and the questions carefully. Some information may be needed for one question but not the other. It is possible that one of the questions cannot be answered using the information.

Here's an Example

■ Frank does piece work at a factory. He is paid by the day for the items he makes. The company pays $1.00 per item for the first 50 items, and $1.25 per item after that. He averages 60 items a day.

1. What are Frank's weekly earnings?

First figure out Frank's daily earnings. You were told that he earns $1.00 per item for the first 50 items. Multiply.

$1.00 \times 50 = 50.00$

You were also told that he earns $1.25 for each item over 50. Since he averages 60 items a day, subtract to find out how many over 50.

$60 - 50 = 10$

So, he makes 10 items over 50. Multiply.

$\$1.25 \times 10 = \$12.50.$

Next, add the earnings for the two rates.

$$\begin{array}{r} \$50.00 \\ +12.50 \\ \hline \$62.50 \end{array}$$

Frank earns $62.50 a day. Multiply by 5 to get his average weekly earnings.

$\$62.50 \times 5 = \$312.50$

2. After six months, Frank's rate will go up to $1.10 for the first 50 and $1.35 for each item over that. How much will he earn per month if he continues to make 60 items per day?

After six months he will be paid more for each item. Figure out what he would earn per day and multiply by 20.

$$\begin{array}{rl} \$55.00\ (50 \times \$1.10) & \text{first 50 items} \\ +\$13.50\ (10 \times \$1.35) & \text{next 10 items} \\ \hline \$68.50 & \text{total daily wages} \end{array}$$

Multiply by 20 to get what he will earn each month.

$$\begin{array}{r} \$ \quad 68.50 \\ \times \quad 20 \\ \hline \$1,370.00 \end{array}$$

He will earn $1,370 a month.

Try It Yourself

Use the information in the paragraph to answer the questions that follow. If there is not enough information to answer a question, choose the answer that reads *Insufficient data is given to solve the problem.*

■ Three friends from 3 different cities went fishing. Fran drove 510 miles due east and used 30 gallons of gas. Lori drove her car due west 675 miles and used 25 gallons. Shawn made a detour of 40 miles and then drove north 280 miles to the meeting. Her compact used just 10 gallons of gas.

1. If they had all paid the same price per gallon for the gasoline they used, what was the price per gallon? ____________

2. Write an expression for the total number of miles the friends drove. ____________

For question 1, you should have answered *Insufficient data is given to solve the problem.* You were not told how much they spent for gasoline.

For question 2, you needed to add the separate mileages to arrive at the total number of miles driven by all 3. Did you remember to add in Shawn's detour? You should have gotten $510 + 675 + 40 + 280 = 1,505$. They drove a total of 1,505 miles.

Warm-up

Solve the following problems.

■ Carla drove to work and parked for free. Gasoline cost her $10 a month (20 working days). She spent $1.50 for lunch at the company cafeteria.

1. What are Carla's out-of-pocket expenses for a year? ______________________

2. How much more would it cost Carla to take the bus to work? ______________________

On the Springboard

Items 1 and 2 are based on the following information.

After the chess tournament the 10 members of the team went to the pier for boating. Their boat costs were as follows.

Boat Hire per hr	
Canoe	$2.00
Rowboat	$2.50
Speedboat	$5.00

1. The co-captains shared a rowboat. Five members of the team each took a canoe. The others shared a speedboat. What was their total cost for one hour of boating?

   (1) $12.50 (2) $17.50 (3) $24.00

   (1) (2) (3)

2. If they had rented their boats during the final hour of the day, everything would have been 50% off except the speedboats. What would the team have spent on boating if they had taken the same boats at the end of the day?

   (1) $8.75 (2) $9.50 (3) $11.25

   (1) (2) (3)

Check your Springboard answers on page 188. If you get these right you're doing well with a very important skill. If you didn't, check over your work carefully to find your mistakes. When you're satisfied that you're ready, go on to "The Real Thing."

Warm-up Answers
1. $480 **2.** Insufficient data is given to solve the problem

The Real Thing

Questions 1 and 2 refer to the following.

During a 2-day craft fair, the DLM Craft Club collected $937.50 for 15 doll houses and $342 for 8 dolls with wardrobes. The RTS Craft Club collected $198 for spinning tops and $1,128.75 for 15 rocking horses.

1. The RTS Craft Club collected how much more money than the DLM Craft Club?

   (1) $47.25 (2) $191.25 (3) $1,279.50
   (4) $1,326.75 (5) $2,606.25

   (1) (2) (3) (4) (5)

2. Joe bought a doll house, a doll, and a rocking horse. How much did he pay for these three items?

   (1) $42.75 (2) $62.50 (3) $75.25
   (4) $180.50 (5) $189.25

   (1) (2) (3) (4) (5)

Questions 3 and 4 refer to the following.

Glenda's jobs pay $8 an hour if she works at home and $12 if she goes into the office. She worked 36 hours at home for the MLK Company. She also went to the DZX Company and worked 7½ hours on each of 15 days.

3. How many times as long did Glenda work for the DZX Company as for the MLK Company?

   (1) $2\frac{2}{5}$ (2) $4\frac{4}{5}$ (3) $3\frac{1}{8}$ (4) $76\frac{1}{2}$
   (5) Insufficient data is given to solve the problem.

   (1) (2) (3) (4) (5)

4. Glenda sets aside 15% of her gross pay for social security and 25% for retirement. Which expression shows how much she should set aside of her pay from the DZX Company?

   (1) $.15(36 + 7.5) + .25(36 + 7.5)$
   (2) $.40(7.5 + 15)$
   (3) $.15(7.5) + .25(7.5)$
   (4) $.15[(7.5 \times 15)12] + .25[(7.5 \times 15)12]$
   (5) Insufficient data is given to solve the problem.

      (1) (2) (3) (4) (5)

Check your answers and record your score on pages 207–208.

LESSON 22

Tables and Meters

Bus schedules, weight charts, price lists, TV guides—all of these are tables that you might need to read on an ordinary day at home, at work, or while traveling. Sometimes you may also need to read dials on meters.

The GED Mathematics Test usually has a few questions about tables or meters. This lesson will help you brush up on how to read them.

Reading Tables

A table has a title that tells you what the table is about. Tables contain information organized into columns that go down the page and rows that go across the page. The columns and rows are usually labeled with headings that tell what's in them.

Here's an Example

Look at this table of unemployment rates for 7 states in the Midwest.

Unemployment Rates
(in percents for months ended Nov. 30)

	Ill.	Ind.	Iowa	Ohio	Mich.	Minn.	Wis.
1981	8.5	10.1	6.9	10.9	12.3	5.6	7.8
1982	11.3	11.9	8.5	14.0	15.5	8.3	10.7
1983	11.4	11.1	8.1	11.0	14.2	6.7	10.4
1984	9.1	8.6	7.0	8.9	11.2	5.4	7.3

The title is "Unemployment Rates." The next line tells you that all the numbers are percents, and that these percents include information through the end of November for each year listed. The column headings are names of states and the row headings are years.

■ In which year did Illinois have its highest rate of unemployment?

Find the column head *Ill.*, then look down the column for the largest number. It's 11.4 and it's in the row labeled 1983, so you've got the answer. Illinois's highest rate of unemployment was in 1983.

Try It Yourself

Use the table of unemployment rates.

■ Which state had the highest unemployment rate in 1982?

____

Did you look across the row for 1982 and see that Michigan had an unemployment rate of 15.5%?

■ Which state had the greatest rise in its unemployment rate from 1981 to 1982?

____

By subtracting the 1981 rate from the 1982 rate for each state, you find that Michigan had the greatest increase, 3.2%

Warm-up

This is a table of how weekly earnings in selected industries went up from 1985 to 1986. Look at it and answer the questions.

Average Weekly Earnings
(nonmanufacturing)*

Industry	1986	1985	% increase
Telephone communications	$508.00	$498.80	1.8
Local transportation	333.60	323.60	3.1
Retail sales	240.00	238.80	0.5
Construction	452.40	447.20	1.2
Hotel/motel	238.80	231.20	3.2

*Based on a 40-hour week.

1. What is the title of the table? __________

____

2. In which industry did workers receive the greatest percent of earnings increase?

____

3. Find the industry with the lowest weekly earnings in 1985 and write down the percent increase for that industry. __________

4. Between industries, what was the greatest dollar difference in earnings in 1985? ________

5. Between industries, what was the greatest dollar difference in earnings in 1986? ________

Warm-up Answers
1. Average Weekly Earnings **2.** Hotel, tourist industry
3. 3.2% **4.** $267.60 **5.** $269.20

On the Springboard

This is a small section of a table that gives distances in miles between major U.S. cities. Answer these questions about the table.

Road Mileage Between Selected U.S. Cities

	Atlanta	Boston	Chicago	Cincin-nati	Cleve-land
Atlanta, Ga	. . .	1,037	674	440	672
Boston, Mass . .	1,037	. . .	963	840	629
Chicago, Ill	674	963	. . .	287	335
Cincinnati, Oh .	440	840	287	. . .	244
Cleveland, Oh .	672	628	335	244	. . .
Dallas, Tex	795	1,748	917	920	1,159
Denver, Col. . . .	1,398	1,949	996	1,164	1,321
Detroit, Mich . . .	699	695	266	259	170
Houston, Tex . .	789	1,804	1,067	1,029	1,273

1. How much farther in miles is it from Atlanta to Chicago than from Atlanta to Cleveland?

   (1) 2 (2) 234 (3) 244

   ① ② ③

2. John drove from Boston to Cincinnati and then from Cincinnati to Atlanta. How many miles did he drive?

   (1) 440 (2) 1,180 (3) 1,280

   ① ② ③

3. Which of the cities listed in the table is farthest from Chicago?

   (1) Boston (2) Denver (3) Houston

   ① ② ③

Check your Springboard answers on pages 188–189. If you got them all correct, go on to "The Real Thing." If you missed any, check over your work. When you're satisfied that you can get them right, go to "The Real Thing."

“The Real Thing”

Questions 1–4 refer to the following table.

Guide to Making Fresh Fruit Sauces

Kind of fruit	Amount of fruit (pounds)	Amount of water (cups)	Amount of sugar (cups)	Cooking time (minutes)	Approximate yield (cups)
Apples	2	⅓	¼	12–15	3
Cherries	1	⅔	½	5	2
Cranberries	1	2	2	15	4 (whole) 3 (strained)
Peaches	1	⅔	½	5–8	2
Rhubarb	1½	¾	⅔	2–5	3

1. Suppose you use 2 pounds of apples to make sauce. How many cups of sugar should you use?

   (1) $\frac{1}{4}$ (2) $\frac{1}{3}$ (3) $\frac{1}{2}$ (4) 2 (5) 3

   ① ② ③ ④ ⑤

2. If you are making 3 cups of strained cranberry sauce, how many minutes should you cook the sauce?

   (1) 1 (2) 2 (3) 2–5
   (4) 5–8 (5) 15

   ① ② ③ ④ ⑤

3. Which kind of fruit sauce takes the least amount of water per pound of fruit?

   (1) Apples (2) Cherries
   (4) Cranberries (4) Peaches
   (5) Rhubarb

   ① ② ③ ④ ⑤

4. Ms. Lindelof used 3 pounds of rhubarb to make sauce. Approximately how many cups of sauce did she get?

   (1) $1\frac{1}{2}$ (2) 2 (3) 3 (4) 4 (5) 6

   ① ② ③ ④ ⑤

Check your answers on page 208.

Reading Meters

To read a dial on a meter, look to see whether the hand on the dial is pointing straight at one of the numbers or if it is between two numbers. If the hand points between two numbers, you choose the smaller number. If the hand points straight at a number, then that's the number you use.

Here's an Example

To read an electric meter, read the dials from left to right. It doesn't matter if the dial runs clockwise or counterclockwise.

■ The dials of an electric meter show these positions. How many kilowatt-hours does the meter show?

Just read the dials from left to right,

Step 1 Start with the dial on the far left. The hand is between 0 and 1, so the reading is 0. Write 0 under that dial.

Step 2 Look at the second dial. The hand points straight at the 5, so write 5 below this dial.

Step 3 Look at the next dial. The hand is not quite to the 9, so the reading is 8. Write 8 below this dial.

Step 4 The hands on the remaining two dials both point directly at 2. So write 2 beneath each of the last two dials.

The meter reads 5,822 kilowatt-hours.

On the GED Mathematics Test you may get a meter with two hands on the dial. If this happens, just read the directions to know for what each hand stands. An altimeter in an airplane is an example of such a meter. An altimeter tells how many feet above sea level the airplane is. The short hand tells thousands of feet and the long hand tells hundreds of feet. To read an altimeter, read the number for the short hand, then the number for the long hand. Write these down and add on two zeros at the end.

■ What altitude does this altimeter show?

Step 1 The short hand points to 5, the long hand points to 2.

Step 2 Write 52 and add on two zeros.

5200.

The altimeter shows 5,200 feet above sea level.

Try It Yourself

■ What is the reading on this gas meter?

_____ _____ _____ _____ _____

Did you remember to write the smaller number when the hand comes between two numbers? Did you get confused on the first dial? Remember 10 comes after 9, so, if the hand is between 9 and 0, 9 is the number recorded. If you read 98,503, you did it correctly.

On the Springboard

4. A small plane crashed on a mountainside. Investigators found that the altimeter looked like this.

What was the altitude of the plane in feet above sea level when it crashed?

(1) 3,500 (2) 5,300 (3) 5,400

1 2 3

5. These are dials from a gas meter. What is the reading in cubic feet of gas?

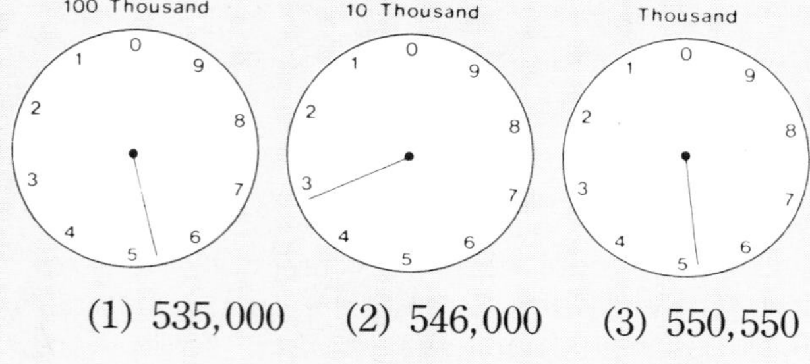

(1) 535,000 (2) 546,000 (3) 550,550

Check your Springboard answers on page 189. If you got these right, go on to "The Real Thing." If you didn't, check your work. Review the section if you wish. When you're sure that you can read these meters, go on.

The Real Thing

5. The dials of a water meter are in the following positions. If the hand is between two numbers use the smaller number. What is the reading on the meter?

(1) 21,362 (2) 21,363 (3) 21,462
(4) 22,472 (5) 22,473

1 2 3 4 5

For questions 6–9, use the dials that follow. These dials on an elecric meter show kilowatt-hours used.

December 2, 1987

6. How many kilowatt-hours should be reported on December 2, 1987.

(1) 54,915 (2) 59,925 (3) 68,915
(4) 69,025 (5) 78,025

1 2 3 4 5

February 3, 1988

7. How many kilowatt-hours should be reported on February 3, 1988?

(1) 50,602 (2) 55,602 (3) 59,612
(4) 60,612 (5) 69,602

1 2 3 4 5

8. How many kilowatt-hours were used between December 2, 1987, and February 3, 1988?

(1) 687 (2) 688 (3) 787
(4) 797 (5) 1,798

1 2 3 4 5

9. A kilowatt-hour costs $0.064. What was the cost of the electricity used between December 2, 1987, and February 3, 1988?

(1) $67.70 (2) $44.96
(3) $43.97 (4) $43.87
(5) Insufficient data is given to solve the problem.

   4 5

LESSON 23

Reading Graphs

You will find graphs in newspapers, magazines, books, and even on television. A graph gives you an easy way to compare information in picture form.

Reading a Bar Graph

The most common kind of graph used in newspapers, magazines, and books is the **bar graph.** A bar graph compares numbers by using bars of different lengths to represent the numbers. The bars may be vertical, as in Graph A, or horizontal, as in Graph B.

Graph A. This is a vertical bar graph.

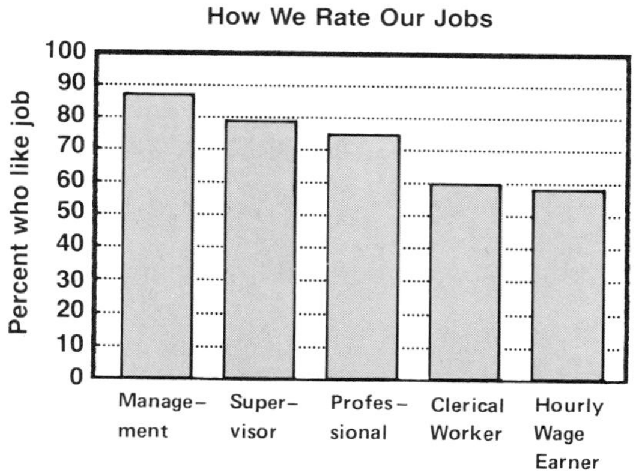

Graph B. This is a horizontal bar graph.

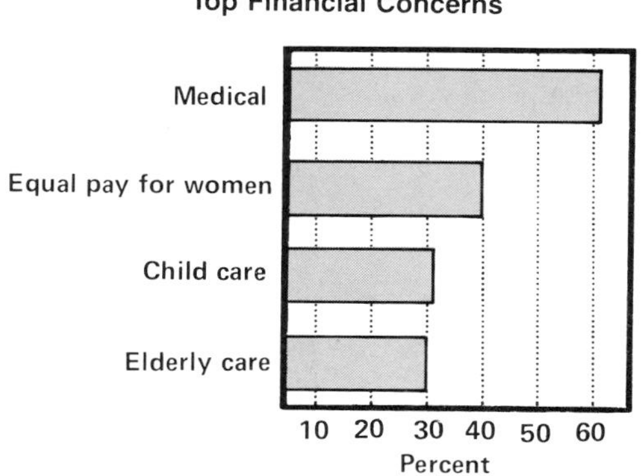

A bar graph may compare several items. Look at Graph C. It compares the low temperature in four different cities and on two separate days.

Graph C. This is a multiple-bar graph.

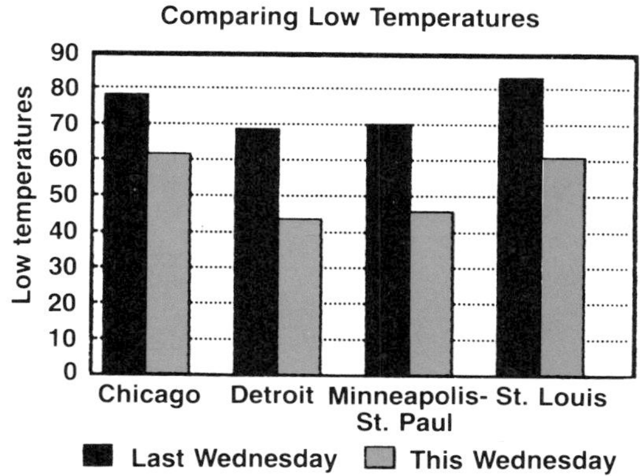

A bar graph may also combine several items. Each stacked bar stands for the *total* of what is named below the bar. In graph D the *total* for both dairy cows and pigs raised on farms is represented. The gray portion of each bar shows the *part* of the total that was cows, the blue portion is the part of the total representing pigs.

Graph D. This is a stacked-bar graph.

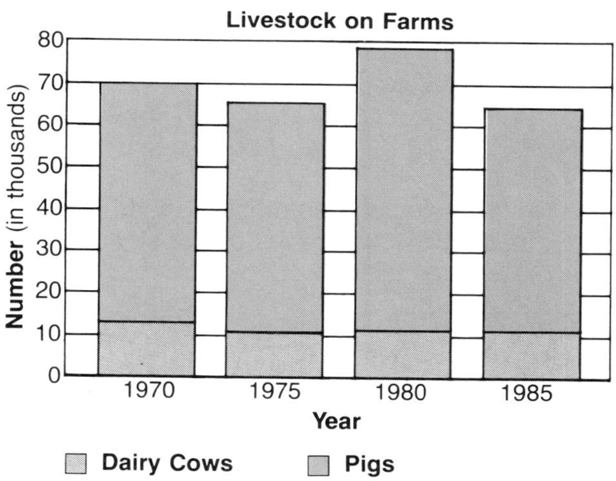

To read a bar graph you should follow four steps.

Step 1 Read the title.

Step 2 Read all headings on columns and rows. Decide what is being compared.

Step 3 Look at the numbers to see how they change (by 10s, 100s, 1,000s, and so on).

Step 4 If the graph uses color or symbols, look for a key to tell you what the symbols mean.

Coming to Terms

bar graph a graph that compares numbers by using bars of different lengths

Here's an Example

Look at Graph A as you work through the 4 steps.

Step 1 The title is "How We Rate Our Jobs."

Step 2 Look at the headings.

Headings of Columns	Headings of Rows
Type of Worker Management Supervisor Professional Clerical Worker Hourly Wage Earner	**Percent Who Like Job** Ranging from 0% to 100% by 10s

The graph compares how people doing different kinds of work feel about their jobs.

Step 3 The numbers along the left side change in 10% jumps, or intervals.

Step 4 Doesn't apply here.

You would need step 4 to read graphs C and D, where the colors stand for special things.

Try It Yourself

Graph B shows financial matters of concern to a group of people. It shows what percent said that a particular thing was important to them.

■ According to Graph B, what percent of all the people polled said that equal pay for women was an important financial concern for them? ______________

Which thing was a concern for more than 60% of the people? ______________

For the first question, if you followed the "equal pay for women" bar to the percent line, you should have said 40%. The only bar representing over 60% is the one labeled "Medical," and this is the answer for the second question.

Graph C compares low temperatures for two Wednesdays in 4 cities. Sometimes a question on the GED Mathematics Test will ask you to find information from a graph and then do something with it.

■ According to Graph C, what is the difference between the low temperatures last Wednesday and this Wednesday in St. Louis?

____

The pair of bars on the far right of the graph are for St. Louis. The difference between the temperatures that they represent is what you need. Maybe you can't tell the exact temperatures from the graph, but with a ruler or a piece of paper you can make a pretty good guess. The temperatures seem to be about 60 degrees and 82 degrees. The difference is about 22 degrees; an answer of 21 degrees is also pretty good. On the GED Mathematics Test, you would find either 21 or 22 as one of the numbers to choose from, so it would be easy to choose the best answer.

On the Springboard

Questions 1–3 refer to the following bar graph.

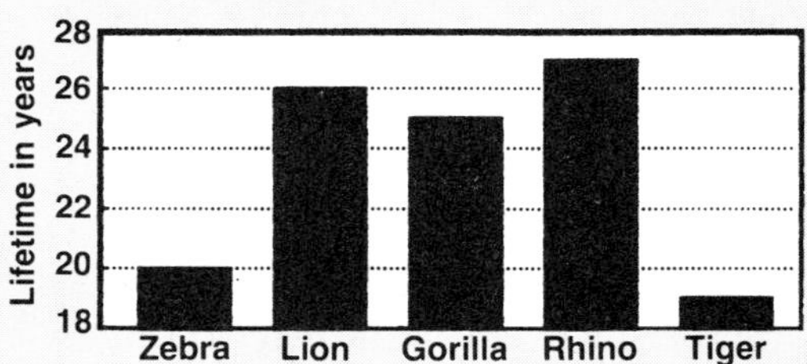

1. How many years does a gorilla usually live?

   (1) 20 (2) 24 (3) 25

   ① ② ③

2. Three of the animals usually live more than 20 years. Of these, which lives longest?

   (1) gorilla (2) rhino (3) tiger

   ① ② ③

3. How many more years does a lion usually live than a tiger?

(1) 6 (2) 7 (3) 8

(1) (2) (3)

Check your Springboard answers on page 189. If you got them right, go on to the section on line graphs. If you missed any, you may want to review bar graphs before you do that.

Reading a Line Graph

Line graphs are often used to show rises and falls or trends over a period of time. When you read a line graph, read the title first. Then read the headings and look carefully at the numbers.

Coming to Terms

line graph a graph that shows change (over a period of time) by a line that connects points on the graph

Here's an Example

Notice how the line on this graph rises.

Graph E. Line graph showing upward trend.

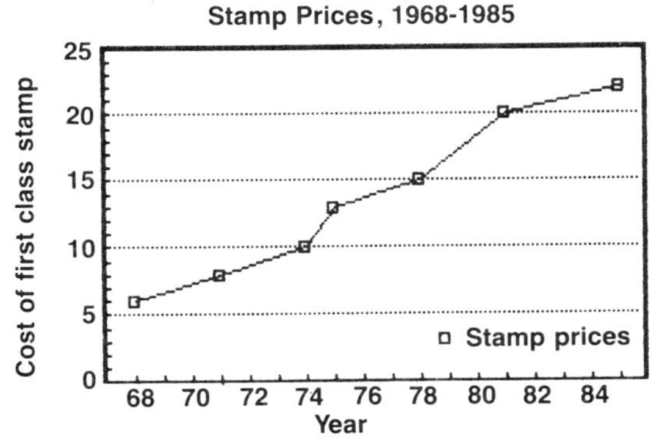

Step 1 The title is "Stamp Prices, 1968–1985."

Step 2 The heading at the bottom of the graph is "Year." The heading on the left is "Cost of first class stamp."

Step 3 The numbers at the bottom represent years. Since there's not enough space to write all the years, only the even-numbered ones are given. The numbers along the left represent the cost (in cents) of one first-class stamp in 5¢ intervals. The missing costs are indicated by short vertical lines called *tick marks.*

■ Find when first-class letter postage was 6¢.

Look up the price scale to the 6-cent level and go across at that level to the line on the graph. When you hit the line, go straight down to the year. You're at 68, so the answer is 1968.

■ What was first-class letter postage in 1978?

Look along the bottom to 78 and go straight up to the line. You hit the line at the 15-cent level, so the answer is 15¢.

Here's an example of a line graph that both rises and falls.

Graph F. Line graph showing rise and fall.

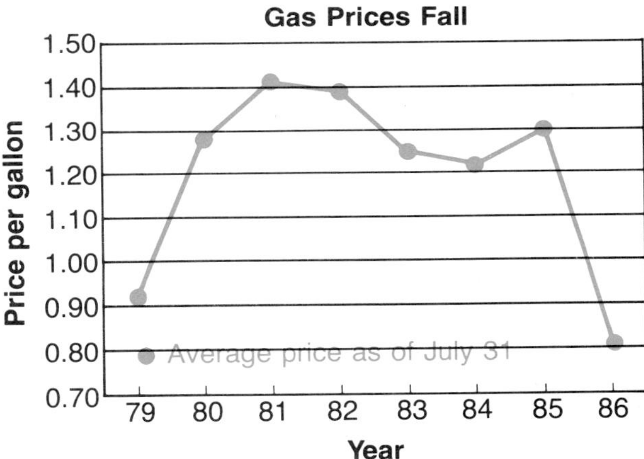

Step 1 The title is "Gas Prices Fall."

Step 2 The heading at the bottom is "Year" and the heading on the left is "Price per gallon."

Step 3 You know from experience that the prices are in dollars and cents, though the graph doesn't say so. The prices are marked in 10-cent intervals. You have to estimate "in-between" prices. The years along the bottom are given in 1-year intervals. The graph shows how the price changed over time.

Step 4 The price check was made on July 31 each year.

■ In what year did the big upward trend in gasoline prices stop?

You can see that the price was going up sharply until it hit about $1.38 a gallon in 1981.

■ What 1-year period saw the greatest price drop, and how much was it?

The greatest drop was from July 31, 1985, to July 31, 1986. The price went from about $1.25 to about $0.78. So the drop in price was about $0.47.

A line graph may have two lines to compare two sets of information.

Graph G. Line graph showing comparison.

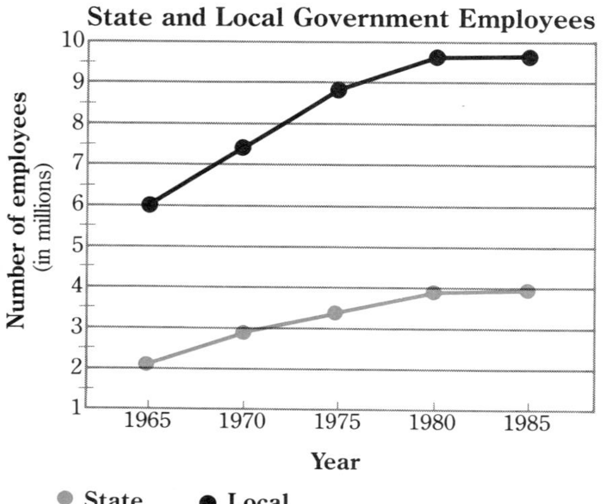

Step 1 The title is "State and Local Government Employees."

Step 2 The headings are "Year" and "Number of employees (in millions)."

Step 3 The numbers up the left side go from 1 to 10, but they stand for *millions.* Many graphs use this system when very large numbers are involved. Not noticing the words in parentheses could lead to serious misunderstanding. The numbers along the bottom are in 15-year intervals.

Step 4 The line for state employees is the blue one. The line for local employees is the black one.

■ How many times more local government employees were there than state employees in 1965?

In 1965, there were 6 million local government employees and 2 million state employees. When you divide 6 by 2, you get the answer. There were 3 times as many local government as state government employees.

Try It Yourself

Look at Graph F.

■ In one or two sentences, describe what happened to gas prices from 1979 to 1984.

____

____

____

____

Your answer should have included the information that gas prices rose sharply from 1979 to 1981 and then fell slowly until 1984.

Now look at Graph G.

■ Over which period did the number of local employees stay the same while the number of state employees rose slightly?

____

There was no increase in the number of local employees from 1980 to 1985. You know this because the graph is level between those years. The state numbers rose very slightly in the same period. No other period fits the description, so 1980 to 1985 is the answer.

On the Springboard

Questions 4–6 refer to the following graph.

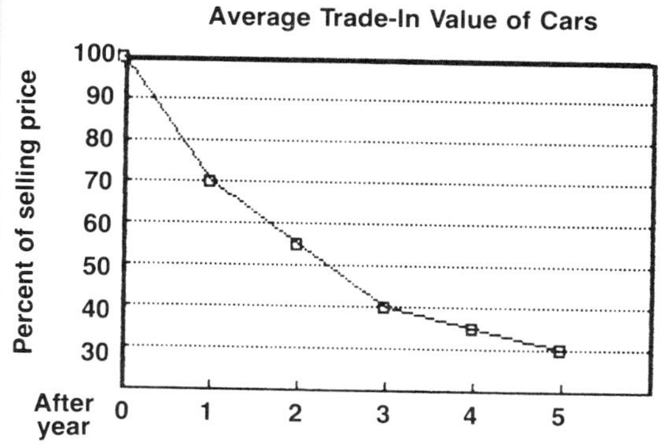

4. How many years old is a car when its trade-in value is only 70% of its original selling price?

   (1) 1 (2) 2 (3) 3

   (1) (2) (3)

5. What percent of the selling price is the trade-in value of a 4-year-old car?

   (1) 30% (2) 35% (3) 40%

   (1) (2) (3)

6. A car cost $8,000 when it was new. What is its trade-in value if it is 3 years old?

   (1) $2,400 (2) $3,200 (3) $4,000

   (1) (2) (3)

Check your Springboard answers on page 189. If you read this line graph correctly, go on to circle graphs. If not, take a few minutes to review the material on line graphs before you go on.

Reading a Circle Graph

A **circle graph** looks like a round pie cut into slices. The entire pie represents 100%, or the whole, of something, and the slices represent parts of the whole.

Coming to Terms

circle graph a circle divided into pie-shaped sections that represent certain percents or parts of a whole

A Test-Taking Tip

When the GED Mathematics Test gives you a problem with a graph, remember that the graph is just a way of picturing information for the problem. If the graph is labeled in percents, you can use your knowledge of solving percent problems to answer the questions.

Here's an Example

Companies often spend a lot of money on business travel. Here is a circle graph that shows how much out of every dollar a certain company spends on different kinds of travel expenses. The slices could be labeled as percents, but, since there are 100 cents in a dollar, the slices are labeled as cents.

Graph H. A circle graph

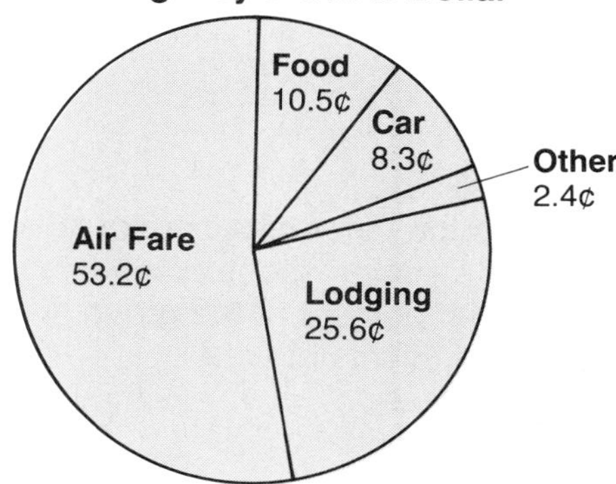

Step 1 is the same for any kind of graph. Read the title. Step 2 is the same too, but for circle graphs you usually have to look at each section of the circle for the headings. For step 3, you'll find a number or percent for each section.

■ How much of each travel dollar was spent on lodging?

The slice labeled "Lodging" has 25.6¢ in it, so that is the answer. Notice that the slice is about a fourth of the pie since 25.6% is very close to 25%, which is one-fourth.

■ If $400 was spent on a business trip, how much of it went for food?

The graph says that 10.5¢ out of every travel dollar was spent on food. Since $400 is 400 times $1 and 10.5¢ equals $0.105, the amount spent on food will be $0.105 × 400 or $42.00. (Check the multiplication on scratch paper if you wish.)

■ How many cents of each travel dollar are spent on car rental and other expenses?

Car rental takes 8.3¢ out of every dollar and the slice called "Other" takes 2.4¢. Add 8.3¢ and 2.4¢ to get the answer. You find that 10.7¢ out of every dollar goes for car rental and other expenses.

The next circle graph shows how many hours people commute each day.

Graph I. Circle graph.

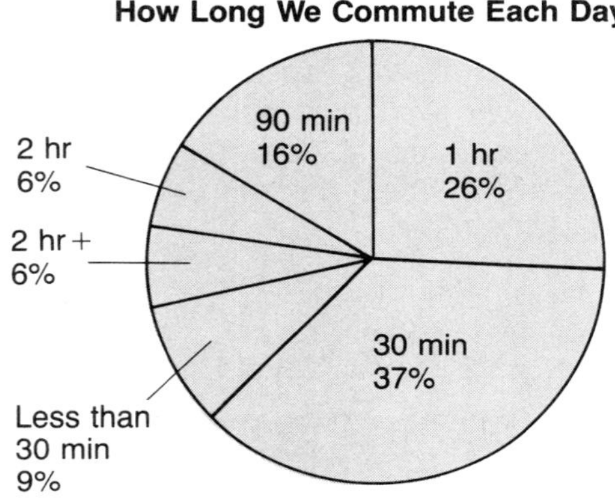

Graph I shows the different amounts of time that people in a survey spend commuting. It also shows what percent of the group spends each of those amounts of time.

Try It Yourself

Use Graph I to answer these questions.

■ What percent of the people commute 2 hours each day? ____________

The 2-hour section says 6%, so that's the answer.

■ What percent travel 30 minutes or 1 hour to work every day? ____________

The 30-minute group is 37% of those surveyed, and the 1-hour group is 26%. Together these two groups make up 37% + 26%, or 63%.

■ If 300 people were surveyed, how many commute more than 2 hours? ____________

The graph says that 6% of the people commute more than 2 hours. That means that 6 out of every 100 people commute more than 2 hours. Out of 300 people that would be 3 × 6, or 18, people.

On the Springboard

Questions 7 and 8 refer to the following graph.

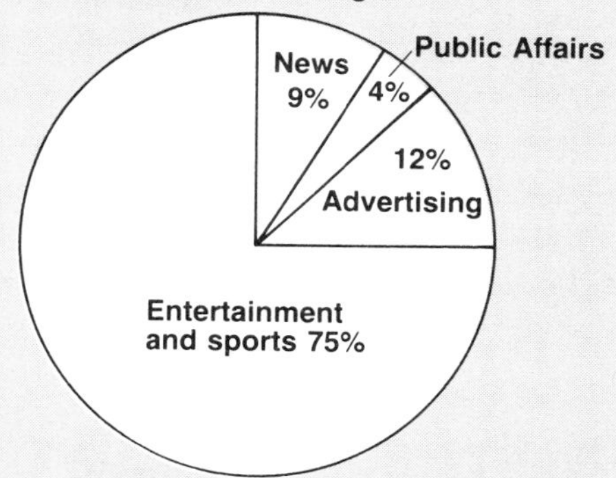

7. Which type of broadcast took up $\frac{3}{4}$ of broadcast time?

(1) public affairs
(2) entertainment and sports
(3) advertising

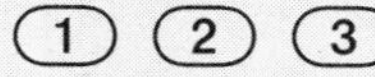

8. What percent of broadcast time was spent on news and public affairs?

   (1) 4% (2) 9% (3) 13%

   ① ② ③

Check your Springboard answers on page 189. If you got these right, go on to the section on pictographs. If you missed either one, you should review this section before continuing.

Reading a Pictograph

Pictographs, as you might guess, use pictures or symbols to show numbers or to compare amounts. A pictograph will always have a key to tell you what the symbols mean. Sometimes you will see a part of a symbol somewhere in the pictograph. This part of a symbol stands for part of the amount that the whole symbol represents.

Coming to Terms

pictograph a graph that uses a small picture or symbol to represent a given amount

Here's an Example

This pictograph tells the number of passengers who fly in or out of the busiest U.S. airports. The key below the graph shows that each small plane symbol stand for 4 million passengers each year.

Busiest Airports in the U.S.

Chicago O'Hare	✈✈✈✈✈✈✈✈✈✈ ½✈
Hartsfield (Atlanta)	✈✈✈✈✈✈✈✈✈ ½✈
Los Angeles	✈✈✈✈✈✈✈✈
John F. Kennedy (NY)	✈✈✈✈✈✈✈
Dallas-Ft. Worth	✈✈✈✈✈✈✈

✈ = 4,000,000 passengers per year.

To find the number of passengers for a certain airport on this pictograph, you must first count the number of planes and then multiply by 4 million. Notice the half planes. Each of these stands for ½ of 4 million, or 2 million, passengers.

Try It Yourself

Answer these questions about the previous pictograph.

■ How many passengers does John F. Kennedy Airport handle each year?

____

Did you see 7 planes drawn after John F. Kennedy? Did you look at the key below the graph to see that each plane stands for 4,000,000 passengers? Then did you multiply $7 \times 4{,}000{,}000 = 28{,}000{,}000$?

John F. Kennedy Airport handles 28,000,000 passengers each year.

Now look at another airport.

■ How many passengers pass through Atlanta each year?

____

Did you find 9½ planes drawn after Hartsfield? Then did you multiply 9½ by 4,000,000? $9\frac{1}{2} \times 4{,}000{,}000 = \frac{19}{2} \times 4{,}000{,}000 = 38{,}000{,}000$, or 38 million.

Atlanta handles 38,000,000 passengers each year.

On the Springboard

Questions 9 and 10 refer to the following pictograph.

National Basketball Association's Final Standings, 1985–86
(Eastern Conference)

Boston Celtics
Milwaukee Bucks
Philadelphia 76ers
Atlanta Hawks
Detroit Pistons
Washington Bullets
NJ Nets
Chicago Bulls
Cleveland Cavaliers
Indiana Pacers
NY Knickerbockers

(one basketball symbol) = 5 games won

9. How many games did the Washington Bullets win?

   (1) 67 (2) 39 (3) 26

   ① ② ③

10. How many times more games did the Celtics win than the Pacers?

   (1) 1.5 (2) 2.5 (3) 3.2

   ① ② ③

Check your Springboard answers on page 189. If you got both correct, go on to "The Real Thing." If you missed either, look at the pictographs again. When you're satisfied with your answers, it might be good to glance through the whole lesson one more time because "The Real Thing" will deal with all the kinds of graphs you have been studying. Good luck!

"The Real Thing"

Questions 1 and 2 refer to the following bar graph.

Planned Advertising Expenditures for 1987

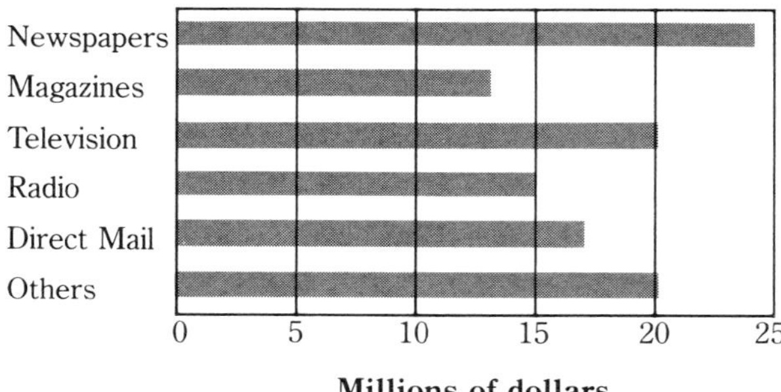

Millions of dollars

1. How much money does the company plan to spend on advertising by direct mail?

   (1) \$13 million (2) \$15 million
   (3) \$17 million (4) \$20 million
   (5) \$24 million

   ① ② ③ ④ ⑤

2. The company plans to spend the least amount of money on what type of advertising?

   (1) direct mail (2) magazines
   (3) newspapers (4) radio
   (5) televison

      ④ ⑤

Questions 3 and 4 refer to the following line graph.

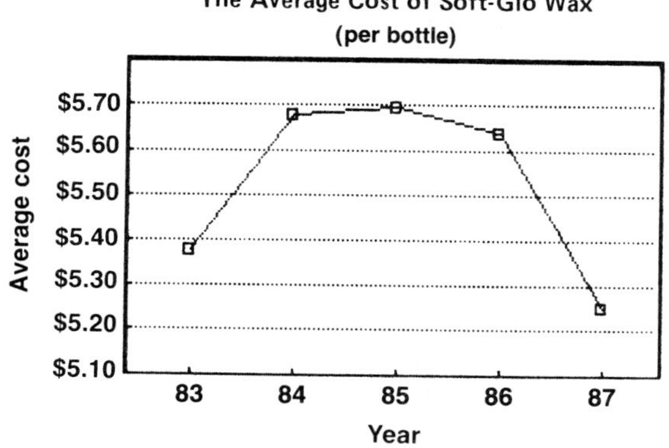

3. Approximately how much did the cost of Soft-Glow Wax increase from 1983 to 1984?

(1) $0.20 (2) $0.30 (3) $0.35
(4) $5.38 (5) $5.68

(1) (2) (3) (4) (5)

4. Between which two years did the cost of Soft-Glow Wax change the least?

(1) 1983–1984 (2) 1984–1985
(3) 1985–1986 (4) 1986–1987
(5) Insufficient data is given to solve the problem.

(1) (2) (3) (4) (5)

Questions 5 and 6 refer to the circle graph.

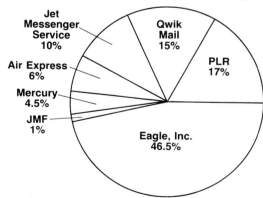

1985 Overnight Express Industry

Jet Messenger Service 10%
Qwik Mail 15%
PLR 17%
Air Express 6%
Mercury 4.5%
JMF 1%
Eagle, Inc. 46.5%

5. What percent of the industry is handled by Jet Messenger Service and Air Express together?

(1) 16% (2) 23% (3) 35.5%
(4) 60% (5) 63.5%

(1) (2) (3) (4) (5)

6. How many times more business do Eagle and Mercury combined get than PLR?

(1) $1\frac{1}{2}$ (2) 2 (3) $2\frac{1}{2}$
(4) 3 (5) 4

(1) (2) (3) (4) (5)

Questions 7 and 8 refer to the pictograph.

Amount of Snow Received in One Year

Deerwood	* * * * * * * * * *
Le Mars	* * * * * * * * * ½
Sun Down	* * * * * *
Danville	* * * ½
Homer	* ½

* = 6 inches

7. Deerwood received how many inches of snow?

(1) 9 (2) 21 (3) 36
(4) 51 (5) 60

(1) (2) (3) (4) (5)

8. Sun Down received approximately how many times as much snow as Homer?

(1) 2.3 (2) 4 (3) 4.5
(4) 5.7 (5) 6.3

(1) (2) (3) (4) (5)

Check your answers and record your score on pages 208–209.

LESSON 24

Finding Means and Medians

The word *average* is used in everyday speech to mean usual or normal. If you say that someone is average height, you mean that person wouldn't stand out in a crowd as being unusually tall or unusually short. Another word that means the same thing is **mean.**

Coming to Terms

mean the number you get when you add several numbers and divide the total by the number of items you added. *Mean* is another word for average.

Here's an Example

Here is a chart of the heights and weights of the members of a boys' basketball team.

■ Find the mean of the players' weights.

Name	Weight (lbs)	Height (in)
Mark	129	66
Joel	139	66
Joshua	141	69
Tim	156	70
Mike	143	69
Jacob	169	72
Tyler	155	69
Brady	146	68
John	157	64
Jeff	140	68
David	131	67

To find the mean, or average, weight add up all the weights and divide the total by 11, the number of players.

Step 1 Add all the weights. You can do this on scratch paper if you wish. You should get 1,606 pounds.

Step 2 Divide by 11. (1,606 ÷ 11 = 146) The mean, or average, weight of the team members is 146 pounds.

Try It Yourself

■ A dieter kept track of all her calories for the week. The results were the following:

Day	Calories
Sunday	1,315
Monday	1,350
Tuesday	1,400
Wednesday	1,825
Thursday	1,230
Friday	1,565
Saturday	1,675

Find the mean daily intake of calories.

You should have added the calories and divided by 7. Did you divide 10,360 by 7 to get 1,480 calories? The mean daily intake was 1,480 calories. Notice an important fact about means. The dieter did not on any day consume the mean number of calories. The mean lets you say what a *typical* individual in a group is like, but it is very possible that no one member of the group exactly matches the description of the *typical* member.

A Test-Taking Tip

If you are finding the mean of a set of numbers on the GED Mathematics Test and one of the numbers is 0, don't forget to count the 0 as a member of the set when you choose your dividing number. The 0 makes no difference to the total, but it is a member of the set of numbers and must be counted.

Warm-up

Find the mean of each of these sets of numbers.

1. 100, 88, 65, 77, 80 ______
2. 89, 73, 77, 81, 90, 88 ______
3. 3, 12, 7, 4, 4, 6, 18, 3, 3, 0 ______
4. 150, 139, 143, 139, 144 ______
5. 1,270, 2,000, 1,575 ______
6. 82, 36, 47, 49 ______

Warm-up Answers
1. 82 **2.** 83 **3.** 6 **4.** 143 **5.** 1,615 **6.** 53.5 or 53½
Be sure you count all the numbers even if one number is zero. For example, did you count ten numbers in Warm-up 3 above?

On the Springboard

1. Find the mean of these test scores: 730, 950, 875, 632, 968.

   (1) 656 (2) 831 (3) 871

   1 2 3

2. Find the mean of these car prices: $6,000; $7,800; $10,750; $8,830.

   (1) $8,000 (2) $8,315 (3) $8,345

   1 2 3

Check your Springboard answers on pages 189–190. If you answered both correctly, you are ready for "The Real Thing." If you missed either one, check your arithmetic or consider reviewing the section before going on to "The Real Thing."

The Real Thing

1. During his 23-year baseball career, Willie Mays had 3,283 hits. What was his average number of hits per year, to the nearest tenth?

   (1) 142.7 (2) 142.9
   (3) 143.2 (4) 186.2
   (5) Insufficient data is given to solve the problem.

   1 2 3 4 5

2. In 1984, there were 3,697,000 births and 2,047,000 deaths in the United States. What was the mean number of births for each of the 50 states?

   (1) 40,940 (2) 71,940 (3) 73,900
   (4) 73,940 (5) 74,900

   1 2 3 4 5

3. In the first 6 six days of January, a seaport had daytime highs of 49, 51, 42, 60, 52, and 58 degrees. The mean daytime high in this seaport was how many degrees?

   (1) 42 (2) 51 (3) 52
   (4) 53 (5) 60

   1 2 3 4 5

4. A customer's utility bills were $117 in January, $139 in February, $85 in March, and $67 in April. What was the average cost of utilities per month?

   (1) $34 (2) $72 (3) $102
   (4) $120 (5) $408

   1 2 3 4 5

Check your answers on page 209.

Finding the Median

If you drive, you may know that the median is a central strip that divides the highway into lanes. In mathematics, the **median** is the central number in a set of numbers written in order of size. If you are given a set of 3, 5, 9, or some other *odd* number of items, there will always be a single number in the center of the spread. If the set has 2, 4, 6, 8, or some other *even* number of items, there will be *two* numbers in the middle of the set. The median in this case is the mean of the two middle numbers.

Here's an Example

■ Here are the weights in pounds of the members of a boys' basketball team. They have been arranged in order for you, from smallest to largest.

Find the median weight. The weights are 129, 131, 139, 140, 141, 143, 146, 155, 156, 157, 169.

There are eleven numbers. So the middle number is the sixth number, or 143. The median weight is 143 pounds.

Here's another example for you to look at.

■ If the 169-pound player in the problem above misses practice, what is the median weight of the people attending practice?

Step 1 Find the middle number. Ten players were at practice, and that means that there isn't just one middle number, but two—141 and 143.

Step 2 To find the mean of 141 and 143, add the two numbers and divide by 2. You will have to calculate 284 ÷ 2. The result is 142. The median weight of the 10 players is 142 lbs.

Coming to Terms

median the middle number in a set of numbers written in order of size

Try It Yourself

■ At the beginning of the lesson, on page 124, there is a chart of the weights and heights of the members of a basketball team. Find the median height of the players.

____

Did you remember to rearrange the heights in order? Did you locate the sixth number in the list? Was it 68? That's the correct median height, 68 inches.

Warm-up

Use the chart to find the medians. Round any answers that are not whole numbers to the nearest tenth.

GED Scores During the Month of May

Name	Sex	Score	Name	Sex	Score
Pao Keu	M	267	Rafael	M	300
May	F	271	Naomi	F	280
Tran	M	255	Leon	M	253
Brendan	M	245	Dana	M	225
Li	F	302	Carla	F	266
Joann	F	288	Joe	M	240

1. Find the median score of the female examinees. _______________

2. Find the median score of the male examinees. _______________

3. Find the median of all the scores. _______________

Warm-up Answers
1. 280 **2.** 253 **3.** 266.5

A Test-Taking Tip

If you multiply the mean of a set of numbers by the number of numbers in the set, you will get the total of all the numbers in the set. For instance, if the mean weight of 10 students is 145 pounds, their total weight is 145 × 10 = 1,450 pounds.

On the Springboard

3. At a carnival, 8 people tried to hit the 3 ducks in the shooting gallery. The number of shots required were 10, 7, 4, 8, 9, 5, 8, and 5. What was the median number of shots required?

   (1) $6\frac{1}{2}$ (2) $7\frac{1}{2}$ (3) $8\frac{1}{2}$

   (1) (2) (3)

Question 4 refers to the following table.

Annual Sales of 10 Largest Retail Companies
(in millions of dollars)

Company	Sales	Company	Sales
Sears Roebuck	41	J.C. Penney	14
K mart	22	Southland	13
Safeway Stores	20	Federated Stores	10
Kroger	17	Lucky Stores	9
American Stores	14	Dayton Hudson	9

4. What was the median amount of the annual sales for these retail companies?

   (1) \$12 million (2) \$13 million
   (3) \$14 million

   (1) (2) (3)

Check your Springboard answers on page 190. If you got them right, you are certainly ready for "The Real Thing." If you missed any, check your work and be sure you can get them right before going on.

The Real Thing

5. Mrs. Hawthorne had bowling scores of 132, 146, 121, 118, and 138. What was her median score?

   (1) 121 (2) 126 (3) 132
   (4) 135 (5) 138

   (1) (2) (3) (4) (5)

6. A basketball team has played eight games so far this year. They had scores of 95, 41, 78, 81, 69, 102, 59, and 75. What is their median score?

   (1) 75.0 (2) 76.5 (3) 77.5
   (4) 78.0 (5) 78.5

   (1) (2) (3) (4) (5)

7. The salaries of the 14 employees of a small company are as follows:

$30,000	$ 9,000	$70,000	$ 8,000
$ 8,000	$ 9,000	$10,000	$ 9,000
$15,000	$10,000	$25,000	$50,000
$ 6,000	$ 7,000		

   What is the median salary in this company?

   (1) $9,000 (2) $9,357 (3) $9,500
   (4) $10,000 (5) $10,500

   (1) (2) (3) (4) (5)

8. Mr. Cantrell is taking a Spanish class that meets 3 times a week from 6:30 PM to 9:30 PM. He has scores of 96, 92, 85, and 87 on his Spanish tests. After his test yesterday, he had a mean score of 91. What was his score on yesterday's test?

   (1) 90.0 (2) 90.2
   (3) 91.0 (4) 95.0
   (5) Insufficient data is given to solve the problem.

   (1) (2) (3) (4) (5)

9. Manuel ran in three marathons during the last 4 months. His times were 2.1 hours, 2.6 hours, and 2.5 hours. What was his mean time?

   (1) 2.16 hr (2) 2.4 hr (3) 6.12 hr
   (4) 7.2 hr (5) 21.6 hr

   (1) (2) (3) (4) (5)

10. During the 18-hole golf season, Trudy's last 10 scores were 95, 86, 88, 102, 86, 89, 93, 101, 87, and 91. What was her median score for these games?

    (1) 90.0 (2) 93.0 (3) 93.5
    (4) 94.0 (5) 94.5

    (1) (2) (3) (4) (5)

11. During his professional football career, Fran Tarkenton completed 3,686 passes. How many passes per year did he complete?

    (1) 204.8 (2) 208.4
    (3) 214.5 (4) 247.8
    (5) Insufficient data is given to solve the problem.

    (1) (2) (3) (4) (5)

12. Sylvia collected data on the number of yards that kickoffs were returned in the National Football League last year. She found that the top kickoff-return specialists had 21.6, 23.0, 24.7, 21.8, 22.6, 22.2, 22.9, and 21.4 yards. What was the median number of yards returned on kickoffs?

    (1) 21.8 (2) 22.2 (3) 22.4
    (4) 22.6 (5) 22.9

    (1) (2) (3) (4) (5)

13. Gabriel had scores of 82, 85, 87, and 92 on his typing tests. What is the lowest score he can get on his next test and have an average of at least 87?

    (1) 86 (2) 87 (3) 88
    (4) 89 (5) 90

    (1) (2) (3) (4) (5)

14. Mrs. Snyder earned $9,840 for each of 2 years and $12,370 for each of 3 years. What was her average yearly income for these 5 years?

    (1) $11,158 (2) $11,358
    (3) $18,930 (4) $28,395
    (5) Insufficient data is given to solve the problem.

    (1) (2) (3) (4) (5)

Check your answers and record all your scores on pages 209–210.

LESSON 25

Ratio, Proportion, and Probability

Ratios and proportions are ways of comparing things. Probability is also a form of comparison; it lets you compare the likelihood that something will happen with the likelihood that it won't. Probabilities are often written as ratios.

Writing a Ratio

A **ratio** is a special kind of comparison between two numbers. For example you may read that the ratio of male dancers to female dancers is 1 to 4. This means that for every male dancer there are four female dancers.

A ratio can be written in different ways.

Here's an Example

■ In a certain shopping mall, 3 out of every 5 people questioned said they drank coffee. Write this as a ratio.

The ratio of coffee drinkers to people questioned can be written in 3 ways:

$$3:5 \text{ or } \frac{3}{5} \text{ or } 3 \text{ to } 5$$

The numbers in a ratio are written in the order in which they come in the sentence. Notice that 2 out of every 5 questioned didn't drink coffee. The ratio of non-drinkers to the number of people questioned is

$$2:5 \text{ or } \frac{2}{5} \text{ or } 2 \text{ to } 5$$

You can see that ratios are a type of fraction. Ratios, like fractions, can be reduced.

■ Ten out of every hundred employed photographers live in the New York metropolitan area. Write this as a ratio.

The ratio can be written as 10:100, but this is the same as $^{10}/_{100}$, which reduces to $^{1}/_{10}$. So 10:100 can be written as 1:10. You always try to reduce fractions to lowest terms in your answers, and the same thing is true of ratios, no matter how they are written.

Coming to Terms

ratio a comparison of two amounts. A ratio may be written with a colon, as a fraction, or with the word *to.*

Try It Yourself

When you answer this question, write the ratio in all three ways.

■ In a large city, 7 out of every 100 dollars earned were paid in taxes. Write the ratio of tax dollars to total dollars earned.

____

You should have written *7:100, 7/100,* and *7 to 100.* You probably noticed that the rate was 7%, and that the ratios expressed that too.

Warm-up

Express the following facts or situations as ratios. Reduce to lowest terms first. Then write each ratio all three ways.

	:	fraction	"to"
1. 36 eggs to 3 eggs	______	______	______
2. 100 years to 1 year	______	______	______
3. 60 inches to 5 inches	______	______	______
4. 1 woman to 3 men	______	______	______
5. 10 fired to 3 hired	______	______	______
6. 15 voters to 45 registered voters	______	______	______

Warm-up Answers

1. 12:1; $\frac{12}{1}$; 12 to 1 **2.** 100:1; $\frac{100}{1}$; 100 to 1

3. 12:1; $\frac{12}{1}$; 12 to 1 **4.** 1:3; $\frac{1}{3}$; 1 to 3

5. 10:3; $\frac{10}{3}$; 10 to 3 **6.** 1:3; $\frac{1}{3}$; 1 to 3

If your answers don't agree, check to see if you reduced the ratios.

Setting Up Ratios in Word Problems

The next thing to do is to recognize ratios in word problems.

Here's an Example

When you use ratios to solve word problems, be sure that the order of the numbers in the ratios is correct. It may not be the same order as in the problem.

■ If 36 men and 63 women are enrolled in an art school, what is the ratio of women to men at this school?

To answer the question, you must set up a ratio and reduce it to lowest terms.

Step 1 Look at the question asked. The ratio should be stated in the same order the question is asked. "What is the *ratio of women to men. . . ?*

Step 2 Find the numbers in the problem that are related to this question: 63 women to 36 men.

Step 3 Express this fact as a ratio and reduce the ratio to lowest terms.

$63{:}36$ or $\frac{63}{36}$ or 63 to 36

Reduce to $7{:}4$ or $\frac{7}{4}$ or 7 to 4

Step 4 State the ratio in a sentence. The ratio of women to men is 7 to 4.

Did you notice that the order in which the numbers were written in the answer is not the order in which they appeared in the problem? Be sure to write the ratio *exactly* as the question asks.

Some ratio problems can be a little more difficult. Read the next problem very carefully before you try to set up the proportion.

■ On a quiz show, a contestant got 16 questions right and 2 wrong. Express the ratio of the number of questions right to the total number of questions.

You are not told the total number of questions, but you can find it by adding the number right to the number wrong. This gives you a total of 18 (16 + 2). The ratio of the number right to the total can be written 16 to 18 or 8 to 9. Careful reading is the key to success on ratio problems like this one.

Try It Yourself

Here you are dealing with a total.

■ A computer store had 4 full-time employees and 6 part-time employees. What was the ratio of full-time employees to the total number of employees?

____

Did you add 4 and 6 to get 10, the total number of employees? Then did you put the number of full-time employees first as the question asked? You should have gotten 4 to 10. Finally, did you reduce the ratio to lowest terms? The ratio is 2 to 5.

Warm-up

Mr. Coty is 42, his wife is 36, his daughter is 15, and his son is 9. Write the following ratios.

1. Mrs. Coty's age to Mr. Coty's age.

____

2. Mr. Coty's age to his son's age.

____

3. Mrs. Coty's age to the sum of the ages of her daughter and son.

____

4. The total ages of the children to the total ages of the parents.

____

5. The son's age to the daughter's age.

____

6. The total ages of the males to the total ages of the females in the family.

____

Warm-up Answers

1. $\frac{36}{42}$ or $\frac{6}{7}$ **2.** $\frac{42}{9}$ or $\frac{14}{3}$ **3.** $\frac{36}{24}$ or $\frac{3}{2}$ **4.** $\frac{24}{78}$ or $\frac{4}{13}$ **5.** $\frac{9}{15}$ or $\frac{3}{5}$ **6.** $\frac{51}{51}$ or $\frac{1}{1}$

Your answers could also have been written with a colon (:) or the word *to*.

On the Springboard

1. The speed of a lion is 50 mph. The speed of a zebra is 40 mph. What is the ratio (in lowest terms) of the speed of the zebra to that of the lion?

   (1) 4:5 (2) 4:9 (3) 5:4

   (1) (2) (3)

2. There were 12,000 registered voters in the first precinct, but only 9,000 of them voted on election day. What was the ratio of those that voted to those that didn't?

   (1) 1:3 (2) 3:1 (3) 3:4

   (1) (2) (3)

3. At a spring gardening sale, the local nursery sold 85 flats of tall marigolds and 35 flats of dwarf marigolds. What was the ratio of dwarf marigolds sold to the total number of marigolds sold that day?

   (1) 7:17 (2) 7:24 (3) 17:24

   (1) (2) (3)

Check your Springboard answers on page 190. If you got these right, you're catching on fine! Go on to "The Real Thing." If you didn't, it may be a good idea to review this section briefly and try again on the questions you missed. Then you should be ready for "The Real Thing."

The Real Thing

1. A billboard is 16 meters long and 5 meters high. What is the ratio, in lowest terms, of the height to the length?

   (1) 1:3 (2) 1:4 (3) 3:1
   (4) 5:16 (5) 16:5

   (1) (2) (3) (4) (5)

2. A recipe calls for 2 cups of sugar and 3 eggs. What is the ratio, in lowest terms, of eggs to sugar?

   (1) 2:3 (2) 3:2 (3) 4:2
   (4) 4:3 (5) 6:3

   (1) (2) (3) (4) (5)

3. Mr. Hagen drove 100 miles on 12 gallons of gasoline. What is the ratio, in lowest terms, of miles to gallons of gasoline?

   (1) 25:3 (2) 25:4 (3) 25:12
   (4) 100:3 (5) 100:36

   (1) (2) (3) (4) (5)

4. A shopkeeper put 25 sweaters on sale. If 15 of the sweaters were sold, what is the ratio of the number of unsold sweaters to the number originally on sale?

   (1) 2:3 (2) 2:5 (3) 3:5
   (4) 5:3 (5) 25:10

   (1) (2) (3) (4) (5)

Check your answers on page 210.

Setting Up Proportions

A ratio compares two numbers. A **proportion** is a statement that two ratios are equal.

Coming to Terms

proportion a statement that two ratios are equal. Since ratios can be written as fractions, the two equal ratios in a proportion can be written as two equal fractions.

Here's an Example

■ Look at the ratios 3:4 and 6:8.

You can write these ratios as $\frac{3}{4}$ and $\frac{6}{8}$. These are equal fractions. (You know this is so because you get $\frac{3}{4}$ if you reduce $\frac{6}{8}$ to lowest terms.) You can write the statement that the ratios are equal like this: $\frac{3}{4} = \frac{6}{8}$. You have then written a proportion. You could also write the proportion with the colons: 3:4 = 6:8.

Problems involving proportions can be solved using a grid. The proportion may be disguised a bit by the words of a problem, but you can usually get the grid filled in if you read carefully.

■ A tree 30 feet high casts a shadow 8 feet long. How long a shadow will a tree 15 feet high cast?

The first ratio compares a tree height to a shadow length. The second ratio must also compare a tree height to a shadow length, since you should make the comparison the same way both times. On a grid, write what you do and do not know.

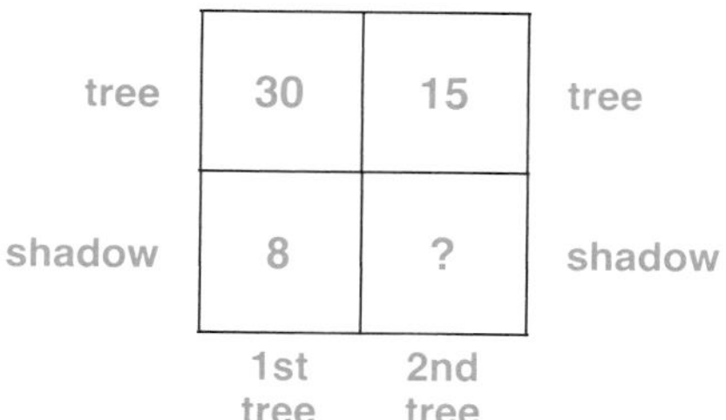

tree
30
15
tree
shadow
8
?
shadow
1st tree
2nd tree

Now you have all the information on the grid and you can solve in the same way as when you put percent problems on a grid.

Step 1 Find the two ratios, including what you need to find.

Step 2 Write the ratios on a grid, with a *?* for what you don't know.

Step 3 Multiply diagonals where possible and divide by the unused number.

$$15 \times 8 = 120 \text{ and } 120 \div 30 = 4$$

The 15-foot high tree will cast a shadow 4 feet long.

See if you can figure out the *change* in temperature from 5:00 A.M. to 6:00 A.M.

■ At 5:00 A.M. yesterday the temperature was 65 degrees. At 7:00 A.M. it was 75 degrees. If the temperature was rising steadily, what was it at 6:00 A.M.?

Step 1 One hour passed from 5:00 A.M. to 6:00 A.M., and 2 hours from 5:00 A.M. to 7:00 A.M. You want to find how much the temperature changed from 5:00 A.M. to 6:00 A.M. You know that it changed 10 degrees from 5:00 A.M. to 7:00 A.M.

Step 2 Put this information on a grid and find what *?* must be.

Time	Temp.	
1	?	Change from 5:00 A.M. to 6:00 A.M.
2	10	Change from 5:00 A.M. to 7:00 A.M.

$$1 \times 10 = 10 \text{ and } 10 \div 2 = 5$$

The temperature change from 5:00 A.M. to 6:00 A.M. was 5 degrees.

Step 3 Add the 5-degree change to the 5:00 A.M. temperature. Use the number you found to answer the question.

$$65 + 5 = 70$$

The temperature at 6:00 A.M. was 70 degrees.

Finding a special in-between number (as in this last example) is called **interpolation.** You interpolate by finding total changes and then using ratios to help find a partial change.

Coming to Terms

interpolation finding a number between two known numbers.

Try It Yourself

■ If a recipe for a loaf of bread calls for ½ cup of shortening, 5 loaves of bread will call for how much shortening?

The ratio of 1 loaf to ½ cup is 1:½. The ratio of the 5 loaves to the needed amount of shortening is 5:? On a grid, the set-up looks like this:

	recipe for 1 loaf	recipe for 5 loaves	
loaves	**1**	**5**	**loaves**
shortening	**1/2**	**?**	**shortening**

When you use the diagonal rule, you get 5 × ½ = 2½ and 2½ ÷ 1 = 2½.

The answer is 2½ cups.

The next problem requires interpolation, finding an in-between number.

■ A plant that was 2 inches tall on Monday was 9 inches tall on the next Saturday. About how tall was the plant on Wednesday?

As in all these interpolation problems, you first find the total changes. Here the total change in time was 6 days, and the total change in height was 7 inches. The in-between time change from Monday to Wednesday was 2 days and the in-between change in height is what you need to find. When you put all of this on a grid, the grid looks like this:

days from Mon. to Wed.	**2**	**?**	**growth from Mon. to Wed.**
days from Mon. to Sat.	**6**	**7**	**growth from Mon. to Sat.**

From the grid you should get 2⅓ inches. Remember this is *not* how tall the plant is on Wednesday. This is how much it *grew*. It was 2 inches + 2⅓ inches, or 4⅓ inches, tall on Wednesday.

Warm-up

Fill in the grids for these problems. Each time, use a ? for the missing number, and then solve the problem.

1. The ratio of men to women on a tour bus was 5 to 3. If there were 9 women on the bus, how many men were there?

____

2. If 350 potatoes were peeled to make mashed potatoes for 200 people, how many potatoes would be needed for a fund-raising dinner for 1,000 people?

____

3. A factory line can produce 50 products every hour. At this rate, how long would it take to produce 250 of these products?

____

4. The temperature in Denver was 25° at 12 noon. At 12 midnight it had dropped to 1°. If the temperature dropped steadily all afternoon and evening, what was the temperature at 6 P.M.?

____

Warm-up Answers

1. 15 men

5	**?**
3	**9**

2. 1,750 potatoes

350	**?**
200	**1,000**

3. 5 hours

50	**250**	number of hrs from 12 n to 6
1	**?**	number of hrs from 12 n to 12 m

4. 13°

6	**?**	change in temp. from 12 noon to 6
12	**24**	change in temp. from 12 noon to 12 midnight

On the Springboard

4. Two record albums are on sale. The minimum purchase is 5 albums, which sell for a total of $45. At this rate, how much will 6 albums cost?

   (1) $54 (2) $63 (3) $72

   (1) (2) (3)

5. It takes 3 gallons of a special paint to treat 200 square feet of steel plate. How many gallons will it take to treat 300 square feet of plate?

   (1) 3½ (2) 4½ (3) 5½

   (1) (2) (3)

6. If 2 packets of mix make 2½ pints of a fruit drink, how many packets would be needed to prepare 15 pints?

   (1) 18 (2) 15 (3) 12

   (1) (2) (3)

7. When the water valve was opened at 8:00 A.M., the tank contained 120 gallons of water. At noon, the tank contained 520 gallons. How many gallons were in the tank at 11:00 A.M.?

   (1) 400 (2) 420 (3) 480

   (1) (2) (3)

Check your Springboard answers on page 190. If you got everything right, then you're ready for "The Real Thing." If you missed any, make sure you find your errors. When you're satisfied with your work, go on to "The Real Thing."

"The Real Thing"

5. A cleaning solution label says to mix 1 part solution to 5 parts water. How many gallons of solution should be mixed with 20 gallons of water?

   (1) 1 (2) 4 (3) 6
   (4) 25 (5) 100

   (1) (2) (3) (4) (5)

6. In a certain school, the ratio of students to teachers is about 12.5 to 1. Approximately how many teachers would there be if the school had an enrollment of 197 students?

   (1) 8 (2) 13 (3) 16
   (4) 24 (5) 36

   (1) (2) (3) (4) (5)

7. A dental assistant earns $45.20 in 8 hours. At this rate, how much does he earn in 40 hours?

   (1) $180.00 (2) $226.00 (3) $361.60
   (4) $404.00 (5) $1,808.00

   (1) (2) (3) (4) (5)

8. After knee surgery, Peter had to lift 8-ounce weights with his legs. He began on the seventh day. He steadily increased the weight each day. On the seventeenth day, he was lifting 3-pound weights. How many pounds was Peter lifting on the twelfth day?

   (1) $1\frac{1}{8}$ (2) $1\frac{1}{4}$ (3) $1\frac{3}{4}$

   (4) $2\frac{1}{2}$ (5) $2\frac{5}{8}$

   (1) (2) (3) (4) (5)

9. The telephone connection for a home computer costs $1.25 for each 3 minutes it is in use. How much does it cost for 21 minutes?

   (1) $3.75 (2) $7.00 (3) $8.45
   (4) $8.75 (5) $26.25

   (1) (2) (3) (4) (5)

Check your answers on page 210.

Finding Probability

"You may be the next winner of the Happy Hopeless Sweepstakes!" That's the message Jeanne Bootz found in her mailbox. In small print, the magazine company sponsoring the contest went on to say that her sweepstakes number was one out of 6,000,000 the company would draw from to select the winner. In this case, the probability of Jeanne's winning is $\frac{1}{6,000,000}$. The **probability** is a ratio written in fraction form. The numerator is 1 (the number of sweepstakes numbers she received) and the denominator is 6,000,000 (the total number of sweepstakes numbers for the whole contest).

Here's an Example

■ Jeanne Bootz decided to subscribe to 2 more magazines. This entitles her to 2 more sweepstakes numbers. What is the probability now that she will win?

To solve, write the probability as a fraction. In the denominator, write the total number of possible winning numbers.

$$\frac{}{6,000,000}$$

In the numerator, write how many chances Jeanne has, 3.

$$\frac{3}{6,000,000}$$

Reduce $\frac{3}{6,000,000}$ to $\frac{1}{2,000,000}$ and you've got the answer. Jeanne's probability of winning is now $\frac{1}{2,000,000}$.

■ Mr. Chen flipped 2 coins, a nickel and a quarter. What is the probability that one of the coins came up heads and the other tails?

First you need to know the total number of possible outcomes. You can find this out by listing all the ways the coins could come up. If you use *H* for heads and *T* for tails, the possible outcomes are these:

	Nickel	Quarter
1	H	H
2	H	T
3	T	H
4	T	T

There are four possible outcomes. There are exactly 2 ways to get one coin heads and the other tails—options 2 and 3 on the table. So the probability of Mr. Chen's getting a heads-tails combination is $\frac{2}{4}$, or $\frac{1}{2}$.

Try It Yourself

■ John has only 2 pairs of blue socks. There are 10 pairs of socks of other colors in his drawer. All the socks are separated. He finds one blue sock and then the lights go out. If he picks out a sock in the dark, what is the probability that it will be blue?

____

There are just 3 ways that John could get a blue sock since there are only 3 blue socks left in the drawer. The number of all possible outcomes is 23 because there are 23 socks left in the drawer (20 of other colors and 3 blue). Therefore, the probability is $\frac{3}{23}$ that he will get a blue sock. Was that your answer?

Coming to Terms

probability The likelihood that a thing will happen written as a ratio in fraction form

$$\text{probability} = \frac{\text{possible ways for a thing to happen}}{\text{total number of possible outcomes}}$$

In the first example with 1 entry, Jeanne had just 1 chance of winning out of a possible 6,000,000 sweepstakes numbers.

Warm-up

1. A different letter of the alphabet is written on each of 26 cards. The cards are mixed in a bowl and one card is drawn at random. What is the probability that the letter on the card drawn is a vowel (A, E, I, O, or U)?

____

2. Here is a spinner that was used in a party game.

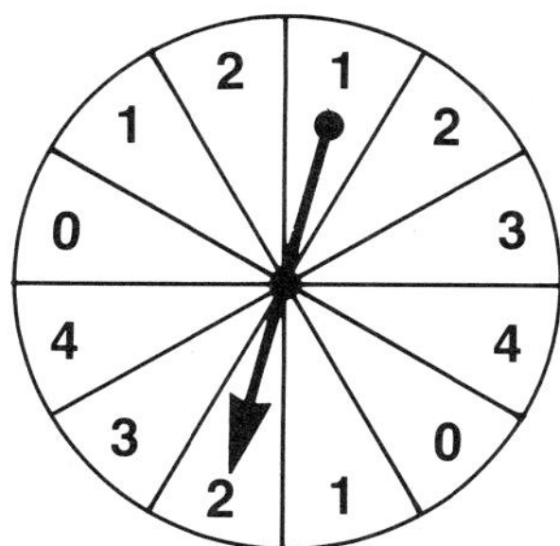

What is the probability that after one spin the pointer will stop in a section with a 2 in it?

---

3. On one roll of a die, what is the probability of getting a number less than 5 (a die has six sides)?

Warm-up Answers

1. $\frac{5}{26}$ **2.** $\frac{1}{4}$ **3.** $\frac{2}{3}$

On the Springboard

8. Marvin borrowed his roomate's suitcase for a trip. It has a 3-digit combination lock. Only after he arrived did Marvin realize that he had failed to memorize or write down the combination. He decided to try 7-2-5. What is the probability that this combination opened the suitcase?

   (1) $\frac{1}{1000}$ (2) $\frac{1}{999}$ (3) $\frac{1}{725}$

   ① ② ③

9. If two pennies are tossed, what is the probability that they will both come up heads?

   (1) $\frac{1}{8}$ (2) $\frac{1}{4}$ (3) $\frac{1}{2}$

   ① ② ③

10. Two couples on vacation in Hollywood, California, went to a TV quiz show. Each member of the audience wrote his or her name on a card, and contestants were chosen by drawing names at random. If there were 350 people in the audience, what is the probability that the first contestant selected was from one of these two couples?

   (1) $\frac{1}{350}$ (2) $\frac{1}{175}$ (3) $\frac{2}{175}$

   ① ② ③

Check your Springboard answers on pages 190–191. If you get all of these correct, go straight to "The Real Thing." If you missed any, review this section briefly and check your work in the ones you missed. When you're ready, go ahead to "The Real Thing."

“The Real Thing”

10. Mrs. Greaves' digital watch has a display for seconds as well as for hours and minutes. What is the probability that the display for seconds will show 35 when Mrs. Greaves suddenly decides to check the time?

(1) $\frac{1}{60}$ (2) $\frac{1}{30}$ (3) $\frac{1}{15}$
(4) $\frac{1}{12}$ (5) $\frac{1}{3}$

① ② ③ ④ ⑤

11. In an experiment, Ed said that the probability of drawing a purple marble was $\frac{2}{3}$. How many times should Ed expect to get a purple marble if he repeats the experiment 180 times?

(1) 60 (2) 90 (3) 120 (4) 270
(5) Insufficient data is given to solve the problem.

① ② ③ ④ ⑤

12. Mr. Howard has 3 red shirts, 6 yellow shirts, 2 blue shirts, and 1 green shirt in a drawer. If he selects a shirt at random and without looking, what is the probability that it will be a blue shirt?

(1) $\frac{1}{12}$ (2) $\frac{1}{6}$ (3) $\frac{1}{4}$
(4) $\frac{1}{3}$ (5) $\frac{1}{2}$

① ② ③ ④ ⑤

13. There are 12 dogs and 15 cats at a pet shop. If one of these animals is chosen at random, what is the probability that it will be a dog?

(1) $\frac{1}{27}$ (2) $\frac{1}{15}$ (3) $\frac{1}{12}$
(4) $\frac{5}{9}$ (5) $\frac{4}{9}$

① ② ③ ④ ⑤

14. Mrs. Rogers spent $14.68 for gasoline. She drove 550 kilometers in 7 hours. At this rate, how many hours, to the nearest tenth, will it take her to drive 680 kilometers?

(1) 5.7 (2) 7.7 (3) 8.5
(4) 8.7 (5) 175.7

① ② ③ ④ ⑤

15. Clarice made 6 free throws out of 11 attempts. What is the ratio, in lowest terms, of free throws made to attempts?

(1) 5:11 (2) 6:11 (3) 11:5
(4) 11:6 (5) 11:11

① ② ③ ④ ⑤

16. Each of the numbers 1 through 10 is written on a separate card. If you select one of the cards without looking, what is the probability that you will select a number greater than 5?

(1) $\frac{1}{10}$ (2) $\frac{1}{4}$ (3) $\frac{1}{2}$
(4) $\frac{2}{5}$ (5) $\frac{2}{1}$

① ② ③ ④ ⑤

17. During a sale, batteries were priced at 6 for $2.10. How much would 9 of these batteries cost?

(1) $1.40 (2) $3.10 (3) $3.15
(4) $3.25 (5) $18.90

① ② ③ ④ ⑤

18. Mrs. Albright gets paid $1.05 for every 5 miles she drives on business trips. On Tuesday, she made a business trip that took her 3.75 hours. How much did she get paid by the company for mileage?

(1) $40.17 (2) $44.72
(3) $53.35 (4) $62.87
(5) Insufficient data is given to solve the problem.

① ② ③ ④ ⑤

19. An orchestra is made up of 28 men and 32 women. If a member from the orchestra is chosen at random, what is the probability of selecting a woman?

(1) $\frac{1}{60}$ (2) $\frac{1}{32}$ (3) $\frac{1}{28}$
(4) $\frac{7}{15}$ (5) $\frac{8}{15}$

① ② ③ ④ ⑤

Check your answers and record all your scores on pages 210–211.

LESSON 26

Signed Numbers

If you live where the winters bring snow and ice, you probably know about below-zero temperatures. If you have ever overdrawn a bank account, you know that it is possible to have less than zero dollars (called an *overdraft*). Numbers that express these less-than-zero quantities are called **negative numbers.** As you would expect, numbers that express quantities greater than zero are called **positive numbers.**

Positive and Negative Numbers

Up to this point, you've worked with numbers larger than zero. Now look at this thermometer and find the temperatures below zero.

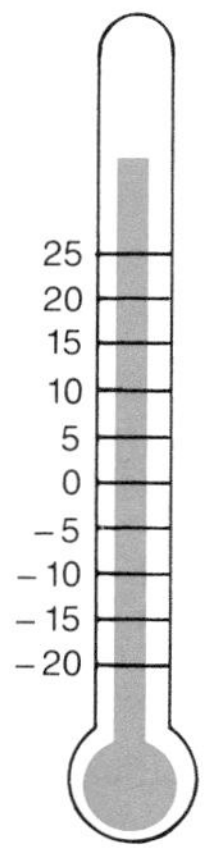

Look at the temperatures below zero. The lowest temperature shown on this thermometer is *negative 20,* or *−20.* The mark just below zero is *negative 1,* or *−1.* The mark just above −20 is *negative 19,* or *−19.* These temperatures below zero are negative numbers and they are written with a minus sign (−). Are you getting the hang of it?

Now look at the numbers above zero on the thermometer. These are positive numbers. They can be written with a plus sign (+). The temperature 15 degrees above zero is called positive 15 and is written +15. The highest temperature marked on the thermometer is positive 25, or +25. Because of these signs (− and +), positive and negative numbers are called **signed numbers.**

Coming to Terms

negative numbers numbers that are less than zero. A negative number is always written with a minus sign (−) in front of it.

positive numbers numbers that are greater than zero. Any positive number is greater than any negative number. A positive number may have a plus (+) sign in front of it or no sign at all. If there is no sign at all, it's understood that the number is positive—except for zero, which is neither positive nor negative.

signed numbers numbers that have a positive or negative sign in front of them (plus or minus sign). Signed numbers may also be fractions or decimals. $+\frac{3}{4}$ is located $\frac{3}{4}$ of a unit to the right of zero. −0.5 is located 0.5 of a unit to the left of zero.

Number Lines

A **number line** is a straight line that shows the position of numbers. Negative numbers are to the left of zero and positive numbers are to the right of zero. The farther to the left a number is, the smaller it is; the farther to the right a number is, the greater it is. A number line continues endlessly in both directions.

Here's an Example

Imagine that the thermometer shown at the beginning of this lesson is turned horizontally instead of vertically.

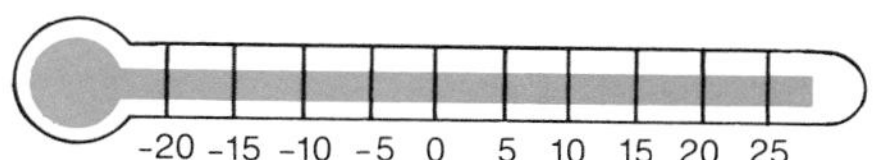

Now, in your mind, replace the thermometer with a straight line. Imagine that zero comes in the center and that the line continues in both directions like this.

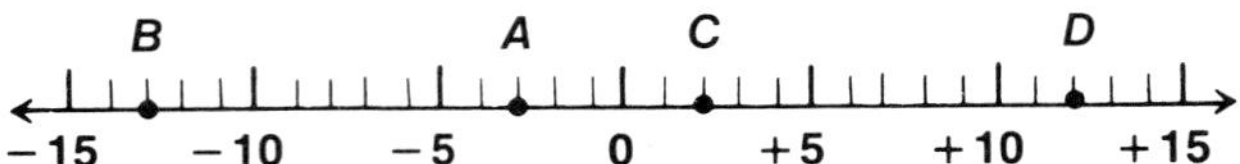

You are now looking at a number line. Look at the letters *A, B, C, and D.*

A is between −5 and zero. It is closer to −5. There is only one mark between *A* and −5. *A* = −3.

B is between −15 and −10. *B* = −13.

C is two units to the right of zero. It is on the positive side of zero. *C* = +2.

D is 2 more than +10. *D* = +12.

On the number line, numbers are greater as you move from left to right and smaller as you move from right to left.

Try It Yourself

For the number line below, write the number for each point marked with a letter. Write each positive or negative number with a positive or negative sign.

A F D B E C
−15 −10 −5 0 +5 +10 +15

A___ B___ C___ D___ E___ F___

Did you see that *A* is between −15 and −10? *A* is −13. *B* is 0. Did you remember that zero has no sign? *C* is +10. Did you locate *D* at −3? Don't forget the negative sign. *D* is −3. Did you notice that *E* is three marks to the right of +5? *E* is +8. *F* is −10.

Coming to Terms

number line a straight line showing the position of numbers

Warm-up

Look at this number line.

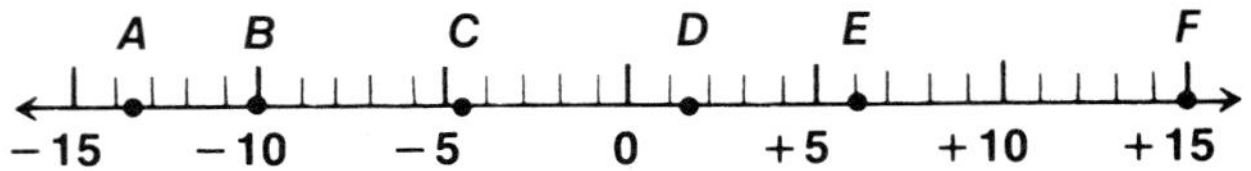

Complete this chart by writing the correct letter below each number. One has been done for you

$-13\frac{1}{2}$	+6	$+1\frac{1}{2}$	$-4\frac{1}{2}$	−10	+15
		D			

Adding When Signs Are the Same

You probably have added positive and negative numbers many times without realizing it. If you make a car payment, you can think of it as adding a positive amount to a negative amount. There are three possibilities when you add signed numbers. You can add two negative numbers, two positive numbers, or one of each. Here you'll deal with the first two cases.

There is a simple rule that will help you get these right. When you add two positive numbers the answer is a positive number, and when you add two negative numbers the answer is a negative number. This makes sense. If you add to your debts, the answer is a bigger debt. If you add to your savings, the result is greater savings.

Here's an Example

You can use the number line here to make things clear.

■ An airplane pilot had to make two trips in one day. She left her base airport in the morning and flew 80 miles due east. Later in the day she had to make a 120-mile trip farther east. Where was she from her base airport at the end of the day?

Warm-up Answers

$-13\frac{1}{2}$	+6	$+1\frac{1}{2}$	$-4\frac{1}{2}$	−10	+15
A	E	D.	C	B	F

To solve, add +80 to +120.

The pilot was flying east on both trips. Think of a number line. What happens when you begin at zero and go 80 units to the right and then 120 more units to the right? Where would you end up?

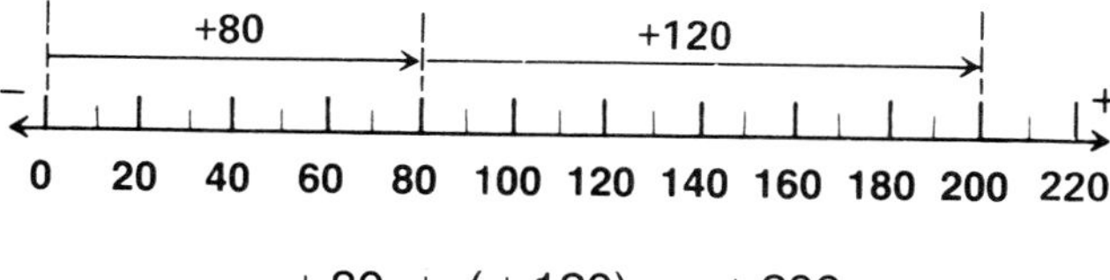

$$+80 + (+120) = +200$$

The +120 is written in parentheses only to separate it from the plus sign *between* +80 and +120. Be sure you understand the difference between the two plus signs. One indicates addition, the other indicates a positive number. The trip would end 200 miles east of the base airport. Can you see that whenever you are adding two positive numbers, you will always get another positive number?

Now look at this example.

■ A deep-sea diver dived 30 feet below the ocean surface. When he got his bearings, he realized he had to go 25 feet farther down. How deep into the ocean did he have to go?

To solve, add −30 to −25.

What would happen if you start at zero (sea level) on the number line and go down 30 units, then down 25 units more? You would end up at 55 below (negative side of) zero. Shown on a vertical number line, the problem would look like this:

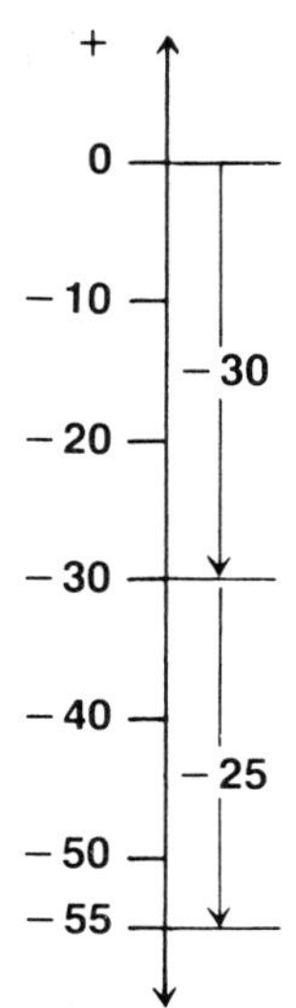

The sum of −30 and −25 is −55, that is, −30 + (−25) = −55.

The diver had to go down 55 feet below sea level.

Try It Yourself

You can use a number line for this if you like.

■ Mr. Brown made two sales in his shoe store one morning. One was for \$12 and the other for \$26. How much money did Mr. Brown take in that morning? ______

Your work should look like this: +12 + (+26) = +38. Mr. Brown took in \$38 that morning.

In this one the word *par* means the average number of strokes for the hole on this golf course.

■ Maria got a golf score of 4 strokes under par. In the next round she was 2 strokes under par. How many strokes over or under par was her score for the two rounds? ______

You should have represented this situation by this addition: −4 + (−2) = −6. Maria was 6 strokes *under* par.

Warm-up

Find these sums.

1. +1 + (+7) = ______

2. −5 + (−4) = ______

3. $-\frac{2}{3} + (-1\frac{2}{3})$ = ______

4. 5 + (+45) = ______

5. −32 + (−23) = ______

6. 65 + 72 = ______

7. +7 + 0 = ______

Warm-up Answers
1. +8 **2.** −9 **3.** $-2\frac{1}{3}$ **4.** +50 **5.** −55
6. 137 **7.** +7

Adding When Signs Are Different

Look at these questions.

How far is −5 from zero? Answer: 5 units
How far is +12 from zero? Answer: 12 units

In order to add positive and negative numbers, you'll have to know how far they are from zero. Think of it as taking a trip. If you are asked how far you went, it doesn't matter which way you went. Whether you went north, south, east, or west, the trip was a certain number of miles. You are only being asked how far you went.

When you add a positive and a negative number, first ask how far from zero each number is.

Here's an Example

You can look at the next problem in two ways. First you'll use the signed numbers.

■ Victor had a checkbook balance of $296. He wrote a check for $49. What was his balance after he wrote the check?

To solve, add +296 and −49. Victor had $296 *in* his account, so call it +296. The check is money to be *taken out* of his account, so call it −49.

+296 + (−49)

On a number line, it would look like this.

Do you see that thinking it through on the number line brings you to +247? You don't want to draw a number line every time you have to add a positive and a negative number though. You can make your work easier if you remember this rule:

To add a positive and a negative number, ask yourself how far each number is from zero. Then subtract these two amounts and use the sign of the number that is farther from zero.

You can look at the same problem using what we knew long ago. When he wrote a check, Victor subtracted money from his account. So the problem is $296 − $49, which is $247. From this, you can see that subtracting a positive number is the same as adding a negative number. It would make sense, then, that subtracting a negative number is the same as adding a positive number. This is so. We deal with it later but you can see that this isn't magic. It boils down to common sense.

Try It Yourself

Set up this problem using signed numbers. Draw a sketch if it helps.

■ Sally bought a black and white TV for $150. When she got home with it, she found that her parents had sent her a color set. She sold the black and white set to a friend for $65. How much did she lose on the set?

____

The problem in signed numbers is −150 + (65). If you used a number line, you could have started at −150, which is what Sally had spent on the black and white set. Then you should have moved 65 units to the right. After 65 units, you get to −85. She lost $85 on the deal. You could also use the rule. First notice that −150 is 150 units from zero and +65 is 65 units from zero. Subtract 65 from 150. You will get 85, and when you use the sign of the number farther from zero, you have −85, the same as before.

Warm-up

Find these sums.

1. −10 + (+3) = ______________
2. +1.7 + −0.9 = ______________
3. 25 + (−2) = ______________
4. (−1.2) + (+1.2) = ______________
5. (−99) + (33) = ______________
6. (+6) + (−6) = ______________
7. 0 + (−19) = ______________

Warm-up Answers
1. −7 **2.** 0.8 **3.** 23 **4.** 0 **5.** −66 **6.** 0 **7.** −19

Subtracting

Do you remember how to add 5 and −2? First ask yourself:

How far is +5 from zero? Answer: 5 units
How far is −2 from zero? Answer: 2 units

What is the difference between 5 and 2? Answer: 3 units

Give this answer (3) the sign of the number farther from zero (+5).

$$5 + (-2) = +3$$

When you subtract signed numbers, remember that every number has an opposite. The opposite of +7 is −7; the opposite of −3 is +3, and so on. When you subtract a number, you are really adding its opposite.

$$5 - 2 = 3$$

is the same as

$$5 + (-2) = +3$$

To subtract signed numbers, change subtraction to addition *and* change the number after the subtraction sign to the opposite of whatever it was. Then add the signed numbers. Some people prefer a slightly different version of this rule. Write the subtraction problem vertically. Put the number you are subtracting at the bottom. Then change the bottom sign and add. The meaning is the same, but some people prefer the second wording. Take your pick.

Here's an Example

■ The temperature at 8:00 A.M. was −18 degrees. At noon, it was −3 degrees. What was the difference between the 8:00 A.M. and noon temperatures?

To solve, subtract the smaller number (−18) from the larger number (−3).

Step 1 Write the subtraction problem.

$$(-3) - (-18)$$

Step 2 Change subtraction to addition and change the number after the subtraction sign to its opposite.

$$-3 + (+18)$$

Step 3 Add the signed numbers. You find that (−3) + (+18) is +15.

The difference in temperature was 15 degrees.

Try It Yourself

Here's another temperature problem.

■ The temperature was 10 below zero at noon. If it dropped 6 degrees in the next hour, what was the temperature then?

____

Did you remember to write *10 below zero* as −10? Then, did you set up the subtraction as −10 − (+6)? Finally, did you add the opposite: −10 + (−6)? The answer is −16°.

If you have to subtract fractions or decimals, use the same method.

Warm-up

Find these differences. Remember that a number without a sign is always a positive number (except, of course, for zero).

1. −10 − (+3) = ________
2. −3 − (−8) = ________
3. +9 − (+6) = ________
4. +16 − (−11) = ________
5. $8\frac{2}{3}$ − 18 = ________
6. 6 − (−17) = ________
7. 0 − (−2) = ________
8. 8 − (−8) = ________
9. +0.13 − (−0.13) = ________
10. −12 − (−12) = ________
11. 5 − (+5) = ________
12. 0 − (+7) = ________

Warm-up Answers
1. −13 **2.** +5 **3.** +3 **4.** +27 **5.** $-9\frac{1}{3}$ **6.** 23
7. +2 **8.** 16 **9.** +0.26 **10.** 0 **11.** 0 **12.** −7

☑ A Test-Taking Tip

If the GED Mathematics Test gives you more than two signed numbers to add, you can do one of two things:

1. you can add the first two numbers and then add this sum to the next number, and so on,

or

2. you can add all the negatives, add all the positives, and then add those two sums.

Use whichever way is easier for you.

Multiplying

"Business is so bad that I'm losing three hundred dollars a day. At that rate I'll be nine thousand dollars in debt by the end of the month!" You hope you'll never find yourself in a situation like this. *Losing* and *in debt* are both negative ideas. The idea of losing $300 each day for a month can be written as a mathematical statement. It is an example of multiplying signed numbers.

(losing $300 per day) (30 days in month)

$$-300 \times 30 = -\$9{,}000$$

(results in loss of $9,000 by end of month)

There is a simple way to remember what the answer is going to be like when you multiply signed numbers. If you multiply two numbers with the same sign, the answer is always a positive number. If you multiply two numbers with different signs, the answer is always a negative number.

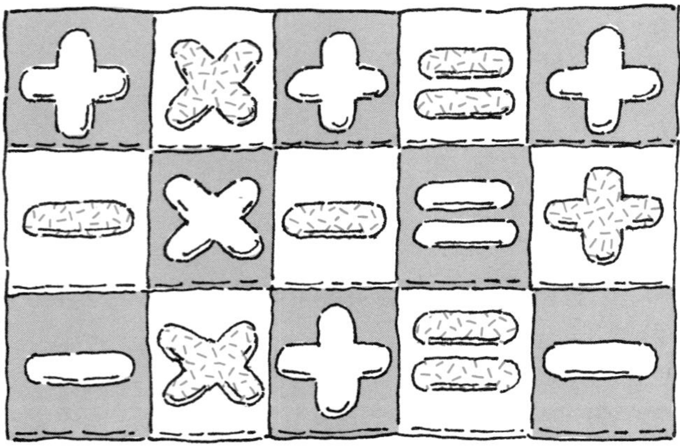

Here's an Example

Let's change our first example slightly.

■ If a successful business person *gains* $300 per day for 30 days, what will be the gain by the end of the month?

To solve, do the multiplication $(+300) \times (+30)$. When you multiply 300 by 30 you get 9,000. Since $+300$ and $+30$ have the same signs, a positive sign goes with the 9,000 for the final answer. The gain is $9,000.

Remember the problem at the very beginning of the lesson?

$$(-300) \times (+30) = -9{,}000$$

The signs were different, so the answer was a negative number.

Try It Yourself

■ Find $(-2) \times (-12)$. 

Both signs were the same kind (negative signs), so the answer is positive, $+24$.

■ Find $(-4) \times (+10)$. ________

Here the signs are different, so the answer is negative, -40.

Warm-up

Multiply these signed numbers.

1. $(-4) \times (-6) =$ ________
2. $(+5) \times (+7) =$ ________
3. $0 \times (-3) =$ ________
4. $(-2) \times (+22) =$ ________
5. $(+8) \times 0 =$ ________
6. $(+9) \times (-6) =$ ________
7. $-\frac{7}{8} \times (-\frac{4}{3})$ ________
8. $2.3 \times (-4.5)$ ________
9. $-\frac{1}{2} \times 2$ ________
10. $-0.8 \times (5)$ ________

Warm-up Answers
1. $+24$ **2.** $+35$ **3.** 0 **4.** -44 **5.** 0 **6.** -54
7. $+1\frac{1}{6}$ **8.** -10.35 **9.** -1 **10.** -4
In 3 and 5 did you remember that zero times any number is zero?

Dividing

The rules for dividing signed numbers are exactly the same as for multiplying signed numbers.

When you divide two numbers with the same signs, the answer is always positive. When you divide two numbers with different signs, the answer is always negative.

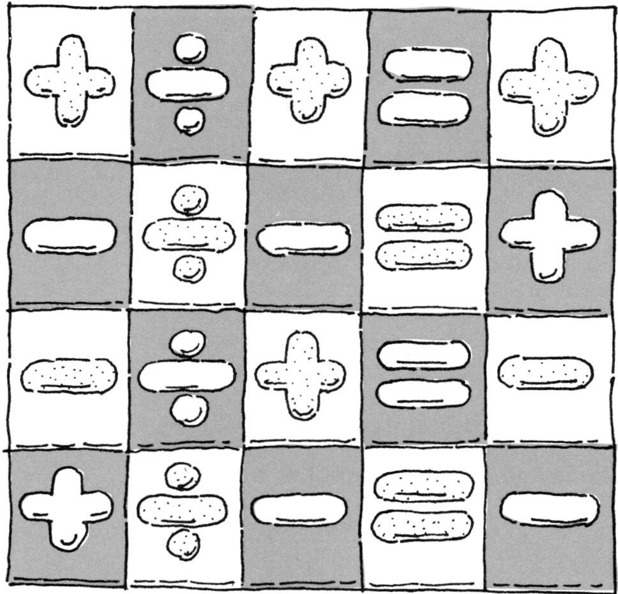

Here's an Example

■ Divide (−63 by (−9).

The signs are alike so (−63) divided by (−9) is positive 7, or +7.

Now let's look at an another example.

■ Divide (+63) by (+9).

The signs are still alike, so the answer is still positive 7, or +7.

Try It Yourself

■ Divide these numbers.

(−63) ÷ (+7) = ________________

(+63) ÷ (−9) = ________________

Did you notice the signs are different in both problems? Did you get a negative answer in both of them? The first is −9 and the second −7.

Even if you're asked to divide signed fractions or decimals, the rules are the same.

Warm-up

Divide these signed numbers.

1. (−72) ÷ (−9) = ________________
2. (−35) ÷ (+5) = ________________
3. (+56) ÷ (−7) = ________________
4. (+45) ÷ (+9) = ________________
5. 32 ÷ (−4) = ________________
6. −81 ÷ 9 = ________________
7. $-2\frac{1}{3}$ ÷ (−8) = ________________
8. 4.8 ÷ (−0.6) = ________________
9. $3\frac{1}{4}$ ÷ $(-\frac{1}{4})$ = ________________
10. (−0.2) ÷ (−5) = ________________

A Test-Taking Tip

The GED Mathematics Test may not specifically ask you to multiply or divide signed numbers. But you will have to know how to do this in order to answer some of the algebra questions on the test. You may prefer to translate some signed-number problems into straightforward arithmetic problems as we did in the section on adding different signs. It's up to you.

Warm-up Answers
1. +8 **2.** −7 **3.** −8 **4.** +5 **5.** −8 **6.** −9 **7.** $\frac{7}{24}$ **8.** −8 **9.** −13 **10.** 0.04
Did you remember that if there is no sign, the number is positive?

LESSON 27

Sequencing

Comparing Decimals

You have learned how to put signed numbers in order using a number line. You know that −57 is actually less than −4 because −57 is a lot further to the left of zero than −4. If you are given any set of numbers, you can put them in the order in which they appear on the number line. Numbers in a particular order are said to be in sequence. You can use the sequence of numbers on the number line to tell at once whether one number is greater than or less than another. When you do that, you are comparing the numbers.

If you add on a zero to 3 to get 30, you have, in effect, multiplied the 3 by 10. This is so because adding on the zero moved the 3 from the ones place to the tens place.

If, however, you add a zero at the right-hand end of 3.0 to get 3.00, you haven't changed the value at all. The 3 is still next to the decimal point in the ones place, so its value is unchanged. You could add a hundred zeros without changing the value of a decimal. It makes sense then that you could take zeros off the end of a decimal without altering the value. You can use this idea to compare decimals.

Here's an Example

Look at these numbers. They all have the same value.

5.4 = 5.40 = 5.400 = 5.4000 = 5.40000

■ Add zeros to .07 to help you compare .07 to .00984. Which is the larger number?

To answer the question, add 3 zeros to .07 to get 5 digits after the decimal point, the same as in the other decimal. You get .07000.

.07000 is seven thousand hundred-thousandths
.00984 is nine hundred eighty-four hundred-thousandths

Now you can see that .07 is the larger number, even though you might have thought it was the other way around at first glance.

Try It Yourself

Remember, when you are comparing decimals, to add on zeros at the right end (after the decimal point) to give all the decimals the same number of digits.

■ Which is the largest decimal number: 0.32, 0.3, or 0.032? ________________

Did you add a zero to the right of 0.32 and add two zeros to the right of 0.3 to make the numbers easier to compare?

0.320 = three hundred twenty thousandths
0.300 = three hundred thousandths
0.032 = thirty-two thousandths

You can see that 0.320 is the largest decimal. So in the original list, 0.32 is the largest number.

Warm-up

Circle the largest number in each group.

1. 0.256 0.3 0.03

2. 0.070 0.007 0.7

3. 0.0025 0.505 0.0505

4. 0.375 0.0375 0.00375

5. 0.125 0.1025 0.025 0.0125

Warm-up Answers
1. 0.3 **2.** 0.7 **3.** 0.505 **4.** 0.375 **5.** 0.125

On the Springboard

1. John put the smallest copper coins from his foreign coin collection into bags. Bag A contained 12 coins, each worth 0.55 cents. Bag B held 11 coins, each worth 0.6 cents, and bag C had 100 coins, each worth 0.065 cents. Which bag held the coins of least total value?

   (1) A (2) B (3) C

   ① ② ③

2. The steel band in the strapping machine was exactly .25 inches wide. The three sizes of the threading slots were exactly 0.245, 0.255, and 0.305 inches long. Which slot size could not be used with the given band?

(1) 0.245 (2) 0.255 (3) 0.305

(1) (2) (3)

Check your Springboard answers on page 191.

Comparing Fractions

An easy way to compare fractions is to make sure they all have the same denominator. If they don't, rename the fractions so that they do have the same denominator. Then compare numerators.

Here's an Example

If you start off with fractions having the same denominator, such as $\frac{6}{16}$ and $\frac{13}{16}$, all you have to do is look at the numerators. Here 13 is greater than 6, so $\frac{13}{16}$ is the larger fraction.

If the denominators are different, find a common denominator. Rename the fractions as needed to give them that same common denominator. Then you can compare. Use the same idea if you have several fractions that you need to put in order. Here's how to put $\frac{3}{18}$, $\frac{5}{6}$, and $\frac{2}{9}$ in order.

Rename to have the common denominator 18.

$$\frac{3}{18} \qquad \frac{5}{6} = \frac{15}{18} \qquad \frac{2}{9} = \frac{4}{18}$$

From least to greatest, the renamed fractions are $\frac{3}{18}$, $\frac{4}{18}$, and $\frac{15}{18}$. So the correct order for the original fractions is $\frac{3}{18}$, $\frac{2}{9}$, $\frac{5}{6}$.

Try It Yourself

■ Which is larger, $\frac{5}{32}$ or $\frac{3}{16}$?

If you used a common denominator of 32 and renamed $\frac{3}{16}$ as $\frac{6}{32}$, you discovered that $\frac{3}{16}$ is the larger of the two original fractions.

Warm-up

Put these fractions in order from least to greatest.

1. $\frac{2}{15}$, $\frac{1}{30}$, and $\frac{5}{12}$ ________________

2. $\frac{2}{9}$, $\frac{2}{3}$, $\frac{5}{6}$, and $\frac{5}{18}$ ________________

On the Springboard

3. A cabinet maker needed some wooden strips to repair a $\frac{9}{40}$-inch gap in the work of an apprentice. He had three strips left. They were $\frac{1}{5}$, $\frac{1}{8}$, and $\frac{1}{10}$ inches thick. Which pair were exactly the size of the crack when put together?

(1) $\frac{1}{5}$, $\frac{1}{8}$ (2) $\frac{1}{8}$, $\frac{1}{10}$ (3) $\frac{1}{5}$, $\frac{1}{10}$

(1) (2) (3)

4. Three prospectors staked out a promising piece of land. A took $\frac{3}{8}$ of it, B took $\frac{5}{12}$ of it, and C took what was left. Write the initials of the prospectors in the order of size of their holdings, from least to greatest.

(1) A, C, B (2) C, A, B
(3) C, B, A

(1) (2) (3)

Check your Springboard answers on page 191. If you got both of these right you did well. If not, make sure where you slipped. To feel really secure, redo what you missed to be sure you're ready to go on.

Warm-up Answers

1. $\frac{1}{30}$, $\frac{2}{15}$, $\frac{5}{12}$ 2. $\frac{2}{9}$, $\frac{5}{18}$, $\frac{2}{3}$, $\frac{5}{6}$

Comparing and Ordering Data

Your practice in comparing and ordering fractions and decimals will help a lot with certain kinds of questions on the GED Mathematics Test.

Here's an Example

■ Doreen is making a tool box and wants to put the sockets for her socket wrench in slots in order of size. She wants the smallest on the left and the largest on the right. If the socket sizes, in fractions of an inch, are $\frac{7}{16}$, $\frac{5}{8}$, $\frac{3}{4}$, and $\frac{1}{2}$, in what order should they be placed in the slots?

To solve the problem, you need to compare the fractions and arrange them in order. First, find a common denominator and rename as necessary so that all the fractions have a common denominator. If you use 16 as the common denominator, you get the following:

$$\frac{7}{16} \qquad \frac{5}{8} = \frac{10}{16} \qquad \frac{3}{4} = \frac{12}{16} \qquad \frac{1}{2} = \frac{8}{16}$$

To arrange the fractions in order, compare the sixteenths. But remember that the final answer must use the sizes *originally* given for the sockets.

	Smallest	⟶		Largest
	$\frac{7}{16}$	$\frac{8}{16}$	$\frac{10}{16}$	$\frac{12}{16}$
original sizes	$\frac{7}{16}$	$\frac{1}{2}$	$\frac{5}{8}$	$\frac{3}{4}$

So from left to right, the sockets should be arranged as follows: $\frac{7}{16}$, $\frac{1}{2}$, $\frac{5}{8}$, $\frac{3}{4}$.

Try It Yourself

In this problem, you'll find it helpful to change all numbers to decimals.

■ Company XYZ asked its 5 sales representatives to report what part of their travel expenses went for entertaining clients. They gave their answers in various ways:

Person A said 17%
Person B said $\frac{1}{5}$
Person C said .23
Person D said $\frac{1}{4}$
Person E said 21.5%

Using the letters A–E, list the sales representatives in order, from the one who spent the least part on entertainment to the one who spent the greatest part on entertainment.

____.

Did you follow the suggestion and change all the numbers to decimals? If you did, you should have gotten the following: 0.17 for A, 0.2 for B, 0.23 for C, 0.25 for D, and 0.215 for E. Did you add zeros to make all the decimals *three*-place decimals? If you did all this and compared correctly, you should have listed the sales representatives in this order (least to greatest): A, B, E, C, D.

On the Springboard

5. In 3 computer classes, attendance on the day after the big blizzard was as follows:

   Class A: 23 present out of 25
   Class B: 19 present out of 20
   Class C: 27 present out of 30

   Which of the following lists the classes in order from best to worst attendance?

   (1) A, C, B (2) B, A, C
   (3) C, A, B

   ① ② ③

6. Family A said it spent 21% of its income on rent, family B said it spent $\frac{1}{5}$, and family C said it spent .205. Which sequence shows the part of family income spent on rent from least to greatest?

(1) A, C, B (2) B, A, C
(3) B, C, A

Check your Springboard answers on page 191. If all of your answers were correct, go right ahead to "The Real Thing." If you missed either question, check your work. Review if you need to do so. When you can correctly answer the ones you missed, go on to "The Real Thing."

 "The Real Thing" 

1. Cold capsules from 5 companies were analyzed to see how much of a certain ingredient they contained. The results were as follows:

Company	Amount of ingredient (grams per capsule)
A	0.015
B	0.0064
C	0.0105
D	0.002
E	0.18

A report indicates that to be safe but effective for people with high blood pressure a capsule should contain no more than 0.007 grams and no less than 0.003 grams of this ingredient. Which company's capsules meet this requirement?

(1) Company A (2) Company B
(3) Company C (4) Company D
(5) Company E

(1) (2) (3) (4) (5)

2. Red and white paint have been mixed to form five different shades of pink. The ratio of red to white paint in each mixture is as follows:

Mixture →	A	B	C	D	E
Ratio of red to white →	$\frac{4}{7}$	$\frac{2}{3}$	$\frac{3}{5}$	$\frac{11}{21}$	$\frac{7}{15}$

Which sequencing lists the mixtures from lightest to darkest shade of pink?

(1) A, B, C, E, D (2) B, A, C, D, E
(3) C, B, A, E, D (4) D, E, A, C, B
(5) E, D, A, C, B

(1) (2) (3) (4) (5)

3. Five different people were asked to guess what $\frac{13}{40}$ would be equal to when expressed as a decimal. Their guesses were as follows:

Jim	0.345
Alice	0.33
Tomas	0.43
Cary	0.275
Seiji	0.425

Which person came closest to the actual decimal value of $\frac{13}{40}$?

(1) Jim (2) Alice (3) Tomas
(4) Cary (5) Seiji

(1) (2) (3) (4) (5)

4. Ellen earns $275 a week. She has earned a raise and has been offered the following options:

A. a raise of $1,200
B. a $55-a-week raise for one year
C. a 15% raise
D. a raise of $105 every two weeks for one year

Which option should she choose?

(1) A (2) B (3) C (4) D
(5) Insufficient data is given to solve the problem.

(1) (2) (3) (4) (5)

5. After a political poll, probabilities were given that the candidates would win the offices for which they were running. The probabilities were expressed as fractions and then changed to decimals with the following results:

A	Wobblies	0.25
B	Senatorials	0.073
C	Workers	0.105
D	Social Democrats	0.3
E	Libertarians	0.08

Which sequencing shows the candidates from most likely to least likely to win?

(1) A, B, C, E, D
(2) C, B, A, E, D
(3) C, A, B, D, E
(4) D, A, C, E, B
(5) D, E, A, B, C

(1) (2) (3) (4) (5)

6. A steel pin must be inserted into a hole that was drilled in a metal plate. The hole measures $\frac{7}{16}$ inch across. The widths of the pins available are as follows:

Pin	→	A	B	C	D	E
Width (in inches)	→	$\frac{15}{32}$	$\frac{3}{8}$	$\frac{13}{32}$	$\frac{5}{8}$	$\frac{9}{16}$

The widest pin possible should be inserted into the hole. Which pin should be used?

(1) Pin A (2) Pin B (3) Pin C
(4) Pin D (5) Pin E

(1) (2) (3) (4) (5)

Check your answers and record all your scores on pages 211 and 212.

LESSON 28

Exponents

What Are Exponents?

In mathematics, you often have to deal with a repeated multiplication such as 2 × 2 × 2, which equals 8, or 10 × 10 × 10 × 10, which equals 10,000. To make it quick and easy to write down such multiplications, you can use exponents as a short cut. In the example 2 × 2 × 2, the number 2 is used 3 times. With an **exponent** this is written 2^3. The 3 is the exponent—it tells how many 2s would be in the multiplication problem if you wrote the problem out in full. The number 2 is called the **base.** It's the number that is repeated in the multiplication.

To write 10 × 10 × 10 × 10 using exponents, you would write 10^4. Here the base is 10 and the exponent is 4. You always write the exponent after the base and a little higher up.

2^3 ← Exponent; 2 ← Base 10^4 ← Exponent; 10 ← Base

To read these numbers aloud, you say "2 to the third" and "10 to the fourth."

Coming to Terms

exponent in an expression such as 5^3, the exponent is 3 and it tells how many 5s are used in the repeated multiplication

base in the expression 5^3, the base is 5, the number that is repeated in the multiplication

A Test-Taking Tip

Many people who take the GED Mathematics Test make the mistake of multiplying the base by the exponent. They say that 5^3 is 5 × 3. Don't make that mistake. Remember that 5^3 means 5 × 5 × 5. There's a big difference—5 × 3 equals 15, but 5^3 equals 5 × 5 × 5, or 125.

Simplifying Numbers

If you are asked to simplify a number with an exponent, you must find the answer for the repeated multiplication. If the base is 10, you can write down the answer at sight.

10^2 means 10×10, which is 100.
10^3 means $10 \times 10 \times 10$, which is 1,000.

If you are dealing with the base 10, the exponent tells how many 0s come after the 1 in the answer. If the base isn't 10, you will have to multiply out, or compute the entire problem.

Here's an Example

■ Simplify 2^3.

Step 1 Write out what the exponent means.

$2^3 = 2 \times 2 \times 2$

Step 2 Compute the answer.

$2 \times 2 \times 2 = 8$

So when you simplify 2^3, you get 8. Sometimes the base is negative, as in $(-2)^3$. Then you have to remember the rules about multiplying signed numbers. Write out the repeated multiplication.

$$(-2)^3 = (-2) \times (-2) \times (-2)$$

Multiply the first two numbers and you get $(-2) \times (-2) = +4$ (both numbers multiplied are negative, so the answer is positive.) Now multiply $+4$ by the last (-2) and you get $(+4) \times (-2) = -8$ (positive times negative gives negative).

If the exponent is 1, you simply get the base. $3^1 = 3$, $10^1 = 10$, $(-5)^1 = -5$, and so on. Sometimes the exponent is zero, as in 2^0, 5^0, $(-8)^0$, and so on. Any number with an exponent of zero is equal to 1. For example, $2^0 = 1$; $5^0 = 1$; $(-8)^0 = 1$, and so on.

A Test-Taking Tip

One quick way to check work with signed numbers on the GED Mathematics Test is to count the negative signs. An even number (2, 4, 6, and so on) of negative signs always gives a positive answer. An odd number of negative signs gives a negative answer.

Try It Yourself

■ Simplify 3^4. ______

Did you first write what 3^4 means? $3^4 = 3 \times 3 \times 3 \times 3$. Then did you compute the answer? Since $3 \times 3 \times 3 \times 3 = 81$, the answer is 81.

Now try another.

■ Simplify $(-5)^2$. ______

Did you write $(-5)^2 = (-5) \times (-5) = +25$?

Multiplying two negatives gives a positive answer. Always keep the negative base in parentheses so you don't get confused with the signs.

Look carefully at the difference between $(-5)^2$ and -5^2. The first one means

$$(-5)^2 = (-5) \times (-5) = +25.$$

The second one means

$$-5^2 = -(5 \times 5) = -25.$$

Warm-up

Simplify.

1. $6^2 =$ ______
2. $(-4)^3 =$ ______
3. $(+7)^3 =$ ______
4. $10^5 =$ ______
5. $(-2)^4 =$ ______
6. $(-10)^2 =$ ______
7. $-4^2 =$ ______
8. $(0.3)^3 =$ ______
9. $+2^6 =$ ______
10. $(\frac{1}{2})^2 =$ ______

Warm-up Answers
1. 36 **2.** −64 **3.** 343 **4.** 100,000 **5.** 16 **6.** 100 **7.** −16 **8.** 0.027 **9.** 64 **10.** ¼
Did 10 give you trouble? Since ½ is the base, it is multiplied by itself. You can tell that the whole fraction is the base because it is all in parentheses.

On the Springboard

1. Simplify 2^5.

   (1) 10 (2) 16 (3) 32

   ① ② ③

2. Simplify $(-9)^3$.

   (1) -729 (2) -27 (3) 729

   ① ② ③

3. $(-1)^6$ equals

   (1) -6 (2) -1 (3) 1

   ① ② ③

4. $(-\frac{2}{3})^3$ equals

   (1) $-\frac{8}{27}$ (2) $\frac{4}{9}$ (3) $\frac{8}{27}$

   ① ② ③

Check your Springboard answers on page 191. If you got all four correct, go on to "The Real Thing." If you missed any, check your work or review this section. Once you've corrected your mistakes, you're ready to go on.

The Real Thing

1. $(-8)^3$ equals

   (1) -512 (2) -24 (3) $+24$
   (4) 64 (5) $+512$

   ① ② ③ ④ ⑤

2. 10^6 equals

   (1) $-1{,}000{,}000$ (2) 60 (3) 600
   (4) 100,000 (5) 1,000,000

   ① ② ③ ④ ⑤

3. $(\frac{2}{3})^2$ equals

   (1) $-\frac{4}{9}$ (2) $\frac{2}{9}$ (3) $\frac{4}{9}$
   (4) $\frac{4}{6}$ (5) $\frac{4}{3}$

   ① ② ③ ④ ⑤

4. What is the value of 5^0?

   (1) -5 (2) 0 (3) 1
   (4) 5 (5) 50

   ① ② ③ ④ ⑤

5. Simplify $(-7)^1$.

   (1) -71 (2) -7 (3) 0
   (4) 1 (5) 7

   ① ② ③ ④ ⑤

6. Simplify $(-\frac{1}{3})^2$.

   (1) -6 (2) $-\frac{1}{6}$ (3) $-\frac{1}{9}$
   (4) $\frac{1}{9}$ (5) $\frac{1}{6}$

   ① ② ③ ④ ⑤

Check your answers on page 212.

Squares

If you are using square tiles for your bathroom, you will notice that you get patterns like these.

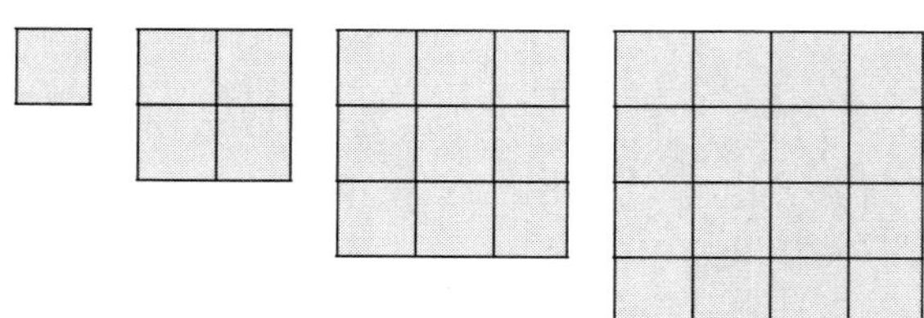

If you have a square shape with 2 tiles down the side, you have 2 × 2 tiles in the shape. If you have a square shape with 3 tiles on one side, you have 3 × 3 tiles in the square shape; a shape with 5 tiles on one side needs 5 × 5 tiles, and so on.

This fact was discovered thousands of years ago. To find how many tiles you need in a square shape you multiply the number on one side by itself. When you multiply a number by itself you **square** it. When you multiply a number by itself you can write it with an exponent of 2. So 5 squared is 5^2 or 5 × 5 or 25; it is the number of tiles in a square shape with 5 tiles to a side. If you are told how many tiles there were in the square, you can say how many are on one side. This is called finding the **square root.** The square root of 25 is 5 and the square root of 16 is 4.

Rather than writing out the words *square root,* we use the sign $\sqrt{\ }$, called the **radical sign.** So $\sqrt{25}$ means the square root of 25, which is 5. In mathematics, the square root of a number is the *positive* number whose square equals the given number. When you multiply -5 by itself, you get 25, and when you multiply 5 by itself you get 25. The *positive* number, 5, is called the square root of 25.

Look carefully at the difference between $(-5)^2$ and -5^2. The first one means: $(-5)^2 = (-5) \times (-5) = +25$. The second one means: $-5^2 = -(5 \times 5) = -25$.

■ What is $\sqrt{100}$? ______________________

Did you think of 10? Both 10×10 and $(-10) \times (-10)$ equal 100. Only the *positive* number, 10, is the square root of 100.

Coming to Terms

square the result of multiplying a number by itself. A square can be written as the number with the exponent 2.

square root the positive number that when multiplied by itself gives the original number. The square root of 49 is 7 because $7 \times 7 = 49$.

radical sign the $\sqrt{\ }$ sign, which means *the square root of*

Here's an Example

■ What is the square of 5?

It is 5^2, or 25, because $5 \times 5 = 25$.

■ Now find the square of -5.

It is $(-5)^2$, or 25, because $(-5) \times (-5) = 25$.

■ What is the square root of 36?

Ask yourself what positive number you multiply by itself to get 36. It is 6, since $6 \times 6 = 36$.

■ What is $\sqrt{49}$?

If you multiply 7 by 7, you get 49. So $\sqrt{49} = 7$.

A Test-Taking Tip

It will help you to memorize the squares of all the numbers from 1 through 12. Here is a table of these squares. Notice that if you read the table from left to right you have the squares, and if you read it from right to left you have the square roots. For example, the square of 11 is 121 and the square root of 121 is 11.

Number →	Square
1	1
2	4
3	9
4	16
5	25
6	36
7	49
8	64
9	81
10	100
11	121
12	144
Square root	← Number

Try It Yourself

■ What is the square of 8? ______________________

Did you remember that the square of a number is the same as the number with an exponent of 2? So the square of 8 means 8^2, or 64. Your answer should have been 64.

■ Simplify $(-5)^2$. ______________________

Did you write $(-5)^2 = (-5) \times (-5) = +25$? Two negatives give a positive answer. Always keep that negative base in parentheses so you won't lose it!

Warm-up

Find these squares and square roots.

1. $17^2 =$ ______________________

2. $300^2 =$ ______________________

3. $4^2 =$ ______________________

4. $\sqrt{25} =$ ______________________

5. $\sqrt{169} =$ ______________________

6. $\sqrt{100} =$ ______________________

Warm-up Answers
1. 289 **2.** 90,000 **3.** 16 **4.** 5 **5.** 13 **6.** 10

On the Springboard

5. The square tiled entrance to a hotel had 37 tiles on each side. How many tiles were in the entrance altogether?

(1) 949 (2) 1,169 (3) 1,369

(1) (2) (3)

6. Jennifer is 6 years old. The square of her age is the same as the total of her age and the age of her father. How old is her father?

(1) 25 (2) 28 (3) 30

(1) (2) (3)

7. John brought three boxes of square floor tiles, each containing 50 tiles. He put down the largest square of tiles he could. How many did he have left over?

(1) 6 (2) 8 (3) 10

(1) (2) (3)

Check your Springboard answers on pages 191–192. If you got these right, go on to "The Real Thing." If not, pause to find out what happened before going on.

The Real Thing

7. Isaac has 157 square tiles to make a square tray. What is the greatest number of tiles that he can put on a side?

(1) 10 (2) 11 (3) 12
(4) 13 (5) 14

(1) (2) (3) (4) (5)

8. Lynn is going on vacation for 2 weeks and has packed 3 blouses: yellow, blue, and red. She packed 3 pairs of slacks of the same colors. How many outfits can she put together (slacks plus blouse) from what she packed?

(1) 2×3 (2) 2^3
(3) 3^2 (4) $(2 \times 3)^2$
(5) Insufficient data is given to solve the problem.

(1) (2) (3) (4) (5)

9. Marco tiled a kitchen floor using square tiles that measured 3 inches on each side. He used 3,600 tiles. How many tiles would he have needed if he had used square tiles that were just 1 inch on each side?

(1) 1,200 (2) 1,800 (3) 7,200
(4) 10,800 (5) 32,400

(1) (2) (3) (4) (5)

Record your scores on page 212.

Scientific Notation

A piece of thin paper may be only 0.00186 inches thick. The distance from the sun to the planet Uranus is about 1,785,000,000 miles. To make it easier to write numbers with so many digits a system called **scientific notation** is used. It uses numbers from 1 up to, but less than 10, multiplied by 10 with an exponent. It's easier to get the idea from actual examples.

Number	Scientific Notation
360	3.6×10^2
3,600	3.6×10^3
36,000	3.6×10^4
360,000	3.6×10^5
3,600,000	3.6×10^6

Notice that the first number in the scientific notation is a decimal number with one digit before the decimal point. This is always so in scientific notation. The digit before the decimal point can be anything from 1 to 9. The second number in the scientific notation is always 10 with an exponent. Look at 36,000 in the left-hand column, then at the scientific notation. If you move the decimal point of 3.6 to the right 4 places, you get 36,000 because you would have multiplied 3.6 by 10,000 or 10^4. The exponent of the 10 is the number of places you move the decimal point to get the original number back again. The exponent can be positive or negative.

Look at these examples.

Number	Scientific notation
0.36	3.6×10^{-1}
0.036	3.6×10^{-2}
0.0036	3.6×10^{-3}
0.00036	3.6×10^{-4}

Notice the pattern. The first number in the scientific notation is 3.6 each time, but now the exponents of 10 are negative. This means that you move the decimal point to the *left* to get back to the original number. As before, the exponent tells you how many places to move the decimal point. Look at 0.0036 in the left-hand column, then at the scientific notation. The exponent of 10 is -3, so by moving the decimal point of the 3.6 to the left 3 places, you get to 0.0036 again. If the original number is less than 1, you will always have a negative exponent for the 10 in the scientific notation for that number.

Here's an Example

■ Write 748,000 in scientific notation.

Step 1 Rewrite the number with the decimal point after the first digit on the left that is not 0. Drop all the unnecessary 0s after the decimal point and write $\times$ 10 at the end.

7.48×10

Step 2 Count the number of places you would have to move the decimal point to get the original number back again. In this example, it is 5 places to the right, so the exponent is positive 5. Write that number as the exponent for the 10.

7.48×10^5

To check that 7.48×10^5 is the correct answer, move the decimal point in the answer 5 places to the right. To do that you have to put three 0s after the 8. Since you get back to 748,000, the answer is correct.

■ Write 0.0000483 in scientific notation.

Step 1 Rewrite the number with the decimal point after the first digit on the left that is not 0. Drop all the unecessary 0s at the left end of the new number. Then write $\times$ 10 at the end as before.

4.83×10

Step 2 Count the number of places you need to move the decimal point to get back to the original number. Here it is 5 places to the left, so the exponent for 10 is -5. Use -5 as the exponent for the 10.

4.83×10^{-5}

To check, move the decimal point 5 places to the left in the answer. Fill in 0s as needed. You get back to 0.0000483.

Coming to Terms

scientific notation a system for writing very large or very small numbers. In scientific notation, the original number is written as a decimal that is multiplied by a 10 with a positive or negative exponent. The original number is always expressed as a decimal greater than or equal to 1.

Try It Yourself

Astronomers have to deal with huge numbers. They often use scientific notation.

■ Mercury is 36,000,000 miles from the sun. Write this number in scientific notation.

You should have rewritten the number with non-zero digits with a decimal point after the first non-zero digit on the left. You should have dropped the unnecessary 0s after the decimal point. After that, you should have put $\times$ 10 to get 3.6×10. Did you count and find that you would have to move the decimal point 7 places to the right to get the original number back? The answer is 3.6×10^7.

Here's one with a small number.

■ A human hair is about 0.00625 cm thick. Write the number in scientific notation.

Did you rewrite the number with the decimal point after the 6? Did you drop unnecessary 0s in front of the decimal and get 6.25×10? To find the exponent for the 10, did you count to find that you'd need to move 3 places to the *left* to get back to the original number? If you used the exponent -3, then you should have 6.25×10^{-3} as your final answer.

Warm-up

Write the following numbers in scientific notation.

1. 7,460,000 ____________

2. 0.00342 ____________

3. 9,000,000 ____________

4. 0.00092 ____________

5. 365 ____________

Write these numbers, which are in scientific notation, as ordinary numbers.

6. 8.15×10^5 ____________

7. 4.78×10^{-3} ____________

8. 3.22×10^4 ____________

9. 1.473×10^{-1} ____________

10. 9.302×10^6 ____________

Warm-up Answers
1. 7.46×10^6 **2.** 3.42×10^{-3} **3.** 9.0×10^6 **4.** 9.2×10^{-4} **5.** 3.65×10^2 **6.** 815,000 **7.** 0.00478 **8.** 32,200 **9.** 0.1473 **10.** 9,302,000

On the Springboard

8. At an average speed of 24,500 miles per hour it would take about 10 hours to get to the moon. What is the distance in miles to the moon?

(1) 24.5×10^4 (2) 2.45×10^5
(3) 2.45×10^6

(1) (2) (3)

9. A scientist wrote 3.67×10^5 as the answer to a problem. He later found that the answer should have been one hundred times bigger than that. What was the correct answer in scientific notation?

(1) 36.7×10^5 (2) 3.67×10^6
(3) 3.67×10^7

(1) (2) (3)

10. A report stated that 6.25×10^{-3} ounces of white lead was found in a certain sample of paint. Written as an ordinary number, how many ounces of lead would that be?

(1) 0.000625 (2) 0.00625
(3) 0.0625

(1) (2) (3)

Check your Springboard answers on page 192. If you got these right, go on to "The Real Thing." If not, why not review a bit to see if you can correct your error before going on.

"The Real Thing"

10. The surface temperature of the sun is estimated to be (5.8×10^3) degrees Celsius. What is another way to write this temperature (in degrees Celsius)?

(1) 0.0058 (2) 0.0580 (3) 5,800
(4) 58,000 (5) 580,000

(1) (2) (3) (4) (5)

11. The length of the light ray that produces the color red is approximately 7×10^{-7} meters. What is another way to write this length (in meters)?

(1) 0.00000007 (2) 0.0000007
(3) 0.000007 (4) 7,000,000
(5) 70,000,000

(1) (2) (3) (4) (5)

12. Which of the following scientific notation expressions equals 0.000000268?

(1) 0.268×10^{-6} (2) 2.68×10^{-7}
(3) 26.8×10^{-8} (4) 268×10^{-9}
(5) 2.68×10

(1) (2) (3) (4) (5)

13. Scientists found a fossil of a fern leaf. They estimated that its age was 360,000,000 years old. What is its age (in years) in scientific notation?

(1) 0.36×10^{-9} (2) 3.6×10^{-8}
(3) 3.6×10^7 (4) 3.6×10^8
(5) 36×10^9

(1) (2) (3) (4) (5)

14. What is the value of $(4/7)^2$?

(1) $\frac{16}{49}$ (2) $\frac{8}{14}$ (3) $\frac{6}{9}$ (4) $\frac{6}{7}$ (5) $\frac{8}{7}$

(1) (2) (3) (4) (5)

15. Which of the following scientific notation expressions equals 1,427,200,000?

(1) 0.14272×10^{10} (2) 1.4272×10^{9}
(3) 14.272×10^{8} (4) 142.72×10^{7}
(5) 14272×10^{5}

(1) (2) (3) (4) (5)

16. If a one-celled animal splits into 2 new animals every hour, how many of these animals would 3 one-celled animals give rise to after 5 hours?

(1) 3×5 (2) $2 \times 3 \times 5$
(3) 3×2^5 (4) $(3 \times 5)^2$ (5) $(2 \times 3)^5$

(1) (2) (3) (4) (5)

17. The earth makes one orbit around the sun in 365 days and travels a distance of approximately 9.344×10^8 kilometers. What is another way to write this distance (in meters)?

(1) 0.00009344 (2) 9,344,000
(3) 93,440,000 (4) 934,400,000
(5) 9,344,000,000

(1) (2) (3) (4) (5)

18. What is the value of $(0.13)^2$?

(1) 0.0169 (2) 0.169 (3) 0.26
(4) 26 (5) 168

(1) (2) (3) (4) (5)

19. Which of the following equals the scientific notation expression of 7.5×10^{-3}?

(1) 0.00075 (2) 0.0075 (3) 0.075
(4) 0.75 (5) 7,500

(1) (2) (3) (4) (5)

Check your answers and record your score on page 212.

LESSON 29

Standard Measurement

You probably have a pretty good feel for the size of a pound, a cup, and a foot. But when you're shopping and you see one bottle of detergent marked 32 ounces and another marked 1 pint 6 ounces, it may not be obvious which is the better buy. Even if you don't shop much, you probably have to deal with measurements when you drive, use recipes, or participate in sports.

In this lesson you will learn how to work with standard measures. Here are some measures you should know for the GED Mathematics Test.

Equivalent Standard Measures			
Length:	1 mile (mi)	=	5,280 feet (ft)
	1 yard (yd)	=	3 feet (ft)
	1 foot (ft)	=	12 inches (in)
Liquid:	1 gallon (gal)	=	4 quarts (qt)
	1 quart (qt)	=	2 pints (pt)
	1 pint (pt)	=	2 cups (c)
	1 cup (c)	=	8 ounces (oz)
Weight:	1 ton (T)	=	2,000 pounds (lb)
	1 pound (lb)	=	16 ounces (oz)
Amount:	1 dozen (doz.)	=	12 items

Take a few minutes to memorize any of these measures you don't already know or of which you aren't quite sure. You'll save time on the GED Mathematics Test if you can change measurements quickly.

Converting from One Unit to Another

Before you add and subtract measures, it's a good idea to practice changing from one type of measurement unit to another if more than one kind is used in the problem. You'll often need to do this in solving measurement problems. There are two rules to use when converting units.

Rule 1 When changing from a larger to a smaller unit, multiply. (You are more inches tall than you are feet tall, so you have to multiply to get from feet to inches.)

Rule 2 When changing from smaller units to larger units, divide. (You are fewer feet tall than you are inches tall, so you have to divide to get from inches to feet.)

Here's an Example

For this example you need to know that there are 12 inches in a foot.

■ Two bookcases are placed side by side. One is 3 feet wide, and the other is 32 inches wide. In inches, what is the total width of wall space that they occupy?

To solve, change 3 feet to inches, then add what you get to 32 inches.

Step 1 Change 3 feet to inches. There are 12 inches in a foot, so multiply by 12.

$$3 \times 12 = 36$$

Step 2 Add 36 inches and 32 inches.

$$36 + 32 = 68$$

The two bookshelves occupy 68 inches of wall space.

If the question had asked for the width in feet, you would have changed inches to feet before you added.

Step 1 Change 32 inches to feet by dividing by 12.

$$32 \div 12 = 2\tfrac{8}{12} \text{ or } 2\tfrac{2}{3}$$

Step 2 Add 3 feet and $2\frac{2}{3}$ feet.

$$3 + 2\tfrac{2}{3} = 5\tfrac{2}{3}$$

The bookshelves take up $5\frac{2}{3}$ feet of wall space.

Notice that the first answer was 68 inches. If you change this to feet by dividing by 12, you get

$$68 \div 12 = \frac{68}{12} = 5\frac{8}{12} = 5\frac{2}{3}$$

So the first answer checks with the second.

✓ A Test-Taking Tip

On the GED Mathematics Test, find out what the problem is asking for. Usually the question itself will help you decide which units to use, but sometimes you need to look at the answer choices to make that decision. Change the measurements to that unit, then solve the problem.

Try It Yourself

■ Which is longer, a 78-inch piece of lumber or one that is 6 feet 5 inches long?

To find out which is longer, you change 6 feet 5 inches to inches. 6 feet = (6 × 12) inches = 72 inches, so 6 feet 5 inches = 72 inches + 5 inches or 77 inches.

You were correct if you said the 78-inch piece is the longer one.

In this next problem, you have to take two steps for the conversion.

■ Which has more detergent, a bottle that contains 32 ounces or one that contains 1 pint 6 ounces?

It is easy to change to ounces. A pint equals 16 ounces, so the second bottle contains 16 + 6, or 22 ounces. The first bottle contains 32 ounces, so it is the larger. If you converted to pints, 32 ounces is exactly 2 pints (32 ÷ 16 = 2). This tells you at once that the 32-ounce bottle is the larger since 1 pint 6 ounces is less than 2 pints.

Warm-up

Change these measures.

1. 4,562 ft = ____ yd ____ ft
2. 18 c = ____ gal ____ oz
3. 3 T 340 lb = ____ lb
4. 3 dozen = ____ items
5. 2 qt 1 pt = ____ oz
6. 7 qt 1 pt = ____ gal ____ qt ____ pt
7. 7 pt 3 oz = ____ oz
8. 128 oz = ____ lb ____ oz
9. 50 yd 2 ft = ____ ft
10. 3 gal = ____ oz

Warm-up Answers
1. 1,520 yd 2 ft **2.** 1 gal 16 oz **3.** 6,340 lb **4.** 36 items
5. 80 oz **6.** 1 gal 3 qt 1 pt **7.** 115 oz **8.** 8 lb 0 oz
9. 152 ft **10.** 384 oz

On the Springboard

1. The floor plan for a house shows that the house is 35 feet 8 inches long. How many inches long is the house?

   (1) 428 (2) 420 (3) 412

   ① ② ③

2. A certain tomato sauce comes only in quart jars. How many jars are needed for a recipe that calls for 32 ounces of sauce?

   (1) 1 (2) 2 (3) 3

   ① ② ③

3. The total distance around the Smiths' property is 193 feet. They want to put a fence around their property. How many yards of fencing should they buy?

   (1) $16\frac{1}{12}$ (2) $32\frac{1}{6}$ (3) $64\frac{1}{3}$

   ① ② ③

Check your Springboard answers on page 192. If you got them all right, go on to the section on adding standard measurements. Adding measurements is a GED-level skill, so if you missed any of the Springboard questions, it's probably a good idea to review the material and try those questions again before going on.

Adding

When you add measurements, you often have to carry numbers in much the same way as when you add whole numbers.

Here's an Example

In this example, keep in mind that there are 12 inches in 1 foot and 24 inches in 2 feet. Any number of inches between 12 and 24 must be more than 1 foot but less than 2.

■ Two windows each measure 3 feet 9 inches wide. They are going to be put side by side in the same opening to form one large window. How wide should the opening be?

To solve, add 3 feet 9 inches and 3 feet 9 inches.

Step 1 Set up the addition vertically. Keep the feet in one column and the inches in another.

$$\begin{array}{r} 3\text{ ft } \ 9\text{ in} \\ +\ 3\text{ ft } \ 9\text{ in} \\ \hline 6\text{ ft } 18\text{ in} \end{array}$$

Step 2 See if the total of the smaller units make up more than 1 of the larger unit. If so, change to the larger unit and add to the total of the larger units.

$$\begin{aligned} 18\text{ in} &= 12\text{ in} + 6\text{ in} \\ &= 1\text{ ft } 6\text{ in} \end{aligned}$$

When you add 6 feet and 1 foot 6 inches, you get 7 feet 6 inches as your answer.

The opening should be 7 feet 6 inches wide. Whenever there are enough smaller units to make a larger unit, change to the larger unit.

Try It Yourself

Remember to change the smaller unit to the larger one in this problem.

■ Two packages weighed 3 pounds 5 ounces and 6 pounds 12 ounces. What was the total weight of the two packages?

Total means add. Did you add pounds to pounds and ounces to ounces? If you did, you got 9 pounds 17 ounces. Did you change 17 ounces to 1 pound 1 ounce? The total of the pounds was 9, and with the other 1 pound 1 ounce you get 10 pounds 1 ounce as the answer.

Warm-up

Do these addition problems on your own paper. Keep like units lined up.

1. 6 yd 2 ft + 4 yd 2 ft
2. 6 yd 2 ft 10 in + 6 yd 1 ft 6 in
3. 1 gal 3 qt + 3 gal 2 qt
4. 2 gal 3 qt 1 pt + 1 gal 3 qt 1 pt
5. 1 lb 15 oz + 1 lb 15 oz

Warm-up Answers
1. 11 yd 1 ft **2.** 13 yd 1 ft 4 in **3.** 5 gal 1 qt **4.** 4 gal 3 qt
5. 3 lb 14 oz

On the Springboard

4. Three recipes called for buttermilk in the following amounts: 2 cups, 1 pint 1 cup, and 1 quart 1 cup. How many quarts of buttermilk are needed for all three recipes?

(1) 2 (2) $2\frac{1}{2}$ (3) 3

5. A costume pattern called for $3\frac{1}{4}$ yards of one fabric and $2\frac{1}{2}$ yards of another. How much fabric was needed altogether?

(1) 5 yd 1 ft 0 in
(2) 5 yd 2 ft 3 in
(3) 6 yd 0 ft 0 in

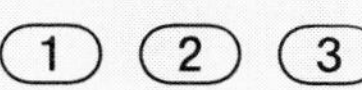

Check your Springboard answers on page 192. If you got both of these right, go on to the section on subtraction. If you got any wrong, check your work to see where you made a mistake before going on.

Subtracting

When you subtract measurements you will often have to borrow from a column with larger units in order to subtract the smaller units.

Here's an Example

To solve this problem you will have to borrow twice.

■ A piece of metal trim was 4 yards 2 feet 3 inches long. A strip 2 yards 2 feet 5 inches was cut off. How much trim was left?

To solve, subtract 2 yards 2 feet 5 inches from 4 yards 2 feet 3 inches.

Step 1 Line up the units in columns.

$$\begin{array}{r} 4\text{ yd } 2\text{ ft } 3\text{ in} \\ -\ 2\text{ yd } 2\text{ ft } 5\text{ in} \\ \hline \end{array}$$

Step 2 Begin subtracting with the smaller units. When you cannot subtract the smaller units, borrow from the next column on the left, as you did with whole numbers. Here you borrow 12 inches from the feet column and 3 feet from the yards column.

```
         4
  3      5        15
  4 yd  2 ft   3 in
– 2 yd  2 ft   5 in
  1 yd  2 ft  10 in
```

There was 1 yard 2 feet 10 inches of trim left.

The next example has a unit missing to show you how to deal with this kind of problem.

■ Two car owners bought antifreeze for the winter. By spring one had used 3 gallons and the other only 1 pint. How much more antifreeze did the first person use?

To solve, subtract 1 pint from 3 gallons.

Step 1 Set up the problem, putting in the missing measures. Here you need everything from gallons through pints in the top line.

```
  3 gal 0 qt 0 pt
–             1 pt
```

Step 2 Borrow first from the gallons column (1 gal = 4 qt).

```
  2      4
  3 gal  0 qt  0 pt
–              1 pt
```

Step 3 Next borrow from the quarts column. (1 qt = 2 pt). Then subtract.

```
         3
  2      4     2
  3 gal  0 qt  0 pt
–              1 pt
  2 gal  3 qt  1 pt
```

The answer is 2 gallons 3 quarts 1 pint. Now you see why you had to put the missing units in the top line before subtracting. The answer contained every unit, though the problem did not.

Try It Yourself

■ In one minute a cheetah ran 1,053 yards. In the same time a garden snail covered 2 feet 8 inches. What is the difference between the distances covered by these two creatures in one minute?

If you put 1,053 yards 0 feet 0 inches as the top line, you started right. Did you borrow a yard and change it to 3 feet? Did you borrow a foot and change it to 12 inches? When you subtracted, you should have gotten 1,052 yards 4 inches as your answer.

Warm-up

Do these subtractions.

1. 5 yd 1 ft 5 in – 2 yd 1 ft 7 in ____________

2. 3 gal 0 qt 1 pt – 1 qt ____________

3. 1 T 10 lb 10 oz – 10 lb 11 oz ____________

4. 2 mi 6 ft – 8 ft ____________

5. 3 qt 1 pt – 1 pt 1 c ____________

On the Springboard

6. The mile track was measured by laser and found to be 8 inches short of a mile. What was the track's true distance in feet and inches?

(1) 5,277 ft 8 in (2) 5,280 ft 4 in
(3) 5,279 ft 4 in

(1) (2) (3)

7. A truck weighed 5 tons after it was loaded. The load was 1 ton 1,000 pounds. What was the weight of the truck?

(1) 3,500 lb (2) 7,000 lb
(3) 4 tons 1,000 lb

(1) (2) (3)

Check your Springboard answers on page 192. If you got them right, go on to "The Real Thing." If you had a wrong answer, make sure that you know how to fix it before you go on.

Warm-up Answers
1. 2 yd 2 ft 10 in **2.** 2 gal 3 qt 1 pt **3.** 1,999 lb 15 oz
4. 1 mi 5,278 ft **5.** 2 qt 1 pt 1 c

The Real Thing

1. A plumber needs two pipes. The pipes are to be 4 feet 9 inches long and 6 feet 7 inches long. How much pipe does the plumber need in all?

   (1) 10 ft 2 in (2) 11 ft 4 in
   (3) 11 ft 6 in (4) 12 ft 2 in
   (5) 12 ft 6 in

   (1) (2) (3) (4) (5)

2. A truck weighs 2 tons 300 pounds when loaded. The load weighs 800 pounds. How much does the truck weigh when it is empty?

   (1) 3 tons 100 lb (2) 2 tons 1,100 lb
   (3) 1 ton 1,500 lb (4) 1 ton 500 lb
   (5) Insufficient data is given to solve the problem.

   (1) (2) (3) (4) (5)

3. A contractor needs to install 186 feet 6 inches of air-conditioning ductwork. One team of workers has installed 45 feet 9 inches and another team has installed 89 feet 10 inches. How much of the ductwork still has to be installed?

   (1) 50 ft (2) 50 ft 9 in (3) 50 ft 11 in
   (4) 60 ft 9 in (5) 60 ft 11 in

   (1) (2) (3) (4) (5)

4. Some material was on sale for $0.79 a yard. Mrs. Ratja bought 3 yards 18 inches of material from one bolt and 4 yards 24 inches from another. She used 2 yards 8 inches to make a skirt and the rest for a dress. How much material did she use for the dress?

   (1) 5 yd 6 in (2) 5 yd 34 in
   (3) 6 yd 8 in (4) 7 yd 4 in
   (5) 7 yd 10 in

   (1) (2)  (5)

Check your answers on page 213.

Multiplying

Multiplying works a lot like adding. You multiply each unit and change the smaller units into larger units if there are enough of them.

Here's an Example

■ A hole in the deck of an old ship was exactly as long as three planks, each 4 feet 9 inches long. What is the length of a single plank that can replace them?

To solve, multiply 4 feet 9 inches by 3.

Step 1 Write down the problem. If there are any missing units between the largest and the smallest, you should fill them in. In this particular case none are missing.

```
  4 ft 9 in
×       3
```

Step 2 Multiply as if you were dealing with ordinary numbers.

```
   4 ft  9 in
×         3
  12 ft 27 in
```

Step 3 Look to see if any column in the answer contains too many units for that column. If so, convert to the next greater unit. Here you have 27 inches in the inches column. Convert the 27 inches to feet and inches and add to the 12 feet already there. 27 inches is 2 feet 3 inches. Add the 12 feet to the 2 feet 3 inches.

The plank should be 14 feet 3 inches long.

In the next example, you have to convert smaller to larger units two times.

■ Three wooden crates are stacked one on top of another. If each crate is 1 yard 1 foot 9 inches high, how high is the stack?

Step 1 Write the multiplication problem and do the calculation.

```
  1 yd 1 ft  9 in
×             3
  3 yd 3 ft 27 in
```

Step 2 In the answer line, change the 27 inches to 2 feet 3 inches and add to the 3 feet already there. You get 3 yards 5 feet 3 inches.

Step 3 Change the 5 feet to 1 yard 2 feet and add to the 3 yards already there. The final answer is 4 yards 2 feet 3 inches.

The stack is 4 yards 2 feet 3 inches high.

All multiplication problems with measurements work basically the same way as in these examples.

Try It Yourself

For this problem you'll need to put in a missing unit.

■ A man bought 11 special offers at a grocery store that had a special sale on milk. Each was for a 1 gallon bottle and a 1 pint carton of milk. In all how much milk did he buy?

If you started with 1 gallon 0 quart 1 pint × 11, you were right on target. After multiplying, you get 11 gallons 0 quart 11 pints. The 11 pints should be changed to 5 quarts 1 pint. The 5 quarts should be changed to 1 gallon 1 quart. Your final answer should have been 12 gallons 1 quart 1 pint.

Warm-up

Do the following multiplication problems on your own paper.

1. 3 gal 3 qt 1 pt × 3
2. 5 yd 2 ft 10 in × 10
3. 1 qt 1 pt 6 oz × 3
4. 1 mi 1,060 ft × 5
5. 3 lb 4 oz × 6

On the Springboard

8. A baby drinks 6 ounces of milk 4 times a day. How much milk must be on hand to feed the baby for a week?

(1) 1 pt 1 c (2) 1 gal 1 qt 1 c
(3) 3 qt 1 pt 1 c

① ② ③

9. A carpenter ordered 10 feet 8 inches of timber. There was a computer error and 10 times the correct amount was delivered. How much timber was delivered?

(1) 33 yd 0 ft 8 in (2) 34 yd 1 ft 4 in
(3) 35 yd 1 ft 8 in

① ② ③

Warm-up Answers
1. 11 gal 2 qt 1 pt **2.** 59 yd 1 ft 4 in **3.** 5 qt 2 oz
4. 6 mi 20 ft **5.** 19 lb 8 oz

Check your Springboard answers on page 192. If you got both right, go on to "The Real Thing."

The Real Thing

5. Kathleen bought 5 pieces of molding. Each piece was 2 feet 9 inches long. How much molding did she buy?

(1) 11 ft 9 in (2) 12 ft 9 in
(3) 13 ft 6 in (4) 13 ft 9 in
(5) 14 ft 5 in

① ② ③ ④ ⑤

6. A cafe opens each day of the week at 7:00 A. M. They use 15 gallons 3 quarts of juice each morning. How much juice do they use in one week?

(1) 100 gal 1 qt (2) 107 gal
(3) 107 gal 1 qt (4) 109 gal 1 qt
(5) 110 gal 1 qt

① ② ③ ④ ⑤

7. A supplier received 60 dozen eggs from a farmer. When the eggs were graded, $\frac{1}{3}$ of them were graded medium and the rest were large. How many eggs were graded large?

(1) 360 (2) 420 (3) 480
(4) 580 (5) 960

① ② ③ ④ ⑤

8. Hamburger was on sale for $0.98 a pound. How much did a customer pay for 2 packages weighing 6 pounds 5 ounces and 5 pounds 11 ounces?

(1) $5.56 (2) $6.18
(3) $10.78 (4) $11.76
(5) Insufficient data is given to solve the problem.

① ② ③ ④ ⑤

Check your answers on page 213.

Dividing

Now that you have seen how to add, subtract, and multiply measurements, dividing should not be a problem.

Here's an Example

■ Over a three-day period a nursery school used 13 gallons 3 quarts 1 pint of milk. What was the average amount used per day?

To solve, divide 13 gallons 3 quarts 1 pint by 3.

Step 1 Set up the problem

```
3)13 gal 3 qt 1 pt
```

Step 2 Begin dividing as you would with numbers. After the subtraction step, change the units you have left as a remainder into the next smaller unit to the right. Add and then continue dividing. Continue this pattern until there are no more numbers left to divide.

```
  4 gal   2 qt    1 pt
3)13 gal  3 qt    1 pt
  12     +4      +2
   1    3)7     3)3
          6       3
          1       0
```

In the first column, you get a remainder of 1 gallon. Change this to 4 quarts and add to the 3 quarts to get 7 quarts. Divide 7 quarts by 3. You get 2. Put the 2 in the answer, above the line. Change the remainder of 1 quart to 2 pints and add to the 1 pint to get 3 pints. Then divide 3 pints by 3 and put the answer, 1, on top.

The average amount of milk used per day was 4 gallons 2 quarts 1 pint.

Sometimes you will get 0 as an answer when you divide. Study this example. Follow the steps carefully.

```
  0 yd   2 ft   11 in
4)3 yd   2 ft    8 in
  0     +9     +36
  3    4)11   4)44
          8     44
          3      0
```

The answer is 2 feet 11 inches.

Try It Yourself

Just follow the pattern.

■ Inez bought 7 yards 1 feet 6 inches of a fabric to make three identical dresses. How much fabric did she allow for each?

If you started out right, you had 2 yards in the answer and 1 yard left over. That 1 yard should be changed to 3 feet and added to the 1 foot already there. Dividing 4 feet by 3 gives 1 foot in the answer and 1 foot left over, which becomes 12 inches. That 12 inches plus the 6 inches in the original problem makes 18 inches. Dividing this by 3 gives you 6 inches. The answer is 2 yards 1 foot 6 inches.

Warm-up

Use your own paper when you perform these divisions.

1. 14 gal 2 qt 1 pt ÷ 3
2. 59 yd 2 ft 2 in ÷ 10
3. 5 qt 5 oz ÷ 3
4. 6 mi 20 ft ÷ 5
5. 25 gal 2 qt ÷ 12

On the Springboard

10. A piece of material 25 yards 6 inches long was divided into 6 equal lengths. How long was each?

    (1) 4 yd $2\frac{1}{2}$ in (2) 4 yd 7 in
    (3) 4 yd 1 ft 6 in

    (1) (2) (3)

11. Three gymnasts were pledged 25¢ a foot for walking on their hands at a charity function. They walked 40 yards, 35 yards, and 30 yards 2 feet. What was the length of their average walk?

    (1) 35 yd 8 in (2) 35 yd 9 in
    (3) 36 yd 2 in

    (1) (2) (3)

Check your Springboard answers on page 193. If you got them right, go on to "The Real Thing." If not, you may want to review this section and check your work before going on.

Warm-up Answers
1. 4 gal 3 qt 1 pt **2.** 5 yd 2 ft 11 in **3.** 1 qt 1 pt 7 oz
4. 1 mi 1,060 ft **5.** 2 gal 1 pt

The Real Thing

9. Mr. Burr purchased a plank that was 7 feet 8 inches long. If he cuts it into 4 pieces of equal length, how long will each piece be?

   (1) 1 foot 8 inches (2) 1 foot $9\frac{1}{2}$ inches
   (3) 1 foot 11 inches (4) 2 feet 3 inches
   (5) Insufficient data is given to solve the problem.

   (1) (2) (3) (4) (5)

10. A stage has a weight limit of 1 ton. What is the maximum number of pianos, each weighing 650 pounds, that the stage can safely support?

    (1) 1 (2) 3 (3) 4 (4) 7 (5) 30

    (1) (2) (3) (4) (5)

11. There are 20 pounds 4 ounces of food, 50 pounds of bedding, and 15 pounds 12 ounces of craft materials to be carried to a campsite. Each camper will carry approximately the same amount. How much will each camper carry?

    (1) 3 pounds $4\frac{4}{5}$ ounces
    (2) 4 pounds 2 ounces
    (3) 5 pounds 8 ounces
    (4) 6 pounds $9\frac{3}{8}$ ounces
    (5) Insufficient data is given to solve the problem.

    (1) (2) (3) (4) (5)

12. Christopher and Manuel bought 3 bags of fruit weighing 2 pounds 12 ounces, 4 pounds 10 ounces, and 3 pounds 8 ounces. What was the average weight of the 3 bags?

    (1) 3 pounds $4\frac{2}{3}$ ounces
    (2) 3 pounds 10 ounces
    (3) 3 pounds 12 ounces
    (4) 4 pounds
    (5) 10 pounds 14 ounces

    (1) (2) (3) (4) (5)

13. A family owned 500 feet of lakefront property. Then they bought 571 feet more of lakefront property. What is the total number of yards of lakefront property that they own?

    (1) $89\frac{1}{4}$ (2) 357 (3) $360\frac{1}{3}$
    (4) 1,071 (5) 3,213

    (1) (2) (3) (4) (5)

14. If 1 pint 2 ounces of applesauce cost \$0.67, what is the approximate cost per ounce?

    (1) \$0.037 (2) \$0.038 (3) \$0.042
    (4) \$0.067 (5) \$0.223

    (1) (2) (3) (4) (5)

15. A group picnic will need 7 tables to seat the 56 people who plan to attend. The picnic tables are going to be covered with strips of butcher's paper, each 3 yards 8 inches long. In all, how much butcher's paper is needed?

    (1) 20 yards 2 feet 10 inches
    (2) 22 yards 1 foot 8 inches
    (3) 24 yards 2 feet 10 inches
    (4) 25 yards 8 inches
    (5) 25 yards 10 inches

    (1) (2) (3) (4) (5)

16. A cook bought packages of ground meat that weighed 7 pounds 8 ounces, 8 pounds 6 ounces, and 9 pounds 2 ounces. The meat cost \$0.96 per pound. He used 5 pounds 8 ounces for meatballs, 6 pounds 8 ounces for meatloaf, and the rest for hamburgers. At lunch, he served the meatballs to 22 persons, the hamburgers to 56 persons, and the meatloaf to the rest. How much meat did the cook use for hamburgers?

    (1) 12 pounds (2) 12 pounds 10 ounces
    (3) 13 pounds (4) 13 pounds 10 ounces
    (5) Insufficient data is given to solve the problem.

    (1) (2) (3) (4) (5)

Check your answers and record all your scores on pages 213–214.

LESSON 30

Metric Measurement

Most people have used, or heard of 35-mm cameras. The Olympic Games have a great many events measured in meters. Most sewing tapes have centimeters and millimeters on one side and inches and feet on the other. Canned goods have two sets of measures on the labels and foreign cars need metric wrenches. Today, all scientific work uses the metric system. Three basic metric units are the *meter, gram,* and *liter.* Other units are based on these three. The following chart gives you some idea how metric units compare to standard units.

Measurement of	Basic Metric Unit	Comparison to Standard
Length:	meter	A little more than a yard (39.4 inches)
Weight:	gram	About the weight of a common paper clip
Liquid:	liter	A little more than a quart (1.057 quarts)

Converting Metric Units

The next chart shows other units, but they are based on the metric units just mentioned. Each unit on the chart is 10 times the one just to the right of it and 1⁄10 of the unit just to the left of it. The metric system uses special prefixes, or beginnings of words, to tell how each of the units is related to one of the three basic units. You can see what they are by studying the chart. *Kilo-* always means thousand, *centi-* means a hundredth, and so on, whether the units are of weight, length, or liquid measure.

If you want to multiply or divide a decimal by 10, 100, or 1,000, you simply move the decimal point to the right or left the same number of places as there are 0s in 10, 100, or 1,000. This makes the conversion of metric units very easy. The system was planned with this in mind.

Here's an Example

You can use the chart to show which way to move the decimal point and how many places.

■ A piece of material is 3 meters long. How many centimeters is that?

Step 1 Compare the units. Meters are larger than centimeters, so you are going from larger to smaller.

Step 2 Find the units on the chart. The space where you find *centimeter* is two spaces to the *right* of the space for *meter.* So move the decimal point 2 places to the *right.*

3.00.

You change from 3 to 300.

The piece of material is 300 centimeters long. If you had been asked to change 300 centimeters to meters, you would have gone 2 places to the *left.* You would have been changing from a smaller to a larger unit.

METRIC EQUIVALENTS

Prefix	kilo-	hecto-	deka-	Basic Unit	deci-	centi-	milli-
Comparison to Basic Units	1000 ×	100 ×	10 ×	1 ×	0.1 ×	0.01 ×	0.001 ×
Length	*kilo*meter km	*hecto*meter hm	*deka*meter dam	meter m	*deci*meter dm	*centi*meter cm	*milli*meter mm
Liquid	*kilo*liter kL	*hecto*liter hL	*deka*liter daL	liter L	*deci*liter dL	*centi*liter cL	*milli*liter mL
Weight	*kilo*gram kg	*hecto*gram hg	*deka*gram dag	gram g	*deci*gram dg	*centi*gram cg	*milli*gram mg

Try It Yourself

Use the table as before.

■ If a bag of flour weighs 11,000 grams, what is its weight in kilograms?______________

Your going from a smaller to a larger unit, so you should have moved the decimal point to the left, towards kilograms. On the chart there are three spaces from gram to kilogram.

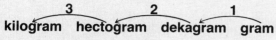

kilogram ←3— **hectogram** ←2— **dekagram** ←1— **gram**

This means that you move the point three places to the left. So 11,000 becomes 11.000, and the answer is 11 kilograms. If you remembered that *kilo-* means 1,000, you could have divided by 1,000 to get the result.

✓ A Test-Taking Tip

If you learn the prefixes in the chart of metric equivalents, you will have the basic facts that you need to do changes from one unit to another on the GED Mathematics Test.

Warm-up

Change these units of metric measurement.

1. 15 mg = _____ g

2. 60 L = _____ mL

3. 40 m = _____ cm

4. 0.3 g = _____ cg

5. 3,020 mm = _____ m

6. 8 mm = _____ m

7. 4 kg = _____ g

8. 2 cg = _____ g

Warm-up Answers
1. 0.015 g **2.** 60,000 mL **3.** 4,000 cm **4.** 30 cg
5. 3.02 m **6.** 0.008 m **7.** 4,000 g **8.** 0.02 g

On the Springboard

1. A package of honey brownies states that each brownie has 13 g of fat. How many milligrams of fat is this?

   (1) 0.013 (2) 130 (3) 13,000

   ① ② ③

2. A miniature horse stands only 1 meter high. How many centimeters high is this horse?

   (1) 0.01 (2) 100 (3) 1,000

   ① ② ③

3. A group of cyclists, after a long day on the road, drank 5 liters of water. How many kiloliters is this?

   (1) 0.005 (2) 0.05 (3) 0.5

   ① ② ③

Check your Springboard answers on page 193. If you got these right, go on to the section on using metric measurements. If you missed any, check your work. It will probably help to refer to the chart to find out where you went wrong.

Using Metric Measurements

Before adding, subtracting, multiplying, or dividing metric measurements, it is usually best to convert all the measurements to the same unit.

Here's an Example

Choosing the best unit to convert to saves time.

■ If 2 liters of soda are mixed with 2,300 milliliters of juice, how many liters of drink are there?

To solve, add 2 liters and 2,300 milliliters.

Step 1 Since the answer is to be in liters, change everything to liters. For 2,300 mL, the change is from smaller to greater, so the decimal point moves to the left. There are 3 spaces between the units on the table, so move the point 3 places to the left.

2.300.

You find that 2,300 mL equals 2.3 L.

Step 2 Add the liters: 2 L + 2.3 L = 4.3 L.

The answer is 4.3 liters of drink.

If you had changed everything to milliliters, you still could have gotten the right answer but with more steps. 2 L = 2,000 mL, and 2,000 mL + 2,300 mL = 4,300 mL, and 4,300 mL = 4.3 L.

Try It Yourself

Remember to change to the same unit before you calculate.

■ Mt. Whitney is 4,414 meters high, and Mt. McKinley is 6.189 kilometers high. By how many meters do their heights differ?

Since the required answer is in meters, you might as well change Mt. McKinley's height to meters as your first step. To do this, you should move the decimal point 3 places to the right. (Since *kilo-* means 1,000, you are multiplying by 1,000.) You get 6,189. Did you subtract 4,414 from 6,189 to get 1,775? The heights differ by 1,775 meters.

Warm-up

Compute the following. Change all amounts to the same units first.

1. 16.8 L + 230 cL = ______ L
2. 250 cm − 0.5 m = ______ cm
3. 0.24 kg − 6 g = ______ kg
4. 48 mL − 2 cL = ______ mL
5. 16,000 mg − 2.6 g = ______ mg
6. 86 L + 23 cL = ______ cL
7. 3,600 g − 0.6 kg = ______ kg
8. 890 mm + 66 cm = ______ cm

Warm-up Answers
1. 19.1 L **2.** 200 cm **3.** 0.234 kg **4.** 28 mL
5. 13,400 mg **6.** 8,623 cL **7.** 3 kg **8.** 155 cm

On the Springboard

4. A mother cat weighed 2.1 kilograms. Each of her 3 kittens weighed 600 grams. What was the total weight of mother and kittens in kilograms?

   (1) 1.8 (2) 2.7 (3) 3.9

   ① ② ③

5. A party began with 3 bottles of soda each containing 2 liters. Six people drank all the soda. What was the average amount in milliliters that each person drank?

   (1) 2,000 (2) 1,000 (3) 666

   ① ② ③

6. A competitor in a 100-kilometer bicycle race rode 76,600 meters before she began having trouble with her front wheel. How many more kilometers did she have to go to finish the race?

   (1) 2.34 (2) 23.4 (3) 66.6

   ① ② ③

7. A packet of artificial sweetener weighs 1 gram. How many packets would it take to make up a package of 1.3 kilograms?

   (1) 1,300 (2) 13,000 (3) 130,000

   ① ② ③

Check your Springboard answers on page 193. If you got them all right, go on to "The Real Thing." If you missed any, it might be a good idea to review this short section and try the problems again. When you're sure that you're ready, go on to "The Real Thing."

The Real Thing

1. A gasoline can holds 12,500 milliliters. How many liters does it hold?

   (1) 0.125 (2) 1.25 (3) 12.5
   (4) 125 (5) 1,250

   ① ② ③ ④ ⑤

2. If a pencil is 18.8 centimeters long, what is its length in millimeters?

   (1) 0.0188 (2) 0.188 (3) 1.88
   (4) 188 (5) 1,880

   ① ② ③ ④ ⑤

3. How many 250-milliliter glasses of juice can Dexter obtain from 1 liter of juice?

   (1) 2.5 (2) 4 (3) 25
   (4) 40 (5) 250

   ① ② ③ ④ ⑤

4. Len takes 2 grams of vitamin C each day. He has run out of vitamin C tablets, but remembers that he has a bottle of multivitamins. He checks the label and finds that each multivitamin tablet contains 200 mg of vitamin C. How many of these tablets would he have to take to get 2 grams of vitamin C?

   (1) 4 (2) 5 (3) 10 (4) 15 (5) 20

   ① ② ③ ④ ⑤

5. Mrs. Swenson bought 1 head of cabbage that weighed 1.5 kilograms, 1 rutabaga weighing 1.45 kilograms, and 3 heads of cabbage that weighed 725 grams each. Altogether, how many grams of cabbage did she buy?

   (1) 3,665 (2) 3,675
   (3) 5,115 (4) 5,125
   (5) Insufficient data is given to solve the problem.

   ① ② ③ ④ ⑤

6. Koyi gained 143 grams the first week after he was born. He gained 115 grams the second week and lost 169 grams the third week. Then he weighed 4.075 kilograms and was 41 centimeters long. How many kilograms did Koyi weigh when he was born?

   (1) 4.272 (2) 4.164 (3) 3.986
   (4) 3.878 (5) 3.996

   ① ② ③ ④ ⑤

7. It takes 8 sheets of typing paper to make a stack 1 mm thick. How many sheets are there in a stack 10 cm thick?

   (1) 80 (2) 125 (3) 250
   (4) 300 (5) 800

   ① ② ③ ④ ⑤

8. A recipe that makes 24 biscuits calls for 14 grams of baking powder. A can contains 2,268 grams of baking powder. This amount of baking powder is enough for how many dozen biscuits?

   (1) 32.4 (2) 81 (3) 162
   (4) 189 (5) 324

   ① ② ③ ④ ⑤

9. A nickel weighs about 5 grams. What is the weight in kilograms of the nickels in a bag that contains $150 worth of nickels?

   (1) 10 (2) 15 (3) 20
   (4) 25 (5) 30

   ① ② ③ ④ ⑤

10. The birth weights of quadruplets were 1.2, 1.5, 2.0, and 1.7 kilograms. What was their average weight in grams?

    (1) 1.6 (2) 6.4 (3) 16
    (4) 1,600 (5) 6,400

    ① ② ③ ④ ⑤

Check your answers and record your score on page 214.

LESSON 31

Dealing with Units of Time

If you use a TV guide, look at a bus schedule, or make an appointment, you are dealing with time measures. Here are the standard measures of time.

1 week (wk) = 7 days (da)
1 day (da) = 24 hours (hr)
1 hour (hr) = 60 minutes (min)
1 minute (min) = 60 seconds (sec)

Converting from One Unit to Another

The same principle holds as for all the other standard measures; you multiply to change larger units to smaller, and you divide to change smaller units to larger.

Here's an Example

Use the table if you need to do so.

■ Stephen spent a total of 3 hours 25 minutes communicating by telephone with the large computer at a company with which he does business. He is charged by the minute. How many minutes of telephone time did he use?

To solve, change 3 hours to minutes and add 25.

Step 1 Multiply 3 hours by 60. Since 3 × 60 = 180, there are 180 minutes in 3 hours.

Step 2 Add the extra 25 minutes.

180 min + 25 min = 205 min

Stephen used 205 minutes of telephone time.

Try It Yourself

This isn't in the table, but it's easy to use the table to get the answer.

■ How many hours are there in a full week?

____

In a full week, there are 7 days. So you should have multiplied 7 by 24 (the number of hours in a day). You get 7 × 24 = 168. There are 168 hours in one week.

Warm-up

Change these time measures.

1. 4 hr 16 min = ______ min

2. 1 hr 1 min = ______ sec

3. 267 hr = ______ da ______ hr

4. 365 da = ______ wk ______ da

5. 1 wk 1 da 1 hr = ______ hr

On the Springboard

1. There is a famous book entitled *Around the World in Eighty Days.* How many hours are there in 80 days?

   (1) 560 (2) 960 (3) 1,920

   ① ② ③

2. Susan said she would split the cost of a long-distance call with her friend. They talked for 1 hour 48 minutes. For how many minutes did she have to pay?

   (1) 53 (2) 54 (3) 56

   ① ② ③

Check your Springboard answers on page 193. If you got both right, go on to the next section. If not, review the section before you go on.

Warm-up Answers
1. 256 min. **2.** 3,660 sec. **3.** 11 da. 3 hr. **4.** 52 wk. 1 da. **5.** 193 hr.

Operations on Time Measurements

The same rules apply to time measurements as to the other measurements. When you do calculations, keep like units lined up in columns and put in a zero for missing units. When you have to carry or borrow, use the table of time units as necessary to change from one unit to another.

Here's an Example

■ Production on a factory line varied one day. The first shift took 3 hours 15 minutes to assemble the complete product. The second shift took 1 hour 55 minutes. Find the difference in the production time.

To solve, subtract 1 hour 55 minutes from 3 hour 15 minutes.

Step 1 Line up like units in columns. Put the greater amount of time on top.

```
   3 hr 15 min
 — 1 hr 55 min
```

Step 2 Begin subtracting with the smaller units. It is not possible to subtract 55 from 15, so you have to borrow. This is where you must know how to change units (1 hour = 60 minutes)

```
   2   (60 + 15)           2   75
   3̶ hr 1̶5̶ min.           3̶ hr 1̶5̶ min
 — 1 hr 55 min          — 1 hr 55 min
                           1 hr 20 min
```

The first shift took 1 hour 20 minutes longer.

If you had been adding the times instead of subtracting them, the first total would have been 70 minutes. You would have changed this to 1 hour and 10 minutes. The 10 minutes would have gone in the minutes column and the 1 hour would have been carried to make 5 hours.

Now let's see a division example.

■ A couple took three trips. The total length of the trips was 7 weeks and 2 days. What was the average length of their trips?

To solve, divide 7 weeks 2 days by 3.

Write the problem as a division problem and divide as usual, changing the left-over units and adding them to the next column before dividing again. Study the calculation to be sure you understand how to work the problem.

```
     2 wk      3 da
  3)7 wk       2 da
    6        → +7
    1          3)9
                 9
                 0
```

The average length of the trips was 2 weeks and 3 days.

You can check by multiplying the answer by 3 and seeing if we get 7 weeks 2 days.

```
   2 wk  3 da
 ×       3
   6 wk  9 da
```

Since 9 days = 7 days + 2 days or 1 week + 2 days, the multiplication answer of 6 weeks 9 days is equal to 7 weeks 2 days. Everything checks.

Try It Yourself

Here you work backwards from an average.

■ Four carpenters worked an average of 2 hours 40 minutes one morning. What was the total number of hours worked that morning? ______________________

You have to multiply the average number of hours by 4 to find the total number of hours worked. Did you multiply 2 hours 40 minutes by 4 and get 8 hours 160 minutes? Did you change 160 minutes to 2 hours 40 minutes? If you did all this, you should have gotten 10 hours 40 minutes as the answer.

Warm-up

Calculate the answers for these problems.

1. 7 hr 30 min − 45 min = ______________

2. 2 da 17 hr 30 min − 1 da 20 hr 50 min =

____

3. 4 hr 20 min + 3 hr 55 min = ___________

4. 3 wk 2 da 5 hr × 5 = ______________

5. 13 wk 1 da ÷ 4 = ________________

Warm-up Answers
1. 6 hr 45 min **2.** 20 hr 40 min **3.** 8 hr 15 min
4. 16 wk 4 da 1 hr **5.** 3 wk 2 da

On the Springboard

3. Alan drives 15 minutes to get to the train each morning when he goes to work. He takes a train trip to town that takes 55 minutes. Going home takes the same amount of time. In a normal five-day work week, how much time in hours and minutes does he spend commuting?

   (1) 5 hr 50 min (2) 11 hr 40 min
   (3) 12 hr 50 min

   ① ② ③

4. Two cyclists competed in a distance race. One took 3 hours and 57 minutes to complete the distance and the other took 231 minutes. By how many minutes did the winner win?

   (1) 6 (2) 37 (3) 108

   ① ② ③

Check your Springboard answers on page 193. If you missed either, check and correct your work before going on to "The Real Thing."

The Real Thing

1. Matthew programmed his microwave oven to cook a turkey. The turkey will cook at high power for 34 minutes and then at medium power for 1 hour 38 minutes. How long will it take the turkey to cook?

   (1) 1 hr 4 min (2) 1 hr 12 min
   (3) 2 hr 12 min (4) 3 hr 24 min
   (5) Insufficient data is given to solve the problem.

   ① ② ③ ④ ⑤

2. An airline flight is scheduled for departure at 2:47 P.M. If its takeoff is 1 hour 25 minutes late, what will be the new departure time?

   (1) 1:22 P.M. (2) 2:22 P.M.
   (3) 3:12 P.M. (4) 4:02 P.M.
   (5) 4:12 P.M.

   ① ② ③ ④ ⑤

3. On a certain compact disc, three songs have playing times of 4 minutes 56 seconds, 2 minutes 30 seconds, and 10 minutes 16 seconds. What is the average playing time of these three songs?

   (1) 5 min 8 sec (2) 5 min 54 sec
   (3) 6 min (4) 6 min 7 sec
   (5) 17 min 42 sec

   ① ② ③ ④ ⑤

4. Mary Lou does part-time work at home. On Monday, she worked 1 hour 40 minutes. On Tuesday she worked three times as long as she did on Monday. On Thursday she worked half as long as she did on Monday. How long did Mary Lou work in all on these three days?

   (1) 4 hr 30 min (2) 5 hr
   (3) 5 hr 30 min (4) 7 hr 30 min
   (5) 10 hr

   ① ② ③ ④ ⑤

Check your answers on pages 214–215.

Motion Problems

If you drove 50 miles per hour for 2 hours, you would go 100 miles. To figure the distance, you multiply your rate of motion (speed) by the time you were driving.

distance = rate × time

You write this more briefly by abbreviating with the first letters of the words *distance, rate,* and *time:* d = distance, r = rate, and t = time.

$$d = r \times t$$

This is a **formula,** a short way of writing a rule that is always true in mathematics. This particular formula is called the *distance formula.* If you are told the values of any two of the letters in the formula, you can find the value of the other letter.

Coming to Terms

formula a short way of using letters instead of words to write a mathematical rule

A Test-Taking Tip

Some formulas are widely used and you will find these on the inside front cover. The GED Mathematics Test includes a page of formulas to which you can refer during the test. You don't have to memorize the formulas.

Here's an Example

You can solve this problem using the distance formula. Just fill in the numbers and compute.

■ The first around-the-world, nonstop jet flight averaged 525 miles per hour and took about 45 hours. What was the approximate distance traveled?

Step 1 Write the formula.

$$d = r \times t$$

Step 2 Replace r with the rate of speed, 525, and t with the time traveled, 45 hours.

$$d = 525 \times 45$$

Step 3 Do the multiplication. The result is $d = 23{,}625$. The approximate distance traveled was 23,625 miles. Formulas are usually written without the multiplication sign, so read $d = rt$ as if it were $d = r \times t$.

Try It Yourself

Use $d = rt$ for this problem.

■ Johnny figured that, if he drove nonstop for 18 hours at 50 miles per hour, he would get to his vacation hotel, which was 945 miles away. Actually he would be a little short of that distance. How many miles short would he be?

If you used $d = rt$ to see how far he would go in 18 hours, you should have gotten $d = 50 \times 18 = 900$ miles.

He needed to go 945 miles, so he was 45 miles short. If you know the distance *(d)* that someone traveled and their rate of speed *(r)*, how would you find the travel time *(t)*? You would divide distance *(d)* by rate *(r)* to get time *(t)*.

$$t = d \div r \quad \text{or} \quad t = d/r$$

Is this really right? Sure. For example, if you drove 150 miles at 50 miles per hour, your trip took $150 \div 50$, or 3 hours.

What if you know distance *(d)* and time traveled *(t)*? You could divide to get the rate.

$$r = d \div t \quad \text{or} \quad r = d/t$$

For example, if you drove at a steady rate of speed and went 110 miles in 2 hours, you were traveling at $110 \div 2$, or 55 miles per hour.

When you are given a problem involving *d, r,* and *t,* knowing any two of the numbers lets you find the third. Just choose the formula you need.

$$d = rt \qquad t = d/r \qquad r = d/t$$

Warm-up

Use $d = rt$, $t = d/r$, or $r = d/t$ to find the distance, rate or time asked for in these problems.

1. $r = 20$ mph, $t = 3$ hours,
$d =$ ____________ miles

2. $d = 450$ miles, $r = 50$ mph,
$t =$ ____________ hours

3. $d = 600$ miles, $t = 12$ hours,
$r =$ ____________ mph

On the Springboard

5. A family on vacation averaged 300 miles a day at 50 miles per hour. On the average, how many hours a day did they travel?

   (1) 6 (2) 8 (3) 10

   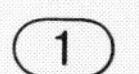 

6. Two travelers were comparing walking tours. The first walked for 4 days, for 6 hours a day. He averaged 5½ miles per hour. The other was in rocky country for 6 days and managed 3 miles an hour for 5 hours a day. How many miles more did the first person walk than the second?

   (1) 30 (2) 42 (3) 56

   (1) (2) (3)

Warm-up Answers
1. 60 **2.** 9 **3.** 50

Check your Springboard answers on pages 193–194. If you missed either one, it may be wise to review this section. When you're satisfied you can solve both correctly, you're ready to go on.

Interest Problems

You can use your knowledge of how to work with a formula to solve interest problems. When you borrow money you are charged a percent of it. The percentage is called the *percent rate* (r). Remember that a percent can be written as a fraction or decimal; for example, 9% is 9/100 or 0.09. The amount you borrow is called the *principal* (p) and the number of years you borrow it for is called the *time* (t). The *interest* is the amount you pay to borrow the money. If you are lending money, the interest is the amount you earn by lending it out. Remember what the letters stand for:

i = amount of interest
p = principal (amount borrowed or loaned)
r = percent rate
t = time in years

The formula that fits all these together is

$$i = p \times r \times t \quad \text{or} \quad i = prt$$

Here's an Example

You can use $i = prt$ to solve this problem.

■ A man got a personal loan of $2,000 for 3 years at 9% interest. What was the total interest charged?

Step 1 Write down the formula. $i = prt$

Step 2 Identify the values of the letters and replace the letters with the numbers. Here p (principal) is $2,000, t (time) is 3 years, and r (percent rate) is 9%, or 0.09. Write the percent as a decimal so that you can multiply. Replacing letters with numbers, you get:

$$i = 2{,}000 \times 3 \times 0.09 = 540$$

The interest is $540.

The formula $i = prt$ contains 4 letters. If you know the values of any three, you can find the value of the one that remains. Simply by moving pairs of letters you can find formulas for p, r, and t.

$$t = i/pr \qquad r = i/pt \qquad p = i/rt$$

Don't forget that the / is a division sign and that when the letters are next to each other, you are to multiply.

In this problem you're finding principal, but it's just as easy to find any of the other values in the formula.

■ The interest on a loan at 8% for 3 years was $48. What was the principal?

Step 1 Write down the formula for what you are looking for. Here it is, the principal

$$p = i/rt$$

Step 2 Identify the values of the other letters.

$$r = 0.08, \quad t = 3, \quad i = 48$$

Step 3 Replacing letters in the formula with numbers, you get $p = 48/(0.08 \times 3)$.

Do the multiplication first, then the division. You get $(48/(0.08 \times 3) = 48/0.24 = 200$. The principal was $200.

Try It Yourself

Here you will have to choose the formula to use.

■ The bank loaned Peggy a certain amount of money for 3 years at a rate of 8%. The total interest she paid was $240. How much did the bank loan her? ______

What the bank loaned her was the principal, so you needed to choose $p = i/rt$. Putting in the numbers and writing 8% as 0.08, you should have come to $240 \div (0.08 \times 3)$, which equals $240 \div 0.24$ or 1,000. The principal was $1,000.

Warm-up

Use the formulas $i = prt$, $p = i/rt$, $r = i/pt$, and $t = i/pr$ to solve these problems. Remember to write the percent (rate %) as a decimal or fraction to do the calculation.

1. principal = $400, time = 4 years, percent rate = 9%; interest = ______

2. principal = $500, time = 3 years; interest = $75; rate = ______

3. principal = $150, percent rate = 8%; interest = $24; time = ______

4. percent rate = 7%, interest = $42, time = 3 years; principal = ______

Warm-up Answers
1. i = $144 **2.** r = 0.05, or 5% **3.** t = 2 years
4. p = $200

On the Springboard

7. John wanted to do some work on his house. One bank offered a loan of $2,000 for 3 years at 8% interest. Another offered $3,000 for 2 years at 9% interest. What was the difference in dollars in the interest on the two loans?

   (1) $50 (2) $60 (3) $70

   ① ② ③

8. Three friends each chipped in $10,000 as a loan to an inventor who promised them a rate of 12% interest until she marketed her invention. How much had she paid in interest to the three friends after 2 years?

   (1) $6,800 (2) $7,100 (3) $7,200

   ① ② ③

Check your Springboard answers on page 194. If you got these right, go on to "The Real Thing." If you missed either, check your work carefully to see where you made your mistake. Did you use the right formula? Was your percentage written as a decimal? When you're satisfied that you're ready, go on to "The Real Thing."

"The Real Thing"

5. Mr. Hayes borrowed $750 for 2 years. The amount of interest for this loan was $93. What was the rate of interest?

   (1) 6.2% (2) 12.4% (3) 16.1%
   (4) 24.8% (5) 62.0%

   ① ② ③ ④ ⑤

6. The McCarthys opened a savings account by depositing $500. After 3 months what was the amount of interest at 5.4% per year?

   (1) $1.25 (2) $6.75 (3) $9.00
   (4) $16.20 (5) $27.00

   ① ② ③ ④ ⑤

7. Some money was invested at 12% for 5 years. If the interest amounted to $840, how much was invested?

(1) $7,000 (2) $4,200 (3) $2,016
(4) $1,400 (5) $1,250

① ② ③ ④ ⑤

8. During a cross-country ski trip, the Boyles averaged $3\frac{3}{4}$ miles per hour. They traveled $7\frac{1}{4}$ miles in the first two hours. How many miles did they travel in $4\frac{2}{3}$ hours?

(1) $\frac{45}{56}$ (2) $1\frac{11}{45}$ (3) $17\frac{1}{2}$
(4) $18\frac{1}{2}$ (5) 35

① ② ③ ④ ⑤

9. A car is traveling at a rate of 90 kilometers per hour. At this rate, how many hours will it take the car to travel 690 kilometers?

(1) $6\frac{1}{3}$ (2) $7\frac{2}{3}$ (3) $7\frac{3}{4}$ (4) $8\frac{1}{4}$
(5) Insufficient data is given to solve the problem.

① ② ③ ④ ⑤

10. Nan borrowed $7,500 for 18 months at an interest rate of 9.9% per year. How much will she have to repay at the end of 18 months?

(1) $1,113.75 (2) $6,386.25
(3) $8,242.50 (4) $8,613.75
(5) $8,625.00

① ② ③ ④ ⑤

11. An employee has to be at work at 8:00 A.M. It takes him 28 minutes to drive to work. Assuming that he wants to be there 15 minutes early, at what time must he leave his home?

(1) 7:17 A.M. (2) 7:27 A.M.
(3) 7:32 A.M. (4) 7:45 A.M.
(5) 7:57 A.M.

① ② ③ ④ ⑤

Check your answers and record all your scores on page 215.

LESSON 32

Linear, Square, and Cubic Measure

Finding the Distance Around

You wouldn't go to the hardware store to buy a new screen door without knowing how big it had to be. Drapery rods must be the correct length or the curtains that go on them won't hang right. You must know how far it is around a garden before you go to buy fencing for it. All these examples involve measuring how long, how wide, or how far around it is. This kind of measurement uses linear measures such as feet, inches, meters, kilometers, and so on.

Here's an Example

When a shape is made up of straight lines, it's usually very easy to find the distance around it.

■ Marion needs to find the distance around the stained glass window in her kitchen because she wants to trim it with lights for a party. The measurements of a window are shown in the diagram.

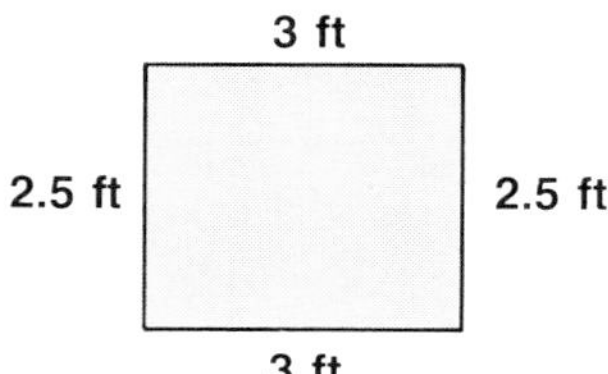

How long will the string of lights be?

To solve, add the lengths of all the sides. You can start anywhere and work your way around. If you start at the left side and work your way around, you get (in feet)

$$2.5 + 3 + 2.5 + 3$$

You can do the addition in your head or on a piece of scratch paper. You get 11. The string of lights will be 11 feet long.

It often helps to make a sketch of the problem and label every side with its length.

■ A sheet of standard, letter-size typing paper is 8½ inches along the short edge and 11 inches along the long edge. What is the distance around the sheet in inches?

A regular sheet of paper has its opposite sides of equal length, just like the pages of a book.

Step 1 Make a sketch. Label all the lengths you know.

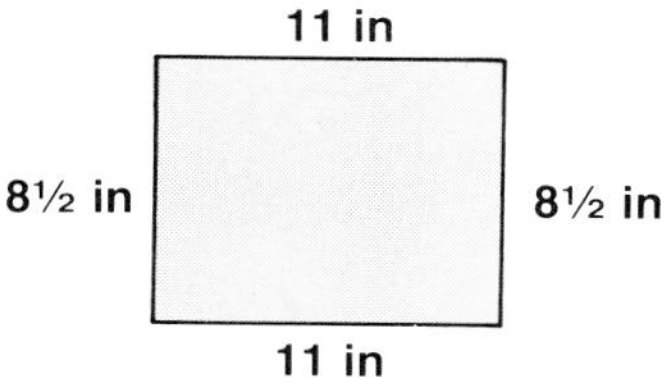

Step 2 Now you have lengths for all the sides, so add the lengths to find the distance around.

$$8\tfrac{1}{2} + 8\tfrac{1}{2} + 11 + 11 = 39$$

In the case of shapes like a sheet of typing paper you know that the short side plus the long side is half the distance you need. So you could add 8½ and 11 and then multiply by 2:

$$(8\tfrac{1}{2} + 11) \times 2 = 19\tfrac{1}{2} \times 2 = 39$$

Again you get 39 inches. The distance around the sheet is 39 inches.

Try It Yourself

■ The rectangular swimming pool in the diagram was given a pool edge path 3 feet wide all around it. How far was it around the outside of the pool edge? (Hint: Sketch what the pool edge looks like and then label its dimensions.)

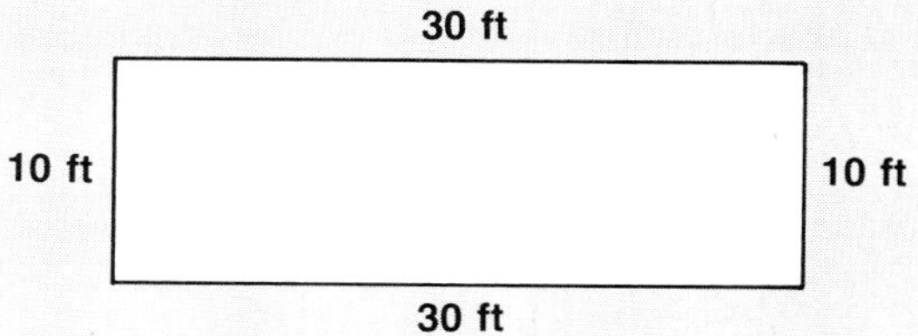

If you sketched the pool edge and labeled its dimensions you should have gotten a diagram that looks like this:

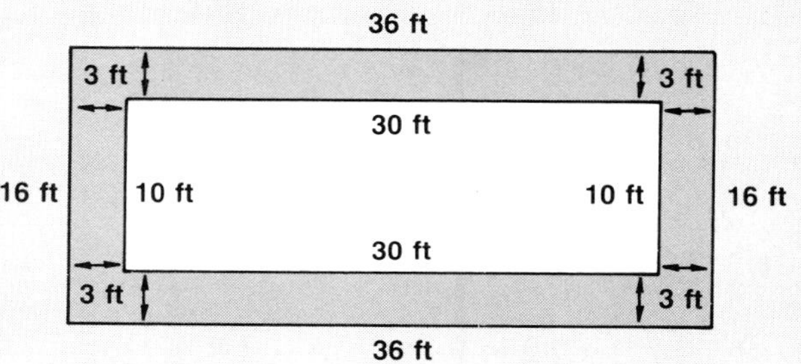

The pool edge adds 3 feet to each short side so it increases the width by 6 feet. The length is also increased by 6 feet. Did you say that the distance around the outer part of the pool edge is 16 + 36 + 16 + 36? If you did, and if you added correctly, you got 104 feet.

Now you see how helpful a sketch can be. Without it many people would have missed the fact that the increase in each dimension was 6 feet and not just 3 feet. When in doubt, make a sketch.

Warm-up

How far around is it for each of these figures?

1. How many yards of fencing would it take to enclose this triangular yard? The dimensions are in feet. ____________

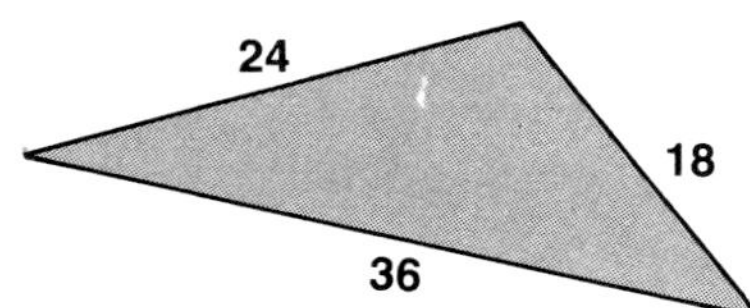

2. The figure in the diagram is made of 6 identical triangular tiles. The tiles have three equal sides, each 1 foot long. How far around is the figure in feet? ____________

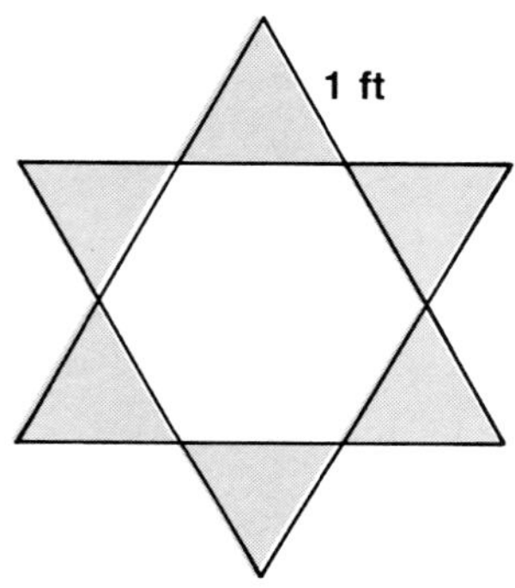

3. If the tiles in number 2 were all turned inward to fill up the space inside and make a solid 6 sided figure, how many yards around would the figure be? ____________

Warm-up Answers
1. 26 yards **2.** 12 feet **3.** 2 yards

On the Springboard

1. If a sheet of typing paper 11 inches long by $8\frac{1}{2}$ inches wide is folded short side onto the opposite short side, how far is it around the folded sheet in inches?

   (1) 24 (2) 26 (3) 28

   (1) (2) (3)

2. If a sheet of typing paper 11 inches long and $8\frac{1}{2}$ inches wide is folded long edge onto the opposite long edge, what is the distance around the folded sheet in inches?

   (1) $28\frac{1}{2}$ (2) $30\frac{1}{2}$ (3) $32\frac{1}{2}$

   (1) (2) (3)

3. A thin wire for a burglar alarm system is between two thick sheets of glass in a jewelry store window. The wire is $2\frac{1}{2}$ inches in from each edge of the window. How long is the wire in inches?

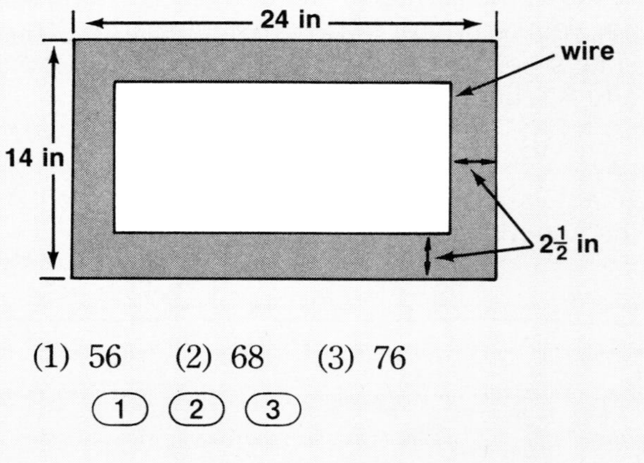

   (1) 56 (2) 68 (3) 76

   (1) (2) (3)

Check your Springboard answers on page 194. If you got these right, you did very well and can go to "The Real Thing." If you missed any, check your sketches and calculations. When you have fixed your work, go on.

The Real Thing

1. The Parthenon is a building with a 4-sided base and square corners. It is approximately 69.5 meters long and 30.9 meters wide. What is the distance in meters around the Parthenon?

   (1) 99.4 (2) 100.4 (3) 198.4
   (4) 200.4 (5) 200.8

   ① ② ③ ④ ⑤

2. A roof has 8 sides. Each side is 4½ feet long. What is the distance around the roof in feet?

   (1) 72 (2) 40 (3) 38
   (4) 36 (5) 32

   ① ② ③ ④ ⑤

3. The base of a pyramid is shaped like a triangle. Each side is the same length. The distance around the pyramid is 417 feet. How many feet long is each side?

   (1) $105\frac{2}{3}$ (2) 138 (3) 139
   (4) 149 (5) 1,251

   ① ② ③ ④ ⑤

4. A rectangular garden has square corners and is 10½ feet wide by 12¾ feet long. Mr. Halloran has 42 feet of fencing. How many more feet of fencing does he need to enclose the garden completely?

   (1) $2\frac{1}{4}$ (2) $4\frac{1}{2}$ (3) $18\frac{3}{4}$ (4) $46\frac{1}{2}$
   (5) Insufficient data is given to solve the problem.

   ① ② ③ ④ ⑤

Check your answers on page 216.

Finding Square Measure

How big will a rug need to be to cover a floor? How much of the wall will one gallon of paint cover? Both of these questions are about how big a surface is. A rug covers the surface of the floor and paint covers the surface of the wall. The amount of surface is called the **area.** The linear units of measure of the last section do not answer how big a surface is, so you have to use other measurement units.

You often see floors covered with tiles and looking like a checkerboard. If the tiles are four-sided and have square corners, you can use them as units to say how much surface is covered. If each side of a tile is 1 foot long, the tile covers an area of 1 square foot. Similarly, a small ceramic bathroom tile with the same shape but having sides 1 inch in length would cover an area of 1 square inch.

Coming to Terms

area the amount of surface that an object has. *Area* is measured in square units.

Here's an Example

For this problem, imagine a tile that's 1 foot on each side. So your unit of area is 1 square foot (1 sq ft). We'll use the floor tile for this one.

■ A room is 10 feet wide by 30 feet long. What is its area in square feet?

Step 1 Find how many 1-square-foot tiles would fill a row along the long side. Each tile takes up 1 foot, so you could fit 30 tiles in the row.

Step 2 Find how many of these long rows of tiles it would take to completely cover the floor. Each tile takes up 1 foot, so you could get 10 rows.

Step 3 Compute how many tiles in all. There are 10 rows with 30 tiles in each row. So in all there are $10 \times 30 = 300$ tiles.

The area of the floor is 300 square feet.

From this example you can see that to find the area of a shape such as the floor, you need only multiply the length by the width. Naturally both length and width need to be in the same units (both in inches, both in feet, and so on). You can write this rule as a simple formula as follows:

Area = length × width or $A = \ell w$

The formula $A = \ell w$ works even if the measurements contain fractions or decimals. But remember—you can use the formula only if both measurements are in the same units. To use the formula to find the area of a floor 30 feet long and 4 yards wide, you couldn't just multiply 30 by 4. You would have to change to one of these:

In feet: 30 × 12 = 360 square feet
In yards: 10 × 4 = 40 square yards

40 square yards is equal to 360 square feet.

Try It Yourself

Finding an area is easy and has lots of practical uses.

■ A metal safety wall 10 feet high and 40 feet long is to be painted with a special paint that covers 200 square feet per gallon. How many gallons will it take to cover the wall?

____

You should have found the area of the wall to be 10 × 40 = 400 square feet. Every 200 square feet needs a gallon of paint. 400 ÷ 200 = 2. The answer is 2 gallons of paint.

Warm-up

All questions are about a shape like a page or an ordinary floor. Find the area, A, in each case. Remember that ℓ = length and w = width. Make sure that your answer is in the units asked for.

1. ℓ = 20 ft, w = 10 ft A = ______ sq ft
2. ℓ = 21 ft, w = 6 yd A = ______ sq yd
3. ℓ = 20½ in, w = 10 in A = ______ sq in
4. ℓ = 30 yd, w = 3 ft A = ______ sq yd
5. ℓ = 11 in, w = 8½ in A = ______ sq in

On the Springboard

4. The outer edge of a picture frame was 24 inches long and 16 inches wide. The frame itself was 2 inches wide. What area of the picture, in square inches, could be seen?

(1) 225 (2) 240 (3) 260

① ② ③

5. A man had two rooms, each 10 feet by 20 feet. He figured that if he told the carpet layers that he had one room 20 feet by 40 feet (doubling each length) he would get enough carpet for both rooms. How many more square feet did he get than he needed?

(1) 100 (2) 300 (3) 400

① ② ③

6. A builder phoned in an order to a lumber yard and told them how many square yards of wall panelling he needed. A few minutes later he called back and told them he meant square feet, not square yards. How many times too large would the shipment have been if he hadn't noticed his mistake in time?

(1) 4 (2) 9 (3) 16

① ② ③

Check your Springboard answers on page 194. If you got them right, go on to "The Real Thing." If you missed any, make sure that you correct your work. When you are satisfied with your understanding, go ahead to "The Real Thing."

Warm-up Answers
1. 200 sq ft **2.** 42 sq yd **3.** 205 sq in **4.** 30 sq yd
5. 93½ sq in

The Real Thing

5. A kitchen floor (four sides, square corners) is 18 feet long and 12 feet wide. How many square feet of floor tile are needed for this kitchen?

   (1) 30 (2) 58 (3) 60
   (4) 108 (5) 216

   (1) (2) (3) (4) (5)

6. The bottom of a baking pan has an area of 117 square inches. If the pan is 9 inches wide, how many inches long is it?

   (1) 13 (2) 24 (3) $49\frac{1}{2}$
   (4) 99 (5) 1,053

   (1) (2) (3) (4) (5)

7. A box-shaped cage for white mice is 20 inches long and 9.5 inches wide. A wall goes between two opposite corners to separate a mother and her babies from the other mice. The floor sections on either side of the wall are the same size. The floor area of one section is how many square inches?

   (1) 29.5 (2) 59.0 (3) 85.0
   (4) 95.0 (5) 190.0

   (1) (2) (3) (4) (5)

Check your answers on page 216.

Finding Cubic Measure

How many cartons of canned peaches will fit into a warehouse space? How many bricks will it take to build a wall of a certain size? Questions like this are about the **volume** of an object—the amount of space taken up by an object.

Linear units and units of area aren't enough to answer questions about volume. You need another kind of unit, one that takes up space, a three-dimensional, or 3-D, unit.

You have seen the small blocks in children's building kits. They look like this.

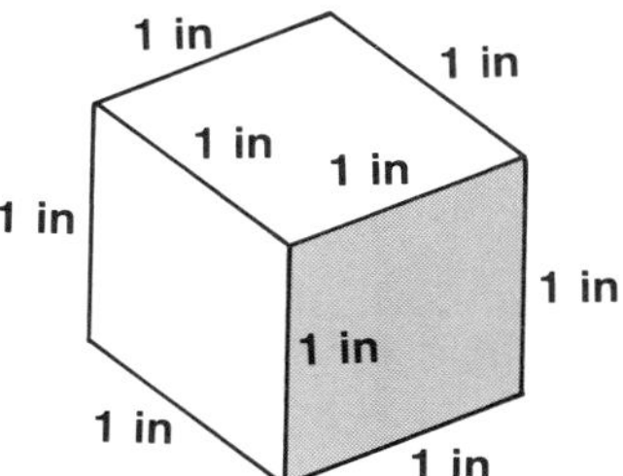

Each edge of the block is 1 inch long. Each side has an area of 1 square inch. The *volume,* or amount of space it takes up is 1 cubic inch. The number of such blocks that will exactly fill a particular space is a measure of the volume of that space.

Here's an Example

Imagine you are systematically filling a box with blocks that each have a volume of 1 cubic inch.

■ The inside of a box is 10 inches long, 5 inches wide, and 6 inches deep. What is the volume of the box in cubic inches?

The volume in cubic inches is the number of 1-cubic-inch blocks that will exactly fill the box.

Step 1 Compute how many blocks would fit into a layer that covers the bottom of the box. The layer can be treated like 1-square-inch tiles. There are 10 rows of 5 blocks each, which gives $10 \times 5 = 50$ blocks.

Step 2 Multiply the number of blocks in the bottom layer by the number of layers it would take to fill the box exactly. The box is 6 inches deep, so you would have 6 layers, each of 50 blocks. To find the total number of blocks, multiply 6 by 50. You get 300. $6 \times 50 = 300$.

The box has a volume of 300 cubic inches.

You could get the same number by just multiplying the three dimensions: $10 \times 5 \times 6 = 300$. You multiply any pair of numbers first and then multiply the answer by the number left.

So to find the volume of a room or a box or anything shaped like a box, just multiply the length, ℓ, by the width, w, by the height, h. The rule can be written as a formula: $V = \ell wh$. Remember that ℓ, w, and h must all be in the same units before you multiply. If they are all in feet, the answer is in cubic feet, if they are all in yards, the answer is in cubic yards, and so on. For something like the volume of the Grand Canyon, you might use cubic miles.

Coming to Terms

volume the amount of space taken up by an object. It is measured in cubic units.

Try It Yourself

The amount of storage space in a big truck can be amazing.

■ The inside of the We-Move-U trailer was 40 feet long by 8 feet wide by 10 feet high. How many cubic feet of storage space did it have? ____________

You can assume that the inside has square corners like a box, so you should have just multiplied the three dimensions together. You should have gotten 40 × 8 × 10 = 3,200. The trailer had 3,200 cubic feet of storage space.

Warm-up

Find the volumes of boxes with these dimensions. Answers should be in the units asked for.

1. ℓ = 10 in, w = 10 in, h = 10 in

V = __________ cu in

2. ℓ = 20 in, w = 20 in, h = 20 in

V = __________ cu in

3. ℓ = 24 in, w = 2½ ft, h = 1 ft

V = __________ cu ft

4. ℓ = 12 in, w = 1 ft, h = 1 ft

V = __________ cu in

5. ℓ = 6 ft, w = 6 ft, h = 1½ ft

V = __________ cu yd

On the Springboard

7. The coal car was carrying 3,000 cubic feet of coal. From it an empty bin 10 feet by 12 feet by 8 feet was filled. How many cubic feet of coal were left in the coal car?

   (1) 2,004 (2) 2,040 (3) 2,400

   (1) (2) (3)

8. How many cartons 2 feet by 3 feet by 1 foot could be packed into a shipping car 20 feet by 30 feet by 10 feet?

   (1) 1,000 (2) 3,000 (3) 8,000

   (1) (2) (3)

Check your Springboard answers on page 194. If you got them right, go on to "The Real Thing." If you didn't get them right, be sure that you know where your error was before going on.

Warm-up Answers

1. 1,000 cu in **2.** 8,000 cu in (if you compare the measurements and answers in 1 and 2, you can see that doubling the length of each side of a box makes its volume 8 times bigger.) **3.** 5 cu ft **4.** 1,728 cu in **5.** 2 cu yd

The Real Thing

8. The box for a computer disk drive is 18 inches long, 10 inches wide, and 6 inches high. The package uses 15 cubic feet of plastic to protect the computer. How many cubic inches of space are in this box?

   (1) 68 (2) 180 (3) 1,008
   (4) 1,080 (5) 1,800

   (1) (2) (3) (4) (5)

9. One of the first electronic computers was in the shape of a huge box. It was 96 feet long and 2.5 feet wide. The amount of space inside was approximately 3,000 cubic feet. How many feet high was the computer?

   (1) 1.25 (2) 12.50 (3) 27.60
   (4) 125.00 (5) 276.00

   (1) (2) (3) (4) (5)

10. Box A is 12 feet long, 10 feet wide, and 6 feet high. Box B is 15 feet long, 8 feet wide, and 11 feet high. How many more cubic feet of space does box B have than box A?

    (1) 576 (2) 580 (3) 600
    (4) 720 (5) 1,320

    (1) (2) (3) (4) (5)

11. Hilary's aquarium will hold $1\frac{1}{2}$ times as much water as Kelly's aquarium. Kelly's aquarium is $2\frac{1}{2}$ feet wide, 4 feet long, and $2\frac{1}{2}$ feet high. How many cubic feet of water will Hilary's aquarium hold?

    (1) 15 (2) $16\frac{3}{5}$ (3) 25
    (4) $37\frac{1}{2}$ (5) $56\frac{3}{10}$

    (1) (2) (3) (4) (5)

12. A cereal box holds 192 cubic inches. The box is 4 inches wide. How many inches tall is this box?

    (1) 2 (2) 4 (3) 12 (4) 48
    (5) Insufficient data is given to solve the problem

    (1) (2) (3) (4) (5)

13. The diagram shows a garden in a city park. A garden committee put a fence around the garden next to the sidewalk. They also put a curb around the outside edge of the sidewalk. How many feet long was the fence around the garden?

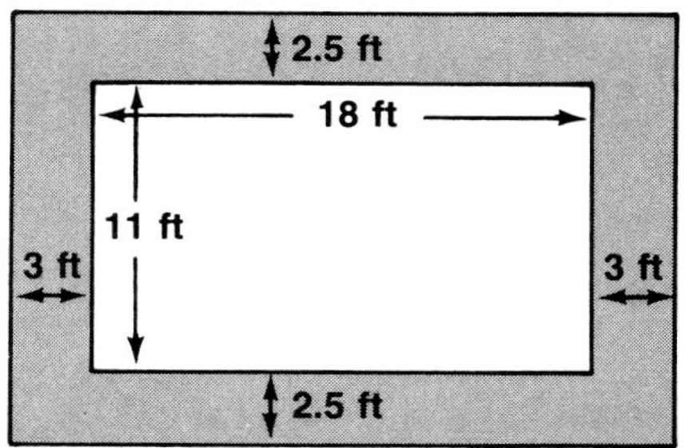

    (1) 36 (2) 56 (3) 58
    (4) 69 (5) 80

    (1) (2) (3) (4) (5)

14. Casey put tape around the triangular flag shown in the diagram. She used 102 yards of binding. She wants to put another tape along side C only. If she allows 0.75 yards of tape for waste, how many yards of the second tape does she need?

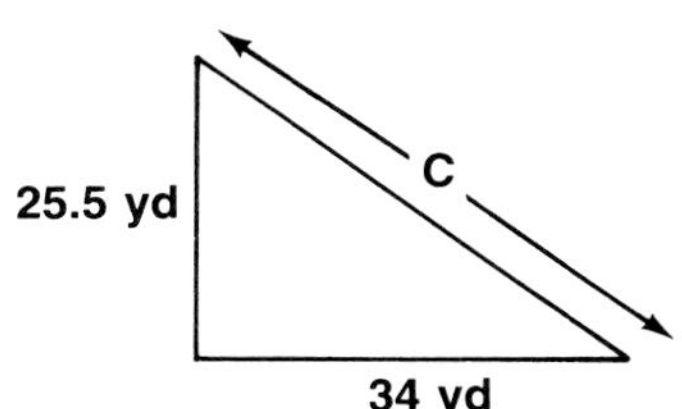

    (1) 42.50 (2) 43.25 (3) 59.50
    (4) 60.25 (5) 68.75

    (1) (2) (3) (4) (5)

15. A feed box is 8 feet long, 3 feet wide, and 30 inches deep. How many cubic feet of feed can the box hold?

    (1) 24 (2) 56 (3) 60
    (4) 240 (5) 720

    (1) (2) (3) (4) (5)

Check your answers and record all your scores on page 216.

Answers: On the Springboard

2 Adding Whole Numbers

(page 23)

1. (2) 85
Add the number of orders filled on the two days.

50 orders on Monday
+ 35 orders on Tuesday
85 in all

2. (3) 697
The words *in all* are your clue to add the numbers of miles she drove on the three days.

275 first day
210 second day
+ 212 last day
697 mi in all

3. (1) 2,879
The word *altogether* is your clue to add. Be careful to line up the digits.

13 box
562 grandstand
+ 2,304 bleacher
2,879 tickets in all

(page 25)

4. (3) 3,140
The word *total* tells you to add. Carry where necessary.

1 1
2,690 during week
+ 450 weekend
3,140 total

5. (2) $1,274
Add to find how much they needed for all three appliances. Carry where necessary.

1 1
$ 595 refrigerator
210 stove
+ 469 washer
$1,274 for all

6. (3) 1,940
Add to find how many calories in all. Carry where necessary.

1 1
560
345 } meals
725
100
+ 210 } snacks
1,940 calories in all

7. (2) 199
The word *total* is your clue to add. Carry where necessary.

1
43 first night
67 second night
+ 89 third night
199 for all nights

3 Subtracting Whole Numbers

(page 27)

1. (2) $116
The word *less* is your clue to subtract.

$148 plane fare
− 32 amount less for train
$116 train fare

2. (2) 5,361
To solve, subtract the number of visiting fans from the number of home fans. Be careful to line up the digits correctly.

5,792 home fans
− 431 visiting fans
5,361 more home fans

3. (2) $72,273
The words *how much more* are your clue to subtract.

$75,698 invested
− 3,425 sales first month
$72,273 more to break even

(page 29)

4. (1) 91,223
The words *how many more* are your clue to subtract. To subtract the tens, you need to borrow one of the hundreds in the top number.

7 10
97,805 senior
− 6,582 junior
91,223 mi more for senior attendant

5. (2) $10,680
Your clue to subtract is the word *difference*. This time you have to borrow 1 thousand (10 hundreds) before you can subtract in the hundreds column.

7 12 17
$88,375 greater price
− 77,695 lesser price
$10,680 difference

(page 30)

6. (1) $166
From the Test-Taking Tip you know that you find net pay by subtracting deductions from gross pay. You have to borrow twice.

14
1 15 15
$2 5 5 gross pay
− 8 9 deductions
$1 6 6 net pay

7. (1) 38
The words *how much more* are your clue to subtract. You have to borrow twice.

10
1 11 7
2 1 7 millions in imports
− 1 7 9 millions in exports
3 8 millions more imports

4 Zeroing In on Word Problems: Two-Step Problems

(page 32)

1. (2) 6
First add to find how many songs in all were published.

11 published by company 1
+ 13 published by company 2
24 published in all

Use this answer. Subtract it from the number of songs she wrote (30) to find how many were *not* published.

2 10
30 songs she wrote
− 24 published
6 not published

2. (1) 250
First add to find how many components Rita installs in one whole day.

```
  1,450  before lunch
+ 1,800  after lunch
  3,250  Rita's total
```

Now subtract the number the average worker installed from Rita's total to find how many more she installed.

```
  3,250  Rita's total
− 3,000  average worker
    250  more for Rita
```

3. (1) $725
Add to find how much in all was taken in on the new record weekend.

```
  $  755
     860
+  1,115
  $2,730  new record
```

To find how much more the new record amount was than the previous record, subtract.

```
     2 10
  $2,730  new record
−  2,005  previous record
  $  725  more for new record
```

5 Multiplying Whole Numbers

(page 34)

1. (3) 3,450
The words in *each of* are your clue to multiply the number of cartons (575) by the number of parts (6) in each carton.

```
    4 3
    575  cartons
×     6  parts per carton
  3,450  parts in all
```

2. (3) 880
Multiply the number of tables (220) by the number of people (4) who can be seated at each table.

```
  220  tables
×   4  people per table
  880  in all
```

(page 36)

3. (3) 636
Multiply the number of dozen (53) by the number in each dozen (12).

```
   53  doz
×  12  eggs per doz
  106
  53
  636  eggs in all
```

4. (2) 504
Multiply the number of students (28) by the number of quizzes that each student took (18).

```
   28  students
×  18  quizzes per student
  224
  28
  504  quizzes to be graded
```

(page 38)

5. (3) 2,100,000
Multiply 2,902 by 675. But only an estimate is asked for, so first round each of the numbers.

```
2,902 →    3,000
  675 → ×    700
```

Multiply 3 by 7 (21) and add 5 zeros on the right (2,100,000). There are about 2,100,000 plants in all.

6. (2) $14,678,820
Multiply 2,091 by $7,020.

```
      2,091
  ×   7,020
     41 820
  14637 0
  14678 820
```

Put in commas and you get $14,678,820.

6 Dividing Whole Numbers

(page 42)

1. (2) $8,569
Your clue to divide is in the words *shared equally.*

```
           $ 8,569  each share
          6)$51,414  total profit
          ↗  48
  partners    34
              30
               41
               36
                54
                54
                 0
```

2. (1) 2,925
Divide the total weight in pounds (14,625) by the number of equal loads (5).

```
         2,925  each load
       5)14,625  total weight
       ↗ 10
  loads   46
          45
           12
           10
            25
            25
             0
```

(page 43)

3. (2) 344
Divide 3,100 gallons (the amount of heating oil the apartment building used) by 9.

```
     344 r4  gal for home
   9)3,100   gal for apartment
     27        building
      40
      36
       40
       36
        4
```

All the answer choices are whole numbers. You were asked for the *approximate* number of gallons used in the home. 344 is the best answer.

4. (1) 58
Divide the number of items packed (350) by the number of items that go to make a full carton (6).

```
           58 r2  cartons
         6)350    items packed
         ↗ 30
  items     50
   per      48
  carton     2
```

58 cartons were packed. The remainder 2 tells you that two items were left over. Since they are not enough to make another carton, the answer is 58 full cartons.

5. (3) 40
Divide the number of pieces of chicken (326) by the number of pieces in a complete basket (8).

```
           40 r6  baskets
         8)326    pieces
         ↗ 32
  pieces     6
   per       0
  basket     6
```

Since 6 pieces is not enough for another *complete* basket, the answer is 40 baskets.

(page 45)

6. (2) 12
Divide the number of feet of lakefront land (780) by the number of feet for each lot (65).

```
        12   number of lots
    65)780   total number of ft
  ↗     65
 ft for 130
each lot 130
          0
```

7. (3) 32
Divide the total number of bushels (6,720) by the number of acres (210).

```
         32   average bushels
   210)6,720  total bushels
   ↗    630
 acres  420
        420
          0
```

7 Zeroing In on Word Problems

(page 48)

1. (3) $320
You need to figure out how much each worker earned and then subtract the garment worker's earnings from the earnings of the auto plant worker. Each worker put in 80 hours of work (40 × 2 = 80). So multiply their hourly wages by 80.

Auto plant worker

```
   80   hours
 × 12   dollars per hour
  160
  80
 $960   total earned
```

Garment worker

```
   80   hours
 ×  8   dollars per hour
 $640   total earned
```

Next subtract their earnings.

```
   $960   auto plant worker
 −  640   garment worker
   $320   more for auto plant worker
```

The answer is $320.

Another way to approach this problem is to first find how much more the auto plant worker was paid each hour. ($12 − $8 = $4) Since he worked 80 hours in all, he earned $4 × 80 more.

```
   80   hours
 ×  4   dollars more per hour
 $320   total amount more for auto
          plant worker
```

2. (2) 2,100
One way to approach the problem is to find how much electricity was saved each month and then multiply by 3 (the number of months they were away). Subtract to find how many kilowatt-hours they saved each month they were away.

```
   1,100   at home
 −   400   away
     700   saved each month
             while away
```

Multiply this result by 3 (the number of months away).

```
   700   saved each month
 ×   3   months away
 2,100   total saved
```

Another approach is to calculate how many kilowatt-hours they used while they were away, how many they would have used if they had been at home, and then subtract the smaller amount from the larger. To do this, multiply and subtract.

Used while away

```
   400   used each month
 ×   3   months away
 1,200   total used (away)
```

Would have used if at home

```
   1,100   used each month
 ×     3   months
   3,300   would have used if
             at home
```

Subtract to find how much was saved.

```
   3,300   would have used
 − 1,200   used (away)
   2,100   total saved
```

3. (3) 200
Figure out the total number of miles he drove each time he drove the route. To do this, add.

```
    5   depot to farm
   15   farm to market
   15   market to farm
 +  5   farm to depot
   40   total mi for route
```

To find how many miles he drove in a week, multiply 40 miles by 5 (the number of times he drove the route).

```
   40   length of route
 ×  5   number of times driven
  200   total mi in 1 wk
```

8 Adding and Subtracting Decimals

(page 55)

1. (3) $468.53
Add all three amounts to find the total. Line up the decimal points and digits correctly.

```
      1 1  1
  $  57.57
    404.16
 +    6.80
  $468.53   total due
```

2. (3) 11.38
Add the weights of all three loads. Put zeros at the ends of the numbers with fewer decimal places as needed.

```
    5.00
    4.03
 +  2.35
   11.38   total weight
```

(page 56)

3. (1) 1.82
Subtract the rainfall at the airport from the rainfall downtown. Add on zeros where needed to make sure the numbers have the same number of decimal places.

```
      14
   2  15 10
   3. 5  0   downtown
 − 1. 6  8   airport
   1. 8  2   more downtown
```

4. (1) $10.01
Subtract the cost of the part ($9.99) from the amount he gave to the clerk ($20) to find his change.

```
       9    9
     1 10   10 10
   $2  0 .  0  0   amount to clerk
 −     9 .  9  9   cost of part
   $1  0 .  0  1   change
```

9 Multiplying Decimals

(page 59)

1. (2) $15.56

Multiply the number of kilowatt-hours used (251) by the cost of each kilowatt-hour ($0.062). Set up the problem vertically. Multiply as usual. Mark off 3 decimal places in the answer.

```
     251   kilowatt-hrs
 × 0.062   cost per kilowatt-hr
     502
   1506
   000
 000
 015.562   amount owed
```

They owe $15.562. Round $15.562 to the nearest cent (hundredth). You get $15.56.

2. (2)1.25

Multiply the distance each ran (0.25 miles) by the number of runners. Mark off 2 decimal places in the multiplication answer.

```
  0.25   mi each ran
×    5   number of runners
  1.25   total mi
```

(page 60)

3. (2) $0.26

Multiply the price per pound ($0.005) by the number of pounds in the bundle. Mark off 3 decimal places in the answer.

```
  0.005   price per lb
×    52   number of lbs
   0010
  0025
 00.260   total to be paid
```

4. (3) 160

Multiply the thickness of a single wire (0.04) by the number of wires (4,000). Mark off 2 decimal places in the answer.

```
   4,000   wires
×   0.04   width each wire
   16000
   0000
  0000
 0160.00   total width
```

Since 0160.00 = 160, the answer is 160 inches.

10 Dividing Decimals

(page 62)

1. (2) $0.51

Divide the total cost ($3.57) by the number of plants in a box.

```
          0.51   cost for each plant
        7)3.57   total cost
        ↗ 3 5
in one     07
box         7
            0
```

The cost of each plant is $0.51.

2. (1) 8.91

Divide the total number of miles (356.4) by the number of working days (40). If you add on 1 zero after the 4 in the number you're dividing, you will come out with no remainder.

```
         8.91   miles each day
   40)356.40    total miles
   ↗  320
days   36 4
       36 0
          40
          40
           0
```

They averaged 8.91 miles a day.

(page 63)

3. (2) $2.75

Divide the price of the package ($15.40) by the weight (5.6 pounds) to get the price per pound. Make the number you divide by a whole number by moving the decimal point 1 place to the right. You must also move the decimal point 1 place to the right in the number you are dividing. Add a zero to make the answer come out without a remainder.

```
             2.75   price per lb
       5.6)15.400   total price
       ↗   11 2
weight      4 20
            3 92
             280
             280
               0
```

4. (3) 21.5

Divide the miles traveled (146.20) by the number of gallons of gas used (6.8). Observe how the decimal point is moved in both numbers before the division is done.

```
               21.5   mi per gal
        6.8)146.20    mi traveled
        ↗    136
gallons      10 2
of gas        6 8
              3 40
              3 40
                 0
```

(page 65)

5. (2) 20

Divide the width of the floor (26 ft) by the width of each tile (1.3 ft).

```
           20.   tiles across width
      1.3)26.0   width of floor
      ↗   26
one        0 0
tile       0 0
width        0
```

There were 20 tiles across the width of the floor.

6. (3) 146

Divide the number of days (365) by the time for completing the process once (2.5 days).

```
              146.   processes a year
        2.5)365.0    days in a year
        ↗   25
days        115
for one     100
process      15 0
             15 0
                0
```

The process can be completed 146 times in 365 days.

(page 66)

7. (1) 15

Divide the cost ($5.65) by the number of pounds (38). Divide till you get a digit in the thousandths place of the answer. Then round to the nearest hundredth (nearest cent).

```
            0.148   cost per pound
         38)5.650   total cost
         ↗  3 8
pounds      1 85
            1 52
              330
              304
               26
```

The number \$0.148 should be rounded to 15 cents.

8. (1) 8.75
Divide the miles traveled (248.5) by the number of gallons of gas used (28.4). Add on zeros as needed in the number you are dividing.

```
          8.75
28.4)248.5,00
     227 2
      21 3 0
      19 8 8
       1 4 20
       1 4 20
            0
```

The vehicle got 8.75 miles per gallon.

11 Zeroing In on Word Problems: Insufficient Data

(pages 68–69)

1. (2) 12
Divide the total number of checks (4,000) by the number of checks that will fit into a file drawer (350).

```
        11   full drawers
350)4,000
    3 50
     500
     350
     150   checks left over
```

The clerk can fill 11 drawers all the way, but there will still be 150 checks left to file. So he needs 12 drawers in all.

2. (3) Insufficient data is given to solve the problem.

The problem tells you neither the number of words per hour nor the number of words per trial. So there is no way you can tell how many words in all might have been taken down.

3. (3) Insufficient data is given to solve the problem.

You could solve the problem if you knew the weight of the oranges. You are not told that. You could *figure out* the weight of the oranges if you knew the weights or total number of packages of the other kinds of fruit. But you are not told either of those things. So there is no way for you to answer the question.

13 Adding Fractions

(page 77)

1. (2) 1 cup
Add the amount of sugar needed for the two recipes. The fractions already have the same denominator. Use that same denominator (4) in the answer and add the numerators.

$$\frac{3}{4} + \frac{1}{4} = \frac{4}{4}$$

Reduce the improper fraction. Since 4 ÷ 4 = 1, the final answer is 1 cup of sugar.

2. (3) $13\frac{3}{16}$ inches

There will be a $2\frac{7}{16}$-inch border on each side, so add $2\frac{7}{16}$ to the width of the photo ($8\frac{5}{16}$ inches) *twice*.

$$8\frac{5}{16} + 2\frac{7}{16} + 2\frac{7}{16}$$

When you add the whole-number parts of the mixed numbers, you get 8 + 2 + 2 or 12. When you add the fraction, add the numerators (5 + 7 + 7) and keep the same denominator (16). You get $\frac{19}{16}$. Reduce $\frac{19}{16}$ to $1\frac{3}{16}$.

$$12\frac{19}{16} = 12 + 1\frac{3}{16} = 13\frac{3}{16}$$

(page 79)

3. (1) $18\frac{11}{12}$ hours

The word *altogether* is your clue to add. Suppose you use 12 as the common denominator.

$$9\frac{3}{4} + 6\frac{5}{6} + 2\frac{1}{3} = 9\frac{9}{12} + 6\frac{10}{12} + 2\frac{4}{12} = 17\frac{23}{12}$$

Since $\frac{23}{12} = 1\frac{11}{12}$, the number $17\frac{23}{12}$ equals $17 + 1\frac{11}{12}$, or $18\frac{11}{12}$.

4. (1) $9\frac{20}{21}$ weeks

Add to find the total time that she worked. You can use 21 as the common denominator when you add the fractions.

$$3\frac{5}{7} + 6\frac{5}{21} = 3\frac{15}{21} + 6\frac{5}{21} = 9\frac{20}{21}$$

14 Subtracting Fractions

(page 82)

1. (2) $5\frac{1}{4}$
Subtract the length of the board that he got ($19\frac{1}{8}$ inches) from the original length ($24\frac{3}{8}$ inches) to find the length of the piece he removed.

$$24\frac{3}{8} - 19\frac{1}{8} = 5\frac{2}{8}$$

Reduce $\frac{2}{8}$ to $\frac{1}{4}$ and you get $5\frac{1}{4}$ as the final answer.

2. (2) $3\frac{3}{5}$

Subtract the amount they used last week ($8\frac{1}{10}$ liters) from the amount they used this week ($11\frac{7}{10}$) to find how much more milk they used this week.

$$11\frac{7}{10} - 8\frac{1}{10} = 3\frac{6}{10}$$

Reduce $\frac{6}{10}$ to lowest terms to get the final answer: $3\frac{3}{5}$.

(page 82)

3. (3) $1\frac{1}{8}$
Subtract the weight of the concrete from the weight of the total load to get the weight of the bricks. When you subtract the fractions, use 8 as the common denominator.

$$2\frac{3}{4} - 1\frac{5}{8} = 2\frac{6}{8} - 1\frac{5}{8} = 1\frac{1}{8}$$

4. (1) $28\frac{1}{4}$

Subtract the amount of chili left from the original amount to find how much was sold.

$$40\frac{3}{4} - 12\frac{1}{2} =$$

$$40\frac{3}{4} - 12\frac{2}{4} = 28\frac{1}{4}$$

(page 84)

5. (1) $1\frac{3}{4}$
Subtract the amount used for the first step from the total amount used. To subtract, you need to change 6 to $5\frac{4}{4}$.

$$6 - 4\frac{1}{4} = 5\frac{4}{4} - 4\frac{1}{4} = 1\frac{3}{4}$$

6. (1) $7\frac{7}{10}$

Subtract the amount used on the first day from the amount they started with. Use 10 as a common denominator.

$11\frac{1}{5} - 3\frac{1}{2} = 11\frac{2}{10} - 3\frac{5}{10}$

You can't subtract 5/10 from 2/10, so borrow 1 from 11 to change 11 2/10 to 10 12/10.

$10\frac{12}{10} - 3\frac{5}{10} = 7\frac{7}{10}$

15 Multiplying Fractions

(page 87)

1. (1) $\frac{5}{12}$

Multiply 5/6 by 1/2 to find what fractions of all the cakes were chocolate cakes with frosting.

$\frac{5}{6} \times \frac{1}{2} = \frac{5 \times 1}{6 \times 2} = \frac{5}{12}$

2. (2) $\frac{2}{15}$

Multiply all three fractions to find what fraction of all the trees could be sold for Christmas trees.

$\frac{5}{6} \times \frac{2}{5} \times \frac{2}{5} = \frac{\overset{1}{\cancel{5}} \times \overset{1}{\cancel{2}} \times 2}{\underset{3}{\cancel{6}} \times \underset{1}{\cancel{5}} \times 5} = \frac{2}{15}$

(page 88)

3. (3) 80

Multiply the number of yards for one sari and shoulder wrap (6 2/3) by the number that are to be made (12). Change both numbers to improper fractions.

$6\frac{2}{3} \times 12 = \frac{20}{3} \times \frac{12}{1} =$

$\frac{20 \times \overset{4}{\cancel{12}}}{\underset{1}{\cancel{3}} \times \underset{1}{1}} \times = \frac{80}{1} = 80$

4. (2) $45\frac{1}{2}$

Multiply the number of pounds in one pack (8 2/3) by the number of packs (5 1/4). In order to multiply, change both mixed numbers to improper fractions.

$8\frac{2}{3} \times 5\frac{1}{4} = \frac{26}{3} \times \frac{21}{4} =$

$\frac{\overset{13}{\cancel{26}} \times \overset{7}{\cancel{21}}}{\underset{1}{\cancel{3}} \times \underset{2}{\cancel{4}}} = \frac{91}{2} = 45\frac{1}{2}$

16 Dividing Fractions

(page 90)

1. (1) $\frac{1}{48}$

In the relay race, the children each ran the same distance. To find how far each child ran, divide the total distance covered (5/16 mile) by the number of children (15).

$\frac{5}{16} \div 15 =$

$\frac{5}{16} \div \frac{15}{1} = \frac{5}{16} \times \frac{1}{15} =$

$\frac{\overset{1}{\cancel{5}} \times 1}{16 \times \underset{3}{\cancel{15}}} = \frac{1}{48}$

2. (3) 13

Divide the number of pints left (16 1/4) by the number of pints each bottle will hold (1 1/4).

$16\frac{1}{4} \div 1\frac{1}{4} =$

$\frac{65}{4} \div \frac{5}{4} = \frac{65}{4} \times \frac{4}{5} =$

$\frac{\overset{13}{\cancel{65}} \times \overset{1}{\cancel{4}}}{\underset{1}{\cancel{4}} \times \underset{1}{\cancel{5}}} = \frac{13}{1} = 13$

17 Fraction and Decimal Names for Numbers

(page 94)

1. (3) $\frac{5}{8}$

If you change 0.625 to a fraction, you get 625/1,000, the number you see in option (1). However, the question asks for a fraction in *lowest terms.* Since you can exactly divide 625 and 1,000 by 5, 625/1,000 is *not* in lowest terms and cannot be the answer. Option (2) is not a fraction in lowest terms either. So (3) must be the correct option.

You can check that 5/8 really is another name for 0.625 by reducing 625/1,000 to lowest terms.

$\frac{625}{1000} = \frac{625 \div 5}{1000 \div 5} = \frac{125}{200} =$

$\frac{125 \div 5}{200 \div 5} = \frac{25}{40} = \frac{25 \div 5}{40 \div 5} = \frac{5}{8}$

2. (3) 0.0625

To change 1/16 to a decimal, change 1/16 to 1 ÷ 16. To do the division, set up the division problem in the usual way. Put a decimal point after the 1 and add several 0s after the decimal point. Then divide.

```
      .0625
16)1.00000
     96
      40
      32      stop when you
       80     get a zero remainder
       80
        0
```

The answer is .0625 or, as it is more commonly written, 0.0625.

18 Zeroing In on Word Problems: Unnecessary Information

(page 96)

1. (3) 4

The information about Sue's 2 days in Arizona is extra, unnecessary information. What you do need to know is how many days of sick leave she had already used.

In March she used 1/4 of her 12 days. So in March she used 12 × 1/4, or 3 days. After her vacation she used 5 days. Since 3 + 5 = 8, she had 12 − 8, or 4, days of sick leave left.

2. (1) 61.7

The information about how much higher above sea level the city in Colorado was than Chicago has nothing to do with figuring out how many meters he threw the discus in Chicago.

You are told that the throw of 65.6 meters (in Colorado) was 1.4 meters more than his previous record. So you can find his previous record by subtracting 1.4 from 65.6.

65.6	Colorado throw
− 1.4	more than previous record
64.2	previous record

His throw in Chicago was 2.5 meters *less* than the previous record. So subtract 2.5 from 64.2 to find how far he threw in Chicago.

64.2	previous record
− 2.5	less in Chicago
61.7	how far in Chicago

20 Working with Percents

(page 104)

1. (2) 25%

Fill in the grid with the information given in the question. The percent is what you need to find in this case, so the ? goes in the percent box.

part of school year	part 45	percent ?	percent of school year
total days in school year	whole 180	100	

Multiply on the diagonal that has two known numbers. Then divide by the unused number.

$45 \times 100 = 4{,}500$
$4{,}500 \div 180 = 25$

So 45 days is 25% of the 180-day school year.

2. (1) 540
The whole collection has 900 marbles, and the part that contains blue makes up 60 percent of the whole collection. Fill in the grid and solve to find how many marbles in the part that contains blue.

part with blue	part ?	percent 60	percent with blue
whole collection	whole 900	100	

$60 \times 900 = 54{,}000$
$54{,}000 \div 100 = 540$

Of the marbles, 540 contain blue.

(page 106)

3. (3) 21
First use the grid to find how many were in the 70% who passed the exam.

number passed	part ?	percent 70	percent who passed
all candidates	whole 120	100	

$70 \times 120 = 8{,}400$
$8{,}400 \div 100 = 84$

So 84 candidates passed. Now use the grid to find how many of these were in the 25% that passed with honors.

honors group	part ?	percent 25	percent passed with honors
number passed	whole 84	100	

$25 \times 84 = 2{,}100$
$2{,}100 \div 100 = 21$

There were 21 honors students.

4. (2) 273
If the storekeeper added 30% to the cost of every article, you need to find 30% of $200.

price increase	part ?	percent 30	percent added
original price	whole 200	100	

$30 \times 200 = 6{,}000$
$6{,}000 \div 100 = 60$

The storekeeper added $60 to the cost of each article, so the price of the article was $260. Next, you need to find the sales tax, or 5% of $260.

dollar amount of sales tax	part ?	percent 5	percent sales tax
total price of article	whole 260	100	

$5 \times 260 = 1{,}300$
$1{,}300 \div 100 = 13$

So the sales tax was $13. Adding $13 to the price of the article gives you 260 + 13 = 273 for the answer.

(page 108)

5. (1) 20
The increase was 36 − 30, or 6, pounds. Set up the grid and solve the problem.

increase in pressure	change 6	percent ?	percent increase
pressure at start	orig. 30	100	

$6 \times 100 = 600$
$600 \div 30 = 20$

There was a 20% increase in the air pressure.

6. (2) 5
The change in the waistband was a 2-inch decrease. Fill in the grid and solve.

decrease	change 2	percent ?	percent decrease
orig. size	orig. 40	100	

$2 \times 100 = 200$
$200 \div 40 = 5$

There was a 5% decrease in the waistband size.

21 Zeroing In on Word Problems: Item Sets

(page 110)

1. (2) $17.50
First, you have to know the total cost for each kind of boat.

rowboat—$2.50
canoe at $2 each—
$10 (5 × 2 = 10)
speedboat—$5

Add the figures for each kind of boat to get the total rental cost.

$ 2.50	rowboat
10.00	canoes
+ 5.00	speedboat
$17.50	

The total boating cost for one hour was $17.50.

2. (3) $11.25
You were told that the cost of the speedboat would remain the same ($5).
Subtract that cost from $17.50 (17.50 − 5 = 12.50).
Next find 50% of $12.50.

$12.50 \times .50 = 6.25$

Subtract from the total hourly cost.

$17.50	regular rental
− 6.25	last hour savings
$11.25	last hour cost

Had they gone boating during the last hour of the day, their cost would have been $11.25.

22 Tables and Meters

(page 112)

1. (1) 2
First look in the row for Atlanta—it's the first row. Find the distance to Chicago (in the column headed *Chicago*). The distance is 674 miles. Next look in the column headed Cleveland. The distance from Atlanta to Cleveland is 672 miles. Subtract.

674	Atlanta to Chicago
−672	Atlanta to Cleveland
2	more to Chicago

It is 2 miles further to Chicago.

2. (3) 1,280
Look in the row for Boston and then over to the column headed Cincinnati. The distance from Boston to Cincinnati is 840 miles. Next look in the row for Cincinnati and over to the column headed Atlanta. The distance from Cincinnati to Atlanta is 440 miles. Add to find the total distance.

840	Boston to Cincinnati
+ 440	Cincinnati to Atlanta
1,280	Boston to Cincinnati to Atlanta

3. (3) Houston
Examine all of the numbers in the *row* for Chicago *and* all of the numbers in the *column* for Chicago. The largest number of all is in the column headed Chicago, the number 1,067. Look to the left in the bottom row, where that number is found, and you see that the city farthest from Chicago is Houston.

(page 114)

4. (2) 5,300
Read the short hand first, then the long hand and add on two 0s. Since the hands are between two numbers in each case, use the smaller number each time.

5, 3 0 0
↑ ↑ ↑ ↑
short hand, long hand, add on 2 0s.

5. (1) 535,000
Read the dials from left to right. When a hand is between numbers, use the smaller number.

500,000	first dial
30,000	second dial
+ 5,000	third dial
535,000	reading

The reading is for 535,000 cubic feet of gas.

23 Reading Graphs

(pages 116–117)

1. (3) 25
Look to see how far up the bar labeled *Gorilla* goes. It stops about halfway between the horizontal lines labeled 24 and 26. So a gorilla lives for about 25 years.

2. (2) rhino
The bars for the lion, the gorilla, and the rhino go above the 20-year line. Of these three bars, the longest is the one for the rhino.

3. (2) 7
The bar for the lion stops at the 26-year line. The bar for the tiger stops at 19 (halfway between the 18-year line and the 20-year line). Subtract and you get $26 - 19 = 7$.

(page 119)

4. (1) 1
Look along the left side of the graph and find 70. Follow the horizontal line across to the graph. You reach the line graph at a point not far from the left-hand side. Stop at this point and then look straight down to see what year number corresponds to this point. The year number is 1.

5. (2) 35%
Look along the bottom of the graph and find the number 4. Go straight up until you come to the point on the line graph directly above the 4. This point is halfway between the horizontal lines for 30% and 40%. Since 35 is the number halfway between 30 and 40, the trade-in value of a 4-year old car is 35% of the selling price.

6. (2) $3,200
The trade-in value of a 3-year old car is 40% of the selling price. To find 40% of $8,000 you can use a grid.

part ?	percent 40
whole 8,000	100

$40 \times 8{,}000 = 320{,}000$
$320{,}000 \div 100 = 3{,}200$

The trade-in value is $3,200.

(pages 120–121)

7. (2) entertainment and sports
Change ¾ to a percent.

$\frac{3}{4} = \frac{?}{100}$
$3 \times 100 = 300$
$300 \div 4 = 75$

So ¾ equals 75%. The circle graph shows 75% for the section representing entertainment and sports, so (2) is the correct choice.

8. (3) 13%
The graph shows that news takes up 9% of the broadcast time and public affairs 4%. Add these and you get 13%.

(page 122)

9. (2) 39
Count the basketball symbols to the right of Washington Bullets. There are 7¾ of these symbols. Since each symbol represents 5 games won, the Bullets won 39 games.

10. (2) 2.5
Count the basketball symbols for each team. For the Celtics, there are 13½ and for the Pacers there are 5¼. Divide.

$13\frac{1}{2} \div 5\frac{1}{4} = \frac{54}{4} \div \frac{21}{4} =$

$\frac{54}{\cancel{4}_1} \times \frac{\cancel{4}^1}{21} = \frac{54}{21} = 2.5$

So the Celtics won 2.6 times more games than the Pacers.

24 Finding Means and Medians

(page 125)

1. (2) 831
To find the mean, add all the scores and divide by 5 (the number of scores in the list).

730
950
875
632
+ 968
4,155

```
        831   average score
      5)4,155 total of scores
     ↗  40
number  15
of scores 15
          05
           5
           0
```

2. (3) 8,345
To find the mean, add all the prices and divide by the number of prices (4).

```
  $ 6,000
    7,800
   10,750
 +  8,830
  $33,380
```

```
   8,345  average
 4)33,380  price
   32
    13
    12
     18
     16
      20
      20
       0
```

(page 126)
3. (2) $7\frac{1}{2}$
First arrange the 8 numbers in order. Then find the two middle numbers.

4, 5, 5, 7, 8, 8, 9, 10
(7, 8: middle numbers)

Find the mean (average) of the two middle numbers, 7 and 8: 7 + 8 = 15, and 15 ÷ 2 = 7½. The median is 7½.

4. (3) $14 million
To find the median, arrange the numbers in order and take the average of the 2 middle numbers. Since the 2 middle numbers are both 14, their average is also 14. The median amount of the annual sales for these companies is thus $14 million.

25 Ratio, Proportion, and Probability

(page 130)
1. (1) 4:5
The speed of the zebra is 40 miles per hour. The speed of the lion is 50 miles per hour. So the ratio is 40:50. The ratio 40:50 can be written as 40/50, which can be reduced to 4/5. So the ratio can be written as 4:5.

2. (2) 3:1
Since 9,000 voted, the number of people who did *not* vote was 12,000 − 9,000, or 3,000. The ratio of those who voted to those who didn't is 9,000:3,000. This ratio can be written as 9,000/3,000, which can be reduced to 3/1. Thus the ratio of voters to non-voters is 3:1.

3. (2) 7:24
First add to find the total number of flats of marigolds sold: 85 + 35 = 120. The ratio of dwarf marigolds to the total sold is 35:120. Since 35/120 = 7/24, the ratio can also be written as 7:24.

(page 133)
4. (1) $54
Set up a grid with the information given in the problem. Then solve.

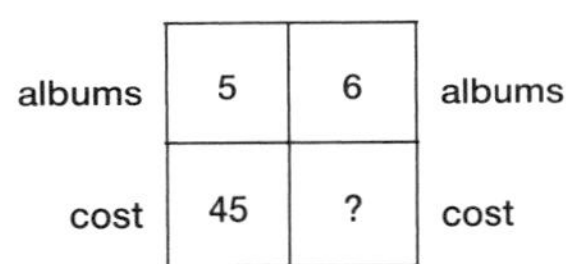

6 × 45 = 270
270 ÷ 5 = 54

The cost of 6 albums is $54.

5. (2) $4\frac{1}{2}$
Put the information in a grid and solve.

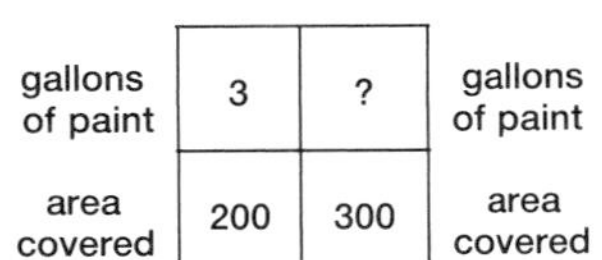

$3 \times 300 = 900$
$900 \div 200 = \frac{900}{200} = \frac{9}{2} = 4\frac{1}{2}$

6. (3) 12
Put the information in a grid and solve.

packets of mix	2	?	packets of mix
pints of drink	$2\frac{1}{2}$	15	pints of drink

$2 \times 15 = 30$
$30 \div 2\frac{1}{2} = \frac{30}{1} \div \frac{5}{2} =$
$\frac{\overset{6}{\cancel{30}}}{1} \times \frac{2}{\underset{1}{\cancel{5}}} = \frac{12}{1} = 12$

7. (2) 420
Set up a grid to find how many gallons of water had been added to the tank in 3 hours. 400 gallons were added in the 4 hours from 8:00 A.M. to noon (520 − 120 = 400).

gallons added	400	?	gallons added
time passed	4	3	time passed

400 × 3 = 1,200
1,200 ÷ 4 = 300

By 11:00 A.M., 300 gallons had been added to the 120 already in the tank. So the total amount of water in the tank at 11:00 A.M. was 120 + 300, or 420 gallons.

(page 135)
8. (1) $\frac{1}{1,000}$
First figure out all the possible combinations that might have been selected for the lock. The first digit could have been any of the 10 digits from 0 to 9, and the same is true for the second and third digits. So in all there are 10 × 10 × 10, or 1,000 possible 3-digit combinations. So there is 1 chance out of 1,000 that Marvin hit the right one when he picked 7-2-5. The probability is 1/1,000

9. (2) $\frac{1}{4}$
There are 4 possible outcomes when you toss two coins.

	First coin	Second coin
outcome 1	H	H
outcome 2	H	T
outcome 3	T	H
outcome 4	T	T

Having both come up heads is 1 of the 4 possible outcomes, so the probability of getting 2 heads is $1/4$.

10. (3) $\frac{2}{175}$

The two couples put 4 cards into the total collection of 350 cards, so the probability that one of the people in the couples is selected first is $4/350$. The fraction $4/350$ can be written in lowest terms as $2/175$.

27 Sequencing

(pages 144–145)

1. (3) C

First multiply to find the total value of the coins in each bag.

Bag	No. of Coins		Value (each coin)		Total Value
A	12	×	0.55¢	=	6.6¢
B	11	×	0.6¢	=	6.6¢
C	100	×	0.065¢	=	6.5¢

Since 6.5¢ is less than 6.6¢, the coins in bag C had the least total value.

2. (1) 0.245

The steel band must be less than (or at the very worst equal to) the size of the threading slot. So compare 0.25 to the widths of the threading slots. First add on a 0 to make 0.25 into the three-place decimal 0.250. This number is less than 0.255 and 0.306, but greater than 0.245. So the slot of width 0.245 could not be used with the band.

(page 145)

3. (2) $\frac{1}{8}, \frac{1}{10}$

You can add the pairs of numbers in each choice to see if the sum you get equals $9/40$.

Do this for option (1).

$$\frac{1}{5} + \frac{1}{8} = \frac{8}{40} + \frac{5}{40} = \frac{13}{40}$$

Too large.

Try option (2).

$$\frac{1}{8} + \frac{1}{10} = \frac{5}{40} + \frac{4}{40} = \frac{9}{40}$$

Just right.

4. (2) C, A, B

First you should find out what part prospector C got. Add the parts that A and B got.

$$\frac{3}{8} + \frac{5}{12} = \frac{9}{24} + \frac{10}{24} = \frac{19}{24}$$

Since all three parts should add up to 1, subtract $19/24$ from 1 to find what part C got.

$$1 - \frac{19}{24} = \frac{24}{24} - \frac{19}{24} = \frac{5}{24}$$

Now write the fractions with a common denominator of 24.

A got $\frac{9}{24}$, B got $\frac{10}{24}$, C got $\frac{5}{24}$

In order from least to greatest:

$$C\left(\frac{5}{24}\right), A\left(\frac{9}{24}\right), B\left(\frac{10}{24}\right)$$

(pages 146–147)

5. (2) B, A, C

You could write fractions that show what part of each class was present and rename using a common denominator of 300.

A had	B had	C had
$\frac{23}{25}$	$\frac{19}{20}$	$\frac{27}{30}$
↓	↓	↓
$\frac{276}{300}$	$\frac{285}{300}$	$\frac{270}{300}$

In order of size (greatest to least) the fractions are $285/300$, $276/300$, $270/300$. So the answer is B, A, C.

6. (3) B, C, A

Write as a decimal the part that each spent on rent, then compare the decimals.

Family A $21\% = \frac{21}{100} = 0.21$
Family B $\frac{1}{5} = 0.2$
Family C 0.205

Add 0s to make the first 2 decimals into 3-place decimals.

Family A 0.210
Family B 0.200
Family C 0.205

The decimals, from least to greatest, are 0.200 (B), 0.205 (C), 0.210 (A).

28 Exponents

(page 150)

1. (3) 32

The exponent 5 tells you that you are to use 2 five times in the repeated multiplication.

$$2^5 = 2 \times 2 \times 2 \times 2 \times 2 = 32$$

2. (1) -729

The exponent 3 tells you that you are to use -9 three times in the repeated multiplication.

$$(-9)^3 = (-9) \times (-9) \times (-9) = -729$$

3. (3) 1

The exponent 6 tells you to use -1 six times in the repeated multiplication. Be careful in using the rules for multiplying positive and negative numbers.

$$(-1)^6 = (-1) \times (-1) \times (-1) \times (-1) \times (-1) \times (-1) = +1, \text{ or } 1$$

4. (1) $-\frac{8}{27}$

The exponent 3 tells you to use $-2/3$ three times in the repeated multiplication.

$$(-\tfrac{2}{3})^3 = (-\tfrac{2}{3}) \times (-\tfrac{2}{3}) \times (-\tfrac{2}{3}) = -\frac{8}{27}$$

(page 152)

5. (3) 1,369

To find how many tiles there are, square the number of tiles that are on each side (37).

$$37^2 = 37 \times 37 = 1{,}369$$

6. (3) 30

The square of Jennifer's age is 36: $6^2 = 6 \times 6 = 36$. This number is supposed to equal the sum of her age (6) and her father's age. Try the choices one at a time. For option (1), is 36 equal to 6 + 25? No. For option (2), is 36 equal to 6 + 28? No. By now you know that (3) has to be the correct choice. You can easily check it out. Is 36 equal to 6 + 30? Yes.

You could also reason this way. 36 equals 6 plus her father's age: 36 = 6 + (father's age)

So her father's age must be 36 − 6, or 30.

7. (1) 6
First multiply the number of boxes of tiles (3) by the number in each box (50) to find how many tiles he had in all.

Try numbers to see how many tiles are in the largest shape he can cover (using the same number of tiles on each side). Does he have enough tiles for a shape that has 10 tiles on each side? Yes, because $10^2 = 10 \times 10 = 100$. What about 11 tiles on a side? Yes, because $11^2 = 11 \times 11 = 121$. Twelve tiles on a side? Yes, because $12^2 = 12 \times 12 = 144$. What about 13 tiles on a side? No, because $13^2 = 13 \times 13 = 169$, and he only has 150 tiles. So the largest shape possible is the one with 12 tiles on each side. For this shape he needed 144 tiles. Subtract to find how many were left over: 150 − 144 = 6.

(page 154)

8. (2) 2.45×10^5
Multiply to find the distance. Multiply the speed by the time (24,500 × 10 = 245,000). To write 245,000 in scientific notation, put a decimal point after the 2: 2.45000

How many places to the *right* would you need to move the decimal point to get back to the original number? (5) So the scientific notation for 245,000 is 2.45000×10^5 or 2.45×10^5.

You could also get the answer by checking out each answer choice. Option (1) is equal to 24,500. Option (2) is equal to 245,000 and is therefore the correct answer.

9. (3) 3.67×10^7
Since the correct answer is 100 times 3.67×10^5, and since $100 = 10 \times 10$ or 10^2, his exponent should not have been 5, but 7: 3.67×10^7.

10. (2) 0.00625
The exponent −3 tells you that to write 6.25×10^{-3} as an ordinary number, you need to move the decimal point 3 places to the left.

.006.25

So the answer is .00625 or, as it is usually written, 0.00625.

29 Standard Measurement

(page 157)

1. (1) 428
Change 35 feet to inches. To do this, multiply 35 by 12 (the number of inches in 1 foot).

35 × 12 = 420

So 35 feet 8 inches = 420 inches + 8 inches, or 428 inches.

2. (1) 1
Since 1 quart equals 2 pints and 2 pints equals 4 cups, you know that 1 quart = 4 × 8 ounces, or 32 ounces. (There are 8 ounces in 1 cup.) So one jar of sauce is exactly the amount he needs.

3. (3) $64\frac{1}{3}$
Divide by 3 (the number of feet in 1 yard) to change the feet to yards.

$193 \div 3 = 64\frac{1}{3}$

(page 158)

4. (2) $2\frac{1}{2}$
Add all the amounts.

```
                 2 cups
         1 pint 1 cup
+1 quart        1 cup
 1 quart 1 pint 4 cups
```

Since 4 cups = 2 pints, the total is 1 quart 3 pints or 2 quarts 1 pint. Since there are 2 pints to a quart, 1 pint = ½ quart. Therefore 2 quarts 1 pint = 2½ quarts.

5. (2) 5 yd 2 ft 3 in
Add 3¼ yard and 2½ yard.

$3\frac{1}{4} + 2\frac{1}{2} = 3\frac{1}{4} + 2\frac{2}{4} = 5\frac{3}{4}$

¾ of a yard is equal to ¾ of 36 inches or 27 inches. But 27 inches, = 24 inches + 3 inches, or 2 feet 3 inches. So 5¾ yards = 5 yards 2 feet 3 inches.

(page 159)

6. (3) 5,279 ft 4 in
First change 1 mile to 5,280 feet. Then subtract. To do this, borrow 1 from the feet and change it to 12 inches.

```
  5279 ft 12 in
−          8 in
  5279 ft  4 in
```

7. (2) 7,000 lb
Change all the measurements to pounds. Use the fact that 1 ton equals 2,000 pounds.

Truck with load: 5 × 2,000 or 10,000 lb
Load: 2,000 + 1,000, or 3,000 lb

Subtract to find the weight of the truck *without* the load: 10,000 lb − 3,000 lb = 7,000 lb.

(page 161)

8. (2) 1 gal 1 qt 1 cup
Multiply 6 ounces by 4 to find the total number of ounces baby drinks in one day.

6 ounces × 4 = 24 ounces

For one week's supply, multiply by 7. 7 × 24 = 168 ounces. Since there are 8 ounces in 1 cup, divide 168 ÷ 8 = 21 cups
2 cups = 1 pint so 21 ÷ 2 = 10 pints 1 cup
2 pints = 1 quart so 10 ÷ 2 = 5 quarts 1 cup
4 quarts = 1 gallon so 5 ÷ 4 = 1 gal 1 qt 1 cup for 1 week.

9. (3) 35 yd 1 ft 8 in
Multiply what was ordered (10 ft 8 in) by 10 to find out how much timber was delivered.

```
   10 ft  8 in
×        10
  100 ft 80 in
```

Change 100 feet 80 inches to yards, feet, inches by using the facts that 1 yd = 3 ft and 1 ft = 12 in.

100 ft = 99 ft + 1 ft
= 33 yd 1 ft
80 in = 72 in + 8 in
= 6 ft 8 in

Add.

```
   33 yd 1 ft
+        6 ft 8 in
   33 yd 7 ft 8 in
```

Since 7 ft = 6 ft + 1 ft = 2 yd 1 ft, the final answer is 35 yd 1 ft 8 in.

(page 162)

10. (2) 4 yd 7 in
Divide the length of the piece of material (25 yd 6 in) by 6. Remember that 1 yd = 36 in.

```
   4 yd 7 in
6)25 yd 6 in
  24     +36
   1     6)42
```

11. (1) 35 yd 8 in
The pledge of 25¢ for each foot walked is extra information. To find the average walk, add the distances that they walked and divide the total by 3.

```
    40 yd
    35 yd
+   30 yd 2 ft
   105 yd 2 ft
```

```
    35 yd 0 ft  8 in
3)105 yd 2 ft
  105    0   → 3)24 in
    0    2       24
                  0
```

30 Metric Measurement

(page 165)

1. (3) 13,000
If you go from grams to milligrams in the chart, you move 3 places to the right. So move the decimal point 3 places to the right in 13.

13.000.

The answer is 13,000 milligrams.

2. (2) 100
When you go from meters to centimeters in the chart, you go 2 places to the right. When you move the decimal in 1. two places to the right, you get 100.

3. (1) 0.005
When you go from liters to kiloliters in the chart, you move 3 places to the left.

.005.

Therefore, 5 liters equals 0.005 kiloliters.

(page 166)

4. (3) 3.9
Multiply to find the weight of all the kittens.

600 g × 3 = 1,800 g
= 1.8 kg

Add the weight of the mother (2.1 kg) and the weight of the kittens (1.8 kg) to get the total weight.

```
  2.1 kg   mother
+ 1.8 kg   kittens
  3.9 kg   total weight
```

5. (2) 1,000
Each of the 3 bottles contained 2L of soda, so there was 3 × 2 or 6L of soda in all. Divide by 6 to find the average amount that the 6 people who drank soda consumed. 6L ÷ 6 = 1L. The problem asks for an answer in milliliters. You need to move the decimal point 3 places to the right to change 1L to milliliters. You get 1,000 mL.

6. (2) 23.4
Change 76,600 meters to kilometers. You get 76.6 kilometers. Subtract from 100 kilometers (the total length of the race route) to find how far she had to go to finish.

```
  100.0 km   length of race
−  76.6 km   already ridden
   23.4 km   yet to go
```

7. (1) 1,300
Change 1.3 kilograms to grams. You need to move the decimal point 3 places to the right.

1.3 kg = 1,300 g

It will take 1,300 packets to make 1.3 kilograms.

31 Dealing with Units of Time

(page 168)

1. (3) 1,920
There are 24 hours in a day, so multiply 80 (the number of days) by 24.

24 × 80 = 1,920

2. (2) 54
Change 1 hour 48 minutes to minutes. Since 1 hour equals 60 minutes, 1 hour 48 minutes = 108 minutes. Susan had to pay for half of this amount of time, or 54 minutes.

(page 170)

3. (2) 11 hr 40 min
Add to find out how long it takes him to get to work: 15 + 55 = 70. It also takes 70 minutes to get from the office back home. So each day he spends 140 minutes commuting. Each 5-day work week the total time is 5 × 140, or 700 minutes. Divide by 60 (number of minutes in 1 hour) to change 700 minutes to hours.

```
    11  ← hours
60)700
   60
   100
    60
    40  ← minutes
```

He commutes for 11 hours 40 minutes each week.

4. (1) 6
First change 3 hours 57 minutes to minutes. 3 hr = 3 × 60 min = 180 min.

3 hr 57 min =
180 min + 57 min = 237 min

Subtract 231 (the minutes the winner took) to find how much less time this cyclist took in the race.

```
  237 min   slower cyclist
− 231 min   winner
    6 min   less for winner
```

(page 171)

5. (1) 6
Here you know the distance (300 miles) and the speed (50 miles per hour). Use the formula $t = d \div r$ to find the time traveled.

$t = d \div r$
$t = 300 \div 50$
$t = 6$ (hours)

6. (2) 42
The first person walked for 4×6 or 24 hours. He averaged $5\frac{1}{2}$ miles per hour. Use $d = rt$ to find how far he walked.

$d = rt$

$d = 5\frac{1}{2} \times 24$

$d = \frac{11}{\not 2} \times \frac{\overset{12}{\not 24}}{1}$

$d = \frac{132}{1}$, or 132 (miles)

The second person walked 6×5, or 30 hours and averaged 3 miles per hour. Use $d = rt$ to find how far this person walked.

$d = rt$

$d = 3 \times 30 = 90$ (mi)

Subtract to find how much more the first person walked.

132	first person
− 90	second person
42	more for first person

(page 173)

7. (2) $60
Use $i = prt$ to calculate the interest for the first loan. Use the fact that 8% = 0.08.

$i = prt$

$i = 2{,}000 \times 0.08 \times 3$

$i = 160 \times 3 = 480$ (dollars)

Now calculate the interest for the second loan. Use 9% = 0.09.

$i = prt$

$i = 3{,}000 \times 0.09 \times 2$

$i = 270 \times 2 = 540$ (dollars)

Subtract to find the difference in interest on the two loans.

$\$540 - \$480 = \$60$.

8. (3) $7,200
Use $i = prt$ to find how much interest he owes each person after 2 years. Use the fact that 12% = 0.12.

$i = prt$

$i = 10{,}000 \times 0.12 \times 2$

$i = 1{,}200 \times 2$

$i = 2{,}400$ (dollars)

Multiply $2,400 by 3 (the number of people he paid interest to).

Total interest $= \$2{,}400 \times 3$
$= \$7{,}200$

32 Linear, Square, and Cubic Measure

(page 176)

1. (3) 28
The long side collapses to ½ of its original 11 inches. So the folded sheet of paper will be $\frac{1}{2} \times 11$, or $5\frac{1}{2}$ inches wide and $8\frac{1}{2}$ inches long. To find the distance around, add the lengths of all 4 sides.

$8\frac{1}{2} + 5\frac{1}{2} + 8\frac{1}{2} + 5\frac{1}{2} =$

$26\frac{4}{2}$, or 28

2. (2) $30\frac{1}{2}$
The short side collapses to ½ of its original $8\frac{1}{2}$ inches.

$\frac{1}{2} \times 8\frac{1}{2} = \frac{1}{2} \times \frac{17}{2} = \frac{17}{4}$, or $4\frac{1}{4}$

So the folded piece of paper has a length of 11 inches and a width of $4\frac{1}{4}$ inches. Add the lengths of all 4 sides to find the distance around.

$11 + 4\frac{1}{4} + 11 + 4\frac{1}{4} =$

$30\frac{2}{4}$, or $30\frac{1}{2}$

3. (1) 56

Subtract $2\frac{1}{2}$ *twice* from the length of the glass ($2\frac{1}{2}$ inches for each end) to get the length of the wire.

$24 - 2\frac{1}{2} - 2\frac{1}{2} = 19$ (inches)

Subtract $2\frac{1}{2}$ *twice* from the width of the glass to get the width of the wire.

$14 - 2\frac{1}{2} - 2\frac{1}{2} = 9$ (in)

Add the lengths of the sides of the wire to find the total amount of wire.

$19 + 9 + 19 + 9 = 56$

(page 178)

4. (2) 240
Subtract 2 inches (the width of the frame) *twice* from 24 inches to find the length of the picture area.

$24 - 2 - 2 = 20$ (in)

Do the same to find the width of the picture area.

$16 - 2 - 2 = 12$ (in)

Multiply 20 by 12 (length by width) to find the area of the picture.

$20 \times 12 = 240$ (sq in)

5. (3) 400
To find the area of each room, multiply length (20 ft) by width (10 ft).

$20 \times 10 = 200$ (sq ft)

He had 2 rooms this size, so the total amount of carpeting needed was 200×2, or 400 sq ft.

The carpet man was told he needed enough carpet for a room 40 feet long and 20 feet wide. He would calculate the number of square feet of carpeting needed by multiplying length (40 ft) by width (20 ft).

$40 \times 20 = 800$ sq ft

To find how much more carpeting was delivered than needed, subtract 400 square feet from 800 square feet. The customer got 400 square feet more carpeting than he needed.

6. (2) 9
A surface that is 1 yard long and 1 yard wide is, in feet, 3 feet long and 3 feet wide. So 1 square yard is 3×3, or 9 square feet. Therefore, 1 square yard is 9 times larger than 1 square foot, so his order would have been 9 times larger than what he really needed.

(page 180)

7. (2) 2,040
First find how many cubic feet of coal the bin will hold. Multiply length times width times height.

$12 \times 10 \times 8 = 960$ cu ft

Subtract 960 cubic feet from the 3,000 cubic feet of coal originally in the coal car.

$3{,}000 - 960 = 2{,}040$ (cu ft)

8. (1) 1,000
Each carton has a volume of 6 cubic feet ($3 \times 2 \times 1 = 6$). The shipping car has a volume of 6,000 cubic feet ($30 \times 20 \times 10 = 6{,}000$). To find the number of cartons that can go into the car, divide.

$6{,}000 \div 6 = 1{,}000$ (cartons)

Answers: "The Real Thing"

4 Zeroing In on Word Problems: Two-Step Problems

(page 32)

1. (3) $118
Add the monthly expenses Ron would have if he did not live with his parents.

$235	rent
23	electricity
7	gas
+ 38	telephone
$303	total expenses for apartment

Subtract $185 (the rent he pays his parents) from $303 (the expenses he would have if he rented an apartment) to find how much he saves by living with his parents.

$303	expenses for apartment
− 185	rent paid to parents
$118	amount saved

2. (2) 769
Add to find how many calories Lynn had for lunch.

268	sandwich
88	milk
+ 75	orange
431	calories for lunch

Subtract 431 from the total number of calories she is allowed each day.

1,200	daily calories
− 431	calories for lunch
769	calories remaining

3. (3) 518
Add the number of students in the four courses.

67	secretarial skills
54	word processing
49	bookkeeping
+ 62	office management
232	students

Subtract 232 from the total enrollment.

750	total enrollment
−232	students in 4 courses
518	students in other courses

4. (5) 524
Add how far she drives going to and returning from the visits.

208	mi to mother's	TO
54	extra mi to sister's	TO
54	extra mi from sister's	FROM
+208	mi from mother's	FROM
524	total mi driven	

5. (2) 1,425
Add to find how many directories were delivered in 4 days.

489	Monday
512	Tuesday
544	Wednesday
+ 530	Thursday
2,075	total in 4 days

Subtract 2,075 from the total number delivered.

3,500	total delivered
−2,075	delivered in 4 days
1,425	delivered on Friday

KEEPING TRACK
Perfect score = 5

Your score ☐

5 Multiplying Whole Numbers

(pages 34–35)

1. (2) $460
Subtract Eric's deductions from his pay to find his take-home pay for 1 week.

$105	weekly pay
− 13	weekly deductions
$ 92	weekly take-home pay

Multiply $92 by 5 to find how much Eric takes home in 5 weeks.

$ 92	weekly take-home pay
× 5	no. of weeks
$460	take-home pay in 5 weeks

If you selected (3) $525 as the answer, you forgot to subtract the deductions from the weekly pay.

2. (2) $79
Multiply to find the total amount Ms. Morales wrote checks for.

$ 25	amt. of each check
× 3	no. of checks written
$ 75	total amt. of checks

Subtract the total amount of checks from the amount in her checking account.

$154	amt. in checking account
− 75	total amt. of checks
$79	balance in account

(pages 36–37)

3. (4) $300
Subtract to find how much more Rosa is paying each month this year.

$405	this year's monthly rent
− 380	last year's monthly rent
$ 25	monthly increase

Multiply $25 by 12 to find how much more she'll pay in 12 months.

$ 25	monthly increase
× 12	number of months
50	
25	
$300	amt. more in 12 months

4. (5) $882
Multiply to find the total cost of the trousers.

$21	cost of 1 pair of trousers
× 14	numbers of pairs
84	
21	
$294	total cost of trousers

Jackets cost 3 times as much as the trousers, and "times as much" tells you to multiply $294 by 3.

$294	cost of 14 pr
× 3	3 times as much
$882	amt. spent for jackets

5. (3) $1,260
Add to find the total amount that the Youngs put into the savings account each month.

$ 45	amt. Mr. Young put in
+ 60	amt. Mrs. Young put in
$105	total amt. saved each month

Multiply $105 by 12 (1 year = 12 months) to find how much the Youngs saved last year.

$ 105	amt. saved each month
× 12	no. of months
210	
1 05	
$1,260	amt. saved last year

(pages 38–39)

6. (2) $480,376

Multiply the number of records sold by the $8 sale price of each record.

60,047	no. of records sold
× $8	price of each record
$480,376	total sales for month

7. (3) 20,355

Multiply the height of the highest point in Florida (345 feet) by 59.

345	highest point in Florida
× 59	59 times higher
3 105	
17 25	
20,355	height of Mt. McKinley

8. (3) 3,516

Multiply to find how many pencils the store had in all on Monday morning.

240	pencils in each box
× 30	boxes the store had
7,200	pencils in all in morning

Subtract the number of pencils the store gave away from the total number of pencils.

7,200	pencils in all in morning
−3,684	pencils given away
3,516	pencils that are left

9. (3) 104,000

Add to find the total number of acres in Parks B and C.

3,500	acres in Park B
+ 1,700	acres in Park C
5,200	acres in Parks B & C

The area of Park A is 20 times 5,200. "Times" tells you to multiply.

5,200	acres in Parks B & C
× 20	number of times
104,000	acres in Park A

10. (1) 9,365

Multiply to find how many miles the employees drove in April.

2,957	mi averaged daily
× 21	no. of days
2 957	
59 14	
62,097	mi driven in April

Subtract the miles driven in April from those driven in May to find how many more miles the employees drove in May than in April.

71,462	miles driven in May
−62,097	miles driven in April
9,365	no. of miles more in May.

11. (5) 39,040

Multiply the average number of papers sold each day in July by 31 (the number of days in July) to find how many newspapers were sold in July.

1,086	newspapers sold/day
× 31	no. of days in July
1 086	
32 58	
33,666	newspapers sold in July

The word *more* tells you to add 5,374 to 33,666 to find the number of newspapers sold in June.

33,666	newspapers sold in July
+ 5,374	how many more in June
39,040	newspapers sold in June

12. (5) 31,104

Multiply to find how many fishhooks are in 1 box.

144	cards in 1 box
× 12	fishhooks on each card
288	
1 44	
1,728	fishhooks in 1 box.

Multiply 1,728 by 18 to find how many fishhooks are in 18 boxes.

1,728	fishhooks in 1 box
× 18	no. of boxes
13 824	
17 28	
31,104	fishhooks in 18 boxes

13. (2) $273

Multiply to find the total cost of 24 boxes of the golf balls.

$11	cost of 1 box
× 24	no. of boxes
44	
22	
$264	total cost of golf balls

You know that Margaret spent $537 in all and that she spent $264 for golf balls. The rest of the money was spent for trophies. Subtract to find how much was spent for trophies.

$537	total amount spent
− 264	amt. spent on golf balls
$273	amt. spent for trophies

Keeping Track

Perfect score = 13

Your score ☐

6 Dividing Whole Numbers

(page 42)

1. (2) 56

Add to find how many thermos bottles there are in all.

235	red thermos bottles
+213	blue thermos bottles
448	no. of bottles in all

Divide the total number by 8 because 8 bottles will be in each carton.

56	number of cartons
8)448	number of bottles
40	↗ bottles in each carton
48	
48	
0	

2. (3) 30,906

Add to find how many cars were produced in all.

98,395	sedans
+ 56,135	station wagons
154,530	total no. of cars

Divide the total number of cars produced by 5 months.

30,906	cars/month
5)154,530	total no. of cars
15	↗ number of months
04	
00	
45	
45	
03	
00	
30	
30	
0	

(page 44)

3. (4) $189

Add to find how much the family spent in all.

$147 gasoline
127 lodging
263 food
\+ 31 admission fees
$568 total spent

Divide the total amount spent by 3 days of vacation.

$189 spent each day
3)$568 spent on vacation
↗ days
3
26
24
28
27
1

Since you have a remainder, you know that they actually spent a little more than $189 each day. All the answer choices are whole numbers, and $189 is clearly the best.

4. (4) 106

Multiply to find the total number of folders.

48 folders in each box
× 20 boxes of folders
960 total no. of folders

Divide the total number of folders by 9 classrooms.

106 r6 folders
9)960 folders in all
↗ classrooms
9
06
00
60
54
6

The answer is 106 folders for each classroom. The remainder tells you that there were 6 folders left over.

(page 45)

5. (4) 266

Add to find how many pounds of apples were sold altogether.

4,468 lbs of red apples
1,342 lbs of yellow apples
\+ 574 lbs of green apples
6,384 total lbs of apples

Divide the total number of pounds by 24 stores.

266 lbs per store
24)6,384 lbs in all
↗ stores
4 8
1 58
1 44
144
144
0

6. (4) 106

Subtract to find the weight of the 27 desks.

4,387 weight of van and desks
−1,525 weight of van
2,862 weight of 27 desks

Divide the total weight of the desks by 27 desks to find the weight of each desk.

106 lbs per desk
27)2,862 lbs in all
↗ desks
2 7
16
00
162
162
0

7. (5) $1,872

Divide the cost of 5 chairs by 5 to find the cost of 1 chair.

$104 cost of 1 chair
5)$520 cost of 5 chairs
↗ number of chairs
5
20
00
20
20
0

Multiply the cost of 1 chair by the number of chairs bought.

$104 cost of 1 chair
× 18 no. bought
832
1 04
$1,872 total cost of chairs

8. (4) 22

First multiply the number of figs per can (8) by the number of cans in a carton (60) to get the number of figs per carton.

60 cans per carton
× 8 figs per can
480 figs per carton

Divide the total number of figs (10,590) by 480 to find the number of full cartons.

22 r30 no. of cartons
480)10,590 total no. of figs
↗ figs per carton
9 60
990
960
30

The answer is 22 full cartons.

9. (2) $164

Add to find the total expenses for the month of January.

$325 rent
\+ 167 utilities and food
$492 total expenses

Divide the total expenses by 3 (the number of people sharing the expenses) to find how much each person should wind up paying.

164 amt. per person
3)492 total expenses
↗ number of people
3
19
18
12
12
0

Each person should spend $164. Since JoAnne had paid none of the expenses, she owes a total of $164.

10. (2) 121

Add to find how many pounds of fish were caught in the two days.

468 pounds caught first day
\+380 pounds caught second day
848 pounds caught in all

Divide total number of pounds by 7, the number of people fishing.

121 r1 per person
7)848 in all
↗ number of people
7
14
14
08
7
1

11. (3) 127

Add to find how much fencing is needed for the two counties.

2,570 ft in Sioux County
\+3,780 ft in Plymouth County
6,350 total no. of ft.

Divide the total number of feet by 50 (the number of feet per roll).

```
        127  rolls
   50)6,350  ft
   ↗  5 0
 ft   1 35
per roll 1 00
        350
        350
          0
```

12. (4) 9
Multiply the number of students by cookies needed for each.

```
   34  students
×   3  cookies for each student
  102  cookies needed in all
```

Divide the total number of cookies (102) by the number of cookies in each box (12).

```
       8 r6  boxes
   12)102    cookies in all
   ↗   96
cookies  6
per box
```

8 boxes plus 6 more cookies are needed, so 1 more full box will have to be bought. This makes 9 boxes in all that are needed.

13. (2) 31
Multiply the number of tiles in each box by the number of boxes.

```
     120  tiles in each box
×     15  boxes
     600
    120   total no. of tiles
   1,800
```

Divide the total number of tiles (1,800) by the number of tiles needed to make each tray (58).

```
        31 r2  trays
   58)1,800    tiles in all
   ↗  1 74
tiles    60
per tray 58
          2
```

A total of 31 trays can be made. The 2 tiles left over are not enough to make another tray.

KEEPING TRACK
Perfect Score = 13
Your Score = ☐

7 Zeroing In on Multistep Problems

(page 49)

1. (3) 20
Multiply the value of 1 green question (2 points) by 3 to find the total points for the 3 green questions.

```
  2  points per green question
×3   number of green questions
  6  total points for green questions
```

Next, add the points for all the questions she answered to get her total points scored.

```
   11  points for 1 yellow
    3  points for 1 red
+   6  points for 3 green
   20  total points
```

2. (3) 84°
Add the temperatures for the 5 days of the week.

```
   84  Monday
   89  Tuesday
   85  Wednesday
   85  Thursday
+  77  Friday
  420
```

Divide by the number of days to get the average. (420 ÷ 5 = 84)

3. (2) 4
Multiply to find how many hours Valerie works on weekdays and on weekends.

Weekdays	**Weekends**
8 hrs per day	3 hrs per day
× 5 days	×2 days
40 hrs weekdays	6 hrs weekends

Add to find the total number of hours Valerie works each week.

```
   40  weekdays
+   6  weekends
   46  Valerie's total hrs
```

Add to find the total number of hours Frida works each day.

```
    7  hrs for day job
+   3  hrs for night job
   10  total hrs per day
```

Multiply by 5 (the number of days she works each week) to find the total number of hours Frida works each week.

```
   10  hrs per day
×   5  days
   50  Frida's total hrs
```

Subtract Valerie's total hours from Frida's total hours to find how many hours more Frida works than Valerie.

```
   50  Frida's total hrs
−  46  Valerie's total hrs
    4  number hrs more for Frida
```

4. (2) 8
He can handle 2 customers per hour (since each takes 30 minutes) when he gives a shampoo and haircut. So multiply by 8 (hours he works per day) to find how many such customers he can handle in a day.

```
    2  shampoos/haircuts per hr
×   8  hrs worked
   16  shampoos/haircuts per day
```

He can handle 3 customers per hour (3 customers, 20 minutes each) if they get only a haircut. Multiply by 8 to find how many in a day.

```
    3  haircuts per hr
×   8  hrs worked
   24  haircuts per day
```

Subtract to find how many more customers he can handle in a day if they all get just haircuts.

```
   24  haircuts per day
−  16  shampoos/haircuts per day
    8  customers more
```

5. (4) phone
Add to find the total of all her bills.

```
 $ 52  department store
  109  car payment
   17  electricity
   28  phone
   35  dentist
 $241  total bills
```

Subtract to see how much more her bills are than her bank balance.

```
 $241  bills
  218  bank balance
 $ 23  amt. more in bills
```

Find the *smallest* bill over $23 that she has. It is the $28 phone bill. This is the one she will have to pay later.

6. (1) $1,050
Multiply regular fare ($210) by number of passengers per flight at regular fare.

$210	regular fare
× 75	passengers
1050	
1470	
$15,750	per flight, regular

Multiply the discount fare by number of passengers to find how much the airline made per flight at that fare.

$150	discount fare
×112	passengers
300	
150	
150	
$16,800	per flight, discount

Subtract to find how much more they made per flight with the discount fare.

$ 16,800	per flight, discount
− 15,750	per flight, regular
$ 1,050	amt. more for discount

7. 2) $178
Multiply to find how much each person earns weekly.

40	hours worked
× $6	Conrad's hourly wage
$240	Conrad's weekly wages

38	Hours worked
×$11	Dorothy's hourly wage
38	
38	
$418	Dorothy's weekly wages

Subtract to find how much more Dorothy earns.

$418	Dorothy's weekly wages
−$240	Conrad's weekly wages
$178	more

8. (4) $390
Multiply the amount of deposit by the deposits per month.

$45	deposit
× 4	deposits per month
$180	deposited per month

Multiply by the number of months.

$180	per month
× 2	months
$360	deposited in 2 months

Add the initial deposit to find his total savings.

$360	deposited in 2 months
+ 30	initial deposit
$390	total savings deposited

9. (3) 186,000
Multiply to find the circulation for Monday through Friday.

26,000	copies sold daily
× 5	days paper is sold
130,000	copies sold weekdays

Add to find the Sunday circulation.

26,000	copies sold daily
+ 30,000	more sold Sun.
56,000	copies sold Sun.

130,000	copies sold weekdays
+ 56,000	copies sold Sun.
186,000	copies sold in all

10. (1) 4 red, 5 yellow, 6 blue
Divide the number of jars of each color by 12.

4	boxes
12)48	jars red paint
48	
0	

5	boxes
12)60	jars yellow paint
60	
0	

6	boxes
12)72	jars blue paint
72	
0	

KEEPING TRACK
Perfect score = 10
Your score ☐

8 Adding and Subtracting Decimals

(pages 56–57)

1. (4) $778.79
Add to find the total amount of the deposits. Be sure to line up the decimal points and fill in zeros where necessary.

$175.00
82.51
125.00
+ 396.28
778.79

2. (3) $59.87
Add the amount the Johnsons saved to the amount they paid.

$49.89	amt. they paid
+ 9.98	amt. they saved
$59.87	original price

3. (4) 2,255.925
Add the lengths of the main span and *each of the two* approaches.

1,158.245	main span
548.840	one approach
+ 548.840	other approach
2,255.925	total length

4. (1) 83.9
Subtract to find how many miles farther she drove on Monday.

337.8	Monday
−253.9	Tuesday
83.9	mi farther on Monday

5. (3) $11.89
Subtract to find the monthly cost.

$86.52	amt. with her son
− 74.63	amt. without her son
$11.89	cost of insuring her son

6. (5) 137.74
Subtract to find how many more points for the winner.

835.65	winner's points
−697.91	2nd place points
137.74	more points for winner

7. (1) 9.75
Add to find how many inches have been completed.

8.40	Monday
7.60	Tuesday
8.25	Wednesday
+ 6.00	Thursday
30.25	total in completed

Subtract the inches completed from the total length.

40.00	total length
−30.25	in completed
9.75	left to be constructed

8. (4) $4.17
Add the cost of the shirt and the tax.

$14.79	cost of shirt
+ 1.04	tax
$15.83	total cost

Subtract the total cost from the amount given to the clerk.

$20.00	given to clerk
− 15.83	amt. of purchase
$ 4.17	change received

9. (2) $38.75
Add to find how much was spent for milk and bread.

$37.50	milk
+ 18.75	bread
$56.25	total for milk & bread

To find how much was spent for meat, subtract the amount for milk and bread from the total spent.

$95.00 total amt. spent
− 56.25 milk and bread
$38.75 meat

10. (5) 2.55
Add to find how much snow has fallen so far.

0.95 December
1.60 January
2.00 February
0.80 March
+4.50 April
9.85 total this year

Subtract.

12.40 usual snowfall
− 9.85 total this year
2.55 less than usual

Keeping Track
Perfect score = 10

Your score ☐

9 Multiplying Decimals

(page 59)

1. (3) 767.125
Multiply kilometers per hour by the number of hours she drove. Mark off 3 decimal places.

80.75 km per hr
× 9.5 hours driven
40 375
726 75
767.125 kilometers

2. (3) $1.64
Change 63 cents to a decimal (63 cents = $0.63). Multiply the cost per pound by the number of pounds. Mark off 3 decimal places.

$0.63 cost per pound
× 2.6 number of pounds
378
1 26
$1.638

The answer should be in dollars and cents, so round to the nearest hundredth. You get $1.64.

(pages 60–61)

3. (1) $1.44
Multiply the cost of 1 yard of tape by the number of yards. Add zeros and mark off 4 decimal places.

$0.015 cost per yard
× 5.5 number yards
75
75
$0.0825 cost of tape

Round $0.0825 to the nearest cent. You get $0.08 for the cost of the tape. Add the cost of the thread to find the total cost.

$0.08 cost of the tape
+ 1.36 cost of the thread
$1.44 total cost

4. (2) 327
Add to find the total number of apartments.

48 building A
56 building B
+114 building C
218 total apartments

Multiply the total number of apartments by the number of parking spaces per apartment. Mark off 1 decimal place.

218 apartments
× 1.5 spaces per apartment
109 0
218
327.0 total spaces needed

5. (3) $0.0058
To find what company B charges per page, multiply what company A charges by 0.8. Mark off 4 decimal places.

$0.029 company A's price
× 0.8
$0.0232 company B's price

You can tell from the answer choices that you should not round. To find the savings per page, subtract company B's price from company A's. You have to add on a zero to A's price before subtracting.

$0.0290 company A's price
−$0.0232 company B's price
$0.0058 savings per page

6. (4) $4.54
Multiply the number of *additional* minutes by the cost per additional minute. 15-minute call = 1 first minute + *14 additional minutes.*

$.29 per additional min
× 14 additional min
116
29
$4.06 for 14 add. min

Add the charge for the first min.

$4.06 14 min
× .48 for first min
$4.54 total for the call

7. (4) $339
Multiply the hourly wage by 2.5 to find the hourly overtime wage. Mark off 3 decimal places.

$5.65 hourly wage
× 2.5 times greater
2 825
11 30
$14.125 hourly overtime wage

Round to the nearest cent: $14.13 an hour for overtime.

48 hrs = 40-hr week +
8 hrs overtime.

Multiply to find how much Tony gets paid for a 40-hour week. Mark off 2 decimal places.

$5.65 hourly wage
× 40 hrs worked
$226.00 pay for 40-hr week

Multiply to find how much Tony gets paid for 8 hours overtime. Mark off 2 decimal places.

$14.13 hourly overtime wage
× 8 overtime hrs worked
$113.04 pay for overtime

Add to find the total pay.

$226.00 pay for 40-hour week
× 113.04 pay for overtime
$339.04 total pay

The answer choices are in whole dollars, so round $339.04 to the nearest dollar. You get $339.

8. (5) 7.0 seconds
To find the time for completing a call the 2nd year, multiply the time for the 1st year by 0.8. Mark off two decimal places.

12.5 seconds for 1st year
× 0.8
10.00 seconds for 2nd year

You can drop the 2 zeros after the decimal point. To find the time for the 3rd year, multiply the time for the 2nd year by 0.7.

Mark off 1 decimal place.

```
  1 0   sec for 2nd yr
 ×0.7
  7.0   sec for 3rd yr
```

KEEPING TRACK

Perfect score = 8

Your score ☐

10 Dividing Decimals

(page 62)

1. (1) $0.12

Divide the total cost of the bars by the number of bars ordered.

```
     $0.12
 48)$5.76
     4 8
       96
       96
        0
```

2. (1) 7.4

Divide the total number of miles by the number of miles the Jacksons rode each hour.

```
       7.4
 16)118.4
    112
      6 4
      6 4
        0
```

(page 64)

3. (4) 87

Divide the total number of kilograms by the number of kilograms in each package.

```
        8 7.
 1.5)130.5
     120
      10 5
      10 5
         0
```

4. (3) 4.6

To find how many times as many, divide the number of grams of carbohydrates in a potato by the number in a carrot.

```
        4.6
 7.2)33.1 2
     28 8
      4 3 2
      4 3 2
          0
```

(page 65)

5. (3) 16

Divide the total number of yards by the number of yards needed for each frame.

```
       1 6.
 3.5)56.0
     35
     21 0
     21 0
        0
```

6. (4) 5,600

Divide the total cost by the cost of each part.

```
           5,600.
 0.075)420.000
       375
        45 0
        45 0
           00
           00
            00
            00
             0
```

(pages 66–67)

7. (1) 0.75

Divide the total number of pounds of fudge by the number of packages.

```
        0.75
 640)480.00
     448 0
      32 00
      32 00
          0
```

8. (4) 37

Divide the total number of inches by the length of each tab.

```
         37.6
 1.25)47.00 0
      37 5
       9 50
       8 75
         75 0
         75 0
            0
```

Only whole tabs should be counted, so 37 is the answer.

9. (5) 25

Divide the total number of acres by the size of each lot.

```
        25.
 .25.)6.25.
      5 0
      1 25
      1 25
         0
```

10. (2) $8.95

Divide the total cost by the number of mixers to find the price of a single mixer.

```
     $   8.95
 12)$107.40
      96
      11 4
      10 8
         60
         60
          0
```

11. (4) $809

Divide the total price by the number of acres to find the price per acre.

```
     $   808.8
 34)$27500.0
     272
       300
       272
        28 0
        27 2
           8
```

Round 808.8 to the nearest whole (809). The price per acre is about $809.

12. (4) 13

Subtract to find how much he had to spend on stemps.

```
  $10.00   total amt. he has
 −  1.75   cost of envelopes
   $8.25   money for stamps
```

Divide what he had left for stamps by 0.22 to find how many 22-cent stamps he will get.

```
          37.
 0.22)8.25
      6 6
      1 65
      1 54
        11
```

He will get 37 stamps, with some change left over. Subtract the number of stamps he will get from the number of envelopes to find how many more envelopes than stamps.

```
  50   envelopes
 −37   stamps
  13   more envelopes
```

13. (3) $31.50

Divide 8.29 (cost of 100 programs) by 100 to find the cost per program.

```
      0.0829
100)8.2900
    0
    8 2
    0 0
    8 29
    8 00
      290
      200
       900
       900
         0
```

Multiply the cost per program ($0.0829) by 380 to find the cost of 380 programs. Mark off 4 decimal places.

```
   $0.0829   cost per program
 ×     380   no. of programs
    0 0000
   06 632
  024 87
  $31.5020   total cost
```

Round to the nearest cent. The final answer is $31.50.

KEEPING TRACK
Perfect score = 13

Your score ☐

11 Zeroing In on Word Problems: Insufficient Data

(page 69)

1. (2) $1,249
Add the two incomes.

```
    $975   first job
 +   743   second job
  $1,718   total monthly income
```

Subtract the deductions.

```
  $1,718   total monthly income
 –   469   monthly deductions
  $1,249   take-home pay
```

2. (5) Insufficient data is given to solve the problem.

The problem does not tell you how many pounds of apples you get for $.68.

3. (5) Insufficient data is given to solve the problem.

The problem does not tell you how much money Ms. Nishimura gave the station attendant.

4. (4) 2,857,250
Multiply the daily attendance by the number of days the park is open.

```
     25,975   daily attendance
 ×      110   days open
     00 000
    259 75
  2 597 5
  2,857,250   people per year
```

5. (1) $1,428.80
Multiply to find the amount of the commission.

```
   $3,245   amt of sales
 ×   0.24   rate of commission
   129 80
   649 0
  $778.80   amt of commission
```

Add the monthly salary and the commission.

```
  $   650.00   monthly salary
 +    778.80   commission
   $1,428.80   total earnings
```

6. (5) Insufficient data is given to solve the problem.

The problem does not tell you how much the chicken costs if not on sale.

7. (5) Insufficient data is given to solve the problem.

The problem does not tell you how much hamburger is used to make each Midget Burger.

KEEPING TRACK
Perfect score = 7

Your score ☐

13 Adding Fractions

(page 77)

1. (2) $1\frac{7}{8}$
Add the three fractions to find how many yards of material he bought.

$$\frac{5}{8} + \frac{7}{8} + \frac{3}{8} = \frac{15}{8}$$

Change the improper fraction $15/8$ to the mixed number $1\,7/8$ and you have the final answer.

2. (4) $6\frac{1}{2}$
Add the miles jogged on Monday and Friday.

$$2\frac{3}{4} + 3\frac{3}{4} = 5\frac{6}{4}$$

Change $6/4$ to $1\,2/4$ or $1\,1/2$.
Then $5\,6/4$ equals $5 + 1\,1/2$ or $6\,1/2$.

3. (4) $1\frac{1}{2}$
Add the amounts of rain for the three days when it rained.

$$\frac{9}{10} + \frac{3}{10} + \frac{3}{10} = \frac{15}{10}$$

Next change $15/10$ to a mixed number in lowest terms.

$$\frac{15}{10} = 1\frac{5}{10} \text{ or } 1\frac{1}{2}$$

The final answer is $1\,1/2$ inches of rain.

(page 80)

4. (5) $4\frac{7}{8}$
All three thicknesses must be added. Change the fractions so that all three fractions have the common denominator 8.

$$3\frac{3}{4} + \frac{3}{4} + \frac{3}{8} = 3\frac{6}{8} + \frac{6}{8} + \frac{3}{8}$$

Add the fractions to get $3\,15/8$. Change the improper fraction to a mixed number to get $3 + 1\,7/8$, or $4\,7/8$.

5. (1) $40\frac{1}{4}$
Add the distance from the floor to table top and the distance from the table top to the shelf. Use a common denominator of 4.

$$27\frac{1}{2} + 12\frac{3}{4} \text{ equals } 27\frac{2}{4} + 12\frac{3}{4}$$

When you add the whole numbers and the fractions, you get $39\,5/4$ or $39 + 1\,1/4$, or $40\,1/4$.

6. (4) $1\frac{1}{16}$
Add to find the total weight. Use a common denominator of 16.

$$\frac{1}{2} + \frac{1}{4} + \frac{1}{16} + \frac{1}{4} \text{ becomes}$$

$$\frac{8}{16} + \frac{4}{16} + \frac{1}{16} + \frac{4}{16}$$

Add and you get $17/16$, which equals the mixed number $1\,1/16$.

7. (4) $3\frac{3}{4}$
Add to find the wingspan. Use a common denominator of 4.

$$2\frac{1}{4} + 1\frac{1}{2} = 2\frac{1}{4} + 1\frac{2}{4}$$

Add to get $3\,3/4$.

8. (2) $1\frac{1}{4}$

Add the thickness of each pane of glass and the thickness of the air space. Use a common denominator of 16.

$\frac{3}{16} + \frac{3}{16} + \frac{7}{8} =$

$\frac{3}{16} + \frac{3}{16} + \frac{14}{16}$ or $\frac{20}{16}$.

The fraction $^{20}/_{16}$ can be changed to the mixed number $1^{4}/_{16}$. Reduce the fraction and you get $1^{1}/_{4}$.

9. (2) $7\frac{5}{12}$

Add to find the total number of hours. Use a common denominator of 12.

$3\frac{2}{3} + 3\frac{3}{4} = 3\frac{8}{12} + 3\frac{9}{12}$, or $6\frac{17}{12}$

Change $^{17}/_{12}$ to $1^{5}/_{12}$ and you get $6 + 1^{5}/_{12}$ or $7^{5}/_{12}$.

10. (1) $1\frac{1}{4}$

Add to find the total length of time. Use a common denominator of 12.

$\frac{1}{3} + \frac{1}{6} + \frac{3}{4} = \frac{4}{12} + \frac{2}{12} + \frac{9}{12}$, or $\frac{15}{12}$

Change to a mixed number and reduce. You get $1^{3}/_{12}$ or $1^{1}/_{4}$.

11. (2) $23\frac{1}{4}$

Add to find the total length. Use a common denominator of 4.

$18\frac{3}{4} + 4\frac{1}{2} = 18\frac{3}{4} + 4\frac{2}{4}$, or $22\frac{5}{4}$

Change $^{5}/_{4}$ to $1^{1}/_{4}$. The final answer is $22 + 1^{1}/_{4}$, or $23^{1}/_{4}$.

KEEPING TRACK

Perfect score = 11

Your score ☐

14 Subtracting Fractions

(pages 84–85)

1. (2) $6\frac{2}{3}$

To make a dozen (12) loaves, the baker needs 12 cups of flour. Subtract the amount he has from the amount he needs. You need to rename 12 as $11^{3}/_{3}$.

$12 - 5\frac{1}{3} = 11\frac{3}{3} - 5\frac{1}{3}$

Do the subtraction and you get $6^{2}/_{3}$.

2. (1) $4\frac{1}{2}$

Subtract to find the difference in the skirt lengths.

$34\frac{1}{4} - 29\frac{3}{4} = 33\frac{5}{4} - 29\frac{3}{4}$

When you subtract, you get $4^{2}/_{4}$. Reduce the fraction to get the final answer, $4^{1}/_{2}$.

3. (3) $4\frac{7}{8}$

Subtract the length used from the original length. Use a common denominator of 8.

$13\frac{1}{2} - 8\frac{5}{8} = 13\frac{4}{8} - 8\frac{5}{8}$

Rename $13^{4}/_{8}$ as $12^{12}/_{8}$, then subtract. $12^{12}/_{8} - 8^{5}/_{8} = 4^{7}/_{8}$.

4. (4) $3\frac{1}{4}$

Subtract to find the difference. Use a common denominator of 4.

$37\frac{1}{2} - 34\frac{1}{4} = 37\frac{2}{4} - 34\frac{1}{4} = 3\frac{1}{4}$

5. (1) $15\frac{1}{4}$

Multiply to find how much material the storekeeper had in all at the start.

```
  25   yards on each bolt
×  3   number of bolts
  75   total yards
```

Subtract the amount sold from the total amount.

$75 - 59\frac{3}{4} = 74\frac{4}{4} - 59\frac{3}{4} = 15\frac{1}{4}$

6. (5) $\frac{5}{12}$

Add the given amounts of varnish and turpentine. Use a common denominator of 12.

$\frac{1}{3} + \frac{1}{4} = \frac{4}{12} + \frac{3}{12} = \frac{7}{12}$

Subtract $^{7}/_{12}$ from the whole amount (1) to find what fraction is tung oil.

$1 - \frac{7}{12} = \frac{12}{12} - \frac{7}{12}$ or $\frac{5}{12}$

7. (2) $12\frac{1}{2}$

Add to find how many miles the Swensons hiked in two days.

$12\frac{1}{4} + 14\frac{5}{8} = 12\frac{2}{8} + 14\frac{5}{8} = 26\frac{7}{8}$

Subtract $26^{7}/_{8}$ from the total miles to find how far they walked on the third day.

$39\frac{3}{8} - 26\frac{7}{8} = 38\frac{11}{8} - 26\frac{7}{8} = 12\frac{4}{8}$

Reduce to get $12^{1}/_{2}$.

8. (3) $4\frac{1}{2}$

Add to find the weight of Diane's fish.

$3\frac{1}{2} + 2\frac{3}{4} = 3\frac{2}{4} + 2\frac{3}{4} = 5\frac{5}{4}$

Since you have to subtract next, it's best to leave $5^{5}/_{4}$ with an improper fraction.

Diane's fish weighed $1^{3}/_{4}$ pounds more than Gwen's. Subtract $1^{3}/_{4}$ from the weight of Diane's fish.

$5\frac{5}{4} - 1\frac{3}{4} = 4\frac{2}{4} = 4\frac{1}{2}$

9. (2) $\frac{7}{8}$

Add to find the weight of the van and its load.

$1\frac{7}{8} + 1\frac{3}{4} = 1\frac{7}{8} + 1\frac{6}{8} = 2\frac{13}{8}$

Change $2\frac{13}{8}$ to $3\frac{5}{8}$.

Subtract the total weight of the van from the weight limit of the bridge.

$4\frac{1}{2} - 3\frac{5}{8} = 4\frac{4}{8} - 3\frac{5}{8} =$
$3\frac{12}{8} - 3\frac{5}{8} = \frac{7}{8}$

KEEPING TRACK

Perfect score = 9

Your score ☐

15 Multiplying Fractions

(page 87)

1. (4) $\frac{1}{4}$

$^{1}/_{3}$ as long tells you to multiply $^{1}/_{3}$ by $^{3}/_{4}$.

$\frac{1}{3} \times \frac{3}{4} = \frac{1 \times 3}{3 \times 4}$

Cancel and multiply in numerator and denominator.

$\frac{1 \times \cancel{3}^{1}}{\cancel{3}_{1} \times 4} = \frac{1}{4}$

2. (1) 36

$^{3}/_{5}$ as fast as tells you to multiply.

First rename 60 as $^{60}/_{1}$.

$\frac{3}{5} \times \frac{60}{1} = \frac{3 \times 60}{5 \times 1}$

Cancel and multiply in numerator and denominator.

$\frac{3 \times \cancel{60}^{12}}{\cancel{5}_{1} \times 1} = \frac{36}{1} = 36$

(page 89)

3. (4) 90

24 tiles each $3\frac{3}{4}$ inches wide are laid side by side. The total width is 24 times $3\frac{3}{4}$ inches.

$$24 \times 3\frac{3}{4} = \frac{24}{1} \times \frac{15}{4} =$$

$$\frac{\overset{6}{\cancel{24}} \times 15}{1 \times \underset{1}{\cancel{4}}} = \frac{90}{1} = 90$$

4. (2) $6\frac{1}{4}$

First find how many times larger the new recipe will be.

$$\frac{30 \text{ pieces}}{8 \text{ pieces}}\ \frac{30}{8} = 3\frac{6}{8} = 3\frac{3}{4}$$

The recipe will be $3\frac{3}{4}$ times larger. Multiply the amount of flour needed ($1\frac{2}{3}$ cups) by how many times larger the new recipe will be.

$$3\frac{3}{4} \times 1\frac{2}{3} = \frac{15}{4} \times \frac{5}{3} =$$

$$\frac{\overset{5}{\cancel{15}} \times 5}{4 \times \underset{1}{\cancel{3}}} = \frac{25}{4} = 6\frac{1}{4}$$

5. (2) $13\frac{1}{4}$

Multiply by 5 to find how many hours he practices on weekdays.

$$5 \times 1\frac{3}{4} = \frac{5}{1} \times \frac{7}{4} = \frac{35}{4} = 8\frac{3}{4}$$

Multiply by 2 to find how many hours he practices during the weekend.

$$2 \times 2\frac{1}{4} = \frac{2}{1} \times \frac{9}{4} =$$

$$\frac{\overset{1}{\cancel{2}} \times 9}{1 \times \underset{2}{\cancel{4}}} = \frac{9}{2} = 4\frac{1}{2}$$

Add $8\frac{3}{4}$ and $4\frac{1}{2}$ to find how many hours he practices in all.

$$8\frac{3}{4} + 4\frac{1}{2} = 8\frac{3}{4} + 4\frac{2}{4} =$$

$$12\frac{5}{4} = 13\frac{1}{4}$$

6. (2) $550

Find how much she puts into the bank by multiplying the income ($1,000) by $\frac{1}{5}$.

$$\frac{1}{5} \times \frac{1{,}000}{1} = \frac{1 \times \overset{200}{\cancel{1{,}000}}}{\underset{1}{\cancel{5}} \times 1} =$$

$$\frac{200}{1} = 200$$

Find how much she uses to pay debts by multiplying the income ($1,000) by $\frac{1}{4}$.

$$\frac{1}{4} \times \frac{1{,}000}{1} = \frac{1 \times \overset{250}{\cancel{1000}}}{\underset{1}{\cancel{4}} \times 1} =$$

$$\frac{250}{1} = 250$$

Add to find how much she saves and uses to pay debts.

$200	saves
+$250	uses to pay debts
$450	total

To find how much she has for current expenses, subtract this total from her income.

$1,000	total income
− 450	saves and pays debts
$ 550	has for current debts

KEEPING TRACK

Perfect score = 6

Your score ☐

16 Dividing Fractions

(page 91)

1. (2) $3\frac{1}{4}$

Divide the total amount of cereal by the number of portions.

$$22\frac{3}{4} \div 7 = \frac{91}{4} \div \frac{7}{1} =$$

$$\frac{\overset{13}{\cancel{91}}}{4} \times \frac{1}{\underset{1}{\cancel{7}}} = \frac{13}{4} = 3\frac{1}{4}$$

2. (4) 2

The $\frac{3}{4}$-yard piece of yarn is being divided into $\frac{3}{8}$-yard pieces.

$$\frac{3}{4} \div \frac{3}{8} =$$

$$\frac{\overset{1}{\cancel{3}}}{\underset{1}{\cancel{4}}} \times \frac{\overset{2}{\cancel{8}}}{\underset{1}{\cancel{3}}} = \frac{2}{1} = 2$$

3. (3) 34

Find the number of working days in 6 months.

20 working days each month × 6 months = 120 working days

Divide the total number of working days by the number of days it takes to make one suit.

$$120 \div 3\frac{1}{2} = \frac{120}{1} \div \frac{7}{2} =$$

$$\frac{120}{1} \times \frac{2}{7} = \frac{240}{7} = 34\frac{2}{7}$$

He can make 34 *complete* suits in 6 months.

4. (2) 13

If two feet are allowed at each end, the length to be divided into $1\frac{1}{3}$ sections is 20 − 4, or 16 feet.

$$16 \div 1\frac{1}{3} = \frac{16}{1} \div \frac{4}{3} =$$

$$\frac{\overset{4}{\cancel{16}}}{1} \times \frac{3}{\underset{1}{\cancel{4}}} = \frac{12}{1} = 12$$

Remember that there will be a plant at the beginning of the 16-foot section and at the end, or 1 more than 12. The number of plants in each row will be 13.

5. (2) $2\frac{5}{8}$

Multiply the weight of 1 bag ($1\frac{3}{4}$) by the number of bags to find the total weight.

$$9 \times 1\frac{3}{4} =$$

$$\frac{9}{1} \times \frac{7}{4} = \frac{63}{4}$$

Leave the answer in fraction form, since you'll need it in that form for the next stage of the solution. Divide the total weight by 6, the number of gift packages.

$$\frac{63}{4} \div \frac{6}{1} =$$

$$\frac{63}{4} \times \frac{1}{6} = \frac{63}{24} = 2\frac{5}{8}$$

6. (4) $2\frac{7}{16}$

Add the four weight losses.

$$2\frac{1}{4} = 2\frac{2}{8}$$
$$1\frac{3}{8} = 1\frac{3}{8}$$
$$3\frac{1}{2} = 3\frac{4}{8}$$
$$+2\frac{5}{8} = 2\frac{5}{8}$$
$$8\frac{14}{8} = 9\frac{6}{8} = 9\frac{3}{4}$$

Divide the total weight loss by the number of people (4).

$$9\frac{3}{4} \div 4 = \frac{39}{4} \div \frac{4}{1} =$$

$$\frac{39}{4} \times \frac{1}{4} = \frac{39}{16} = 2\frac{7}{16}$$

7. (2) $32\frac{2}{5}$

Add to find how many pounds of apples there were in all.

45	Jonathan apples
+36	Winesap apples
81	pounds of apples in all

Divide the pounds of apples in all (81) by the number of pounds used for each quart ($2\frac{1}{2}$).

$81 \div 2\frac{1}{2} = \frac{81}{1} \div \frac{5}{2} =$

$\frac{81}{1} \times \frac{2}{5} = \frac{162}{5} = 32\frac{2}{5}$

8. (5) 72
First find how many books will fit in each box by dividing the total depth of a box (15 in) by the thickness of each book (5/8 in).

$\frac{15}{1} \div \frac{5}{8} =$

$\frac{\overset{3}{\cancel{15}}}{1} \times \frac{8}{\underset{1}{\cancel{5}}} = \frac{24}{1} = 24$

Each box will hold 24 books. Multiply 24 by the number of boxes.

24	books in each box
× 3	boxes
72	books in shipment

9. (3) 8
To find the amount of unused material, subtract the amount used (9¼) from the original amount (16½).

$$\begin{array}{rcr} 16\frac{1}{2} & = & 16\frac{2}{4} \\ -\ 9\frac{1}{4} & = & 9\frac{1}{4} \\ \hline & & 7\frac{1}{4} \end{array}$$

To find how many pillows can be made, divide the amount of unused material (7¼) by the amount needed for each pillow (7/8).

$7\frac{1}{4} \div \frac{7}{8} = \frac{29}{4} \div \frac{7}{8} =$

$\frac{29}{\underset{1}{\cancel{4}}} \times \frac{\overset{2}{\cancel{8}}}{7} = \frac{58}{7} = 8\frac{2}{7}$

He can cover 8 pillows, with 2/7 of a yard left over.

10. (5) Insufficient data is given to solve the problem.
To find how many aprons Alma can make, you would need to know how many yards of material she needs for each apron. The problem does not give you that information.

KEEPING TRACK
Perfect score = 10
Your score ☐

17 Fraction and Decimal Names for Numbers

(pages 94–95)

1. (1) 8
Change 12½ cents to a decimal. First change ½ to a decimal.

$\frac{1}{2} = 2\overline{)1.0}$ = .5

$\quad 1.0$
$\quad \overline{\ \ 0}$

$12\frac{1}{2}¢ = 12.5¢ = \$.125$

Then divide the total amount ($1) by the cost of each tile ($.125).

$.125.\overline{)1.000.}$ = 8.

$\quad 1\ 000$
$\quad \overline{\quad 0}$

2. (3) 79.5
Multiply the amount of meat for each hamburger (¼ pound) by the number of hamburgers (318).

$\frac{1}{4} \times 318 = \frac{1}{\underset{2}{\cancel{4}}} \times \frac{\overset{159}{\cancel{318}}}{1} =$

$\frac{159}{2} = 79\frac{1}{2}$

Option (3) is the only one that has 79 as the whole-number part of the answer, so it must be the answer. (You could change 79½ to a decimal, just to be sure. You would get 79.5.)

3. (2) 2.4
First add the 3 times.

$$\begin{array}{rcl} 2\frac{1}{10} & = & 2\frac{1}{10} \\ 1\frac{3}{5} & = & 1\frac{6}{10} \\ +3\frac{1}{2} & = & 3\frac{5}{10} \\ \hline & & 6\frac{12}{10} = 7\frac{2}{10} = 7\frac{1}{5} \end{array}$$

Next divide by 3 days.

$7\frac{1}{5} \div 3 = \frac{36}{5} \div \frac{3}{1} =$

$\frac{\overset{12}{\cancel{36}}}{5} \times \frac{1}{\underset{1}{\cancel{3}}} = \frac{12}{5} = 2\frac{2}{5}$

Then change the answer to a decimal.

$2\frac{2}{5} = \frac{12}{5}$

$5\overline{)12.0}$ = 2.4

$\quad 10$
$\quad \overline{\ 2\ 0}$
$\quad\ \ 2\ 0$
$\quad \overline{\quad 0}$

4. (4) $54.25
Multiply ¼ of the monthly income ($2,604) to find how much is budgeted for housing costs.

$\frac{1}{\underset{1}{\cancel{4}}} \times \frac{\overset{651}{\cancel{\$2,604}}}{1} = \frac{651}{1} = \$651$

Multiply 1/12 of the housing budget ($651) to find how much is budgeted for electricity.

$\frac{1}{\underset{4}{\cancel{12}}} \times \frac{\overset{217}{\cancel{\$651}}}{1} = \frac{217}{4} = \$54\frac{1}{4}$

Change $\frac{1}{4}$ to a decimal.

$4\overline{)1.00}$ = .25

$\quad 8$
$\quad \overline{\ 20}$
$\quad\ 20$
$\quad \overline{\ \ 0}$

$\$54\frac{1}{4} = \$54.25$

5. (4) 0.5
First find the total number of laps each swimmer completed.

Swimmer 1	Swimmer 2
$3\frac{1}{2} = 3\frac{2}{4}$	$2\frac{1}{2} = 2\frac{2}{4}$
$+2\frac{3}{4} = 2\frac{3}{4}$	$+4\frac{1}{4} = 4\frac{1}{4}$
$5\frac{5}{4} = 6\frac{1}{4}$	$6\frac{3}{4}$

Then find the difference between the totals of the first and second swimmers. The second swam farther, so put that number on top.

$$\begin{array}{r} 6\frac{3}{4} \\ -6\frac{1}{4} \\ \hline \frac{2}{4} = \frac{1}{2} \end{array}$$

Change ½ to a two-place decimal.

$\frac{1}{2} = 0.50$ or 0.5

6. (2) $213.75
First find out how much Teresa earned for the first 40 hours. Multiply her regular hourly wage by 40.

$$\begin{array}{r} \$4.50 \\ \times \quad 40 \\ \hline 180.00 \end{array} = \$180.00$$

Next find her overtime hourly wage. Change 1½ to a decimal and multiply it by $4.50.

$$1\frac{1}{2} = \frac{3}{2} \qquad \begin{array}{r} 1.5 \\ 2\overline{)3.0} \\ \underline{2} \\ 1\,0 \\ \underline{1\,0} \\ 0 \end{array}$$

$$\begin{array}{r} \$4.50 \\ \times \quad 1.5 \\ \hline 2250 \\ 450 \\ \hline 6.750 \end{array} = \$6.75$$

Teresa worked 5 hours overtime (45-hour week = 40 regular hr + 5 overtime hr). Multiply her overtime hourly wage ($6.75) by 5 hours to find her overtime pay.

$$\begin{array}{r} \$6.75 \\ \times \quad 5 \\ \hline 33.75 \end{array} = \$33.75$$

Add her regular pay and her overtime pay.

$$\begin{array}{r} \$180.00 \\ + \quad 33.75 \\ \hline \$213.75 \end{array}$$

KEEPING TRACK
Perfect score = 6
Your score ☐

18 Zeroing In on Word Problems

(pages 96–97)

1. (1) $0.20
Extraneous information: How much money Kevin had to spend.

Divide the cost by the number of issues to find the cost per issue.

For 8 issues:
$8.00 ÷ 8 = $1.00 per issue

For 16 issues:
$12.80 ÷ 16 = $0.80 per issue

Subtract to find how much less he pays if he buys the 16-issue subscription.

$1.00	per issue for 8 issues
− 0.80	per issue for 16 issues
$0.20	less

2. (5) $250
Extraneous information: how long the vacation lasted.

Multiply the total amount ($1,200) by 5⁄12 to find how much was spent on motels.

$\$1,200 \times \frac{5}{12} = \$500$

Multiply the total amount ($1,200) by 3⁄8 to find what was spent for gas and entertainment.

$\$1,200 \times \frac{3}{8} = \$450$

Add the costs of these items.

$500	motels
+ 450	gas and entertainment
$950	cost

Subtract the cost of motels, gas, and entertainment from the total spent ($1200) to find how much was spent for food.

$1,200	total cost of vacation
− 950	motels, gas, entertainment
$ 250	food

3. (2) 2.2
Extraneous information: how long the driver worked each day.

Multiply the miles per gallon by the total amount of gasoline.

26.7	miles per gallon
× 12.4	number of gallons
10 68	
53 4	
267	
331.08	total mi cab can go

Divide total miles by 150, the number of miles driven in 1 day.

$$\begin{array}{r} 2.20 \\ 150\overline{)331.08} \\ \underline{300} \\ 310 \\ \underline{300} \\ 108 \end{array} \quad \text{total mi cab can go}$$

Stop dividing here and round to the nearest tenth, 2.2.

4. (1) 9
Extraneous information: number of times class meets a week and the cost of the class.

Multiply the total number in the class (48) by 3⁄4 to find how many had never had sailing lessons.

$48 \times \frac{3}{4} = 36$

Out of the 36, 1⁄4 had never sailed. Multiply 36 by 1⁄4.

$36 \times \frac{1}{4} = 9$

5. (5) Insufficient data is given to solve the problem.

The problem does not tell you the cost of vitamin E capsules.

KEEPING TRACK
Perfect score = 5
Your score ☐

20 Working with Percents

(page 105)

1. (4) $4,368.00
Use the grid to find the gross sales.

amt. earned in commission	part $218.40	percent 5	% commission
gross sales on which commission is figured	whole ?	100	

$218.40 × 100 = $21,840
$21,840 ÷ 5 = $4,368.00

2. (1) $41\frac{2}{3}\%$
Use the grid to find what percent of the weekend was used biking.

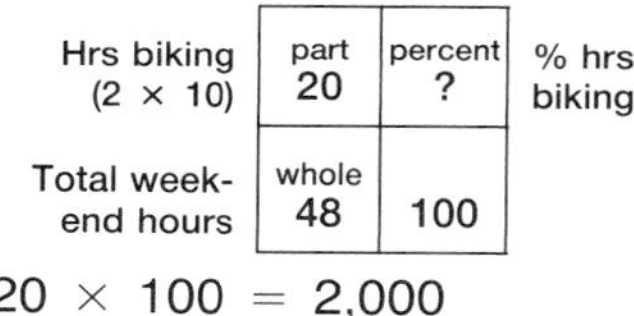

20 × 100 = 2,000
2,000 ÷ 48 = 41⅔

3. (3) $4,000
Use the grid to find what her gross sales must be.

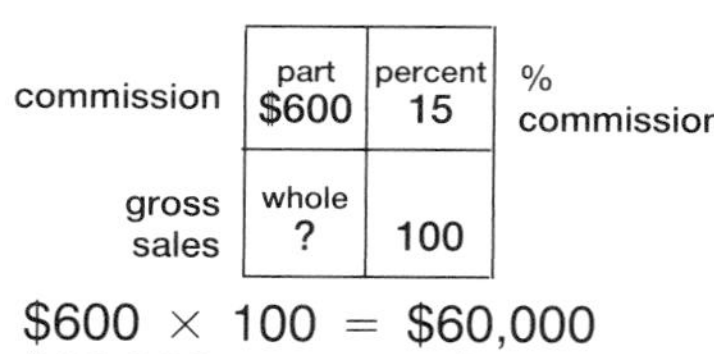

$600 × 100 = $60,000
$60,000 ÷ 15 = $4,000

(page 106)

4. (3) $546.00
If the Andersons bought a sofa marked 30% off, they paid 70% of the original price.

sale price	part ?	percent 70	% of original price
original price	whole 750	100	

$750 × 70 = $52,500
$52,500 ÷ 100 = $525

The sale price is \$525. They had to pay 4% of \$525 to have it delivered.

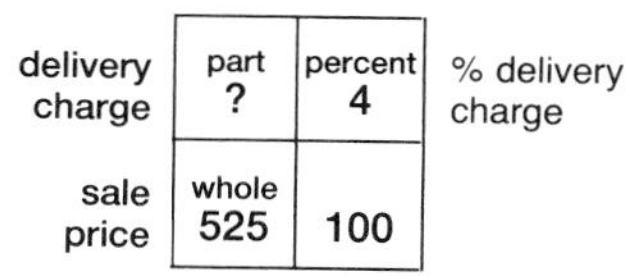

delivery charge
part ?
percent 4
% delivery charge
sale price
whole 525
100

\$525 × 4 = \$2,100
\$2,100 ÷ 100 = \$21

Add to find the total cost.

	\$525.00	sale price
+	21.00	delivery charge
	\$546.00	total cost

5. (5) \$7,560
Use grids to find 25% of \$7,560 and 20% of \$7,560.

rent	part ?	percent 25	% for rent
monthly salary	whole 1,400	100	

food	?	20	% for food
monthly salary	1,400	100	

Add to find how much she spends each month for rent and food.

	\$350	rent per month
+	280	food per month
	\$630	monthly rent and food

Multiply by 12 to find how much she will spend in one year for rent and food.

\$630 × 12 = \$7,560

You can solve this problem in another way. First add the two percents (20% + 25% = 45%) and solve on one grid. Then multiply by 12.

(page 108)

6. (1) 340%
Subtract to find the amount of increase.

\$1.76 − 0.40 = \$1.36

Use a grid to find the percent that the price has increased.

amt. of increase	change \$1.36	percent ?	% of increase
starting amt.	orig. \$0.40	100	

\$1.36 × 100 = \$136.00
\$136.00 ÷ \$0.40 = 340

The price increased 340%.

7. (4) $33\frac{1}{3}$%
Subtract to find the amount of decrease.

60 − 40 = 20

Use a grid to find the percent of decrease.

amt. of decrease	change 20	percent ?	percent of decrease
starting amt.	orig. 60	100	

20 × 100 = 2,000
2,000 ÷ 60 = $33\frac{1}{3}$

8. (4) 103.5

lbs. water	part ?	percent 75	% water
total weight	whole 138	100	

138 × 75 = 10,350
10,350 ÷ 100 = 103.5

9. (3) 8%
Find the dollar amount of pay increase by subtracting.

	\$972	new salary
−	900	old salary
	\$ 72	more per month

amt. of increase	change \$72	percent ?	% increase in pay
original salary	orig. \$900	100	

\$72 × 100 = \$7200
\$7200 ÷ \$900 = 8

10. (1) 45%
Find the number of orange-leaved trees.

	5	red		20	total
+	6	gold	−	11	
	11			9	orange

no. of orange trees	part 9	percent ?	% orange trees
total trees	whole 20	100	

9 × 100 = 900
900 ÷ 20 = 45

11. (1) \$3,500.00

amt. of commission	part \$245	percent 7	% commission
total sales	whole ?	100	

\$245 × 100 = \$24,500
\$24,500 ÷ 7 = \$3,500

12. (5) Insufficient data is given to solve the problem.

The problem does not tell you the length of the boards purchased.

KEEPING TRACK
Perfect score = 12
Your score ☐

21 Zeroing In on Word Problems: Item Sets

(page 110)

1. (1) \$47.25
Add to find how much each club collected.

DLM Club

	\$937.50
+	342.00
	\$1,279.50

RTS Club

	\$ 198.00
+	1,128.75
	\$1,326.75

Subtract to find how much more the RTS Club collected.

	\$1,326.75	RTS Club
−	1,279.50	DLM Club
	\$ 47.25	more for RTS Club

2. (4) \$180.50
To find the cost of a doll house, divide \$937.50 by 15. (937.50 ÷ 15 = 62.50) Next find the cost of 1 rocking horse. To do so, divide \$1,128.75 by 15. (1,128.75 ÷ 15 = 75.25) Then, divide \$342 by 8 to find the cost of 1 doll. (342 ÷ 8 = 42.75)

Add to find the total cost.

	\$ 75.25	rocking horse
	62.50	doll house
+	42.75	doll
	\$180.50	total cost

3. (3) $3\frac{1}{8}$
Multiply $7\frac{1}{2}$ and 15 to find how long Glenda worked at DZX.

$7\frac{1}{2} \times 15 = 112\frac{1}{2}$

To find *times as long,* divide the time she worked for DZX by the time she worked for MLK.

$112\frac{1}{2} \div 36 = 3\frac{1}{8}$

4. (4) .15 [(7.5 × 15)12] + .25 [(7.5 × 15)12]
First, find the amount of time Glenda worked for the DZX Company.

7.5 × 15

Find her pay at the DZX Company.

(7.5 × 15)12

She saves 15% of her gross pay for social security.

.15 [(7.5 × 15)12]

She saves 25% of her gross pay for retirement.

.25 [(7.5 × 15)12]

Add the two.

.15 [(7.5 × 15)12] +
.25 [(7.5 × 15)12]

KEEPING TRACK
Perfect score = 4

Your score ☐

22 Tables and Meters

(page 112)

1. (1) $\frac{1}{4}$
Use the chart. Find the row labeled "Apples." The amount of fruit is 2 pounds. Go across the row to the column labeled "Amount of sugar (cups.)" It says ¼.

2. (5) 15
Find the row labeled "Cranberries." Go across the row to the column labeled "Cooking time (minutes)." It says 15.

3. (1) Apples
You need to know how much water for 1 pound of apples and 1 pound of rhubarb. Look in the row labeled "Apples." You see that 2 pounds of apples take ⅓ cup of water. So 1 pound would take ½ of ⅓ (that is, ½ × ⅓) or ⅙ cup.

Look in the row labeled "Rhubarb." You see that 1½ pounds of rhubarb takes ¾ cup of water. To find how much water 1 pound would take divide ¾ by 1½: You get ¾ ÷ 3/2 = ¾ × ⅔ or ½. Of all the amounts of water for 1 pound of fruit, the ⅙ cup for apples is the least.

4. (5) 6
Use the row for rhubarb in the chart. 1½ pounds will yield 3 cups of sauce. Ms. Lindelof used 3 pounds of rhubarb, or twice as much (3 ÷ 1½ = 2). Since she used twice as much rhubarb, she will get twice as much sauce (2 × 3 = 6). She got 6 cups of sauce.

(page 114)

5. (1) 21,362
On the first dial, the hand is directly on 2. On all the others, the hand is between the numbers, so you use the smaller number.

6. (1) 54,915
The hand is between the numbers on dials 1, 2, 3, and 4, so you use the smaller number. It is directly on 5 on dial 5. On dial 3, it is between 0 and 9. When it is in this position, you read the dial as 9.

7. (2) 55,602
On dials 1, 2, and 4, the hand is between the numbers, so you use the smaller number. On 3 and 5, it is directly on the number.

8. (1) 687
Use your answers from numbers 8 and 9. Subtract.

55,602	Feb. 3 reading
− 54,915	Dec. 2 reading
687	kw-hrs used

9. (3) $43.97
Use your answers from number 8. Multiply

$0.064	cost of kw-hr
× 687	kw-hrs used
448	
5 12	
38 4	
$43.968	

Round 43.968 to the nearest cent (hundredth): $43.97.

KEEPING TRACK
Perfect score = 9

Your score ☐

23 Reading Graphs

(pages 122–123)

1. (3) $17 million
Find the horizontal bar for direct mail advertising and follow it to the right. Go down to determine the number of millions. The bar stops a little less than halfway between 15 and 20 million, so the estimated number of dollars is $17 million.

2. (2) magazines
Look at the horizontal bars to determine which is shortest. Then go to the left and read the label for this bar. The label for the shortest bar is "Magazines."

3. (2) $0.30
In 1983 the cost of Soft-Glow Wax was approximately $5.38. In 1984 the cost was approximately $5.68. Subtract to find the increase.

$5.68	1984
− 5.38	1983
$0.30	increase

4. (2) 1984–1985
If there were no change in cost between two years, the line graph would be horizontal (straight across). The least amount of change in prices is shown between 1984 and 1985, where the line connecting the two prices is almost straight across.

5. (1) 16%
Add together the percents for Jet Messenger Service and Air Express.

10%	Jet Messenger Service
+ 6%	Air Express
16%	Total

6. (4) 3
Add the percentages for Eagle and Mercury.

46.5%	Eagle
+ 4.5%	Mercury
51.0%	combined Eagle and Mercury

Divide the combined percentage by what PLR gets (17%) to find how many times more the other two get.

51 ÷ 17 = 3

7. (5) 60
Deerwood shows 10 flakes. Each flake stands for 6 inches. Multiply to find the total amount of snow.

10 × 6 = 60

8. (2) 4
Sun Down shows 6 flakes and Homer shows 1½ flakes. Divide 6 by 1½ to find how many times as much.

$6 \div 1\frac{1}{2} = 6 \div \frac{3}{2} = 6 \times \frac{2}{3} = 4$

You can work the problem another way. Multiply to find how much snow each town received and then divide to find how many times as much.

KEEPING TRACK
Perfect score = 8

Your score ☐

24 Means and Medians

(page 125)
1. (1) 142.7
The total number of hits (3,283) is given. Divide this total by 23, the number of years Mays played.

3,283 ÷ 23 = 142.73

The answer choices are one-place decimals, so you only divide to hundredths. Round 142.73 to the nearest tenth: 142.7.

2. (4) 73,940
Extraneous information: 2,047,000 deaths.

Divide the total number of births (3,697,000) by 50 states.

3,697,000 ÷ 50 = 73,940

3. (3) 52
Add to find the total number of degrees.

49 + 51 + 42 + 60 + 52 + 58 = 312

Divide the total number of degrees (312) by 6 days.

312 ÷ 6 = 52

4. (3) $102
Add the monthly utility bills.

$117	January
139	February
85	March
+ 67	April
$408	total for 4 months

Divide this total by 4 months.

$408 ÷ 4 = $102

(page 127)
5. (3) 132
Arrange the bowling scores in order.

118, 121, 132, 138, 146

There are 5 scores, so the median is the score in the middle.

6. (2) 76.5
Arrange the scores in order.

41, 59, 69, 75, 78, 81, 95, 102

There are 8 scores, so find the two middle numbers, 75 and 78. Find their mean by adding them and then dividing by 2.

75 + 78 = 153
153 ÷ 2 = 76.5

7. (3) $9,500
Arrange the salaries in order.

$6,000; $7,000; $8,000; $8,000; $9,000; $9,000; $9,000; $10,000; $10,000; $15,000; $25,000; $30,000; $50,000; $70,000

There are 14 salaries, so find the two middle numbers, $9,000 and $10,000. Find their mean. Add and then divide by 2.

$9,000 + $10,000 = $19,000
$19,000 ÷ 2 = $9,500

8. (4) 95.0
You know that the mean of 5 scores is 91. Multiply 5 and 91 to get the total number needed to find a mean of 91.

5 × 91 = 455

Add the 4 test scores that you know.

96	
92	
85	
+ 87	
360	total of 4 scores

Subtract to find the unknown score.

455	total number needed
−360	total of 4 scores
95	score of yesterday's test

9. (2) 2.4 hr
To find his mean time, add the 3 times and then divide.

2.1
2.6
+2.5
7.2   7.2 ÷ 3 = 2.4

His mean time was 2.4 hr.

10. (1) 90.0
Arrange the scores in order: 86, 86, 87, 88, 89, 91, 93, 95, 101, 102. Because there are 10 scores, find the mean of the two middle scores, 89 and 91. Add and divide by 2.

89 + 91 = 180
180 ÷ 2 = 90

11. (5) Insufficient data is given to solve the problem.

The problem does not tell you the number of games in which he played.

12. (3) 22.4
Arrange the averages in order: 21.4, 21.6, 21.8, 22.2, 22.6, 22.9, 23.0, 24.7. Because there are 8 scores, find the mean of the two middle averages, 22.2 and 22.6. Add and divide by 2.

22.2 + 22.6 = 44.8
44.8 ÷ 2 = 22.4

13. (4) 89
You know that the mean of his 5 scores must be at least 87. Multiply 5 and 87 to get the total of his scores for the 5 tests (the total needed to get a mean of 87).

5 × 87 = 435

Add the 4 scores that you know.

82	
85	
87	
+ 92	
346	total of 4 scores

Subtract to find the unknown score.

435	total number needed
−346	total of 4 scores
89	least score needed on next test

14. (2) $11,358
Multiply to find how much she earned in 2 years.

$9,840 × 2 = $19,680

Multiply to find how much she earned in 3 years.

$12,370 × 3 = $37,110

Add to find the total amount.

$19,680	earned in 2 yrs
+ 37,110	earned in 3 yrs
$56,790	total earnings in 5 yrs

To find the average, divide $56,790 by 5, the number of years.

$56,790 ÷ 5 = $11,358

KEEPING TRACK
Perfect score = 14
Your score ☐

25 Ratio, Proportion, and Probability

(page 130)

1. (4) 5:16
Find the ratio of height to width. The billboard is 5 meters high and 16 meters long, so the ratio is 5:16.

2. (2) 3:2
Find the ratio of eggs to sugar. The recipe calls for 3 eggs and 2 cups of sugar, so the ratio is 3:2.

3. (1) 25:3
Find the ratio of miles to gallons. The car went 100 miles on 12 gallons. The ratio is 100:12. Divide each number by 4 to write the ratio in lowest terms. You get 25:3.

4. (2) 2:5
Subtract to find the number of unsold sweaters.

25	original no. of sweaters
− 15	sweaters sold
10	unsold sweaters

Write the ratio of unsold sweaters (10) to the original number (25).

The ratio is 10:25. Divide each number by 5 and you get 2:5.

(page 133)

5. (2) 4
Set up a grid and solve.

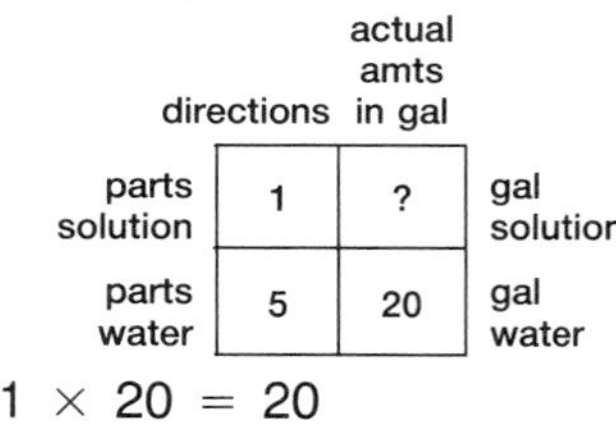

	directions	actual amts in gal	
parts solution	1	?	gal solution
parts water	5	20	gal water

1 × 20 = 20
20 ÷ 5 = 4

6. (3) 16
Set up a grid and solve.

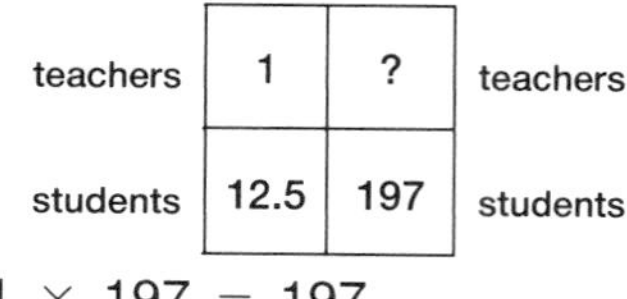

teachers	1	?	teachers
students	12.5	197	students

1 × 197 = 197
197 ÷ 12.5 = 15.76

You cannot have 15.76 teachers, so round the answer to the nearest whole number (16).

7. (2) $226.00
Set up a grid and solve.

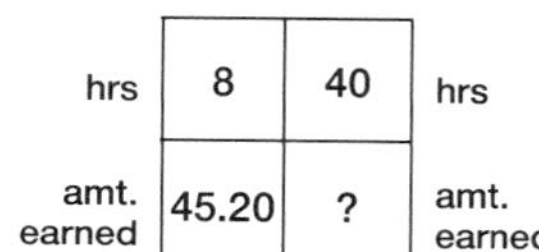

hrs	8	40	hrs
amt. earned	45.20	?	amt. earned

$45.20 × 40 = $1,808
$1,808 ÷ 8 = $226.00

8. (3) $1\frac{3}{4}$
There are two kinds of measure, weight and days. You are to find a weight between two other weights. You know a beginning day, a middle day, and a last day.
Set up an interpolation table:
(Change 3 lbs to 48 ounces.)

		Weights	Day		
40	x	8	7	5	10
		?	12		
		48	17		

From the differences, set up a grid and solve:

	weights	Days	
wt. diff. day 7–12	x	5	no. days from 7–12
wt. diff. day 7–17	40	10	no. days from 7–17

40 × 5 = 200
200 ÷ 10 = 20

The difference in the weights from day 7 to day 12 is 20 ounces. This is not the answer. The question asks how much is being lifted on day 12. Add to find this number.

8	beginning weight
+20	increase in weight
28	weight lifted on day 12

Divide by 16 to change ounces to pounds.

$28 \div 16 = 1\frac{12}{16} = 1\frac{3}{4}$

9. (4) $8.75
Set up a grid and solve.

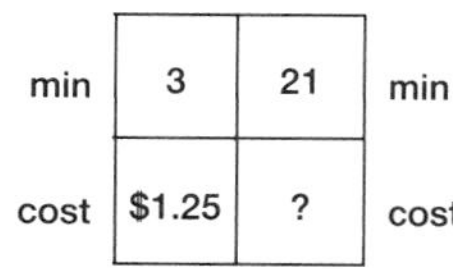

min	3	21	min
cost	$1.25	?	cost

$1.25 × 21 = $26.25
$26.25 ÷ 3 = $8.75

(page 136)

10. (1) $\frac{1}{60}$
The seconds display will show 60 different numbers. 35 is one of these numbers. The probability is 1 in 60 or $\frac{1}{60}$.

11. (3) 120
Multiply the number of times the experiment was repeated by the probability.

$$\frac{2}{\cancel{3}_1} \times \frac{\cancel{180}^{60}}{1} = 120$$

12. (2) $\frac{1}{6}$
Add to find the total number of shirts in the drawer.

3 + 6 + 2 + 1 = 12

There are 12 shirts in the drawer and 2 of them are blue. The probability of getting a blue shirt is $\frac{2}{12}$ or $\frac{1}{6}$.

13. (5) $\frac{4}{9}$
Add to find how many dogs and cats are in the pet shop.

12 + 15 = 27

There are 27 cats and dogs in the pet shop and 12 of them are dogs. The probability of getting a dog is $\frac{12}{27}$, or $\frac{4}{9}$.

14. (4) 8.7
Set up a grid and solve.

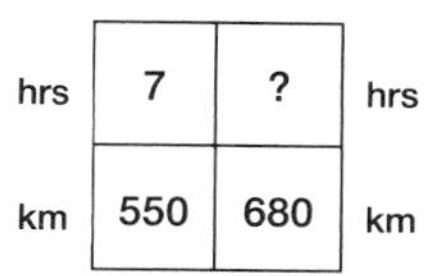

$7 \times 680 = 4{,}760$
$4{,}760 \div 550 = 8.65$

Stop dividing when you reach hundredths because the problem says to give the answer in tenths. Round 8.65 to the nearest tenth: 8.7.

15. (2) 6:11
Find the ratio of free throws made to attempts. In all, 6 free throws and 11 attempts were made, so the ratio is 6:11.

16. (3) $\frac{1}{2}$
There are 10 cards and 5 of these cards (6, 7, 8, 9, 10) have numbers greater than 5. The probability of drawing a number greater than 5 is $5/10$ or $1/2$.

17. (3) $3.15
Set up a grid and solve.

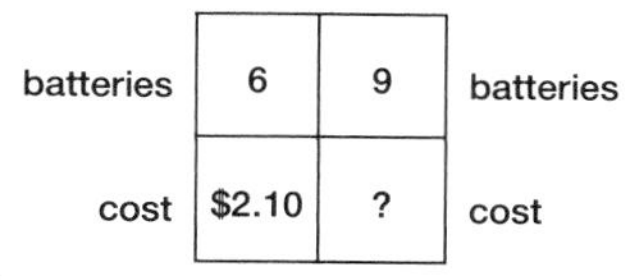

$2.10 × 9 = $18.90
$18.90 ÷ 6 = $3.15

18. (5) Insufficient data is given to solve the problem.

The problem does not tell you how far Mrs. Albright drove on the trip.

19. (5) $\frac{8}{15}$
Add to find the number of members.

28	men
+32	women
60	members in all

There are 60 members and 32 of these are women. The probability of selecting a woman is $32/60$, or $8/15$.

KEEPING TRACK
Perfect score = 19

Your score ☐

27 Sequencing

(pages 147–148)

1. (2) Company B
You must compare the numbers for the companies listed to 0.003 and 0.007 to see which of the numbers is *between* these two. Start with company A. You know that 0.015 is greater than both 0.003 and 0.007, so (1) is not the correct answer. What about company B? To compare 0.0064 to 0.003 and 0.007, add zeros to make all the decimals four-place decimals. You get

0.0030   0.0070

Since 0.0064 is greater than 0.0030 and less than 0.0070, company B's capsules meet the requirements. There's only one correct answer, so you can stop here.

2. (5) E, D, A, C, B
You must list the fractions in order from least to greatest. (The smaller the fraction, the lighter the shade of pink, the greater the fraction, the darker the shade of pink.) To compare the fractions, rename to have a common denominator of 105.

You get the following:

A	B	C	D	E
$\frac{4}{7}$	$\frac{2}{3}$	$\frac{3}{5}$	$\frac{11}{21}$	$\frac{7}{15}$
↓	↓	↓	↓	↓
$\frac{60}{105}$	$\frac{70}{105}$	$\frac{63}{105}$	$\frac{55}{105}$	$\frac{49}{105}$

Arrange the fractions in order from least to greatest and below them write the matching letters.

$\frac{49}{105}$	$\frac{55}{105}$	$\frac{60}{105}$	$\frac{63}{105}$	$\frac{70}{105}$
E	D	A	C	B

3. (2) Alice
Change $13/40$ to a decimal by dividing.

$$
\begin{array}{r}
0.325 \\
40\overline{)13.000} \\
\underline{12\ 0} \\
1\ 00 \\
\underline{80} \\
200 \\
\underline{200} \\
0
\end{array}
$$

Calculate the *difference* between each guess and 0.325.

List these.

Person →	Jim	Alice	Tomás
Guess →	0.345	0.33	0.43
Diff. →	0.020	0.005	0.105

Person →	Cary	Seiji
Guess →	0.275	0.425
Diff. →	0.050	0.100

The differences tell *by how much* each person missed. So look for the smallest difference. It is 0.005. So person 2's answer comes closest.

4. (2) B
Work through the options to see which offers the most.

Option (1) is a flat rate of $1,200.

Option (2) = $55 a week for one year, or 55 × 52 = $2,860.

Option (3) = $275 × .15 for 1 week. To compare you must multiply 275 × .15 × 52 = $2,145.

Option (4) = $105 × 26 = $2,730.

Option (2) is obviously the best choice.

5. (4) D, A, C, E, B
You need to compare all the decimals. So make them all 3-place decimals by adding on zeros where necessary. You get

A 0.250
B 0.073
C 0.105
D 0.300
E 0.080

List these in order from greatest to smallest, and write the corresponding letters below.

0.300	0.250	0.105	0.080	0.073
D	A	C	E	B

6. (3) Pin C
To make comparisons easy, rename the fractions to have a common denominator of 32.
The hole is $14/32$ inch across. The new fractions for the widths of the pins are as follows:

A	B	C	D	E
$\frac{15}{32}$	$\frac{12}{32}$	$\frac{13}{32}$	$\frac{20}{32}$	$\frac{18}{32}$

Only pins *less* than 14⁄32 inch across can fit inside the hole. There are pins B and C. You want the larger pin, So choose C.

KEEPING TRACK
Perfect score = 6

Your score ☐

28 Exponents

(page 150)

1. (1) − 512

$(-8)^3 = (-8) \times (-8) \times (-8) = (+64) \times (-8) = -512$

2. (5) 1,000,000

$10^6 = 10 \times 10 \times 10 \times 10 \times 10 \times 10 = 1,000,000$

3. (3) $\frac{4}{9}$

$(\frac{2}{3})^2 = \frac{2}{3} \times \frac{2}{3} = \frac{4}{9}$

4. (3) 1

$5^\circ = 1$ (by definition)

5. (2) − 7

Any number with an exponent of 1 is equal to that number. So $(-7)^1 = -7$.

6. (4) $\frac{1}{9}$

Remember that when you multiply two negative numbers you get a positive answer.

$(-\frac{1}{3})^2 = (-\frac{1}{3}) \times (-\frac{1}{3}) = +\frac{1}{9}$, or $\frac{1}{9}$

(page 152)

7. (3) 12

Start with the least number in the list of choices and calculate their squares. Stop when you get to a number whose square is *greater than* 157.

Choices	*Squares*	
10	100	
11	121	
12	144	
13	169	← Greater than 157
14		

Isaac can make a tray with 12 tiles on a side, since it requires only 144 tiles. But he can't make one with 13 tiles on a side. (It would require 169 tiles and he has only 157.

8. (3) 3^2

You could list all possible outfits (yellow slacks/red blouse, and so on) and then count the possibilities. You'll get 9, which equals 3^2.

You can also think of it this way: For any color blouse she picks, there are 3 outfits she can make, depending on the color of the slacks. There are 3 possible colors for the blouses. The total number of possibilities is 3 × 3 (blouse possibilities × slacks possibilities). 3 × 3 equals 3^2.

9. (5) 32,400

Each tile measuring 3 inches on a side would need 9 of the 1-inch tiles to exactly cover it. So he would need 9 times as many of the smaller tiles. Multiply 9 by 3,600 to get the total number of smaller, 1-inch tiles that he would need.

(pages 154–155)

10. (3) 5,800

In 5.8×10^3, the 10^3 tells you to move the decimal point 3 places to the right.

11. (2) 0.0000007

In 7×10^{-7}, the 10^{-7} tells you to move the decimal point 7 places to the left. Recall that the decimal point is directly to the right of 7.

12. (2) 2.68×10^{-7}

Recall that, when you write a number in scientific notation, there is always one number to the left of the decimal point. For 0.000000268, the 2 will be at the left of the decimal point. If you have 2.68 and need to get to 0.000000268, you'd move the decimal point 7 places to the *left.* So the exponent you need for the scientific notation is −7. The answer is 2.68×10^{-7}.

13. (4) 3.6×10^8

For 360,000,000, the 3 will be at the left of the decimal point, so you move the decimal point 8 places to the left. To go back to the original number, you move 8 places to the right. The exponent for 10 is 8: 3.6×10^8.

14. (1) $\frac{16}{49}$

$(\frac{4}{7})^2 = \frac{4}{7} \times \frac{4}{7} = \frac{16}{49}$

15. (2) 1.4272×10^9

For 1,427,200,000, the 1 will be at the left of the decimal point, so you move the decimal point 9 places to the left. To get back to the original number from 1.4272, you need to go *right* 9 places. The exponent for 10 in the scientific notation is 9: 1.4272×10^9.

16. (3) 3×2^5

In 5 hours, each one-celled animal will yield

$2 \times 2 \times 2 \times 2 \times 2 = 2^5$ animals

1st hr.
2nd hr.
3rd hr.
4th hr.
5th hr.

So 3 one-celled animals will yield 3 times this number, or 3×2^5 one-celled animals.

17. (4) 934,400,000

In 9.344×10^8, the 10^8 tells you to move the decimal point 8 places to the right.

18. (1) 0.0169

$(0.13)^2 = 0.13 \times 0.13 = 0.0169$

19. (2) 0.0075

In 7.5×10^{-3}, the 10^{-3} tells you to move the decimal point 3 places to the left.

KEEPING TRACK
Perfect score = 19

Your score ☐

29 Standard Measurement

(page 160)

1. (2) 11 ft 4 in
Add to find how much pipe is needed in all.

$$\begin{array}{rr} & 4\text{ ft } 9\text{ in} \\ + & 6\text{ ft } 7\text{ in} \\ \hline & 10\text{ ft } 16\text{ in} \end{array} = 11\text{ ft } 4\text{ in}$$

2. (3) 1 ton 1,500 lbs
Subtract to find the weight of the empty truck.

$$\begin{array}{rr} \overset{1}{\cancel{2}}\text{ tons} & \overset{2300}{\cancel{300}}\text{ lbs} \\ & -800\text{ lbs} \\ \hline 1\text{ ton} & 1500\text{ lbs} \end{array}$$

3. (3) 50 ft 11 in
Add to find how much ductwork has been installed.

$$\begin{array}{rr} & 45\text{ ft } 9\text{ in} \\ + & 89\text{ ft } 10\text{ in} \\ \hline & 134\text{ ft } 19\text{ in} \end{array} = 135\text{ ft } 7\text{ in}$$

Subtract the ductwork that was installed from 186 feet 6 inches, which is the total ductwork.

$$\begin{array}{rr} & \overset{185}{\cancel{186}}\text{ ft } \overset{18}{\cancel{6}}\text{ in} \\ - & 135\text{ ft } 7\text{ in} \\ \hline & 50\text{ ft } 11\text{ in} \end{array}$$

4. (2) 5 yds 34 in
Add to find the total amount of material.

$$\begin{array}{rr} & 3\text{ yds } 18\text{ in} \\ + & 4\text{ yds } 24\text{ in} \\ \hline & 7\text{ yds } 42\text{ in} \end{array} = 8\text{ yds } 6\text{ in}$$

Subtract the amount used for the skirt from the total amount. The difference is how much she used for the dress.

$$\begin{array}{rr} & \overset{7}{\cancel{8}}\text{ yds } \overset{42}{\cancel{6}}\text{ in} \\ - & 2\text{ yds } 8\text{ in} \\ \hline & 5\text{ yds } 34\text{ in} \end{array}$$

(page 161)

5. (4) 13 ft 9 in
Multiply the length of each piece by 5, the number of pieces.

$$\begin{array}{rr} 2\text{ ft} & 9\text{ in} \\ & \times\ 5 \\ \hline 10\text{ ft} & 45\text{ in} \\ 13\text{ ft} & 9\text{ in} \end{array}$$

6. (5) 110 gal 1 qt
Multiply the amount used daily by 7, the number of days the cafe is open.

$$\begin{array}{rr} 15\text{ gal} & 3\text{ qt} \\ & \times\ 7 \\ \hline 105\text{ gal} & 21\text{ qt} \end{array} = 110\text{ gal } 1\text{ qt}$$

7. (3) 480
To find how many eggs there are in all, multiply the number of dozens (60) by 12, the number in a dozen.

$60 \times 12 = 720$ eggs in all

To find how many medium eggs there are, multiply 720 by ⅓.

$\frac{1}{3} \times 720 = 240$ medium eggs

To find the number of large eggs, subtract the medium eggs from the total.

$720 - 240 = 480$ large eggs

8. (4) $11.76
Add to find the total weight of the packages.

$$\begin{array}{rr} & 6\text{ lbs } 5\text{ oz} \\ + & 5\text{ lbs } 11\text{ oz} \\ \hline & 11\text{ lbs } 16\text{ oz} \end{array} = 12\text{ lbs}$$

The meat cost $0.98 a pound. To find the total cost, multiply $0.98 by 12.

$\$0.98 \times 12 = \$11.76$

(page 163)

9. (3) 1 foot 11 inches
Divide 7 feet 8 inches by 4.

$$\begin{array}{rrr} & 1\text{ ft} & 11\text{ in} \\ 4\overline{)} & 7\text{ ft} & 8\text{ in} \\ & \underline{4} & \underline{+36} \\ & 3 \nearrow & 44 \\ & & \underline{44} \\ & & 0 \end{array}$$

10. (2) 3
Change 1 ton to 2,000 pounds. To find how many pianos can be safely placed on the stage, divide 2,000 by 650, the weight of 1 piano.

$2{,}000 \div 650 = 3\frac{50}{650} = 3\frac{1}{13}$

The stage will hold 3 pianos; 4 pianos would be too heavy.

11. (5) Insufficient data is given to solve the problem.

The problem does not tell you how many campers there are.

12. (2) 3 pounds 10 ounces
Add to find the total weight.

$$\begin{array}{rr} & 2\text{ lbs } 12\text{ oz} \\ & 4\text{ lbs } 10\text{ oz} \\ + & 3\text{ lbs } 8\text{ oz} \\ \hline & 9\text{ lbs } 30\text{ oz} \end{array} = 10\text{ lbs } 14\text{ oz}$$

To find the average weight, divide the total weight by 3 (the number of bags).

$$\begin{array}{rrr} & 3\text{ lbs} & 10\text{ oz} \\ 3\overline{)} & 10\text{ lbs} & 14\text{ oz} \\ & \underline{9} & \underline{+16} \\ & 1 \nearrow & 30 \end{array}$$

13. (2) 357
Add to find the total length of the lake-front property in feet.

$500 + 571 = 1{,}071$

There are 3 feet in 1 yard. To change the feet to yards, divide 1,071 by 3.

$1{,}071 \div 3 = 357$

14. (1) $0.037
Change 1 pint 2 ounces to ounces. 1 pint equals 2 cups, and each cup equals 8 ounces. So 1 pint equals 16 ounces.
1 pint 2 ounces = 18 ounces

To find the cost per ounce, divide $0.67 by 18, the number of ounces.

$\$0.67 \div 18 = 0.0372$

Round to the nearest thousandth because all answers are given in thousandths: $0.037.

15. (2) 22 yards 1 foot 8 inches
Multiply the amount of butcher paper needed for each table (3 yards 8 inches) by the number of tables (7).

$$\begin{array}{rr} & 3\text{ yds } 8\text{ in} \\ \times & 7 \\ \hline & 21\text{ yds } 56\text{ in} \\ = & 22\text{ yds } 20\text{ in} \\ = & 22\text{ yds } 1\text{ ft } 8\text{ in} \end{array}$$

16. (3) 13 pounds
Add to find out how much meat was bought.

```
    7 lbs  8 oz
    8 lbs  6 oz
 +  9 lbs  2 oz
   24 lbs 16 oz = 25 lbs
```

Add to find how much meat was used for meatballs and meat loaf.

```
    5 lbs  8 oz
 +  6 lbs  8 oz
   11 lbs 16 oz = 12 lbs
```

Subtract 12 pounds from the total amount to find how much was used for hamburgers.

25 − 12 = 13

KEEPING TRACK
Perfect score = 16
Your score ☐

30 Metric Measurement

(page 167)

1. (3) 12.5
Divide to change a smaller unit to a larger unit. There are 1,000 milliliters in 1 liter. Divide 12,500 milliliters by 1,000 milliliters.

12,500 ÷ 1,000 = 12.5 liters

2. (4) 188
Multiply to change a larger unit to a smaller unit. There are 10 millimeters in 1 centimeter. Multiply 18.8 centimeters by 10.

18.8 × 10 = 188

3. (2) 4
Change 1 liter to milliliters.

1 liter = 1,000 milliliters

Divide 1,000 milliliters by 250 milliliters to find the number of glasses.

1,000 ÷ 250 = 4

4. (3) 10
First find how many milligrams (mg) of vitamin C he takes each day. There are 1,000 mg in 1 gram, so there are 1,000 × 2, or 2,000 mg in 2 grams.

Next divide the 2,000 mg that he takes each day by 200 mg (the number of milligrams of vitamin C in 1 multivitamin tablet).

2,000 ÷ 200 = 10

5. (2) 3,675
Three heads of cabbage weigh 725 grams each. Multiply 725 by 3 to find the total weight of these cabbages.

725 × 3 = 2,175 grams

Multiply 1.5 by 1,000 to find how many grams the 1.5 kg cabbage weighed

1.5 pts 1,000 = 1,500 grams

Add to find the total weight of the cabbages.

```
  2,175 g
 +1,500 g
  3,675 g
```

6. (3) 3.986
Add to find change in weight at the end of the second week because Koyi gained weight the first 2 weeks.

```
  143  gain first week
 +115  gain second week
  258  change in 2 weeks
```

He lost weight the third week, so subtract to find the change.

```
  258  change in 2 weeks
 −169  loss third week
   89  change in 3 weeks
```

Koyi had gained 89 grams since birth. Divide by 1,000 to change the gain to kilograms.

89 ÷ 1,000 = 0.089 kg

Subtract to find his weight at birth.

```
  4.075  weight at end of 3 weeks
 −0.089  weight gain in 3 weeks
  3.986  weight at birth
```

7. (5) 800
Multiply by 10 to find how many millimeters in 10 centimeters.

10 × 10 = 100 mm

Each millimeter (mm) of thickness contains 8 sheets of paper. Multiply by 100 to find the total number of sheets of paper.

8 × 100 = 800

8. (5) 324
Each recipe uses 14 grams of baking powder. Divide the amount of baking powder (2,268) in one can by 14 to find how many recipes can be made.

2,268 ÷ 14 = 162

Each recipe makes 2 dozen biscuits, and 162 recipes can be made. Multiply 162 by 2 to find how many dozen biscuits can be made.

162 × 2 = 324

9. (2) 15
There are 20 nickels in $1. Multiply 20 by 150 to find the number of nickels in $150.

20 × 150 = 3,000 nickels

Multiply 3,000 by 5 (the weight of 1 nickel in grams) to find how many grams 3,000 nickels weigh.

3,000 × 5 = 15,000 g

The question asks how many *kilograms* the nickels weigh, so divide 15,000 by 1,000 (the number of grams in 1 kilogram).

15,000 ÷ 1,000 = 15

10. (4) 1,600
Add to find the total weight of the quadruplets in kilograms.

1.2 + 1.5 + 2.0 + 1.7 = 6.4

Change 6.4 kilograms to grams by multiplying by 1,000.

6.4 × 1,000 = 6,400 g

To find the average weight, divide the total weight in grams by 4 (the number of babies).

6,400 ÷ 4 = 1,600 g

KEEPING TRACK
Perfect score = 10
Your score ☐

31 Dealing with Units of Time

(page 170)

1. (3) 2 hrs 12 min
Add to find the total cooking time.

```
        34 min  high power
 + 1 hr 38 min  medium power
   1 hr 72 min  total time
 = 2 hrs 12 min
```

2. (5) 4:12 P.M.
Add the delay to the departure time. Think of 2:47 as 2 hours 47 minutes.

	2 hrs 47 min	scheduled time
+	1 hr 25 min	delay
	3 hrs 72 min	
=	4 hrs 12 min	

Think of 4 hours 12 minutes as 4:12.

3. (2) 5 min 54 sec
Add the times to find the total.

	4 min	56 sec
	2 min	30 sec
+	10 min	16 sec
	16 min	102 sec = 17 min 42 sec

Change 17 minutes 42 seconds to seconds. To do this, first multiply 17 by 60.

$17 \times 60 = 1{,}020$

Then add the other seconds (42).

$1{,}020 + 42 = 1{,}062$

To find the average, divide the total time in seconds (1,062) by the number of songs (3).

$1{,}062 \div 3 = 354$ sec

Change the seconds back to minutes by dividing by 60.

```
    5    ← min
60)354
   300
    54   ← sec remaining
```

So 354 seconds equals 5 minutes 54 seconds, which is the final answer.

4. (4) 7 hr 30 min
To find how long Mary Lou worked on Tuesday, multiply 1 hour 40 minutes (the time worked on Monday) by 3.

	1 hr	40 min
		×3
	3 hr	120 min
=	5 hr	

To find how long she worked on Thursday, first change 1 hour 40 minutes to minutes (60 + 40 = 100). Multiply 100 by ½.

$100 \times \frac{1}{2} = 50$ minutes

Add to find the total times.

	1 hr 40 min	Monday
	5 hrs	Tuesday
+	50 min	Thursday
	6 hrs 90 min	
=	7 hrs 30 min	total time

(pages 173–174)

5. (1) 6.2%
Use the formula $r = {}^{i}/_{pt}$ and replace *i, p,* and *t* with numbers.

$r = \frac{i}{pt}$

$r = \frac{93}{750 \times 2}$

$r = \frac{93}{1{,}500}$

Divide 93.00 by 1,500. You get $r = 0.062$.

You have to change 0.062 to a percent by moving the decimal point two places to the right: 6.2%.

6. (2) $6.75
Use $i = prt$ and put in the numbers you know for *p, r,* and *t.* Change 5.4% to a decimal: 5.4% = 0.054. Express 3 months as a decimal part of a year: 0.25.

$i = prt$
$i = 500 \times 0.054 \times 0.25$
$i = 27 \times 0.25$
$i = 6.75$

7. (4) $1,400
Use $p = {}^{i}/_{rt}$ and put in the numbers you know for *i, r,* and *t.* Change 12% to a decimal: 0.12.

$p = \frac{i}{rt}$

$p = \frac{840}{0.12 \times 5}$

$p = \frac{840}{0.6}$

Divide 840 by 0.6. You get $p = 1{,}400$.

8. (3) $17\frac{1}{2}$
Use $d = rt$ and put in the numbers you know for *r* and *t.*

$d = rt$

$d = 3\frac{3}{4} \times 4\frac{2}{3}$

$d = \frac{\cancel{15}^{5}}{\cancel{4}_{2}} \times \frac{\cancel{14}^{7}}{\cancel{3}_{1}} = \frac{35}{2} = 17\frac{1}{2}$

9. (2) $7\frac{2}{3}$
Use $t = {}^{d}/_{r}$ and put in the numbers you know for *d* and *r.*

$t = \frac{d}{r}$

$t = \frac{690}{90}$

$t = 7\frac{2}{3}$

10. (4) $8,613.75
First find the interest by using $i = prt$. Change 9.9% to 0.099 and 18 months to 1.5 years.

$i = prt$
$i = \$7{,}500 \times 0.099 \times 1.5$
$i = \$742.50 \times 1.5$
$i = \$1{,}113.75$

Add the interest to the principal to find the total amount due.

	$7,500.00	principal
+	1,113.75	interest
	$8,613.75	total amount due

11. (1) 7:17 A.M.
Add the travel time and the early arrival.

	28 min	travel time
+	15 min	early arrival
	43 min	before starting time

Subtract the amount of time he must leave early from 8:00 AM. Think of 8:00 AM as 8 hours 0 minutes.

	$\overset{7}{8}$ hrs	$\overset{60}{0}$ min	starting time
−		43 min	before starting
	7 hrs	17 min	departure

Think of 7 hours 17 minutes as 7:17 AM.

KEEPING TRACK
Perfect score = 11
Your score ☐

32 Linear, Square, and Cubic Measure

(page 177)

1. (5) 200.8
The base has 4 sides with lengths of 69.5, 30.9, 69.5, and 30.9 meters. Add to find the total distance around.

$69.5 + 30.9 + 69.5 + 30.9 = 200.8$

2. (4) 36
Multiply the length of each side by number of sides to find the distance around.

$8 \times 4\frac{1}{2} = 36$

You could also add (using 4½ eight times) and get the same answer.

3. (3) 139
A triangle has 3 sides and the distance around this triangle is 417 feet. Divide the distance around by the number of sides to find the length of each side.

$417 \div 3 = 139$

4. (2) $4\frac{1}{2}$
Add to find the distance around the garden. Remember to use each length twice.

$10\frac{1}{2} + 12\frac{3}{4} + 10\frac{1}{2} + 12\frac{3}{4} = 46\frac{1}{2}$

Mr. Halloran needs 46½ feet of fencing in all and he has 42 feet. Subtract to find how much more he needs.

$46\frac{1}{2}$	needs
-42	has
$4\frac{1}{2}$	how much more needed

(page 179)

5. (5) 216
Multiply length by width.

$18 \times 12 = 216$ (sq ft)

6. (1) 13
Since the length times the width (9) equals the number of square inches (117), divide 117 by 9 to find the length.

$117 \div 9 = 13$ (in)

7. (4) 95.0
First multiply to find how many square inches are in the whole floor.

$20 \times 9.5 = 190$ sq in

Divide by 2 to find how many square inches are in half of the floor.

$190 \div 2 = 95$ (sq in)

(page 181)

8. (4) 1,080
Multiply the length, width, and height.

$18 \times 10 \times 6 = 1{,}080$ (cu in)

9. (2) 12.50
You know 2 dimensions and you have to find the third. Multiply 96 by 2.5.

$96 \times 2.5 = 240$ ft

Now you know that 240 times some number is equal to 3,000. Divide to find this number.

$3{,}000 \div 240 = 12.5$ ft

10. (3) 600
Multiply to find how many cubic inches are in each box.

box A
$12 \times 10 \times 6 = 720$ (cu ft)

box B
$15 \times 8 \times 11 = 1{,}320$ (cu ft)

Subtract to find how many more cubic feet are in box B.

1,320	box B
− 720	box A
600	cu ft more in box B

11. (4) $37\frac{1}{2}$
Multiply to find how many cubic feet of water Kelly's aquarium will hold.

$2\frac{1}{2} \times 4 \times 2\frac{1}{2} = 25$ (cu ft)

Hilary's aquarium holds 1½ times as much. Multiply 25 by 1½.

$1\frac{1}{2} \times 25 = 37\frac{1}{2}$ (cu ft)

12. (5) Insufficient data is given to solve the problem.

The problem does not tell you how long the box is.

13. (3) 58
The fence will have 4 sides that measure 18, 11, 18, and 11 feet. Add to find the distance around the garden.

$18 + 11 + 18 + 11 = 58$ (ft)

14. (2) 43.25
First find the length of side C. You know the total distance around and the lengths of two sides. Add the lengths of these two sides.

$34 + 25.5 = 59.5$

Subtract to find the length of side C.

102.0	total distance around
− 59.5	total length of two sides
42.5	length of side C

Add to find how much more tape Casey needs.

42.50	length of side C
+ 0.75	allowed for waste
43.25	amount needed

15. (3) 60
First change 30 inches to feet, because the question asks for an answer in cubic feet. Divide by 12.

$30 \div 12 = 2.5$

Multiply to find how many cu feet are in this box.

$8 \times 3 \times 2.5 = 60$ (cubic feet)

KEEPING TRACK
Perfect score = 15
Your score ☐

Keeping Track

Now enter all your scores from the Keeping Track boxes on the lines below.

Lesson	Perfect Score	Your Score
Zeroing In on Word Problems: Two Step Problems	5	_______
Multiplying Whole Numbers	13	_______
Dividing Whole Numbers	13	_______
Zeroing In on Multistep Problems	10	_______
TOTAL **Whole Numbers**	**41**	_______
Adding and Subtracting Decimals	10	_______
Multiplying Decimals	8	_______
Dividing Decimals	13	_______
Zeroing In on Word Problems: Insufficient Data	7	_______
TOTAL **Decimals**	**38**	_______
Adding Fractions	11	_______
Subtracting Fractions	9	_______
Multiplying Fractions	6	_______
Dividing Fractions	10	_______
Fraction and Decimal Names for Numbers	6	_______
Zeroing In on Word Problems: Unnecessary Information	5	_______
TOTAL **Fractions**	**47**	_______
Working with Percents	12	_______
Zeroing In on Word Problems: Item Sets	4	_______
TOTAL **Percents**	**16**	
Tables and Meters	9	_______
Reading Graphs	8	_______
Finding Means and Medians	14	_______
Ratio, Proportion, and Probability	19	_______
TOTAL **Data Analysis**	**50**	_______
Sequencing	6	_______
Exponents	19	_______
TOTAL **Number Relationships**	**25**	_______
Standard Measurement	16	_______
Metric Measurement	10	_______
Dealing with Units of Time	11	_______
Linear, Square, and Cubic Measure	15	
TOTAL **Measurement**	**52**	_______
TOTAL	**269**	

In which lessons did you get perfect scores? Were there some sections that gave you trouble? If so, review those lessons before you try the *Extra Practice* that follows.

Extra Practice: Arithmetic

1. Carla had \$50. She spent \$29.76 for groceries and \$12.50 for gasoline. How much did she have left?

   (1) \$6.74 (2) \$7.74 (3) \$7.84
   (4) \$32.74 (5) \$42.26

   1 2 3 4 5

2. A student obtained an educational loan for business school. He borrowed \$2,400 and repaid it by paying \$112 a month for 2 years. What was the rate of interest?

   (1) 2% (2) 6% (3) 12%
   (4) 24% (5) 28%

   1 2 3 4 5

3. It took 40 minutes for a driver to travel to Chicago. The distance Milwaukee to Chicago is half the distance Milwaukee to Indianapolis. How many miles per hour was he going?

   (1) 60 (2) 55 (3) 50 (4) 45
   (5) Insufficient data is given to solve the problem.

   1 2 3 4 5

4. David and Don took a trip. They spent \$267.84 for meals, \$309.52 for gas, \$71.10 for tolls, and \$32 for campsite fees. If they began their trip with \$800, how much did they have left when they got home?

   (1) \$680.46 (2) \$429.06
   (3) \$151.22 (4) \$120.46
   (5) \$119.54

   1 2 3 4 5

5. How much milk does a family of 4 need in a day if each person drinks 20 ounces?

   (1) $2\frac{1}{2}$ cups (2) 2 qt $\frac{2}{3}$ cups
   (3) 2 qt 2 cups (4) 2 qt $2\frac{1}{2}$ cups
   (5) 10 qt

   1 2 3 4 5

6. What is the length AB in the floor plan below?

   A, B, C, D, E; 20; 10; 5; 5

   (1) 10 (2) 15 (3) 20 (4) 30
   (5) Insufficient data is given to solve the problem.

   1 2 3 4 5

7. How many gallons of water are needed to fill a waterbed measuring 6 feet long, 5 feet wide and 10 inches deep if 7.5 gallons will fill 1 cubic foot?

   (1) 40 (2) 182.5 (3) 187.5
   (4) 1,875 (5) 2,250

   1 2 3 4 5

8. A car rental agency charges \$24.95 a day for a subcompact with 100 free miles. If Ms. Ruiz drives 219 miles, how much will she pay?

   (1) \$21.90 (2) \$24.95 (3) \$29.00
   (4) \$32.16 (5) Insufficient data is given to solve the problem.

   1 2 3 4 5

9. A container of carpet cleaner is enough to clean a 10-foot × 14-foot rug. How many cans of cleaner are needed for two rugs, one 9 × 12 feet and one 13 × 15 feet?

   (1) 2 (2) 3 (3) 4 (4) 5 (5) 6

   1 2 3 4 5

10. The cost of a single-family home increased 75% between 1976 and 1980. If a home was selling for \$42,000 in 1976, what was its approximate selling price in 1980?

    (1) \$98,000 (2) \$73,500 (3) \$56,000
    (4) \$45,150 (5) \$31,500

    1 2 3 4 5

11. Two horses ran in a steeplechase. One horse took 2 hours 47 minutes to finish the race and the other took 173 minutes. By how many minutes did the faster horse beat the other horse?

(1) 2 (2) 3 (3) 6 (4) 12 (5) 20

① ② ③ ④ ⑤

12. What is the area, in square yards, of a rectangular piece of property measuring 1,350 feet long and 450 feet wide?

(1) 13,500 (2) 67,500
(3) 85,278 (4) 202,750
(5) 607,500

① ② ③ ④ ⑤

13. A soccer team won 17 games. This was 85 percent of the games they played. How many games did they play in all?

(1) 15 (2) 20 (3) 22
(4) 39 (5) 200

① ② ③ ④ ⑤

14. A plumber has a section of tubing 72 meters long. How many pieces can he cut from it if each piece must be $\frac{3}{4}$ meter long?

(1) 32 (2) 64 (3) 86
(4) 88 (5) 96

① ② ③ ④ ⑤

15. A business office uses an average of 4 reams of paper per month. The paper comes in cartons of 6 reams each. How many cartons of paper does this office use per year?

(1) 2 (2) 8 (3) 18 (4) 120
(5) Insufficient data is given to solve the problem.

① ② ③ ④ ⑤

16. In 1985 the U.S. Postal Service increased the price of a postage stamp from 20¢ to 22¢. The increase was what fraction of the old price?

(1) $\frac{1}{11}$ (2) $\frac{1}{10}$ (3) $\frac{10}{11}$

(4) $1\frac{1}{10}$ (5) $\frac{10}{1}$

① ② ③ ④ ⑤

17. An empty container will hold 6 cups of liquid. A person pours in $2\frac{1}{2}$ cups of orange juice and $1\frac{1}{3}$ cups of pineapple juice. How many more cups of liquid will the container hold?

(1) $2\frac{1}{6}$ (2) $2\frac{4}{5}$ (3) $\frac{1}{6}$

(4) $3\frac{1}{5}$ (5) $3\frac{5}{6}$

① ② ③ ④ ⑤

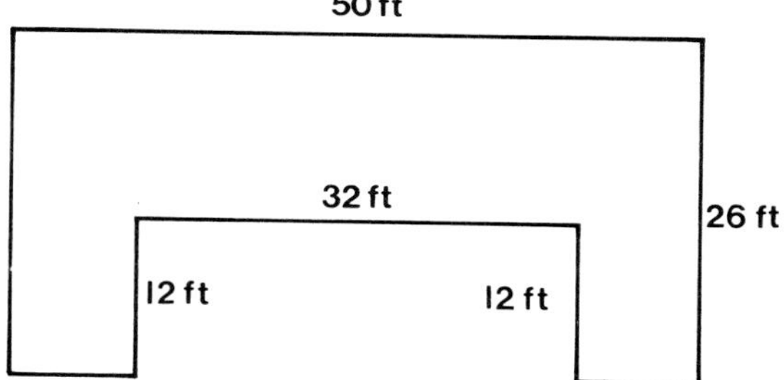

18. The area, in square feet, of the figure above is

(1) 916 (2) 1,156 (3) 1,300
(4) 1,600 (5) 2,216

① ② ③ ④ ⑤

19. A weekday telephone call costs 52¢ for the first minute and 34¢ for each additional minute. A weekend call to the same town costs 21¢ for the first minute and 14¢ for each additional minute. How much would a person save on a 5-minute call by calling on the weekend instead of on a weekday?

(1) \$1.11 (2) \$1.31 (3) \$1.88
(4) \$2.04 (5) \$2.55

① ② ③ ④ ⑤

20. In April a family used 210 kilowat-hours of electricity. In May they used $\frac{1}{7}$ less than in April. How many kilowatt-hours did they in May?

(1) 30 (2) 170
(3) 180 (4) 240
(5) Insufficient data is given to solve the problem.

① ② ③ ④ ⑤

Answers to Extra Practice begin on page 359.
Record your score on the Progress Chart on the Inside Back Cover.

Geometry

Of the 56 problems on the GED Mathematics Test about 11 are geometry questions. This is one of every five questions, or ⅕, or 20%. To prepare for the GED Mathematics Test, you should study three kinds of geometry—plane geometry, solid geometry, and coordinate geometry. *Plane geometry* deals with figures like triangles, squares, rectangles, and circles. Plane figures have only two dimensions: length, or height, and width. *Solid geometry,* on the other hand, is three-dimensional. It deals with cubes, cylinders, and rectangular and triangular solid figures. These figures have height, width, and depth. *Coordinate geometry* involves finding points on a grid made up of two number lines. Your work with number lines in the *Arithmetic* section of this book will provide you with a good foundation for your study of coordinate geometry.

If you feel uneasy just hearing the word *geometry,* you may discover you actually have been doing real-life geometry problems for years. The following problem is an example of how geometry is all around us. See if you can solve it.

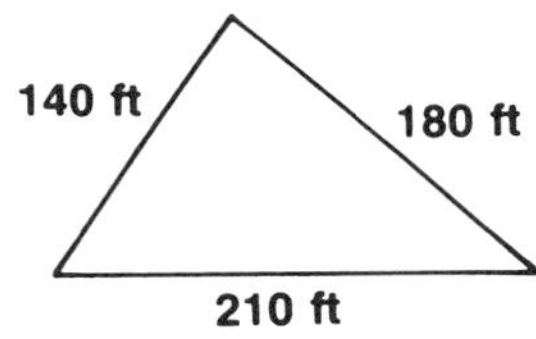

1. How much fencing will be needed to enclose the triangular park shown above?

   (1) 420 ft (2) 440 ft (3) 530 ft
   (4) 540 ft (5) 630 ft

   ① ② ③ ④ ⑤

Did you choose option (3)? If so, you found the perimeter of a triangle. Now try another problem. You may want to draw a picture to help yourself.

2. How many pieces of leather 2 inches by 2 inches can be cut from a larger piece that measures 16 inches by 20 inches?

   (1) 1,280 (2) 100 (3) 80
   (4) 60 (5) 18

   ① ② ③ ④ ⑤

If you chose option (3), 80 pieces, you have just solved a plane geometry problem involving area. This problem was not as simple as the one about the perimeter of the triangle.

Have you ever had to find the volume of a container? Try this problem that asks for volume.

3. A storage building is 200 feet by 300 feet by 13 feet. In cubic feet, what is its volume of storage space?

   (1) 78 (2) 2,600 (3) 3,900
   (4) 60,000 (5) 780,000

   ① ② ③ ④ ⑤

If you followed the formula, length × width × height, and multiplied the three numbers correctly, you got 780,000 cubic feet and you've just solved a solid geometry problem. Remember, you don't have to memorize formulas, they will be given to you on the formula page of the GED Mathematics Test. All the formulas you will be given on that page are printed on the inside front cover of this book. When you solve the problems in *"The Real Thing"* sections, the *Extra Practice* and the *Posttests*, don't look back in the lessons to find geometric formulas. Instead, use the formula page on the inside front cover. That way, you'll be working with formulas the way you'll have to when you take the actual GED Test.

Now see if you can solve one more problem by counting units on a grid.

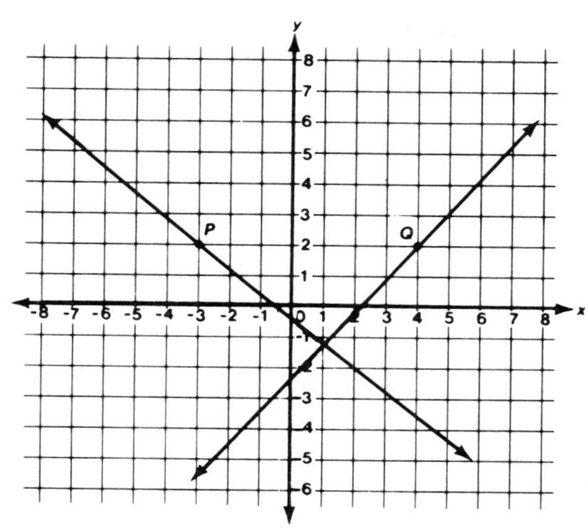

4. The distance in units between points P and Q on the figure above is

(1) 6 (2) 7 (3) 10
(4) 11 (5) 16

(1) (2) (3) (4) (5)

If you said option (2), you have just solved a coordinate geometry problem. Probably your knowledge of positive and negative numbers helped you to reach the solution. But it's likely that your common sense helped you just as much.

If you got any of these problems right, you probably could answer some geometry problems on the GED Mathematics Test correctly if you took the Test right now. Even if you got some wrong, by the time you finish this *geometry* section in your book, you'll be prepared to make a good showing on the geometry problems you find on the GED Mathematics Test.

LESSON 33

Plane Geometry

A carpenter uses a plane to smooth a piece of wood so that there aren't any bumps or dips on it. The Great Plains of the United States are enormous areas of flat land that stretch for miles in all directions. Both *plane* and *plain* come from a word that means "to make smooth." In plane geometry you study shapes and figures that lie on a perfectly smooth, flat surface called a **plane.**

Every special subject has special words and you must know what they mean in that particular subject. Plane geometry uses many words that you use in everyday language, but in plane geometry these words take on special meanings—**point**, **line** and **ray**.

In Lesson 32, you dealt with some figures from plane geometry. When you found the distance around, you were working with shapes made up of **line segments.** In plane geometry, there is a special name for closed figures that are made up of line segments. These figures are called **polygons,** a word which means "many sides."

Coming to Terms

point a dot like that made with a very sharp pencil. In plane geometry, a *point* is the smallest dot imaginable.

line a straight series of connected points that goes on endlessly in two opposite directions. A line has no starting or ending points.

line segment on a line, a piece of the line with a beginning and an end.

ray on a line, a part that consists of a starting point and all of those points on the line that lie to one side of that point. A ray has no ending point. You can imagine a ray easily by thinking of a very powerful flashlight pointing up at the sky at night. The starting point (end point) of the ray is the flashlight bulb. The rest of the ray is the beam of light that goes off endlessly into the sky.

polygon a closed figure made up of three or more line segments.

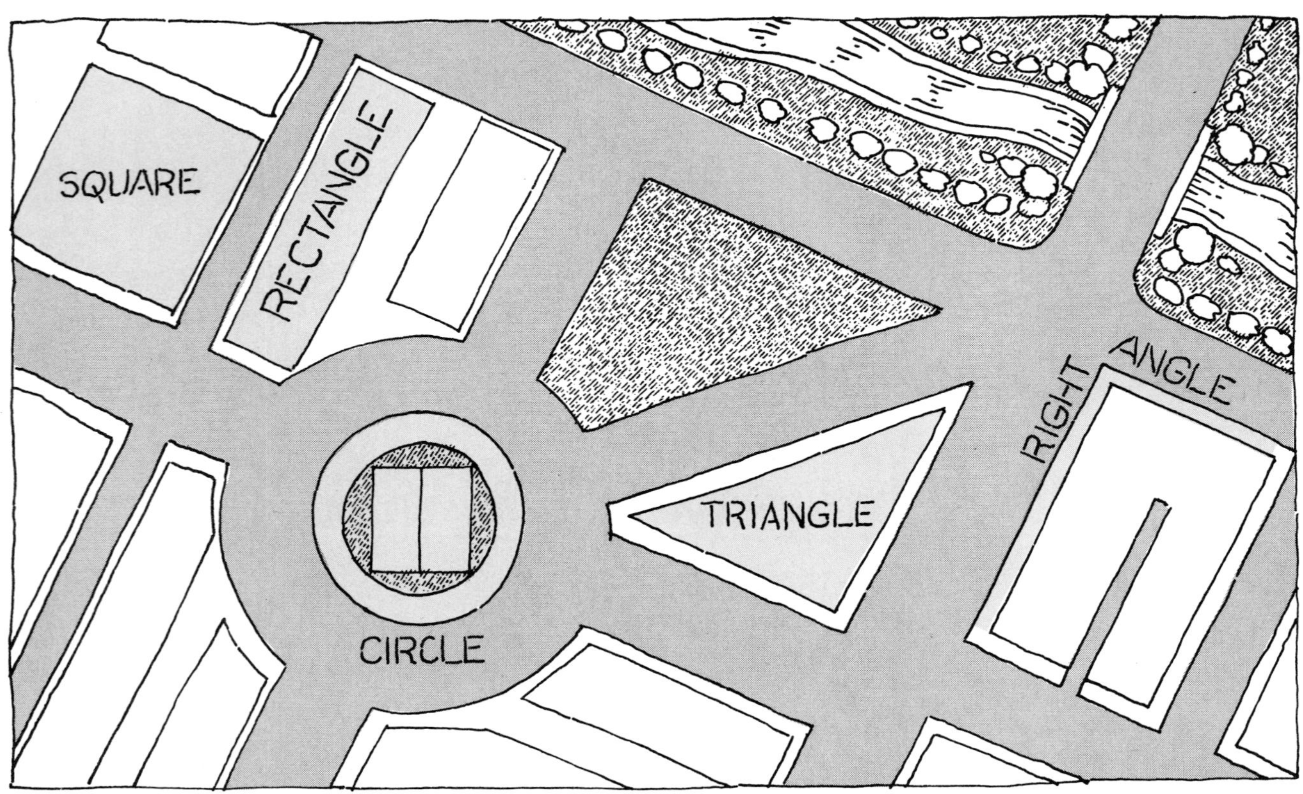

Angles

If two rays start from the same point, you have an **angle.** The point where the rays start is called the **vertex of the angle.** The two rays are called the **sides of the angle.**

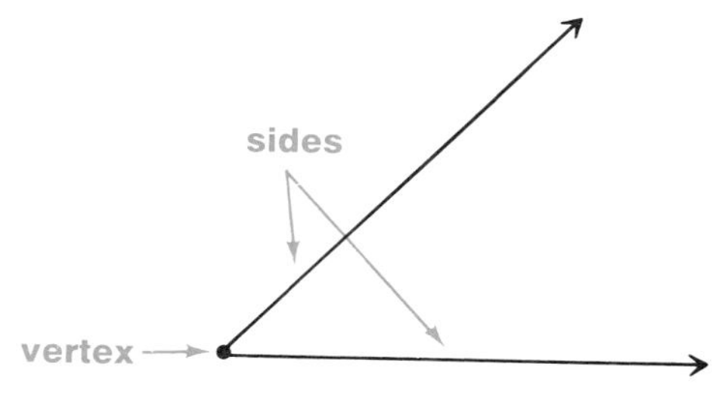

Coming to Terms

angle a figure formed by two rays that start at the same point

sides of an angle the two rays that form an angle

vertex of an angle the point shared by the two sides of an angle. The plural of *vertex* is *vertices.*

Labeling Angles

Angles can be named by using letter labels.

Here's an Example

Look at this angle.

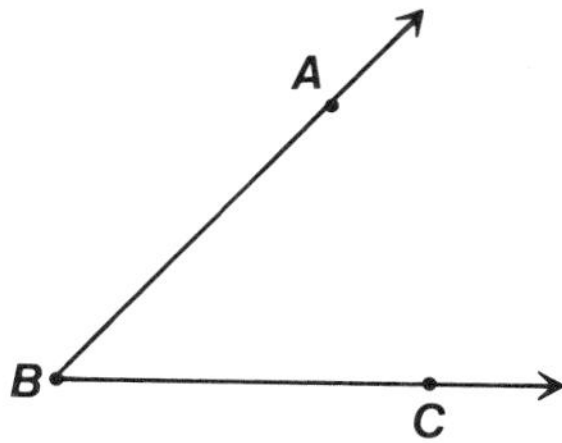

Notice the three points marked on the angle, one on each side and one at the vertex. Any angle can be named by labeling three such points. It is important to know that the *vertex is always named between* the points on the sides. This angle can be named either $\angle ABC$ or $\angle CBA$. The little angle symbol in front of the letters tells that you are referring to an angle. You read $\angle ABC$ as "angle ABC" and $\angle CBA$ as "angle CBA."

A second way to name an angle is by the vertex alone. The angle in the preceding figure could be referred to simply as $\angle B$. (If the figure showed more than one angle with the vertex *B,* you would have to use some other way of naming the angle to make it clear which angle you are talking about.)

A third way to name angles is to use a small letter inside the angle like this:

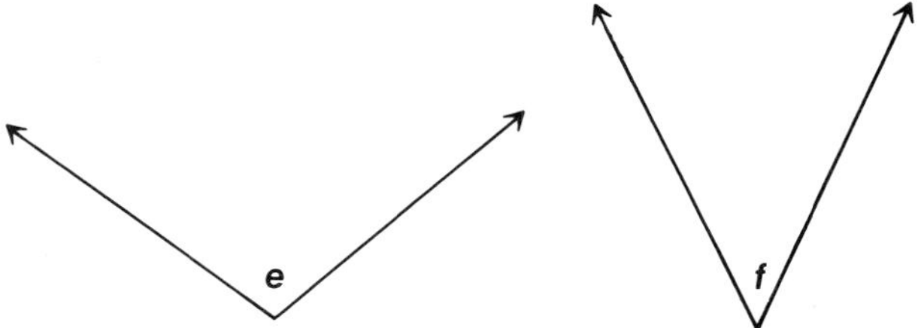

These angles can be referred to as $\angle e$ and $\angle f$.

Try It Yourself

■ Write four names for this angle.

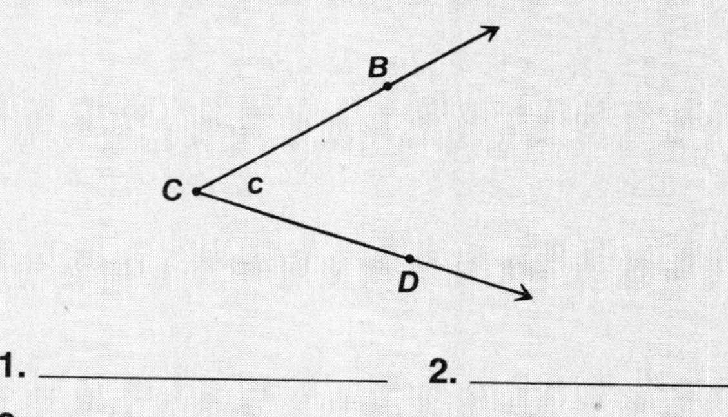

1. ____________ 2. ____________

3. ____________ 4. ____________

Using the points on the sides, you should have $\angle BCD$ and $\angle DCB$. Since the figure shows only one angle with the vertex *C,* you can use $\angle C$. You can also use the small letter *c* and name the angle $\angle c$.

Measuring Angles

Think of the sides of an angle as being like the hands of a clock. Suppose you held the hour hand steady. The measure of the angle tells how far you would need to turn the minute hand to move it to the same position as the hour hand.

The unit for measuring angles is the **degree,** and the instrument for measuring angles is the **protractor,** a curved scale marked off into 180 equal parts. Even though you will not have to measure angles on the GED Mathematics Test, it may help to look at a diagram to remind yourself how it is done. Here's how you place the protractor to measure ∠*ABC*.

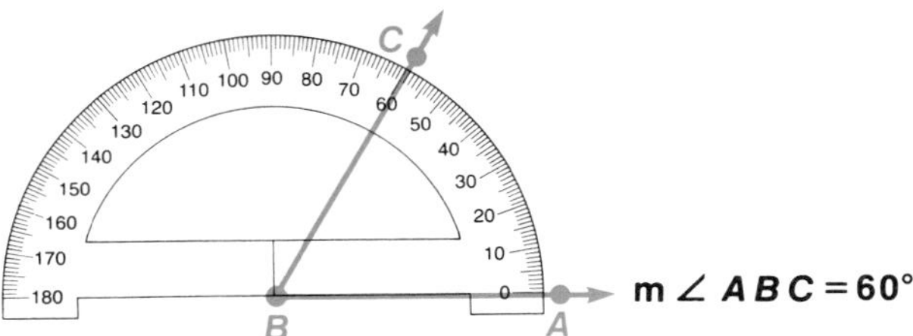

Notice that the center point of the protractor is at the vertex of the angle. The protractor is fixed so that one side of the angle crosses the scale at the 0 mark. The place where the other side crosses the scale gives you the measure of the angle—in this case 60 degrees. The symbol for *measure* is the letter *m,* and the symbol ° stands for *degree.* You can say that the measure of ∠*ABC* is 60° and you can write m∠*ABC* = 60°.

A **right angle** is a special angle. It's an angle whose sides form a square corner. A right angle measures exactly 90°. In diagrams you can show that an angle is a right angle by writing 90° inside the angle. Another way is to put the right angle symbol ⌉ inside the angle.

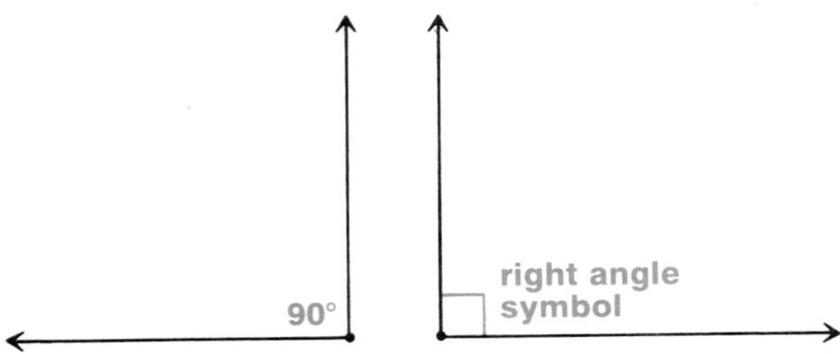

A **straight angle** is another special angle. Its sides point in exactly opposite directions and form a straight line. If you measure a straight angle with a protractor, you will find that it measures 180°—the largest measure that an angle can possibly have.

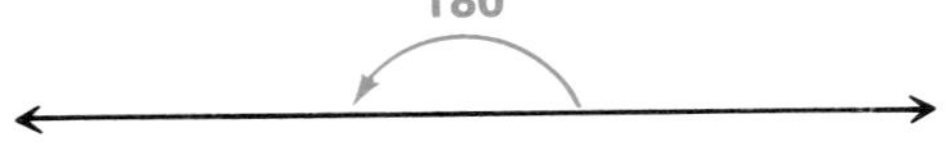

Coming to Terms

degree the unit used for measuring angles

protractor the instrument used for finding the degree measure of an angle

right angle an angle whose measure is exactly 90°. A right angle forms a square corner.

straight angle an angle whose measure is exactly 180°. A straight angle forms a straight line.

Try It Yourself

■ Give one name for each angle and tell its measure. If the angle is a special kind of angle because of its measure, tell what kind of angle it is.

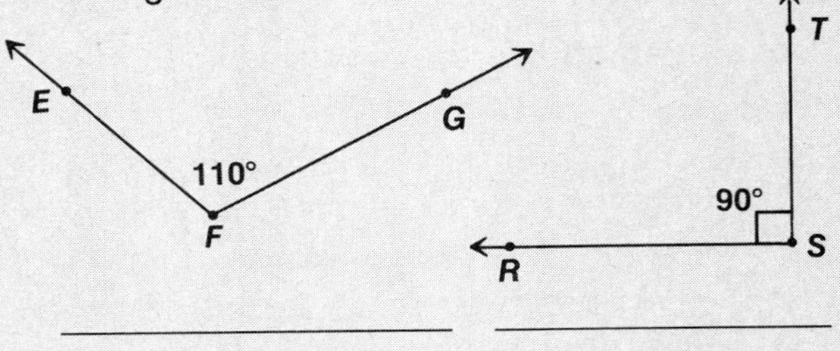

Did you name the angle on the left ∠*F* or ∠*EFG* or ∠*GFE*? It measures 110°. Did you say that the angle on the right is ∠*S* or ∠*RST* or ∠*TSR* and m∠*RST* = 90°? The angle on the right is called a *right angle* because it measures 90 degrees.

Special Pairs of Angles

Some angles are often found in pairs. Three of the most important pairs are dealt with here.

Here's an Example

If two angles measure a total of 90°, they are called **complementary** angles. If two angles measure a total of 180°, they are called **supplementary** angles.

■ Look at these complementary angles.

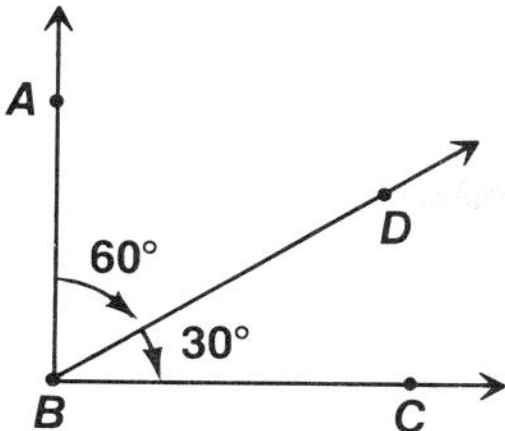

Together their measures total 90°.

$m\angle ABD + m\angle DBC = 60° + 30° = 90°$

The angles in this diagram are side by side, but angles can be complementary without being side by side. Any two angles whose measures equal 90° when added together are complementary (no matter how they are positioned in relation to each other).

■ Here are two supplementary angles.

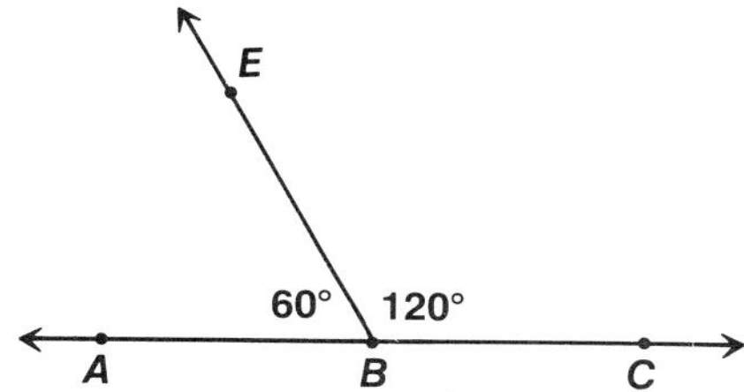

Together their measures total 180°.

$m\angle ABE + m\angle EBC = 60° + 120° = 180°$

Since 60° + 120° = 180°, the angles are supplementary.

Another important pair of angles are called **vertical angles.** Vertical angles are the *opposite* angles formed where two straight lines cross. Vertical angles always have the same measure.

■ Study this diagram. Do you see that the two 15° angles ($\angle c$ and $\angle d$) are vertical angles?

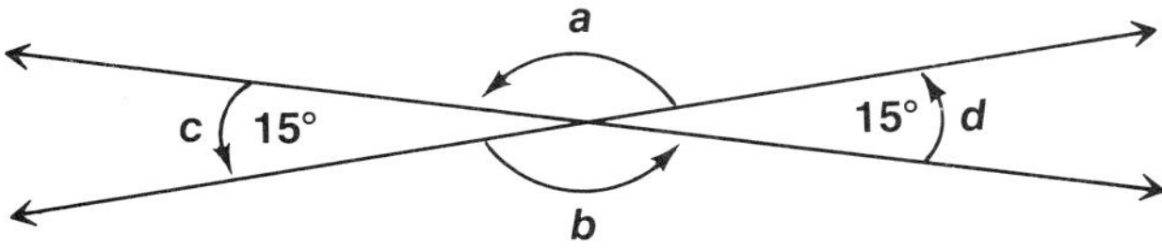

Do you see that $\angle a$ and $\angle b$ form another pair of vertical angles?

Try It Yourself

■ What size angle is complementary to a 47° angle?

____

Did you remember that the measures of complementary angles add up to 90°? An angle complementary to a 47° angle must measure 43°.

Try another.

■ What size angle is supplementary to a 105° angle?

____

Did you remember that the measures of supplementary angles must add up to 180°? An angle supplementary to a 105° angle must measure 75°.

■ Study this diagram.

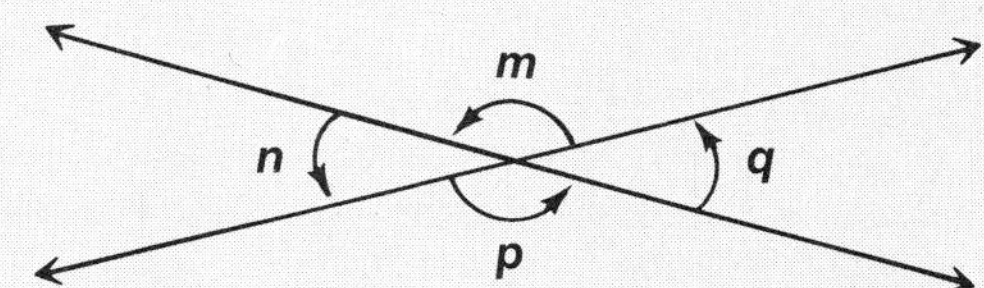

Name two vertical angles.

____________ and ____________

Did you answer $\angle m$ and $\angle p$? Or $\angle n$ and $\angle q$? Both sets of answers are correct.

A Test-Taking Tip

Sometimes it's hard to keep from confusing the terms *complementary* and *supplementary.* One easy way is to use the first letters of the words. Remember that *c* comes before *s* in the alphabet and that 90 comes before 180.

Coming to Terms

complementary angles a pair of angles whose measures total 90°. Each angle is called a **complement** of the other.

supplementary angles a pair of angles whose measures total 180°. Each angle is called a **supplement** of the other.

vertical angles a pair of opposite angles formed when two lines cross. Two such pairs of angles are formed when two lines cross. Vertical angles always have the same degree measure.

Warm-up

Name the complement of each of the following angles.

1. 36° ______ **2.** 10° ______ **3.** 89° ______

Name the supplement of each of the following angles.

4. 100° ______ **5.** 1° ______ **6.** 90° ______

7. Explain in your own words how vertical angles are formed and what is true about the measures of any pair of vertical angles.

____

____

Warm-up Answers

1. 54° **2.** 80° **3.** 1° **4.** 80° **5.** 179° **6.** 90° **7.** Vertical angles are formed by two lines crossing. They are opposite each other and are always equal in measure.

On the Springboard

For 1–5, find the measure in degrees of the angle labeled x.

1.

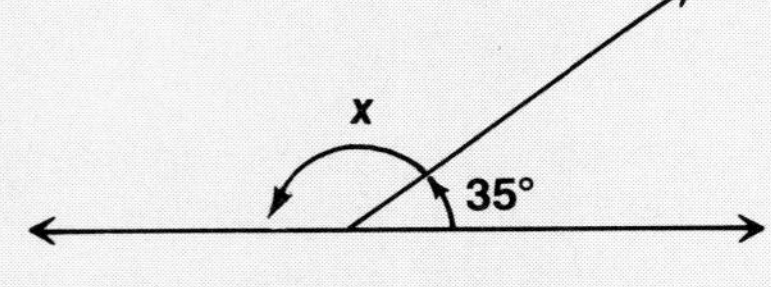

(1) 55° (2) 145° (3) 155°

(1) (2) (3)

2.

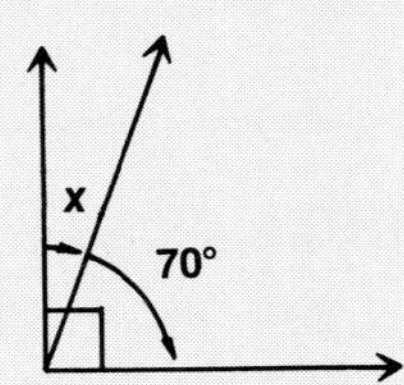

(1) 20° (2) 70° (3) 110°

(1) (2) (3)

3.

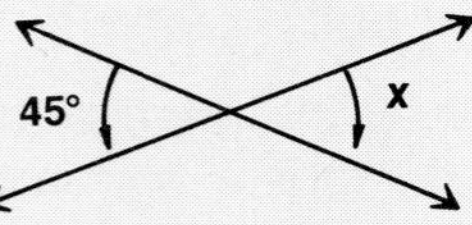

(1) 45° (2) 135°
(3) Insufficient data is given.

(1) (2) (3)

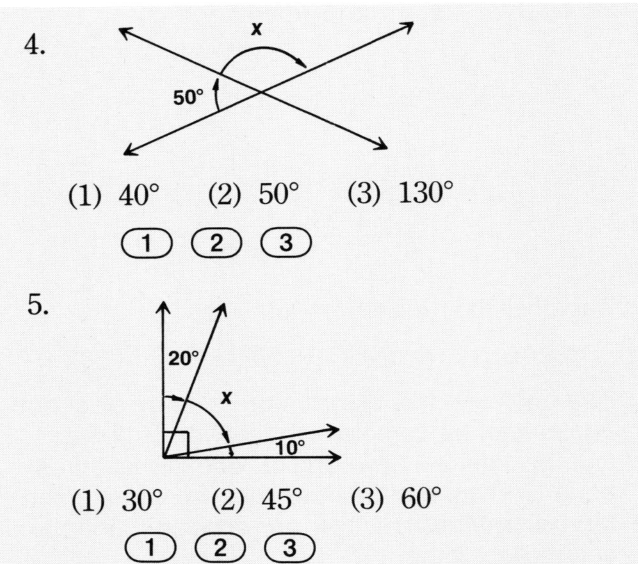

4\.

(1) 40° (2) 50° (3) 130°

(1) (2) (3)

5\.

(1) 30° (2) 45° (3) 60°

(1) (2) (3)

Check your Springboard answers on page 283. If you got them all right, go on to "The Real Thing." If you missed any, check your work. Review if necessary. When you are confident that you can answer all of these correctly, you're ready for "The Real Thing."

"The Real Thing"

1. What angle is shown below?

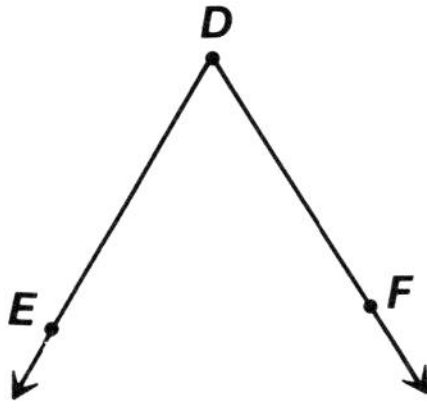

(1) $\angle EFD$ (2) $\angle DEF$ (3) $\angle FDE$

(4) $\angle DFE$ (5) $\angle FED$

(1) (2) (3) (4) (5)

2. If $\angle G$ measures 60° and $\angle H$ measures 120°, what kind of angles are $\angle G$ and $\angle H$?

(1) complementary (2) obtuse (3) right
(4) supplementary (5) vertical

(1) (2) (3) (4) (5)

3. In the figure below, which pair of angles are vertical angles?

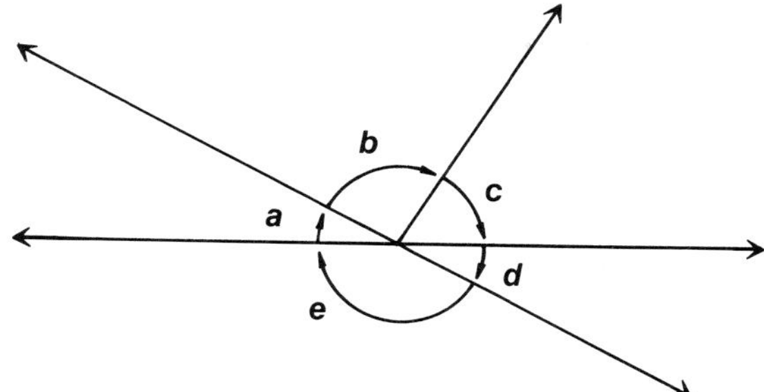

(1) $\angle a$ and $\angle c$ (2) $\angle b$ and $\angle e$
(3) $\angle c$ and $\angle e$ (4) $\angle d$ and $\angle a$
(5) $\angle e$ and $\angle d$

(1) (2) (3) (4) (5)

4. In the figure below, $\angle RXT$ measures 90°. Which other angle(s) are complements of $\angle a$?

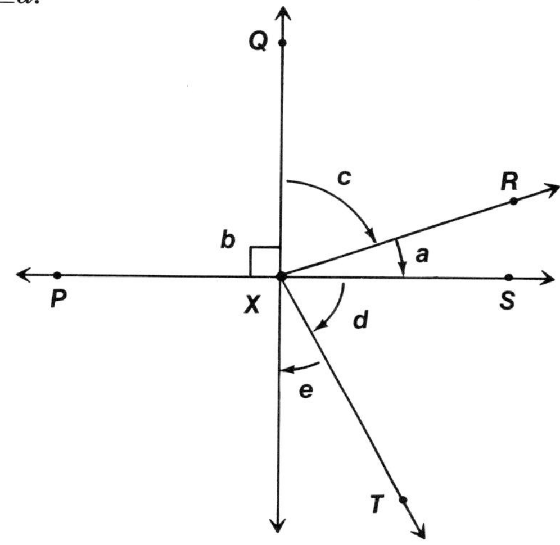

(1) $\angle b$ only (2) $\angle c$ only
(3) $\angle d$ only (4) $\angle c$ and $\angle d$
(5) $\angle e$ and $\angle c$

(1) (2) (3) (4) (5)

Check your answers on page 288.

Angles Formed by Parallel Lines and Transversals

Parallel lines are like the ruled lines on a piece of notebook paper or straight railroad tracks going across flat land. Parallel lines stay the same distance apart and never meet. Here are two parallel lines:

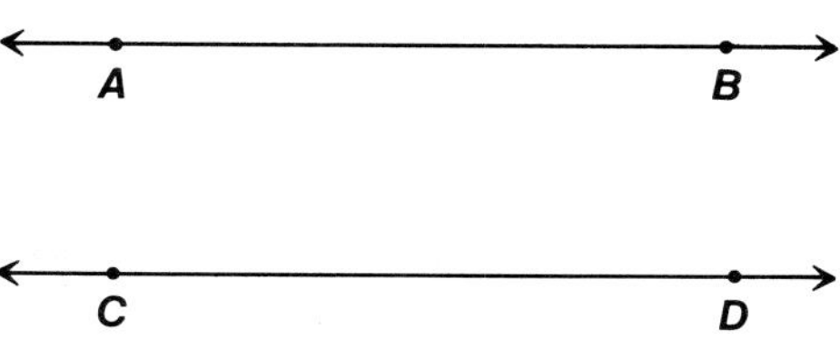

A **transversal** is a line that cuts across two or more parallel lines, like this:

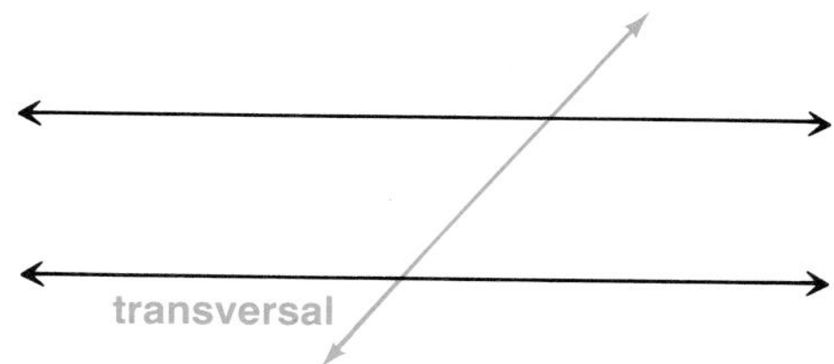

Special pairs of angles are formed when two parallel lines are cut by a transversal.

Coming to Terms

parallel lines lines that run next to each other and are always the same distance apart, like railroad tracks

transversal a line that cuts or crosses parallel lines

Here's an Example

In the following examples, you will see three special pairs of angles that are formed when parallel lines are cut by a transversal. The first kind is a pair of **corresponding angles.** The diagram shows that $\angle a$ and $\angle e$ are corresponding angles.

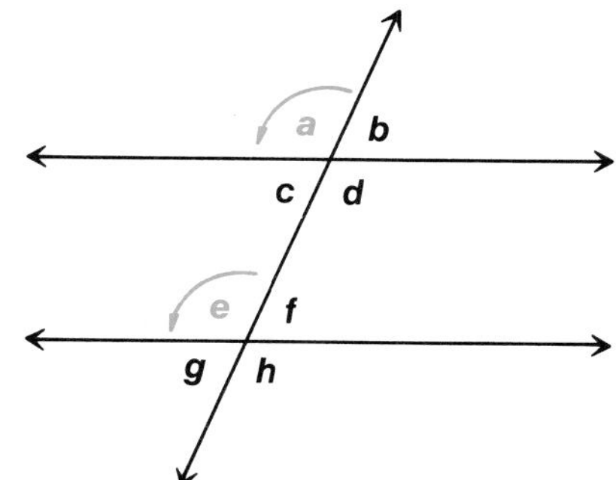

Corresponding angles are on the same side of the transversal, they face in the same direction, and they have the same degree measure. There are three other pairs of corresponding angles in the diagram above: $\angle c$ and $\angle g$, $\angle b$ and $\angle f$, and $\angle d$ and $\angle h$.

In the diagram below, $\angle a$ and $\angle h$ are **alternate exterior angles.**

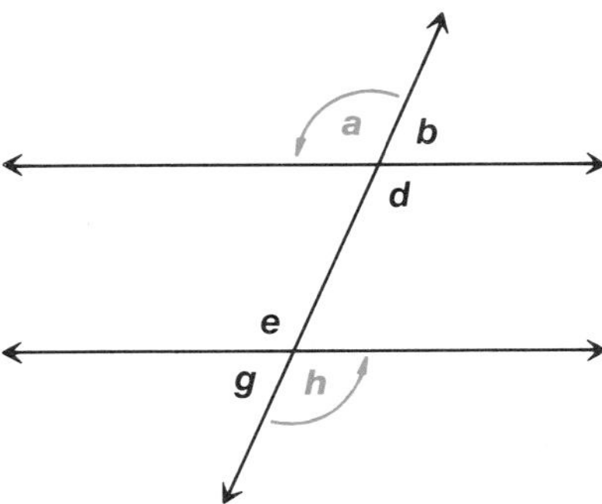

Exterior means outside the parallel lines, and *alternate* means on opposite sides of the transversal. Alternate exterior angles face the space outside (not between) the two parallel lines. They are on opposite sides of the transversal and they have the same degree measure. When two parallel lines are cut by a transversal, there are two pairs of these angles. In the diagram above, the other pair of alternate exterior angles is $\angle b$ and $\angle g$.

Here is a pair of **alternate interior angles,** $\angle c$ and $\angle f$.

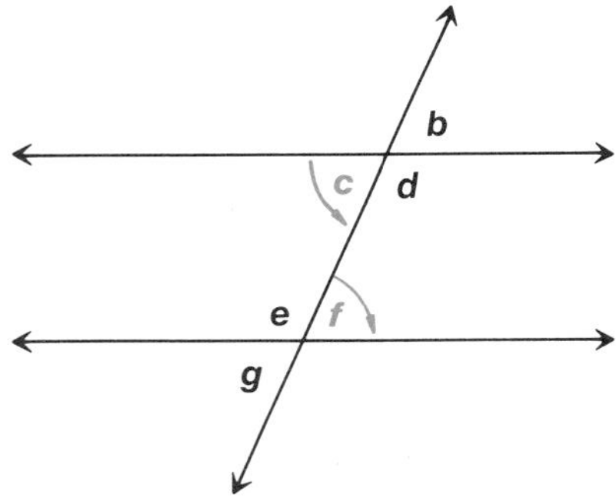

Interior means inside, and *alternate* means on opposite sides of the transversal. *Alternate interior angles have the same degree measure.* There are two pairs of these angles. In the diagram above, the other pair is $\angle d$ and $\angle e$.

Here is a special case. The transversal cuts the parallel lines at right angles. All the angles are equal to 90°.

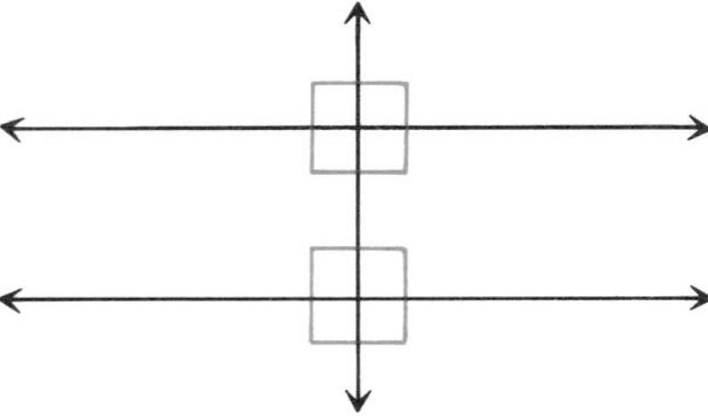

When one line cuts or meets another to form right angles, the lines are called **perpendicular lines.** *If a transversal is perpendicular to one of two parallel lines, it is also perpendicular to the other.* Here is another useful fact: *If two lines are both perpendicular to the same third line, then the first two lines are parallel to each other.*

Coming to Terms

corresponding angles two angles on the same side of a transversal and facing in the same direction. Corresponding angles are always equal.

alternate exterior angles two angles on opposite sides of a transversal and facing to the outside of the parallel lines. Two alternate exterior angles have the same measure.

alternate interior angles two angles on opposite sides of a transversal and facing the inside space between the parallel lines. Alternate interior angles have the same measure.

perpendicular lines two lines that meet at right angles (like many street intersections)

Try It Yourself

■ This diagram shows two parallel lines cut by a transversal and the eight angles that are formed.

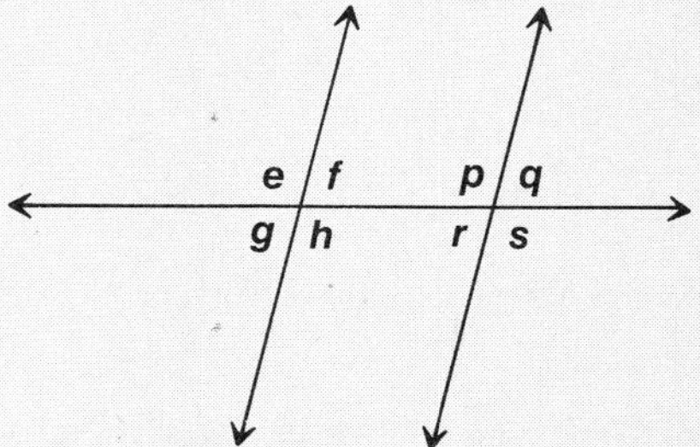

Identify the following:

A pair of corresponding angles _____ _____

A pair of alternate exterior angles _____ _____

A pair of alternate interior angles _____ _____

You could have several different choices for each answer. Did you remember that corresponding angles are on the same side of the transversal and face in the same direction? Did you choose $\angle e$ and $\angle p$, or $\angle f$ and $\angle q$, or $\angle g$ and $\angle r$, or $\angle h$ and $\angle s$?

Did you choose angles facing outside the parallel lines and on opposite sides of the transversal? Alternate exterior angles are $\angle e$ and $\angle s$, $\angle g$ and $\angle q$.

Did you pick those angles on opposite sides of the transversal but facing between the parallel lines? Alternate interior angles are $\angle f$ and $\angle r$, $\angle h$ and $\angle p$.

Warm-up

Look at the diagram below. The horizontal lines are parallel. If $\angle a$ is 35°, give the measures of these angles.

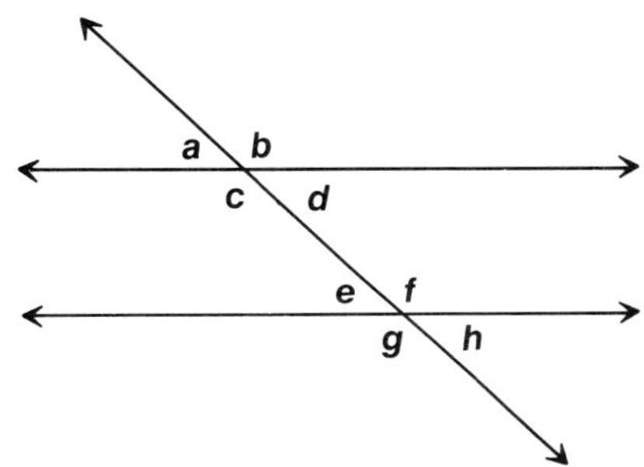

1. $\angle b$ ______ **2.** $\angle f$ ______

3. $\angle c$ ______ **4.** $\angle g$ ______

5. $\angle d$ ______ **6.** $\angle h$ ______

7. $\angle e$ ______

Warm-up Answers
1. 145° **2.** 145° **3.** 145° **4.** 145° **5.** 35° **6.** 35° **7.** 35°

On the Springboard

6. What is true about all these pairs of angles: vertical, corresponding, alternate interior, and alternate exterior?
   (1) They are formed by two lines crossing.
   (2) Each pair is always equal in measure.
   (3) Exactly two such pairs are formed when two parallel lines are cut by a transversal.

   (1) (2) (3)

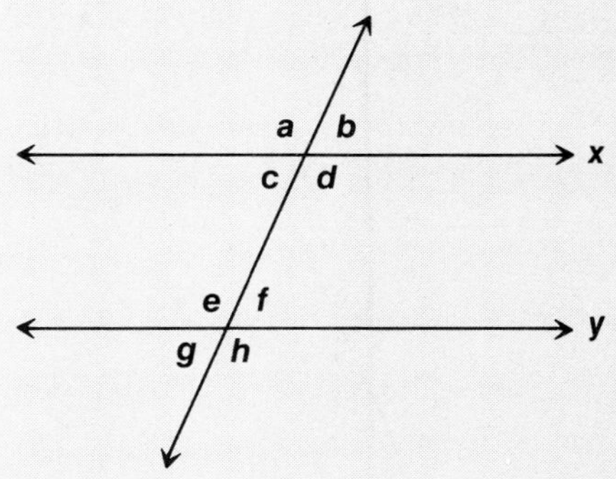

7. In the figure above, line x is parallel to line y. If $\angle g$ is 30°, then $\angle b$ is
   (1) 15° (2) 30° (3) 60°

   (1) (2) (3)

8. In the diagram below, what is the measure of $\angle 3$?

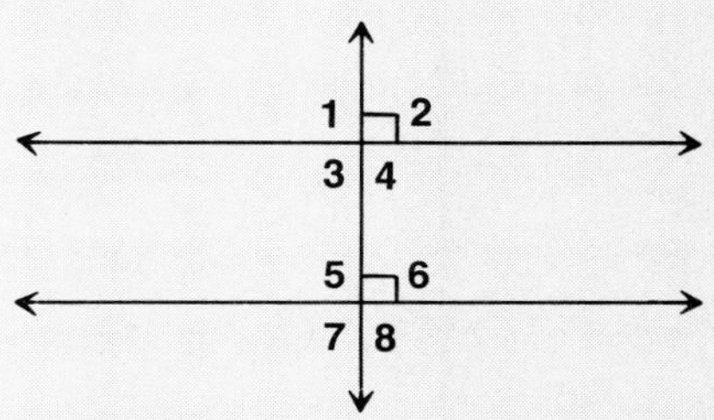

   (1) 30° (2) 60° (3) 90°

   (1) (2) (3)

Check your Springboard answers on page 283. If you got them all right, go to "The Real Thing." If you missed any, make sure of where your error was. It may pay you to review the section before you go on.

"The Real Thing"

5. Angles a and b are supplementary angles. Angles b and c are vertical angles. If the measure of $\angle c$ is 40°, what is the measure of $\angle a$?
   (1) 40° (2) 50° (3) 90°
   (4) 140° (5) 150°

   (1) (2) (3) (4) (5)

Questions 6–8 refer to the diagram below. Line RS is parallel to line TU.

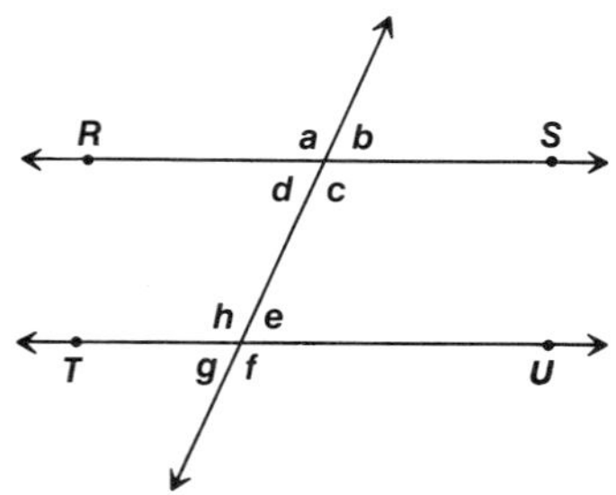

6. Angles b and g are what kind of angles?
   (1) alternate exterior
   (2) alternate interior
   (3) complementary
   (4) corresponding
   (5) vertical

   (1) (2) (3) (4) (5)

7. If the measure of $\angle e$ is 55°, what is the measure of $\angle g$?
   (1) 35° (2) 55° (3) 125°
   (4) 145° (5) 155°

   (1) (2) (3) (4) (5)

8. The sum of the measures of angles $e, f, g,$ and h is how many degrees?
   (1) 90° (2) 140° (3) 180°
   (4) 240° (5) 360°

   (1) (2) (3) (4) (5)

9. In the diagram below, line WX is parallel to line YZ. How many degrees is $\angle SRT$?

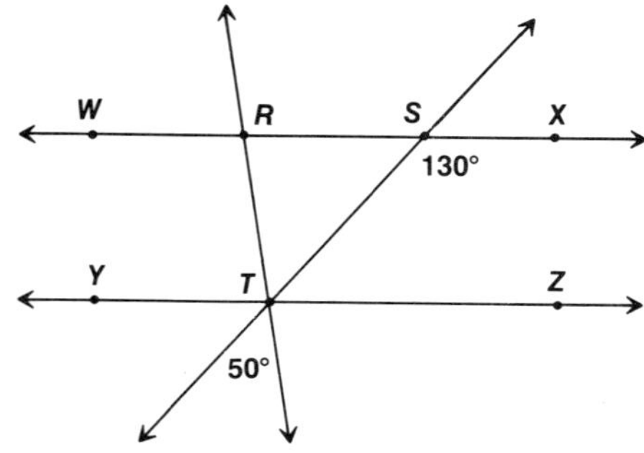

   (1) 30 (2) 50 (3) 80
   (4) 130 (5) 180

   (1) (2) (3) (4) (5)

Check your answers and record all your scores on pages 288–289.

LESSON 34

Quadrilaterals

The first part of the word quadrilateral is *quadri-*, which means *four.* The second part, *-lateral,* means *side.* So a **quadrilateral** is a 4-sided figure—a closed, 4-sided figure made up of line segments.

In this lesson you will concentrate on the special kinds of quadrilaterals shown in Coming to Terms on the next page. You will *not* be dealing with figures such as this:

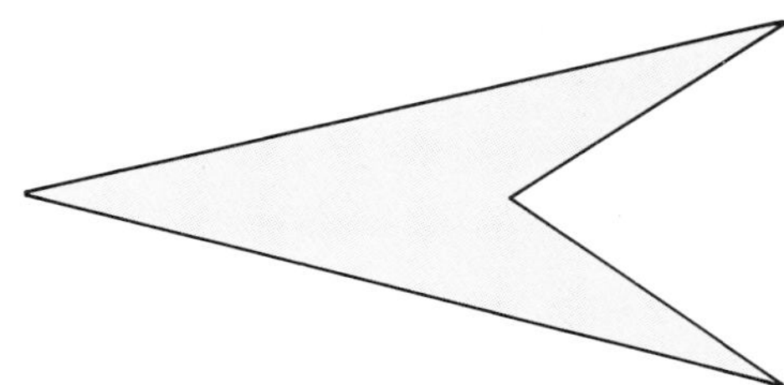

Strictly speaking, this figure is a quadrilateral. After all, it is a closed figure with four straight sides. *This lesson will not discuss quadrilaterals with dents.*

All quadrilaterals have four sides and four angles. The angles of a quadrilateral have an interesting property: *If you add the measures of all four angles of a quadrilateral, you get 360°.*

A *parallelogram* is a special kind of quadrilateral. It is made up of two pairs of parallel sides. The opposite sides of a parallelogram are equal, and the opposite angles of a parallelogram are equal. You can use these facts to find angles in a parallelogram quite easily.

Here's an Example

■ In this figure, XY is parallel to WZ and XW is parallel to YZ. Find the measure of $\angle X$.

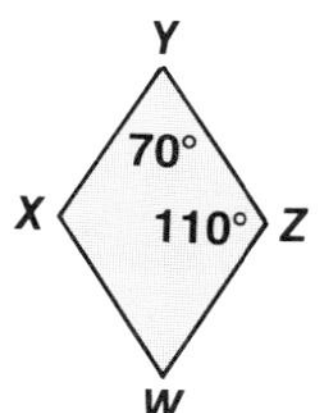

To solve, find an angle equal to $\angle X$. Since the quadrilateral is a parallelogram, $\angle X$ is equal to the angle opposite angle $\angle Z$. The figure shows that m$\angle Z$ is 110°. So m$\angle X$ is also 110°.

Coming to Terms

Here are some special quadrilaterals you should be able to identify.

QUADRILATERALS
(four sides)
angles total 360°

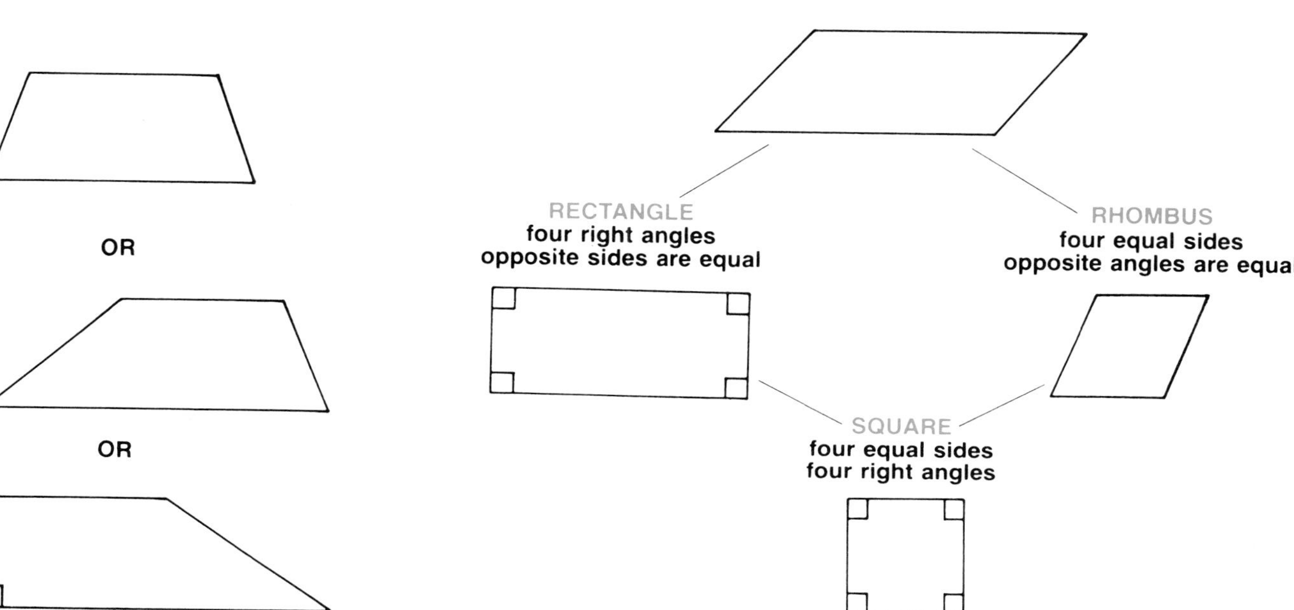

Try It Yourself

■ In parallelogram *ABCD*, m∠*B* = 50°. Find the measure of ∠*D*.

ABCD is a parallelogram. Did you notice that ∠*D* is opposite ∠*B* and therefore equal to it? If you did, you got the correct answer: m∠*D* = 50°.

Warm-up

Use the information in the figures to fill in the blanks.

1.

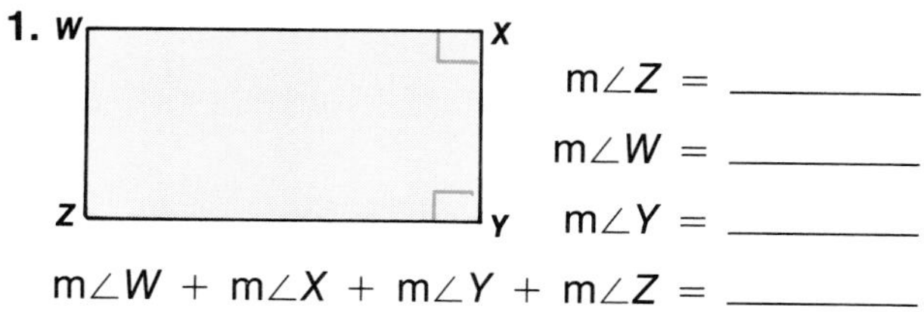

m∠*Z* = ________

m∠*W* = ________

m∠*Y* = ________

m∠*W* + m∠*X* + m∠*Y* + m∠*Z* = ________

2.

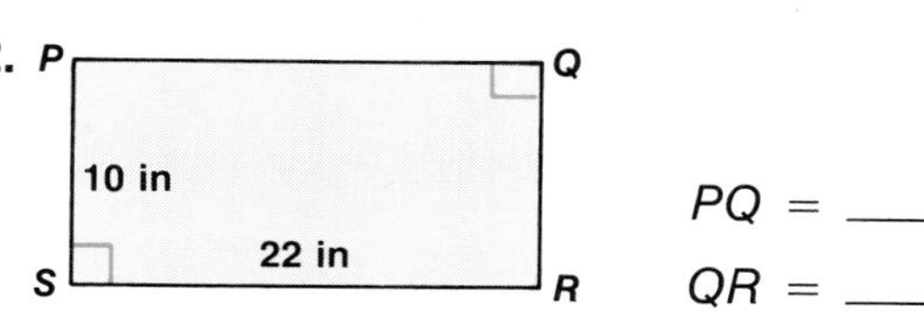

PQ = ________

QR = ________

3.

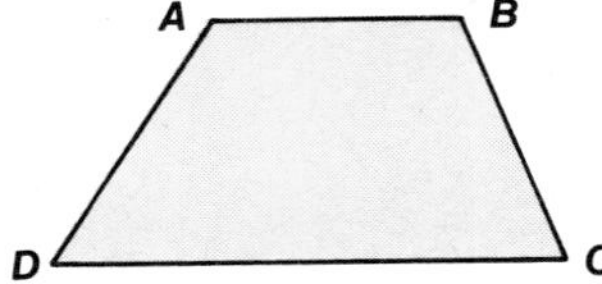

$m\angle A + m\angle B + m\angle C + m\angle D =$ ________

4.

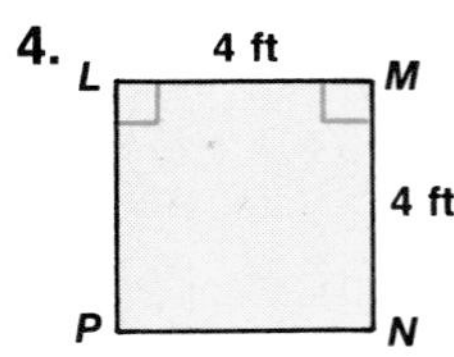

$NP + PL =$ ________

Warm-up Answers
1. 90°, 90°, 90°, 360° **2.** 22 in, 10 in **3.** 360° **4.** 8 ft

On the Springboard

1. This figure is what kind of quadrilateral?

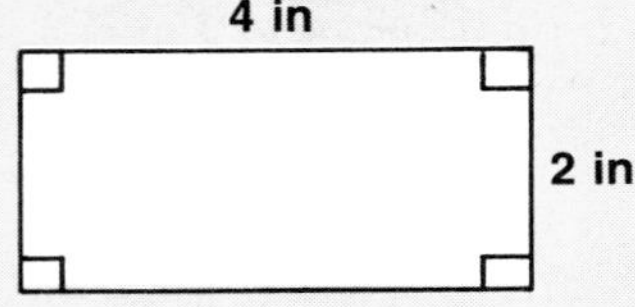

(1) rectangle (2) square (3) triangle

(1) (2) (3) (4) (5)

2. In figure *ABCD, AB* is parallel to *CD* and *AD* is parallel to *BC*. What kind of figure is *ABCD?*

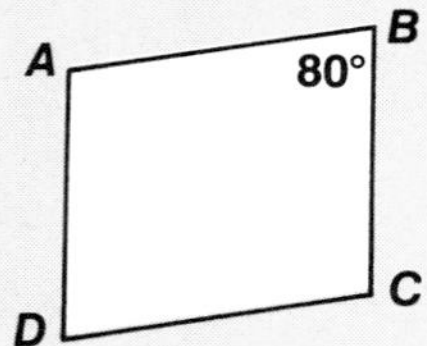

(1) parallelogram (2) rectangle
(3) square

(1) (2) (3)

3. What is the measure of side *XY* of square *WXYZ*?

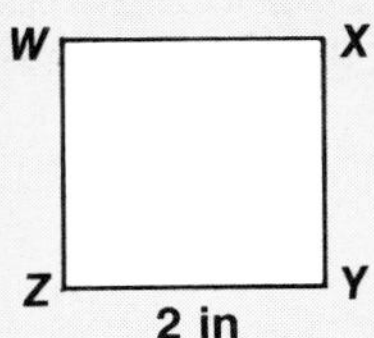

(1) 2 in (2) 4 in
(3) Insufficient data is given to solve the problem.

(1) (2) (3) (4) (5)

4. In the figure, what is the measure of $\angle R$?

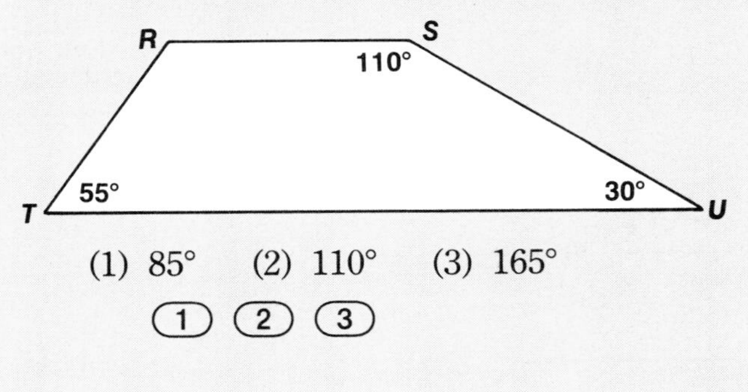

(1) 85° (2) 110° (3) 165°

(1) (2) (3)

Check your Springboard answers on page 283. If your got these right, go on to "The Real Thing." If you missed any, decide whether you should review the section. When you're satisfied that you understand, go on.

Finding Perimeters

If you want to put a fence around a garden or a plot of land, you have to know how far it is around. The distance around a plot of land, a figure, and so on, is called the **perimeter.** You found perimeters in Lesson 32 when you studied linear measure. It is easy to find perimeters of many figures if you have a little knowledge of geometry.

Coming to Terms

perimeter the distance around the outside of a closed figure such as a polygon

Here's an Example

Here you use the knowledge of geometry to solve a problem that straight arithmetic can't solve.

■ Lucas walked his dog around the rectangular city block he lives on. The block is 100 yards long and 50 yards wide. How far did Lucas walk?

Draw a picture showing what you know about the situation:

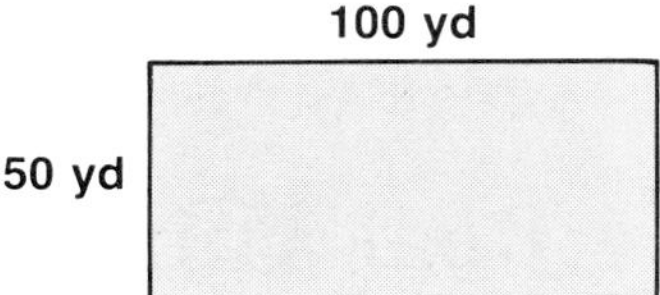

There are four sides to the rectangular block, but only two measurements are given. You have to apply what you know about rectangles to find the other two. By definition, rectangles have two pairs of equal sides. If one side measures 100 yards, the side opposite it must also be 100 yards. The side opposite the 50-yard side is 50 yards.

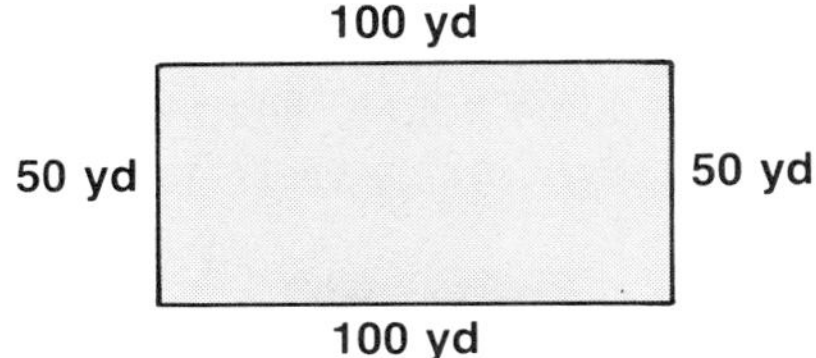

Now you can find the perimeter by adding all four sides.

$$100 + 100 + 50 + 50 = 300$$

Instead of adding 100 + 100, you can also write 2×100 or 2(100). Then, 50 + 50 can be written 2(50). You can solve the problem like this: $2(100) + 2(50) = 200 + 100 = 300$. The answer is the same. Lucas walked 300 yards.

So, to find the perimeter of a rectangle, you can add the length twice and the width twice. You recall how you used formulas as a shorthand system of writing information. Here, if you let P stand for perimeter, ℓ for length, and w for width, you get the formula $P = \ell + \ell + w + w$, or $P = 2\ell + 2w$. This tells you that if you add twice the length to twice the width you will get the perimeter of a rectangle. Of course, all the measurements must be in the same unit.

Try It Yourself

■ The Huynhs placed edging around their garden, which is a square 3 feet on each side. How much edging did they need?

____

Did you remember that a square has four equal sides? If one side is 3 feet long, the other three sides are each 3 feet long too. Did you add all four sides to find the perimeter: $3 + 3 + 3 + 3 = 12$? Or maybe you just multiplied 3 by 4. Either way is correct. To find the perimeter of a square, you can add all four sides ($P = s + s + s + s$), or you can multiply one side by 4 to get 12: ($P = 4s$). Whichever method you used, the answer is 12 feet.

Sometimes it's helpful to first sketch the problem. Try this one.

■ A model train club set up two pieces of 4×8 plywood with the 8-foot sides meeting. Then they attached a wood strip around the outside of the plywood. How many feet of wood were used around the edge?

____

Did you draw a sketch with the 8-foot sides of the two plywood pieces meeting? If you had the 4-foot sides meeting, you got the wrong answer. Check your sketch against the wording of the problem to be sure your mental picture is accurate. The wood strips form the perimeter of the plywood platform for the model trains. Did you come up with 32 feet?

☑ A Test-Taking Tip

The formulas $P = 2\ell + 2w$ and $P = 4s$ are shorthand for the perimeters of rectangles and squares. You don't have to memorize them. They will be on the formula page for the GED Mathematics Test. The important thing is to know what they are and how to use them.

Warm-up

Find the perimeters of each of these polygons.

1.

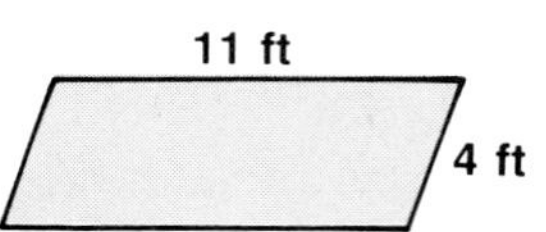

$P =$ ________

2.

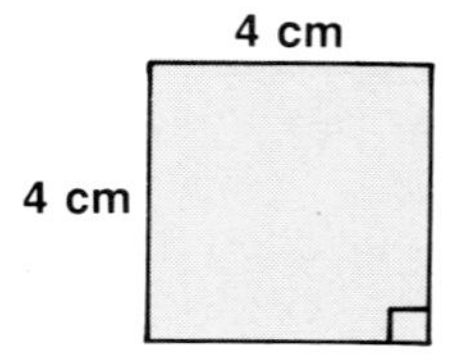

$P =$ ________

3.

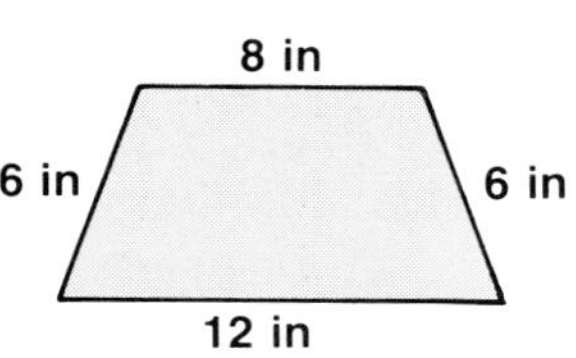

$P =$ ________

Warm-up Answers
1. 30 ft **2.** 16 cm **3.** 32 in

On the Springboard

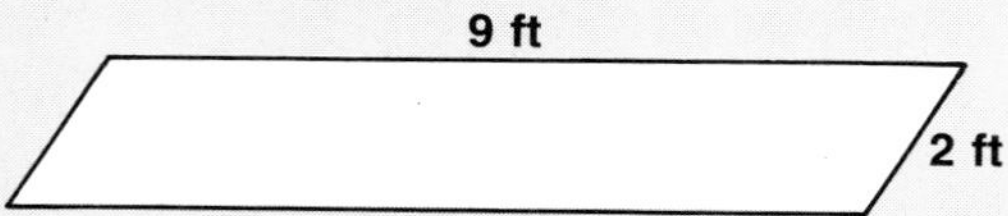

5. What expression will give you the perimeter *in feet* of this parallelogram?

   (1) $2 \times 2 \times 9 \times 9$ (2) $2 + 9$
   (3) $2(2) + 2(9)$

   ① ② ③

6. What is the length *in feet* of one side of a square whose perimeter is 64 feet?

   (1) 8 (2) 16 (3) 32

   ① ② ③

7. Find the perimeter *in inches* of a trapezoid with sides whose measures are 3.17, 5.50, 3.33, and 4.00 inches.

   (1) 8.67 (2) 12.00 (3) 16.00

   ① ② ③

8. Find the perimeter *in inches* of a rectangle whose length is 10 inches and whose width is 3 inches.

   (1) 23 (2) 26 (3) 30

   ① ② ③

Check your Springboard answers on page 283. If you answered all these correctly, go on to "The Real Thing." If you had some wrong, find out where you made your errors. Then go on.

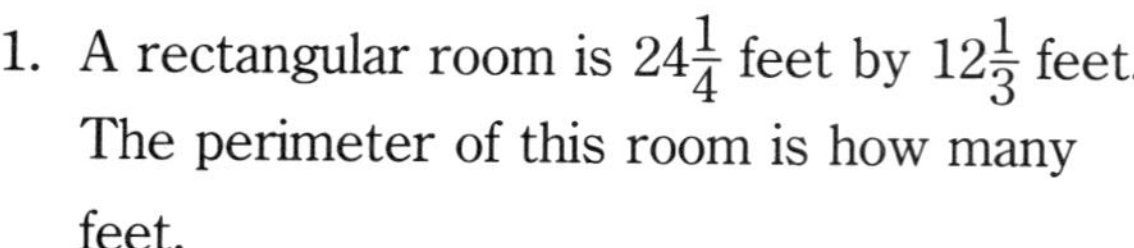

The Real Thing

1. A rectangular room is $24\frac{1}{4}$ feet by $12\frac{1}{3}$ feet. The perimeter of this room is how many feet.

   (1) $36\frac{7}{12}$ (2) $62\frac{3}{4}$ (3) $63\frac{1}{6}$
   (4) $72\frac{3}{4}$ (5) $73\frac{1}{6}$

   ① ② ③ ④ ⑤

2. A patio is shown below. The perimeter of this patio is how many feet?

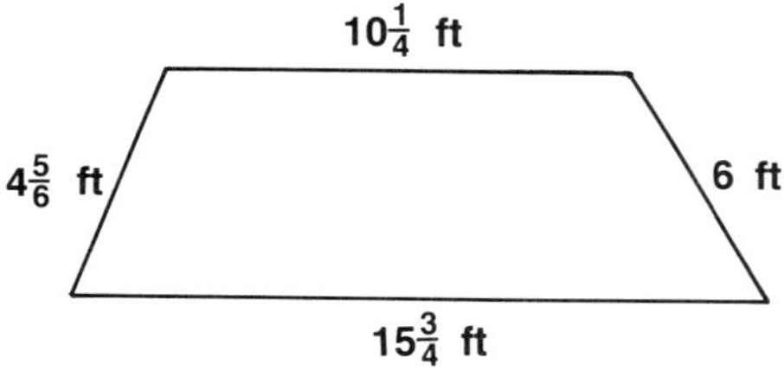

   (1) $29\frac{1}{3}$ (2) $31\frac{1}{3}$ (3) $35\frac{5}{6}$
   (4) $36\frac{5}{6}$ (5) $37\frac{1}{6}$

   ① ② ③ ④ ⑤

3. The perimeter of a square building is 3,026 meters. How many meters long is each side of the building?

   (1) 378.25 (2) 504.30 (3) 706.5
   (4) 756.5 (5) 12,104

   ① ② ③ ④ ⑤

4. A rectangular room is 12 feet by 9 feet. If you allow a total of 6 feet for the doorways, how many feet of baseboard are needed for this room?

(1) 30 (2) 36 (3) 42
(4) 48 (5) 318

① ② ③ ④ ⑤

5. A certain type of low fencing costs \$0.36 per running foot. How much will it cost to put this fencing around a rectangular lot that is 24 feet by 12 feet?

(1) \$12.96 (2) \$21.82
(3) \$25.92 (4) \$32.40
(5) Insufficient data is given to solve the problem.

① ② ③ ④ ⑤

Check your answers on page 289.

Area

To measure the perimeter (distance around) a quadrilateral, you use linear units such as inches, feet, centimeters, and so on. To measure the surface space inside a quadrilateral, you need **square units** such as *square inches, square feet, square centimeters,* and so on. In Lesson 32 you learned to do this for rectangles and squares. Although you did not use the words *rectangle* and *square* for the surfaces you were working with, that is, in fact, what they were.

To find how many 1-square-foot tiles are needed to cover a rectangular section of floor such as this, you simply find the length and width of the floor and multiply. When you do this, you are finding the **area** of the floor.

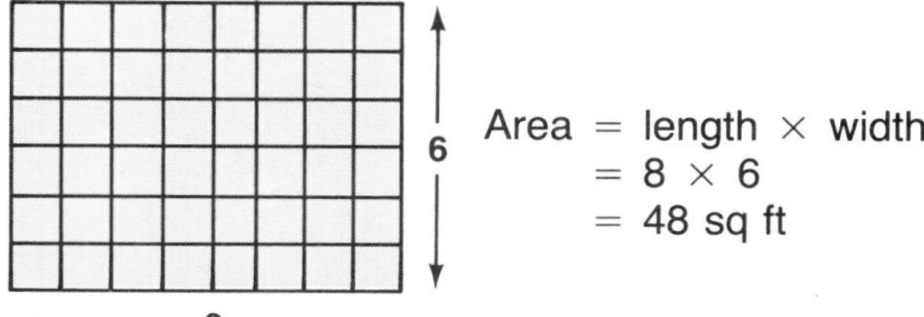

Area = length × width
= 8 × 6
= 48 sq ft

The **length** (ℓ) of a rectangle is the length of the longer side, and the **width** *(w)* is the length of the shorter side. So the area *(A)* is found by multiplying ℓ by *w:*

$$A = \ell w$$

A square is a special kind of rectangle that has four sides of equal length. This square measures 10 units on each side. The length and width are the same.

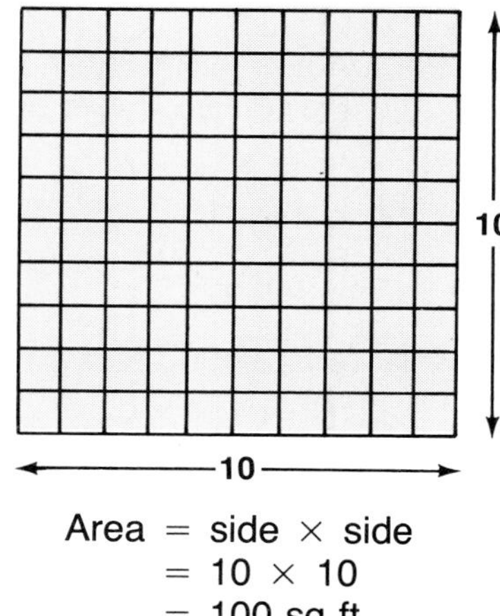

Area = side × side
= 10 × 10
= 100 sq ft

If you use the letter *s* to stand for the length of each side of a square, then the area of the square is length *(s)* times width *(s):*

$$A = s \times s \quad \text{or} \quad A = s^2$$

Every quadrilateral—in fact, every polygon—has an area, and, given full information on a quadrilateral or polygon, it would be possible to find its area. But the most basic and most important quadrilaterals are rectangles and squares.

Coming to Terms

area the number of square units that can be fitted inside a polygon

length (of a rectangle) the number of units in the longer side of a rectangle

square unit the amount of surface covered by the inside of a square that measures 1 unit on each side

width (of a rectangle) the number of units in the shorter side of a rectangle

Here's an Example

■ A store owner found her store measured 60 feet by 50 feet. Find the area of the store in square feet.

To solve, use the formula for the area of a rectangle, $A = \ell w$. Here ℓ is 60 feet and w is 50 feet.

$$A = \ell w$$
$$= 60 \times 50$$
$$= 3{,}000$$

The area of the store is 3,000 square feet.

■ What is the area in square feet of a square parking lot that measures 100 yards on each side?

To solve, change 100 yards to feet; then use the formula for the area of a square, $A = s^2$.

Step 1 There are 3 feet in 1 yard. So 100 yards is 3×100, or 300, feet.

Step 2 Use $A = s^2$, with 300 as the value of s.

$$A = 300^2$$
$$= 300 \times 300$$
$$= 90{,}000$$

The area of the parking lot is 90,000 square feet.

■ The area of a rectangular store is 3,000 square feet. If it is 60 feet long, how wide is it?

To solve, put the known information into the formula for the area of a rectangle.

Step 1 Write the formula.

$$A = \ell w$$

Step 2 Put in the numbers you know.

$$3{,}000 = 60 \times w$$

Step 3 You now need to know what you'd multiply 60 by to get 3,000. To find out, divide 3,000 by 60.

$$3{,}000 \div 60 = 50$$

The store is 50 feet wide.

Try It Yourself

■ Find the area in square feet of the blue shaded portion of the diagram. ________

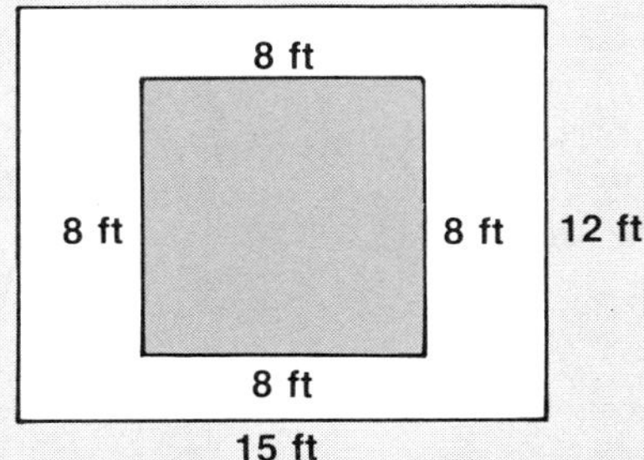

Did you see that the shaded portion is a square because all sides are equal? Whichever formula you used, $8 \times 8 = 64$ square feet.

■ Now find the area of the white portion of the same figure. ________

Did you find the area of the whole rectangle first? $A = \ell w$, or $12 \times 15 = 180$ square feet. You found the area of the blue part before. It is 64. Then did you see you had to subtract the center shaded portion? $180 - 64 = 116$ square feet.

Warm-up

Fill in the missing measurements. Assume that the figures are rectangles or squares.

	Length	Width	Area
1.	12	10	______
2.	9	______	81
3.	______	8	192
4.	16	11	______
5.	______	50	2,500
6.	18	13	______

Warm-up Answers
1. 120 **2.** 9 **3.** 24 **4.** 176 **5.** 50 **6.** 234

A Test-Taking Tip

The GED Mathematics Test may contain questions that ask you to find areas of polygons other than rectangles or squares. For example, you may find a question that requires you to find the area of a parallelogram. You do not need to memorize formulas for the areas of such figures. The GED Mathematics Test includes a formula page that you are free to use to look up special formulas you may need.

On the Springboard

9. Lynn plans to open her own business and needs to find commercial space with 1,100 square feet of floor space. She finds ads for space of the following sizes:

   Space A: 50 ft by 25 ft
   Space B: 40 ft by 30 ft
   Space C: 35 ft by 35 ft

   Which space comes closest to meeting her needs?

   (1) A (2) B (3) C

   (1) (2) (3)

10. A square pool is going to be put into a rectangular plot. The area outside the pool is to be covered with white gravel.

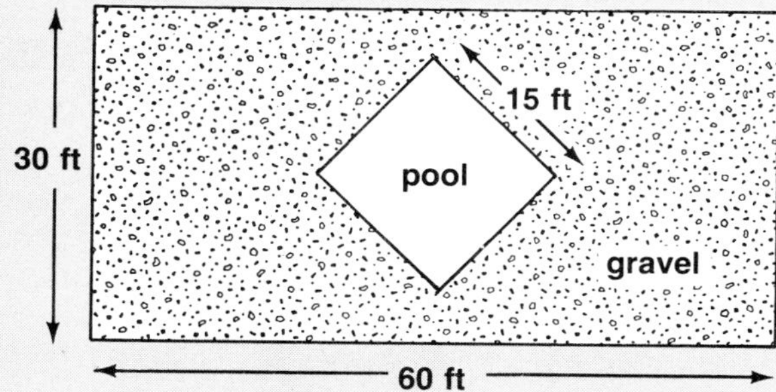

   How many square feet of space will be covered with gravel?

   (1) 675 (2) 1,575
   (3) Insufficient data is given to solve the problem.

   (1) (2) (3)

11. The flat roof of a building is a rectangle measuring 50 yards wide and 30 yards long. The building is 6 yards high. How many square feet of tarpaper are needed to cover the roof?

   (1) 1,500 (2) 9,000 (3) 13,500

   (1) (2) (3)

Check your Springboard answers on pages 283–284. If you answered all of them correctly, you're ready for "The Real Thing." If you missed any, review this section before you go on.

The Real Thing

6. A dollar bill is about 6 inches long and $2\frac{1}{2}$ inches wide. What is its area in square inches?

   (1) 7.5 (2) 12.0 (3) 15.0
   (4) 24.0 (5) 30.0

   (1) (2) (3) (4) (5)

7. The area of a parallelogram is $22\frac{1}{2}$ square yards. The height is 15 feet. What is the length of the base in yards? Use the formula $A = bh$ (A is area, b is base, h is height).

   (1) $4\frac{1}{2}$ (2) 9 (3) $17\frac{1}{2}$
   (4) $56\frac{1}{4}$ (5) $112\frac{1}{2}$

   (1) (2) (3) (4) (5)

8. A wall in Ms. Jones's apartment is shown below. What is the area of this wall in square feet?

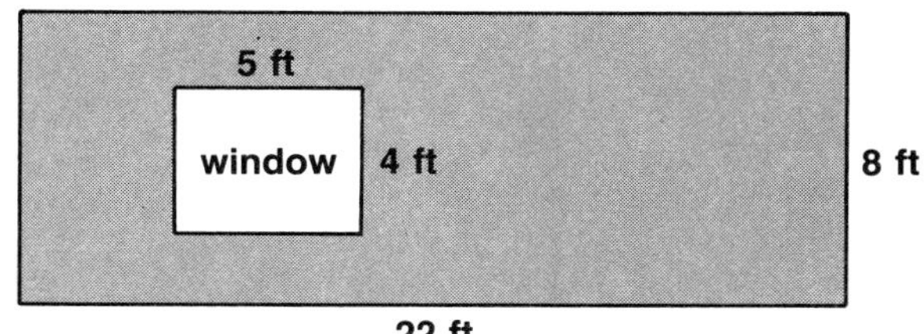

   (1) 42 (2) 60 (3) 156
   (4) 176 (5) 196

   (1) (2) (3) (4) (5)

9. Alan's rectangular table is 18 inches by 32 inches. He has 7 feet of edging. How many more feet of edging does he need to put it completely around the table?

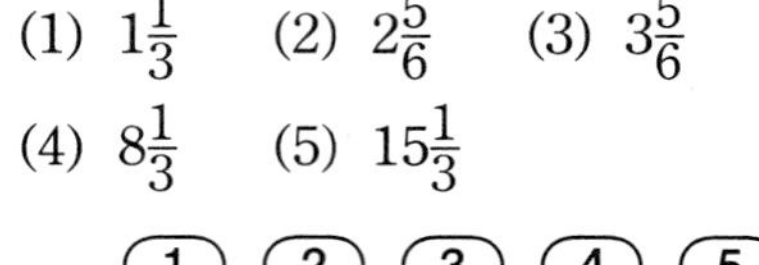

(1) $1\frac{1}{3}$ (2) $2\frac{5}{6}$ (3) $3\frac{5}{6}$
(4) $8\frac{1}{3}$ (5) $15\frac{1}{3}$

① ② ③ ④ ⑤

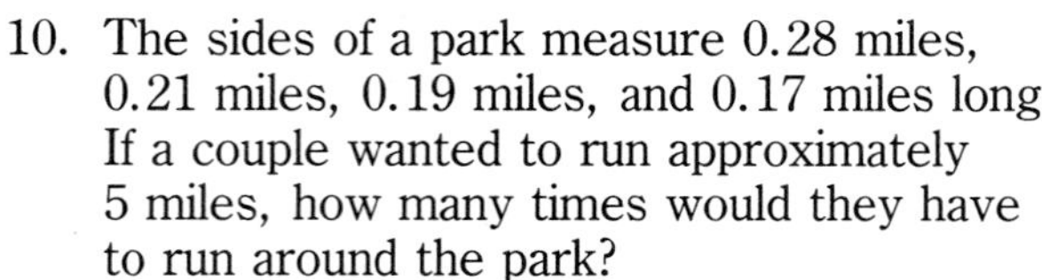

10. The sides of a park measure 0.28 miles, 0.21 miles, 0.19 miles, and 0.17 miles long. If a couple wanted to run approximately 5 miles, how many times would they have to run around the park?

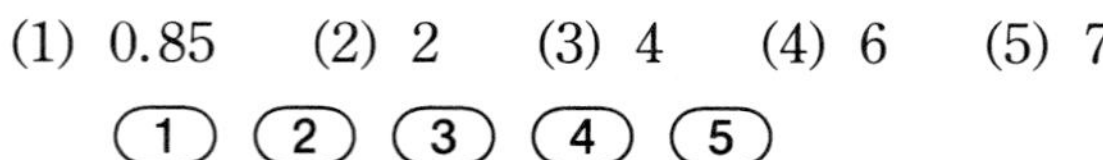

(1) 0.85 (2) 2 (3) 4 (4) 6 (5) 7

① ② ③ ④ ⑤

Questions 11 and 12 refer to the diagram below.

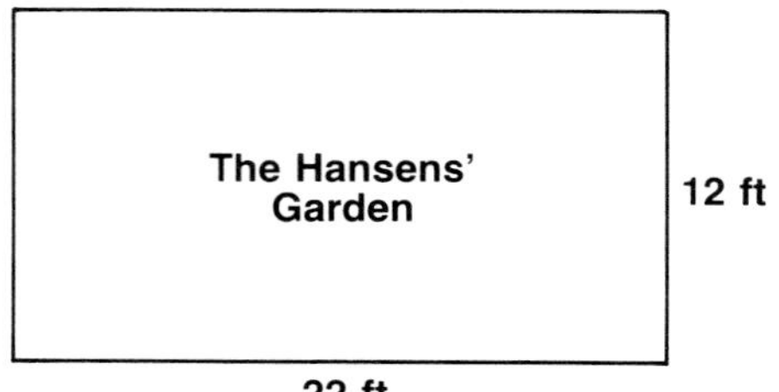

11. The Hansens are going to put cement curb around the garden. The curbing costs $2.60 a running foot. How much will the curbing cost?

(1) $17.68 (2) $36.40 (3) $99.40
(4) $176.80 (5) $360.00

① ② ③ ④ ⑤

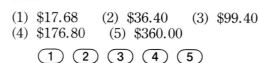

12. Certain plants cost $0.89 apiece and need 2.4 square feet per plant. Approximately how many of these plants can the Hansens put in their garden?

(1) 10 (2) 30 (3) 100
(4) 200 (5) 250

① ② ③ ④ ⑤

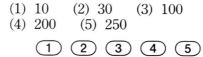

13. A building has eight sides. Each of four sides is 54.5 meters long. Each of three sides is 46.7 meters long. How many meters is the perimeter of this building?

(1) 186.8 (2) 218.0
(3) 359.3 (4) 394.8
(5) Insufficient data is given to solve the problem.

① ② ③ ④ ⑤

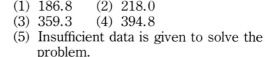

14. The Garcias' living/dining room is shown below. They want to carpet the entire room with carpet that costs $9.50 a square yard. How much will the new carpet cost?

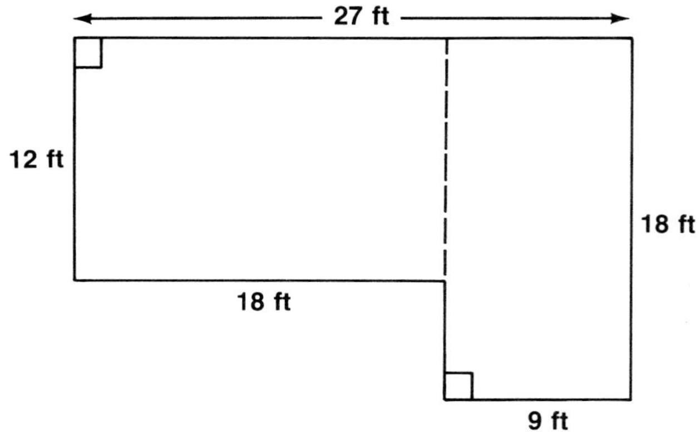

(1) $190 (2) $228 (3) $399
(4) $570 (5) $3,591

① ② ③ ④ ⑤

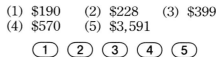

15. The perimeter of a square flower bed is 100 feet. What is the area of the flowerbed in square feet?

(1) 25 (2) 100 (3) 400 (4) 625
(5) Insufficient data is given to solve the problem.

① ② ③ ④ ⑤

Check your answers and record all your scores on page 290.

LESSON 35

Triangles

The first part of the word *triangle* is *tri-*, which means *three,* so a *tri*angle is a polygon with three angles. Triangles also have three sides. In geometry the symbol for triangle is △. With information about the sides of a triangle you can tell a great deal about its angles, and with information about a triangle's angles you can tell a great deal about its sides.

The Sides of Triangles

Here are two special triangles named for their sides.

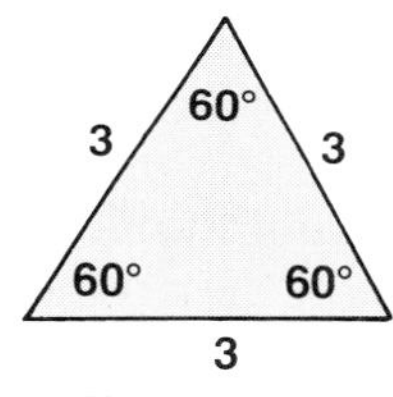

Equilateral **Isosceles**

Equilateral triangles have three equal sides, and three equal angles. Each angle of an equilateral triangle measures 60°. *This is true of every equilateral triangle.*

Isosceles triangles have two equal sides. The angles opposite the equal sides are also equal. Equal sides are always opposite equal angles. Some triangles have no equal sides and no equal angles. When two sides of a triangle are *not* equal, the larger side is always opposite the larger angle and the smaller side opposite the smaller angle.

Coming to Terms

equilateral triangle a triangle with three equal sides. The angles of an equilateral triangle are all equal and each measures 60°.

isosceles triangle a triangle with two equal sides. The angles opposite the equal sides are equal angles.

Here's an Example

■ In △*ABC,* ∠*A* = ∠*B* = ∠*C.* If side *AB* = 4 cm, how long is side *CA*?

To solve, use the fact that sides opposite equal angles are equal. Here all the angles are equal, so all the sides are equal. Since all the sides are equal, the length of *CA* must be 4 cm.

Try It Yourself

■ In triangle *ABC,* ∠*A* = ∠*C* and *AB* = 5 feet. What is the length of *BC?*

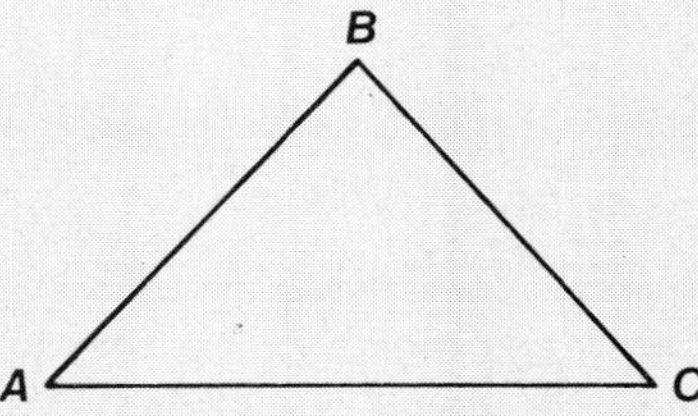

Did you remember that the sides opposite equal angles are equal? Since *AB* is 5 feet long, so is *BC.*

Warm-up

Answer the questions about these triangles.

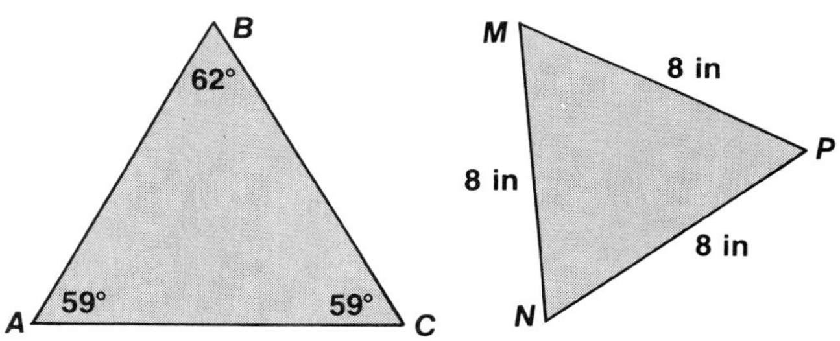

1. In △ *ABC,* what is the longest side? ______

2. In △ *ABC,* if *BC* is 7 cm long, how long is *BA?* ______

3. In △ *MNP,* how many degrees are there in ∠*P?* ______

4. In △ *MNP,* what is the *sum* of the measures of ∠*M* and ∠*N* in degrees? ______

Warm-up Answers
1. *AC* **2.** 7 cm **3.** 60° **4.** 120°

The Angles of Triangles

You know that all three angles of an equilateral triangle are equal to 60°, so the angles of an equilateral triangle add up to 180°. But did you know that *the angles of every triangle add up to 180°*—even if the triangle is not equilateral?

If you draw *any* triangle, you can tear off the angles and fit them together like this.

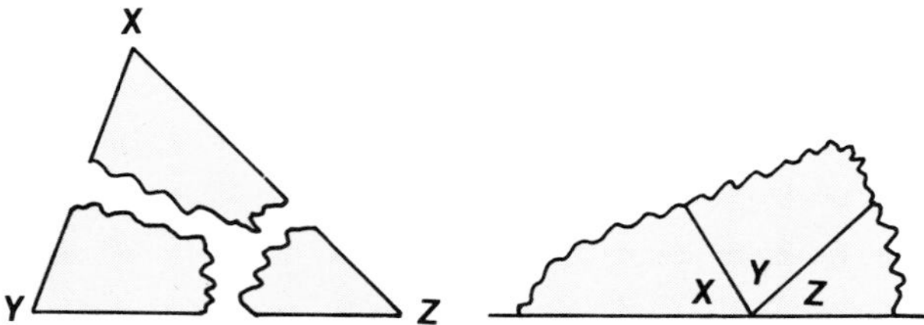

You will always be able to piece them together to make a straight angle, which is 180°. No matter what the triangle, its angles add up to 180°. This is a very important fact about triangles.

Here's an Example

■ Find $\angle Y$ in triangle XYZ.

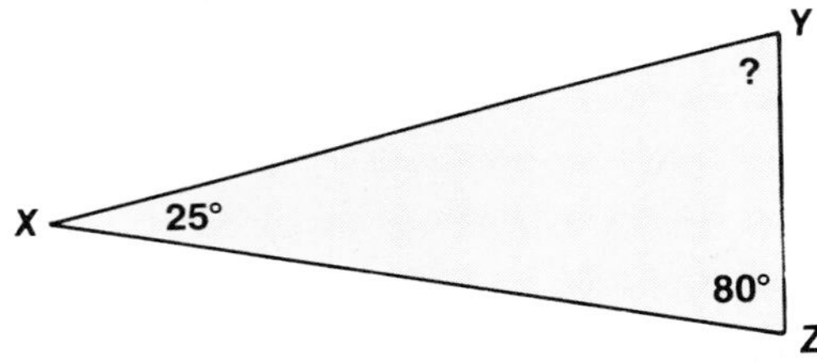

To solve, use the fact that the sum of all the angles is 180°.

Step 1 Add the angles you know.

$$25° + 80° = 105°$$

Step 2 To find how many degrees more you need for a total of 180°, subtract 105° from 180°.

$$180° - 105° = 75°$$

So $\angle Y = 75°$.

Try It Yourself

This problem takes two steps.

■ Find $\angle D$ in this triangle. The measure of $\angle E$ is ⅓ the measure of $\angle C$. ____________

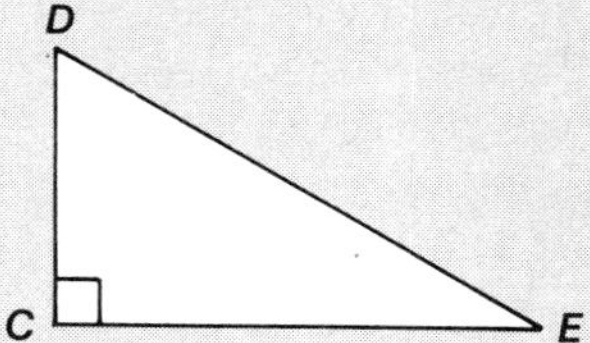

Did you recognize $\angle C$ as being a right angle? It measures 90°. If $\angle E$ is ⅓ of $\angle C$, then $\angle E = 30°$. Then did you add 90° and 30°, and subtract from 180° to get 60°? The measure of $\angle D$ is 60°.

A Test-Taking Tip

If you are given information about a triangle in a sentence or short paragraph, draw a diagram and put in the information. This can help you picture the information and get the facts straight in your mind before you start to use the information to answer questions.

Warm-up

Find the missing angle measure in each triangle.

1. 2.

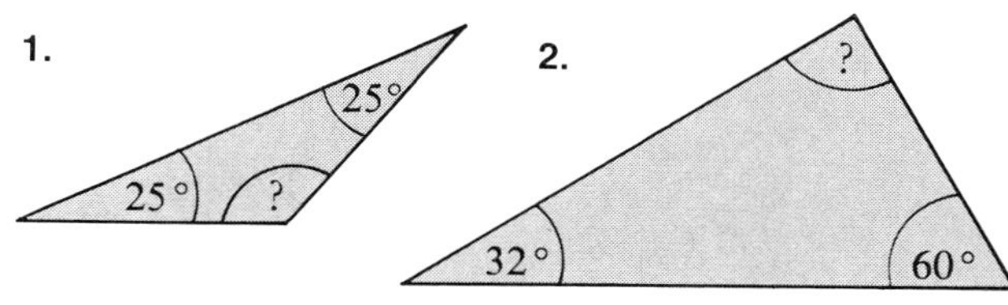

_______ _______

3.

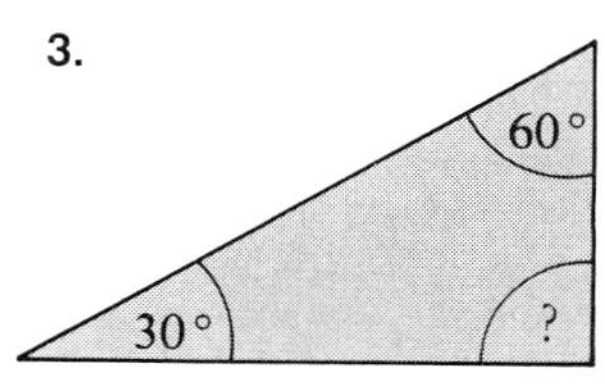

4.

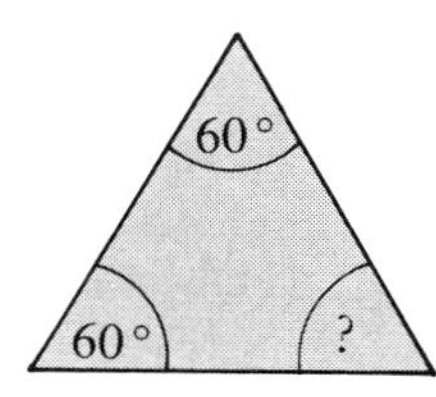

_______ _______

5. 6.

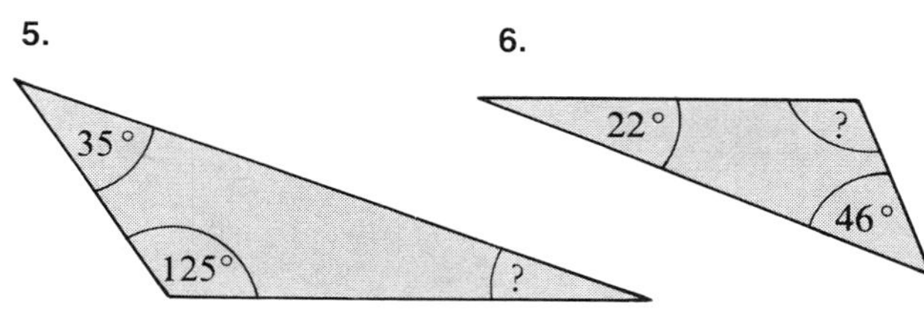

_______ _______

7.

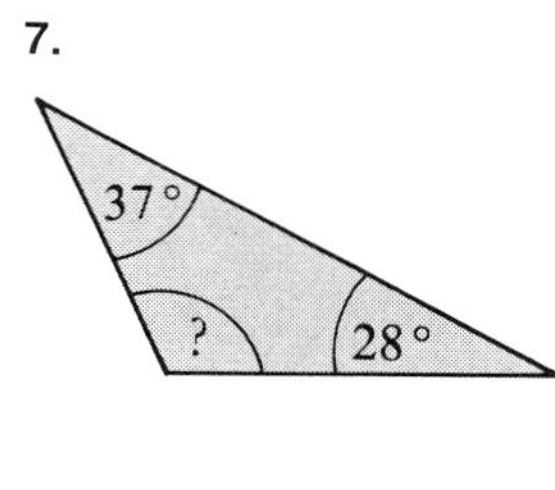

8.

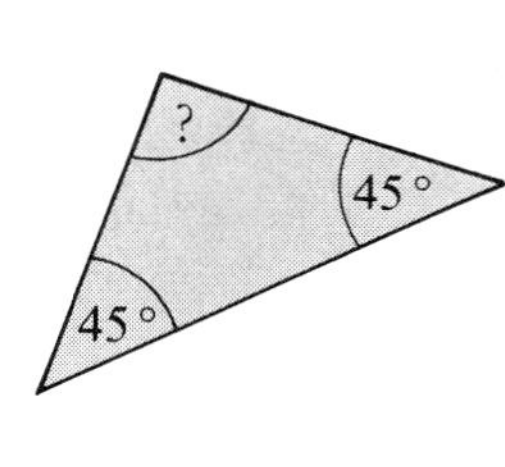

_______ _______

Warm-up Answers
1. 130° **2.** 88° **3.** 90° **4.** 60° **5.** 20° **6.** 112°
7. 115° **8.** 90°

On the Springboard

1. In triangle ABC, $\angle C = 90°$ and $\angle A = \angle B$. What is the measure of $\angle B$?

   (1) 30° (2) 45° (3) 60°

   ① ② ③

2. Which of the following can represent the measures of the three angles of a triangle?

   (1) 30°, 90°, 61° (2) 30°, 110°, 40°
   (3) 55°, 45°, 90°

   ① ② ③

3. If two angles of a triangle measure 42° and 86°, what does the third angle measure?

   (1) 26° (2) 52° (3) 94°

   ① ② ③

Check your Springboard answers on page 284. If you got them right, go on to "The Real Thing." If you missed any, check your work before going on.

The Real Thing

1. In the diagram below, how many degrees are in $\angle B$?

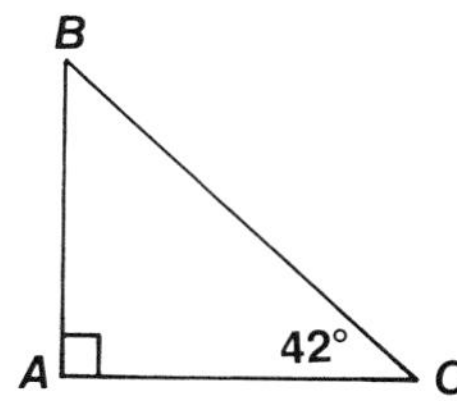

   (1) 42° (2) 48° (3) 58°
   (4) 138° (5) 180°

   ① ② ③ ④ ⑤

2. In triangle RST, $RS = ST$. If $\angle S = 40°$, how many degrees are in $\angle T$?

   (1) 40° (2) 70° (3) 80°
   (4) 90° (5) 140°

   ① ② ③ ④ ⑤

3. In the diagram below, line m is parallel to line n. How many degrees are in $\angle BAC$?

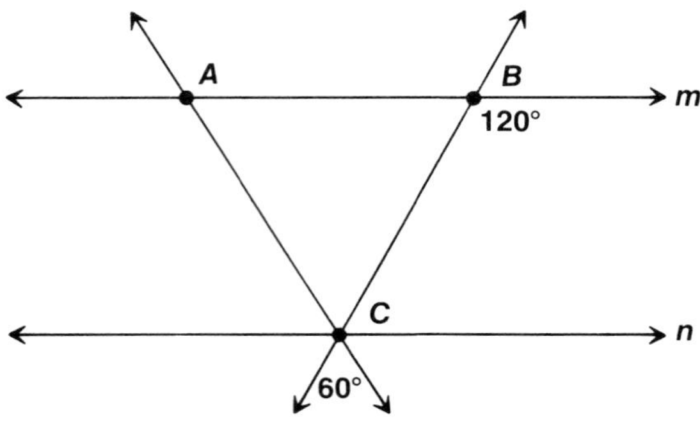

(1) 30° (2) 45° (3) 60°
(4) 120° (5) 180°

① ② ③ ④ ⑤

4. If the measure of $\angle X$ in triangle XYZ is twice the measure of each of the other two angles, how many degrees are in $\angle X$?

(1) 22.5° (2) 45° (3) 72°
(4) 90° (5) 180°

① ② ③ ④ ⑤

Check your answers on page 290.

Finding Perimeters

You've had some practice in finding perimeters in Lessons 32 and 34. The perimeter of a polygon is the distance around it. To find the perimeter of a triangle, you have to know the measures of the three sides and then add them.

Here's an Example

With what you now know about triangles, these examples should be easy to understand.

■ Find the perimeters of triangles *A* and *B*.

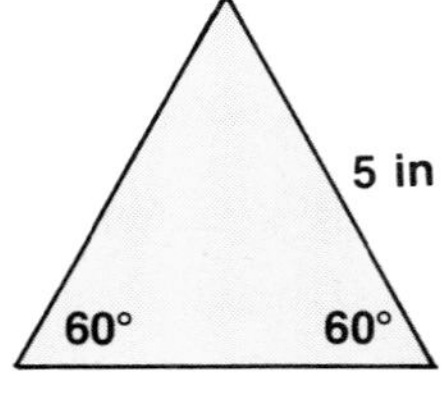

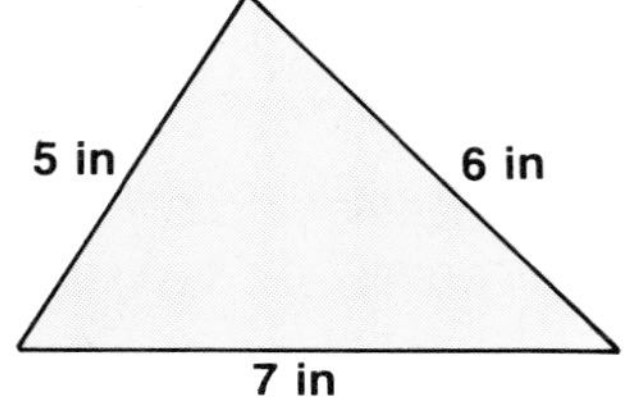

The triangle on the left has two angles of 60°, so the third angle also must be 60°. (Remember, the angle sum must be 180°.) This means that the triangle is equilateral. All its sides are, therefore, equal to 5 inches. Add the three sides. (5 + 5 + 5 = 15) The perimeter of the triangle is 15 inches.

For the second triangle, the measures of all the sides are given right in the diagram. All you have to do is add. (5 + 6 + 7 = 18.) The perimeter of this triangle is 18 inches.

Try It Yourself

You may want to draw a diagram to help answer this question.

■ A farmer wanted to fence off a triangular half of his rectangular field, which measures 120 yards by 50 yards. The fence from one corner to the opposite corner was 130 yards long. How many yards of fencing did he need? ______________________

The triangle had sides that measured 50 yards, 120 yards, and 130 yards. The amount of fencing he needed is the total of these sides. Since 50 + 120 + 130 = 300, the farmer needed 300 yards of fencing.

Warm-up

Find the perimeters of these triangles in inches.

1.

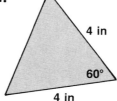

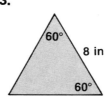

4 in
60°
4 in

____

2.

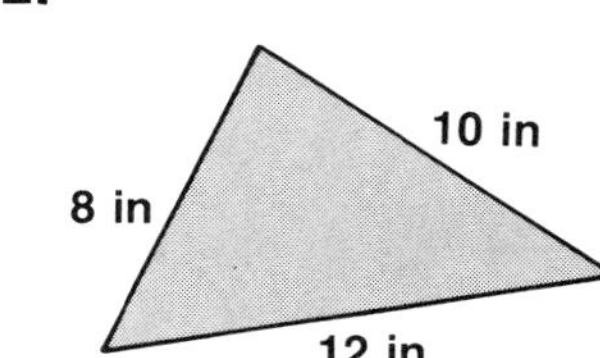

____

3.

60°
8 in
60°

____

4.

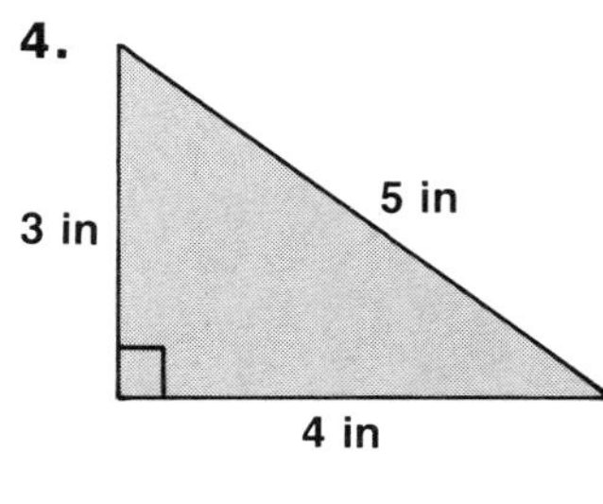

____

Warm-up Answers
1. 12 **2.** 30 **3.** 24 **4.** 12

A Test-Taking Tip

In case you forget what is meant by the phrase *perimeter of a triangle,* there will be a formula for the perimeter of a triangle on the formula page of the GED Mathematics Test.

On the Springboard

4. Which of these triangles has the greatest perimeter? All lengths are in inches.

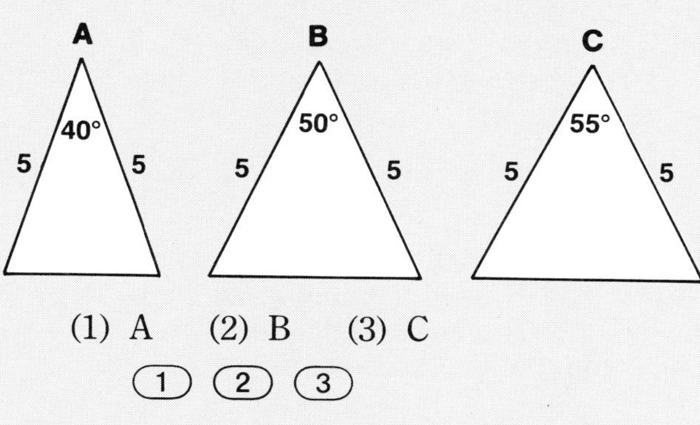

(1) A (2) B (3) C

(1) (2) (3)

5. Which of these triangles has the greatest perimeter? All lengths are in inches.

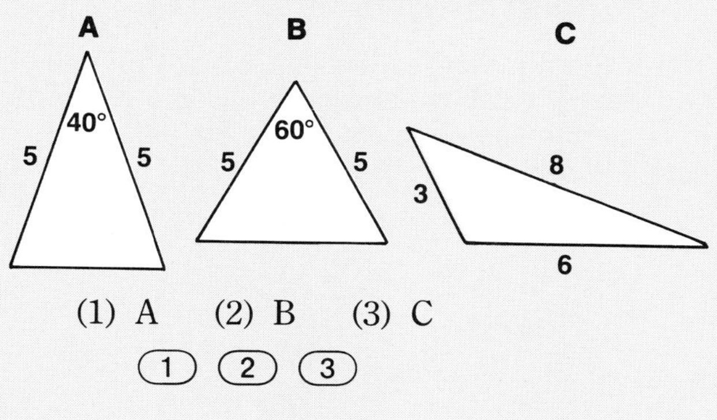

(1) A (2) B (3) C

(1) (2) (3)

Check your Springboard answers on page 284. If you got both correct, you're on the right track and ready to go on to "The Real Thing." If you missed either, you may want to check your knowledge of the two special triangles and the way sides vary with angles. Try again on the ones you missed and, when you're confident that you're ready, go on.

"The Real Thing"

5. The sides of a triangular-shaped garden are $3\frac{1}{2}$ yards, $4\frac{3}{4}$ yards, and $5\frac{1}{4}$ yards. How many yards in the perimeter of the garden?

(1) $12\frac{1}{2}$ (2) 13 (3) $13\frac{1}{4}$
(4) $13\frac{1}{2}$ (5) $14\frac{3}{4}$

(1) (2) (3) (4) (5)

6. A poster has the shape of an equilateral triangle. Each side is 4.6 meters long. How many meters long is the perimeter?

(1) 10.58 (2) 12.50 (3) 12.80
(4) 13.50 (5) 13.80

(1) (2) (3) (4) (5)

7. A table top is an isosceles triangle. The perimeter is 78.4 centimeters. The length of the longest side is 32 centimeters. The length of each of the equal sides is how many centimeters?

(1) 14.4 (2) 23.2
(3) 23.35 (4) 45.70
(5) Insufficient data is given to solve the problem.

(1) (2) (3) (4) (5)

8. Mrs. Polson drove 38 miles from town A to town B. Next she drove 59 miles to town C. After driving directly back to town A, she found that her total mileage was 167 miles and that it took her 3.5 hours. If all the roads are straight roads, how many miles is it from town C directly to town A?

(1) 43 (2) 67 (3) 70
(4) 107 (5) 127

(1) (2) (3) (4) (5)

Check your answers on page 290.

Finding Area

Look at these triangles:

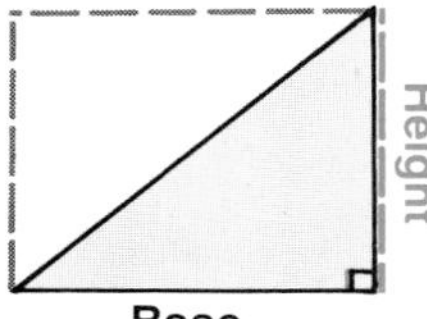

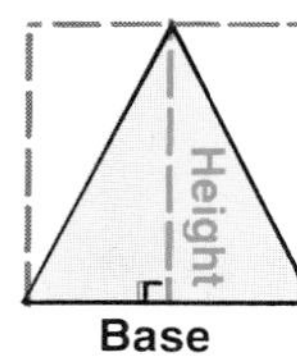

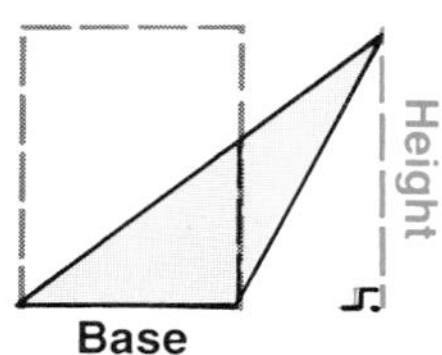

Instead of calling the dimensions of a triangle *length* and *width,* they are called **base** and **height** (or **altitude**). The base is one side of the triangle. The height forms a right angle with the base. Look for the dotted blue line in each of the triangles shown. You can find the area of a triangle by using the formula.

$$A = \frac{1}{2}(b \times h) \quad \text{or} \quad A = \frac{1}{2}bh$$

Before you use the formula to calculate the area of a triangle, be sure that the base and height are in the same units. If both are in inches, the area will be in square inches; if both are in centimeters, the area will be in square centimeters, and so on. Note that the metric system has its own special symbols for square units:

Metric Unit of Area	Symbol
Square meter	m^2
Square centimeter	cm^2
Square millimeter	mm^2
Square kilometer	km^2

Coming to Terms

base the side of a triangle to which the perpendicular height (altitude) is dropped

height (or **altitude)** the perpendicular distance in a triangle from the angle opposite the base *to* the base. The height can be inside, outside, or along one side of a triangle.

Here's an Example

■ A designer wanted to tile a triangular floor space. What is the area if the base is 8 feet and the height is 10 feet?

To solve, use the formula $A = \frac{1}{2}bh$.

Step 1 Multiply the base by the height.

$8 \times 10 = 80$

Step 2 Take ½ times the answer.

$$\frac{1}{\not{2}_1} \times \frac{\not{80}^{40}}{1} = 40$$

The area of the triangle is 40 square feet.

Try It Yourself

When a builder says "straight up" or "straight down," he is talking about a perpendicular.

■ Find the area of the triangle forming the front side of an A-frame home that measures 20 feet across the base of the house and 16 feet from the peak of the roof straight down to the ground. __________

Did you recognize 20 and 16 as the base and height of a triangle that forms the front side of the house? Then, did you multiply the base by the height? The answer, 320, multiplied by ½ gives you the final answer, 160 square feet.

Warm-up

The following triangles are labeled so that you can identify the base and height. Use the letters to name the line segment that is the base and height in each drawing.

1.

A, B, C, D

Base = __________

Height is = __________

2.

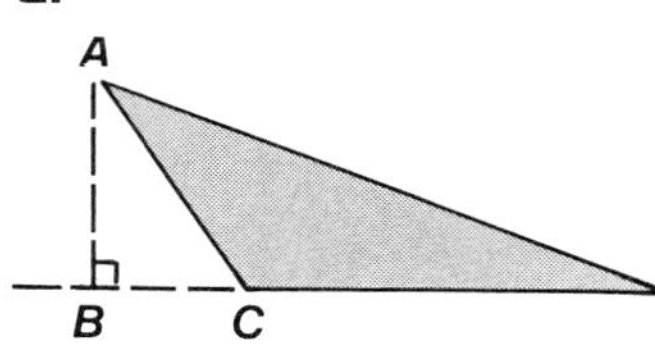

Base = __________

Height = __________

3.

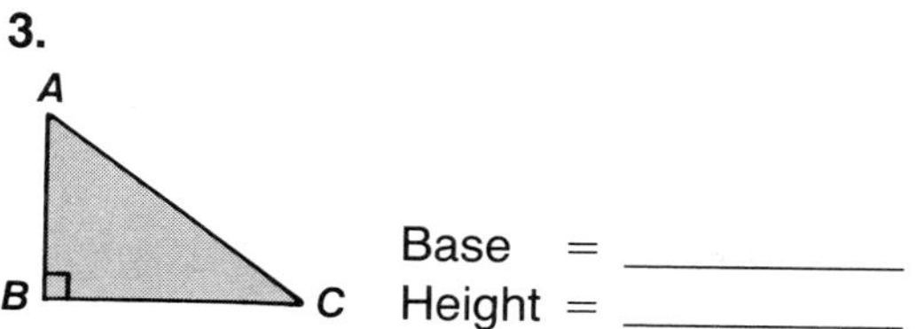

Find the area of the following triangles. Be sure to include the area unit of measurement (sq in, sq ft, and so on).

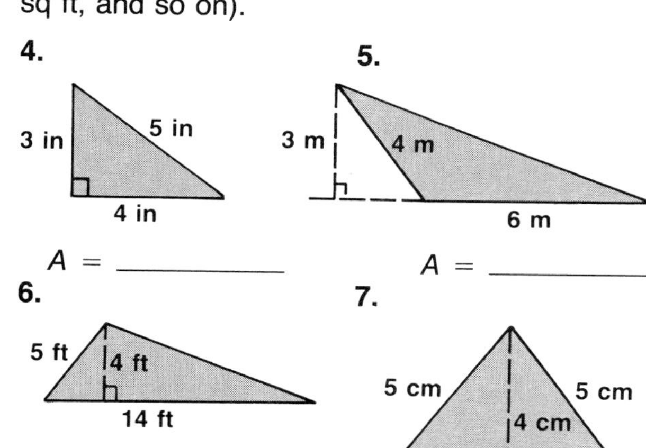

A = ______________ A = ______________

Find the base or height of the following triangles.

8. Area = 30 sq yd
Base = 6 yd

Height = ______________

9. Area = 15 cm^2

Base = ______________
Height = 15 cm

10. Area = 72 sq in

Base = ______________
Height = 12 in

If you disagree with any answers, make certain you understand that the height of a triangle is the perpendicular distance from any angle of the triangle to the opposite side.

Warm-up Answers
1. b = BD and h = AC **2.** b = CD and h = AB
3. b = BC and h = AB or b = AB and h = BC
4. 6 sq in **5.** 9 m^2 **6.** 28 sq ft **7.** 12 cm^2 **8.** 10 yd
9. 2 cm **10.** 12 in

On the Springboard

6. If a triangle has a base of 12 inches and a height of 12 inches, what is the area in square inches?

(1) 24 (2) 72 (3) 144

① ② ③

7. A triangle with an area of 32 square centimeters has a height of 16 centimeters. How many centimeters is the base?

(1) $\frac{1}{2}$ (2) 4 (3) 8

① ② ③

Check your Springboard answers on page 284. If you got both right, go on to "The Real Thing."

"The Real Thing"

9. The height of a sail shaped like a right triangle is 59 feet. The base of the sail is 38 feet. What is the sail's area in square feet?

(1) 324.5 (2) 621.0 (3) 1,071.0
(4) 1,121.0 (5) 1,142.0

① ② ③ ④ ⑤

10. A window is shaped like an isosceles triangle and has a base of 19 inches. The length of the equal sides is $14\frac{1}{2}$ inches. What is the area of the window in square inches?

(1) $13\frac{1}{2}$ (2) 56

(3) $195\frac{3}{4}$ (4) 391

(5) Insufficient data is given to solve the problem.

① ② ③ ④ ⑤

11. The floor of a room is shaped like a right triangle. It has a base of 24 yards and a height of 18 yards. If carpeting costs $11.63 per square yard, how much will it cost to carpet this room?

(1) $1,930.58 (2) $2,502.08 (3) $2,512.08
(4) $3,831.16 (5) $5,024.16

① ② ③ ④ ⑤

12. Debby is wallpapering a ceiling shaped like a right triangle. It has a base of 20 feet and a height of 15 feet. A roll of paper covers 30 square feet. If the paper costs $7.50 a roll, how much will the paper cost?

(1) $18.75 (2) $36.50 (3) $37.50
(4) $75.00 (5) $112.50

(1) (2) (3) (4) (5)

Questions 13 and 14 refer to the following diagram.

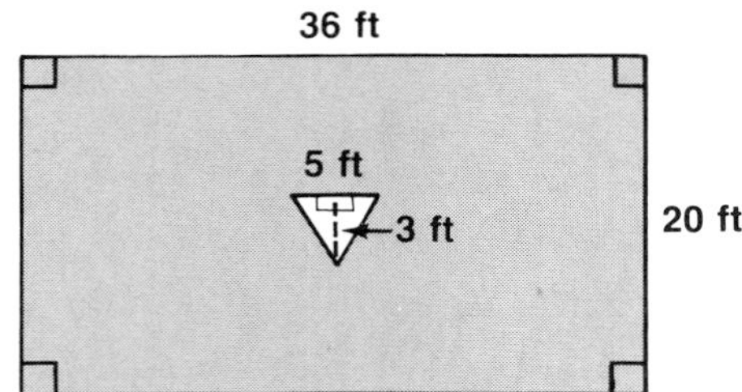

13. A triangular flagpole base was set in a courtyard. What is the area of this base in square feet?

(1) 720.0 (2) 712.5 (3) 15.0
(4) 8.0 (5) 7.5

(1) (2) (3) (4) (5)

14. What is the distance in feet around the courtyard?

(1) 56 (2) 92 (3) 112
(4) 127 (5) 224

(1) (2) (3) (4) (5)

15. The Kershaws are going to paint part of their den. Three walls are rectangular with a length of 15 feet and a height of 8 feet. Part of the ceiling is shaped like a triangle that has a base of 15 feet and a height of 12 feet. If one quart of paint covers approximately 100 square feet, how many quarts of paint will the Kershaws have to buy to paint this much of the den?

(1) 3 (2) $3\frac{1}{2}$ (3) 4

(4) $4\frac{1}{2}$ (5) 5

(1) (2) (3) (4) (5)

Questions 16 and 17 refer to the diagram.

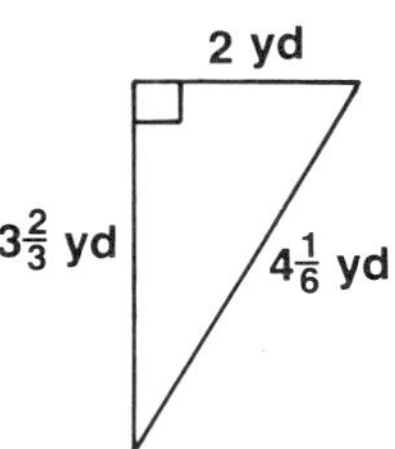

16. The patio costs $1.85 per square foot when it is installed. About how much will it cost to have this patio installed?

(1) $50.00 (2) $60.00 (3) $80.00
(4) $120.00 (5) $280.00

(1) (2) (3) (4) (5)

17. A family is going to put an edge around the patio. The edging costs $1.68 a yard. About how much will the edging cost?

(1) $5.00 (2) $10.00
(3) $16.00 (4) $26.00
(5) Insufficient data is given to solve the problem.

(1) (2) (3) (4) (5)

18. A club flag has the shape of an isosceles triangle. The perimeter of the flag is 120 inches. The length of one of the equal sides is 42 inches. How many inches long is the shortest side of the flag?

(1) 36 (2) 78
(3) 126 (4) 162
(5) Insufficient data is given to solve the problem.

(1) (2) (3) (4) (5)

19. The diagram below shows a building lot. What is the area of this lot in square feet?

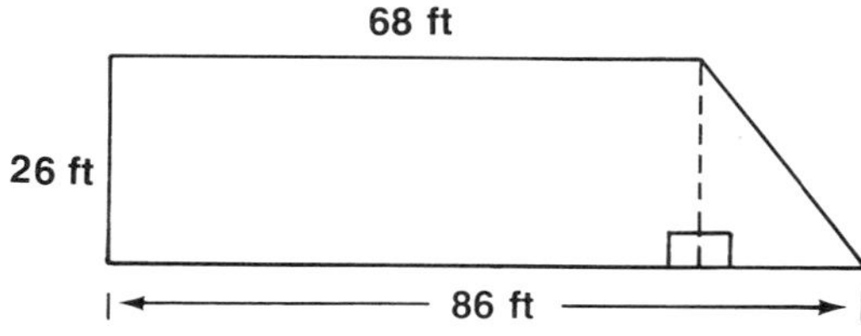

(1) 1,012 (2) 2,002
(3) 2,236 (4) 2,652
(5) Insufficient data is given to solve the problem.

(1) (2) (3) (4) (5)

Check your answers and record all your scores on page 291.

LESSON 36

Congruent and Similar Triangles

You've already learned a lot about triangles. When you have two triangles in a problem and know something about one of them, you can often figure out several things about the other. This lesson will show you some special pairs of triangles and certain things that are true about their sides and angles.

Congruent Triangles

Two triangles that are exactly the same are called **congruent triangles.** One triangle can be placed directly on top of the other, so that all their points match exactly.

Coming to Terms

congruent triangles identical triangles, the same size and same shape

Here's an Example

■ Look at these two congruent triangles.

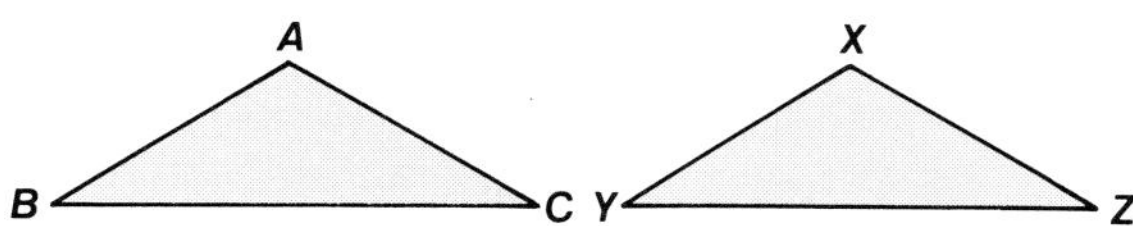

The measures of corresponding parts of congruent triangles are equal. So, in the example just given, m∠*A* = m∠*X*, m∠*B* = m∠*Y*, and m∠*C* = m∠*Z*. You can also see that the measures of corresponding sides are equal: *AB* = *XY*, *BC* = *YZ*, and *AC* = *XZ*.

Try It Yourself

■ The two following triangles are congruent. If *AC* = 5 inches, *AB* = 10 inches, and *BC* = 12 inches, what does RS equal?

____

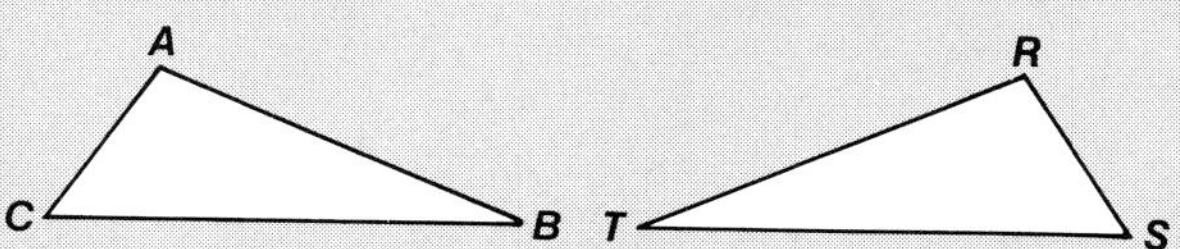

Did you flip one of the triangles in your mind to see that *RS* corresponds to *AC?* The triangles are congruent, so *RS* equals *AC.* If *AC* = 5 inches, *RS* = 5 inches.

Warm-up

Look at these two congruent triangles. Name the parts that correspond to those listed.

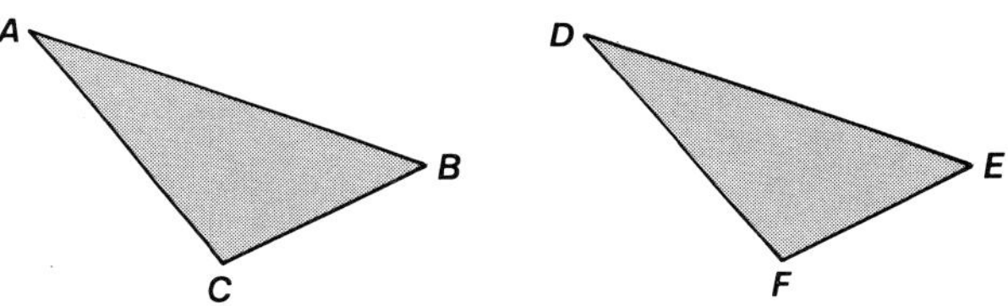

1. *AB* = ______________________
2. ∠*A* = ______________________
3. *BC* = ______________________
4. ∠*E* = ______________________
5. *FD* = ______________________
6. ∠*C* = ______________________

Warm-up Answers
1. *DE* **2.** ∠*D* **3.** *EF* **4.** ∠*B* **5.** *CA* **6.** ∠*F*

On the Springboard

1. The triangles in the diagram below are congruent. If $\angle A = 36°$, what is the measure of $\angle E$?

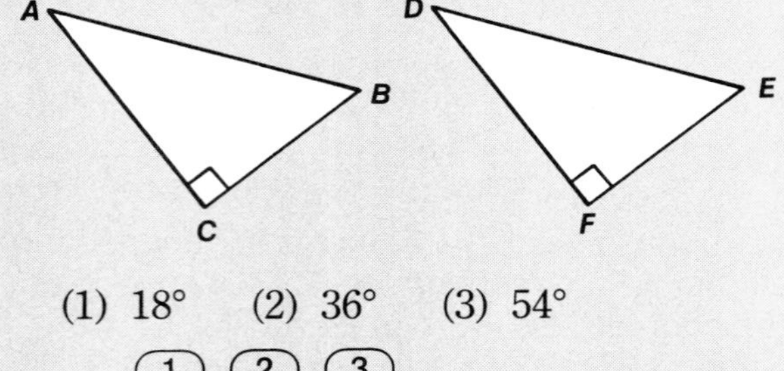

(1) 18° (2) 36° (3) 54°

1 2 3

2. *ABCD* is a parallelogram. What is the measure of $\angle C$?

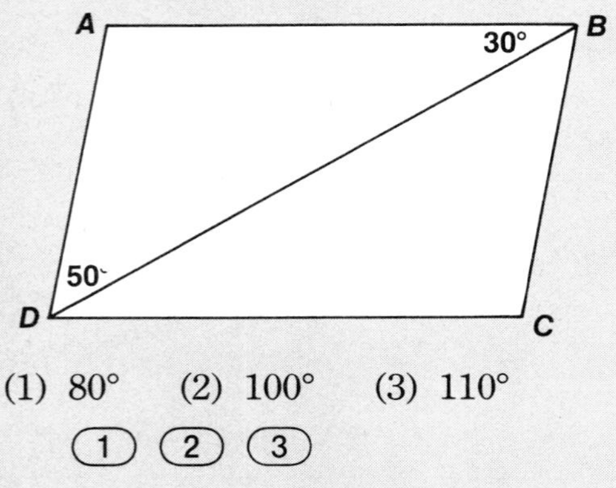

(1) 80° (2) 100° (3) 110°

1 2 3

Check your Springboard answers on page 285. If you got these right, go on to "The Real Thing." If you didn't, check your thinking carefully. Which parts of the triangles are corresponding parts? When you're sure you can answer both questions correctly, you're ready to go on.

"The Real Thing"

Questions 1 and 2 refer to the two triangles shown in the diagram. Triangle *JKL* is congruent to triangle *RST*.

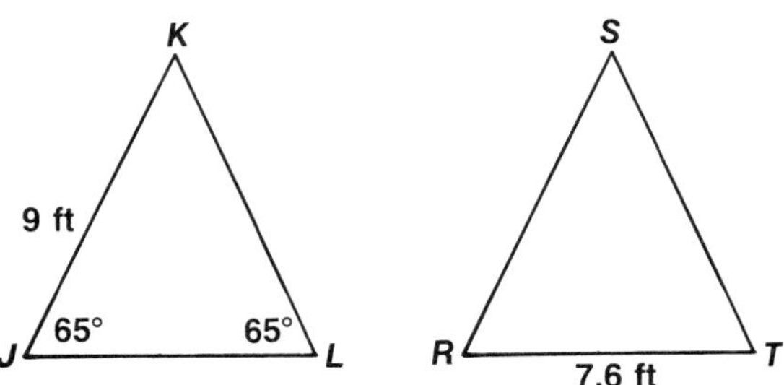

1. What is the measure of $\angle S$ in degrees?

   (1) 50° (2) 65° (3) 115° (4) 130°
   (5) Insufficient data is given to solve the problem.

   1 2 3 4 5

2. What is the perimeter of triangle *JKL* in feet?

   (1) 16.6 (2) 18.0 (3) 25.6 (4) 34.2
   (5) Insufficient data is given to solve the problem.

   1 2 3 4 5

Check your answers on page 291.

Similar Triangles

If you look at a triangle through a magnifying glass or through the wrong end of a pair of binoculars, you see what looks like a different triangle with the same shape but a different size. These are **similar triangles.** All the corresponding sides of similar triangles are in proportion and all the corresponding angles are equal.

Coming to Terms

similar triangles triangles with the same shape but not necessarily the same size

Look at these two triangles. Notice that they have the same angle measurements, and, therefore, the same shape. Although their shapes are the same, their sizes are not. The triangles are not congruent, but they are similar.

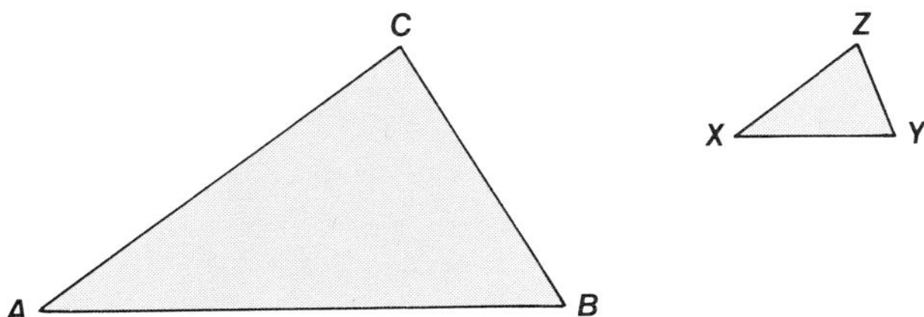

The corresponding sides of similar triangles are in proportion to each other. The longest side of the larger triangle corresponds to the longest side of the smaller triangle. The shortest side of the larger triangle corresponds to the shortest side of the smaller triangle.

In these two similar triangles, the corresponding sides are named as follows: *AB* corresponds to *XY, AC* corresponds to *XZ, BC* corresponds to *YZ.*

You can also make a proportion statement about the two triangles:

$$AB:XY = AC:XZ = BC:YZ$$

This means, for example, that if *AB* is double *XY,* then *AC* is double *XZ,* and *BC* is double *YZ.* If *AB* is ⅓ of *XY*, then *AC* is ⅓ of *XZ,* and *BC* is ⅓ of *YZ.*

You can write a proportion using any two of the three ratios. *AB*:*XY* = *BC*:*YZ* is a proportion. In such a proportion, if you know three of the four measures, you can find the fourth measure by using a grid.

Here's an Example

■ Triangles *ABC* and *XYZ* are similar. Find the length of *XZ.*

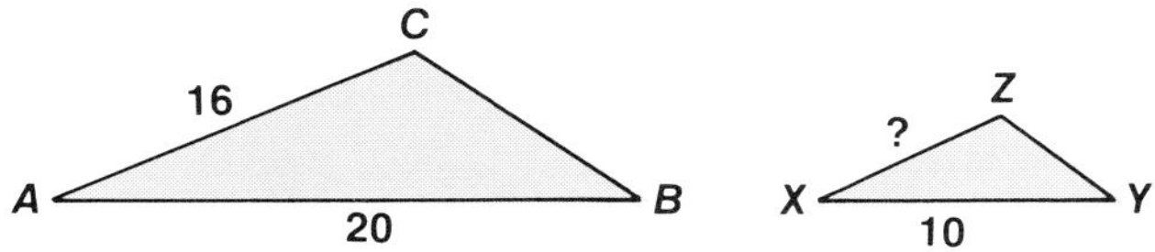

To solve, set up a proportion and use a grid to find the missing number.

Step 1 Write a proportion with *XZ* and the three sides whose measures you know (*AB*, *AC*, and *XY*).

$$AB:XY = AC:XZ$$

Step 2 Replace the names of the sides with the lengths given in the diagram.

$$20:10 = 16:?$$

This is the same as 20/10 = 16/?, so you can use the grid to solve it.

Step 3 Fill in the grid. Multiply the known diagonals and divide by the unused number.

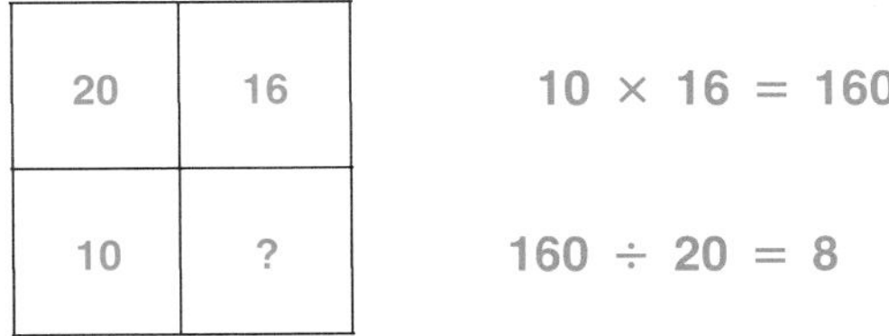

The length of *XZ* is 8 units.

Try It Yourself

Don't let the positions of similar triangles in a diagram confuse you. Remember that corresponding sides are always opposite corresponding angles.

■ How many feet does side *RT* measure?

____

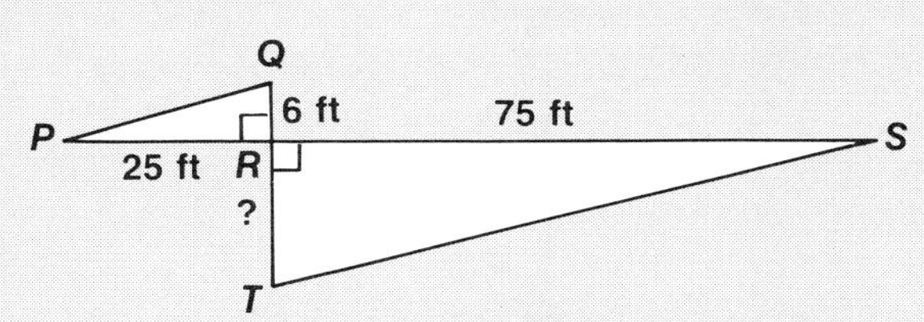

Did you set up one of the right proportions? The proportion 25:75 = 6:? will do fine, but this is not the only possible proportion. You might also have had 75:25 = ?:6. No matter which way you set it up, side *RT* is 18 feet.

✓ A Test-Taking Tip

On the GED Mathematics Test, you may be given a problem with similar triangles drawn on a map. Similar triangles are often used to measure distances to places not easily reached. For example, you may have to find the distance across a river.

Warm-up

Write proportions for pairs of corresponding sides in each exercise as indicated. Do not solve for the ?.

1. 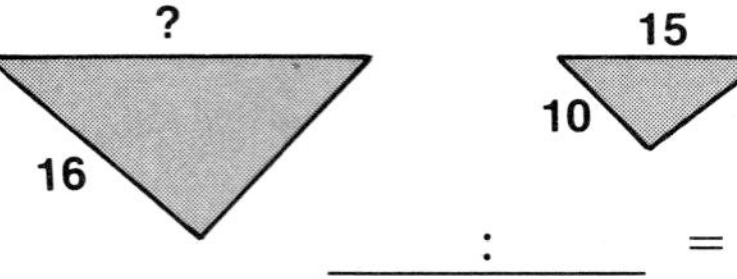

_____ : _____ = _____ : _____

2. 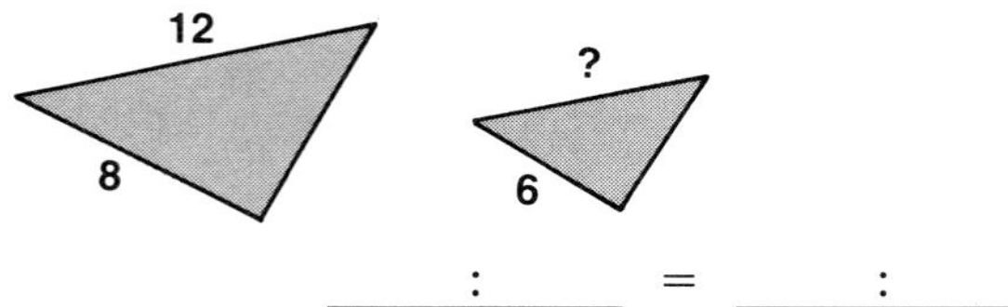

_____ : _____ = _____ : _____

3. 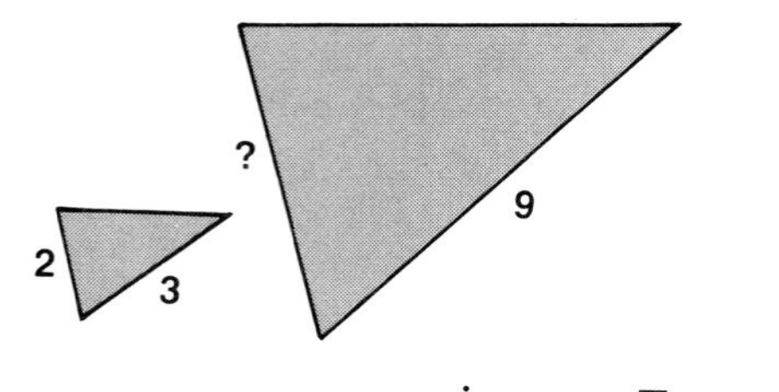

_____ : _____ = _____ : _____

4. 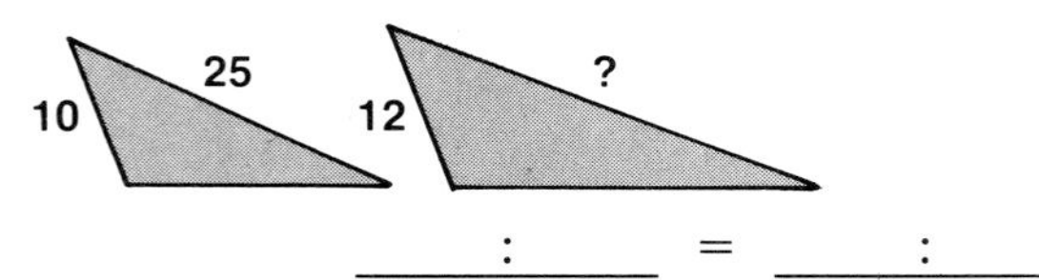

_____ : _____ = _____ : _____

5. 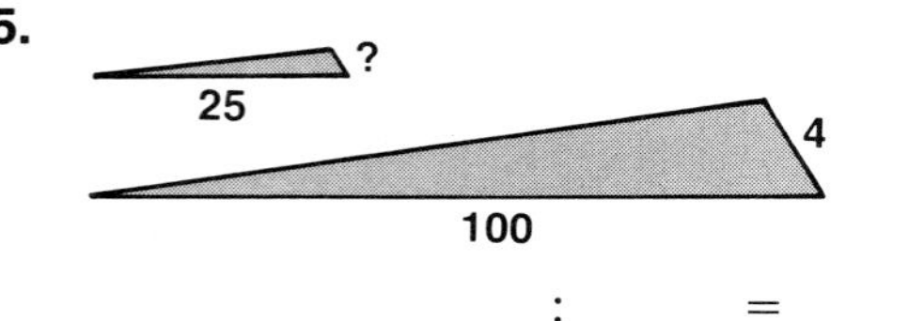

_____ : _____ = _____ : _____

Now, using the same figures, find the missing part.

6. Use #1. ________ **7.** Use #2. ________

8. Use #3. ________ **9.** Use #4. ________

10. Use #5. ________

On the Springboard

Questions 3–5 refer to these similar triangles.

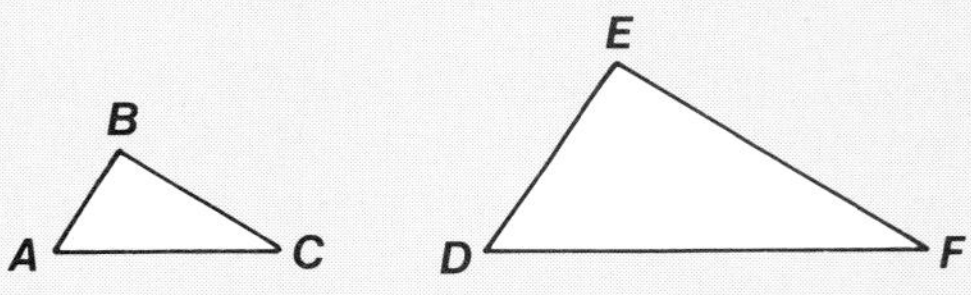

3. Angle *ABC* corresponds to angle
   (1) *BCA* (2) *CAB* (3) *DEF*
   ① ② ③

4. Side *DF* corresponds to side
   (1) *AB* (2) *AC* (3) *BC*
   ① ② ③

5. If *AB* is 4 feet, *BC* is 6 feet, and *DE* is 8 feet, how long is *EF*?
   (1) 3 ft (2) 8 ft (3) 12 ft
   ① ② ③

Check your Springboard answers on page 285. If you got them wrong, go back over this lesson and read the parts that gave you trouble. If you got them right, go ahead to "The Real Thing."

Warm-up Answers
1. 15:? = 10:16 or ?:15 = 16:10 **2.** 6:8 = ?:12 or 8;6 = 12:? **3.** 2:? = 3:9 or ?:2 = 9:3 **4.** 10:12 = 25:? or 12:10 = ?:25 **5.** ?:4 = 25:100 or 4:? = 100:25 **6.** 24 **7.** 9 **8.** 6 **9.** 30 **10.** 1

The Real Thing

3. Margaret and her shadow and a tree and its shadow form similar triangles. Margaret is 5.6 feet tall and her shadow is 12 feet long. The tree's shadow is 66 feet. How many feet tall is the tree?

   (1) 30.80 (2) 29.80 (3) 27.50
   (4) 14.14 (5) 3.80

   ① ② ③ ④ ⑤

4. Kevin is making a scale drawing of a flowerbed that is a right triangle. On the scale drawing, the height is 12 centimers and the base is 22 centimeters. The actual base of the garden is 55 meters. In meters, what is the actual height of the garden?

   (1) 7.5 (2) 18.3
   (3) 30.0 (4) 100.8
   (5) Insufficient data is given to solve the problem.

   ① ② ③ ④ ⑤

5. When a man stands next to a flagpole, his shadow measures 4.5 meters and the pole's shadow is 22.5 meters. How many meters tall is the flagpole?

   (1) 6.0 (2) 9.0
   (3) 10.8 (4) 12.6
   (5) Insufficient data is given to solve the problem.

   ① ② ③ ④ ⑤

6. Chris used measurements from a surveyor's map to make the drawing below. In meters, what is the distance across the lake from point C to point D?

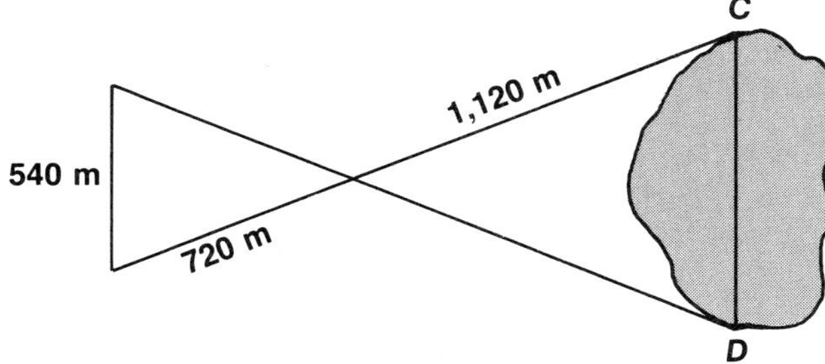

   (1) 347.1 (2) 840.0 (3) 853.9
   (4) 884.2 (5) 1,400.0

   ① ② ③ ④ ⑤

7. The drawing below shows part of a roof. The dimensions are actual measurements. How many feet tall should post KJ be?

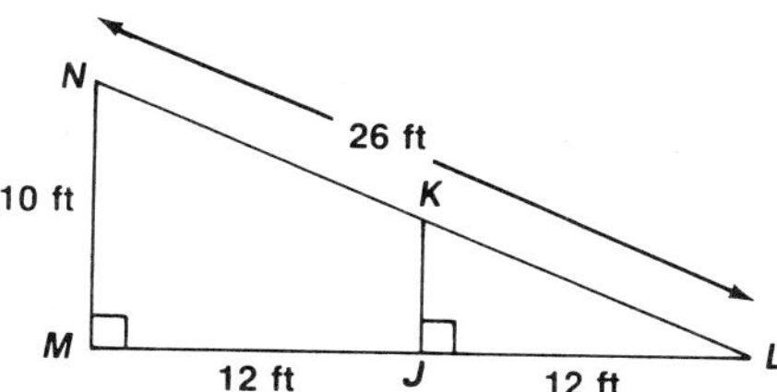

   (1) 5 (2) 10 (3) 15
   (4) 20 (5) 25

   ① ② ③ ④ ⑤

8. A surveyor drew the map shown below to find the distance across the stream at DE. How many feet is it from D to E?

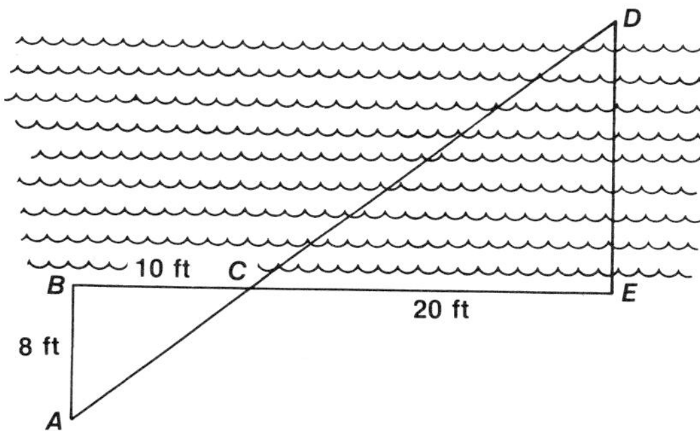

   (1) 4 (2) 16 (3) 22
   (4) 25 (5) 200

   ① ② ③ ④ ⑤

9. In the figure, triangle ABC is congruent to triangle EDF. If DE and DF each measure 10 inches, what is the measure of $\angle D$?

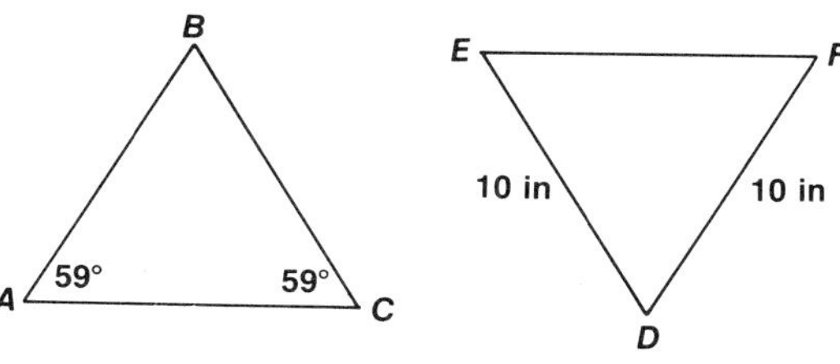

   (1) 59° (2) 62° (3) 121° (4) 131°
   (5) Insufficient data is given to solve the problem.

   ① ② ③ ④ ⑤

Check your answers and record all your scores on page 292.

LESSON 37

Pythagorean Theorem

A **right triangle** is a triangle that contains a right angle—an angle of 90°. The **Pythagorean Theorem** is a fact about right angles that was discovered about 2,500 years ago by a Greek mathematician named Pythagoras. This fact is called a *theorem* because Pythagoras was able to prove that it is true for *all* right triangles, with no exceptions.

Before you look at what the Pythagorean Theorem says, you need to know what the **legs** and **hypotenuse** of a right triangle are. Look at this right triangle.

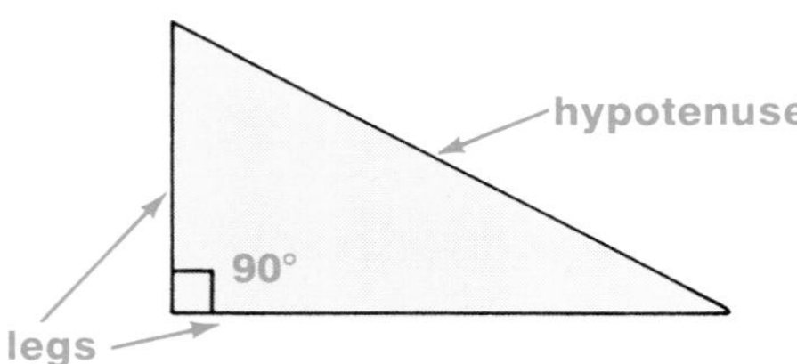

The *legs* are the sides that form the right angle. The *hypotenuse* is the side opposite, or across from, the right angle. Since the right angle is always the largest angle in any right triangle, the hypotenuse is always the longest side of the right triangle.

Now let's go on to the Pythagorean Theorem. Suppose *a* and *b* stand for the lengths of the legs and *c* for the length of the hypotenuse in a right triangle. (So *a, b,* and *c* tell you how many units you get if you measure the sides of the triangle.) The Pythagorean Theorem says that, if you square these numbers, the sum of the squares of the legs is equal to the square of the hypotenuse.

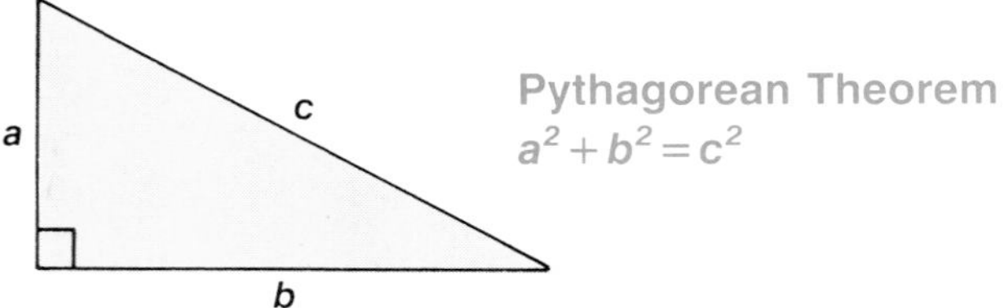

Look at this right triangle.

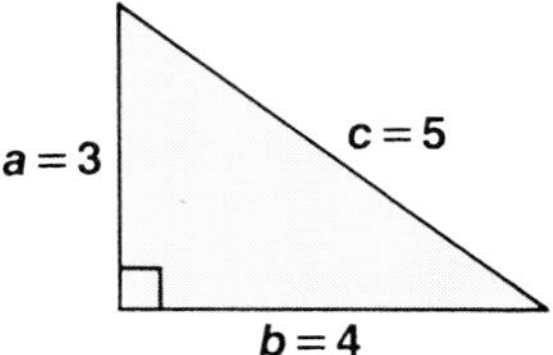

This right triangle has legs that measure 3 units and 4 units. The hypotenuse measures 5 units. Is it true that $a^2 + b^2 = c^2$? Yes, because

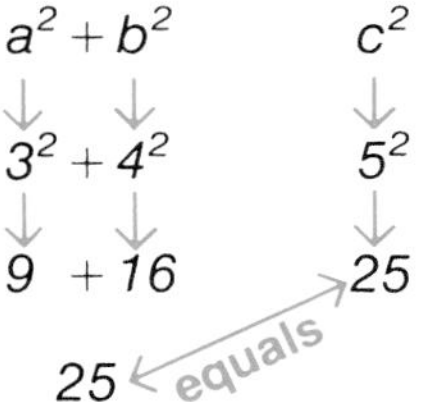

The amazing thing about the Pythagorean Theorem is that it is true for *all* right triangles. You do not need to memorize the formula version ($a^2 + b^2 = c^2$) of the Pythagorean Theorem. It will be on the formula page that you will receive when you take the GED Mathematics Test. However, it is important that you understand what the Pythagorean Theorem means and how to use it.

Coming to Terms

right triangle a triangle that has a right angle

Pythagorean Theorem The fact that, in every right triangle, the sum of the squares of the legs equals the square of the hypotenuse. In mathematical symbols, this theorem is written $a^2 + b^2 = c^2$, where *a* and *b* are the lengths of the legs of the right triangle and *c* is the length of the hypotenuse.

leg in a right triangle, one of the sides that form the right angle

hypotenuse the side opposite the right angle in a right triangle. Because it is opposite the largest angle, the hypotenuse is the longest side of the right triangle

Because of the Pythagorean Theorem, if you are told the lengths of any two sides of a right triangle, you can find the third side.

Here's an Example

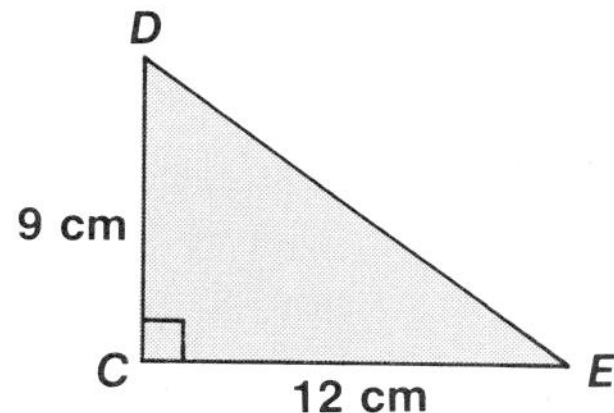

■ In triangle *CDE,* leg *CE* is 12 centimeters and *CD* is 9 centimeters. Find the length in centimeters of hypotenuse *DE.* To solve, use $a^2 + b^2 = c^2$.

Step 1 Fill in as much as you can of the formula from the sketch.

$$9^2 + 12^2 = c^2$$

Step 2 Compute what you can.

$$81 + 144 = c^2$$
$$225 = c^2$$

Step 3 Find the number that you can square to get 225.

$$15^2 = 225; \text{ so } c = 15$$

Since c is the hypotenuse *DE,* you get $DE = 15$ cm.

If you know one leg and the hypotenuse, you can find the other leg.

■ Find the other leg of a right triangle that has one leg 10 units long and a hypotenuse of 26 units.

To solve, use the formula $a^2 + b^2 = c^2$.

Step 1 Draw a sketch to help you picture the information you are given.

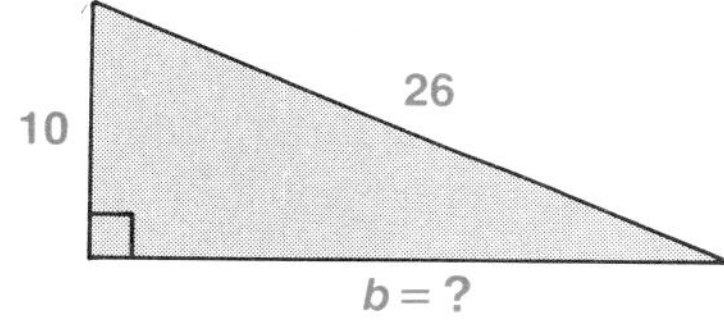

Step 2 Fill in as much of the formula as you can from the sketch.

$$10^2 + b^2 = 26^2$$

Step 3 Compute what you can.

$$100 + b^2 = 676$$

Step 4 Reason out what b^2 must equal. Since you have to add 100 to b^2 to get 676, it makes sense that b^2 is equal to 676 minus 100. So, $b^2 = 576$.

Step 5 Try numbers to see what number squared gives 576. The answer is 24. So, $b = 24$.

The other leg is 24 units long.

☑ A Test-Taking Tip

Remember that the GED Mathematics Test will have five choices for answers, so you can always try each of the five choices until you find the one that works.

Try It Yourself

Use the Pythagorean Theorem to solve this problem.

■ A right triangle has a hypotenuse of 20 millimeters and a leg that measures 16 millimeters. What is the length of the other leg?

____

Here $16^2 + b^2 = 20^2$, so $256 + b^2 = 400$. As in last example, you can find b^2 by reasoning that it must equal $400 - 256$, or 144. Since $12^2 = 144$, $b = 12$. The other leg is 12 millimeters long.

■ The diagram shows an end view of a roof. $AC = 8$ yards and $BD = 3$ yards. If $\triangle BAD$ is congruent to $\triangle BCD$, what is the length of *BC?*

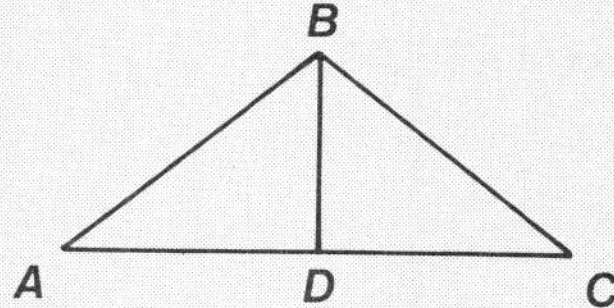

Before you started work on this question, did you ask yourself why you were told that the two triangles are congruent? Since the triangles are the same in every way, the two angles at *D* are equal. Therefore each must be 90°. $\angle ADC$ is a straight angle and measures 180°. Also $AD = CD$, so *DC* is half of 8, which is 4. Now you have a right triangle with two legs of 3 and 4 yards and you need the hypotenuse. Using the formula and computing, you get $3^2 + 4^2 = 25 = BC^2$. The length of *BC* is 5 yards.

A Test-Taking Tip

Before you try to use the Pythagorean Theorem to solve a problem about a triangle, be sure that you have enough evidence to know that the triangle *is* a right triangle. The Pythagorean Theorem is true for all right triangles, but *only for right triangles.*

Warm-up

Using figures 1–3, replace the letters in the formula $a^2 + b^2 = c^2$.

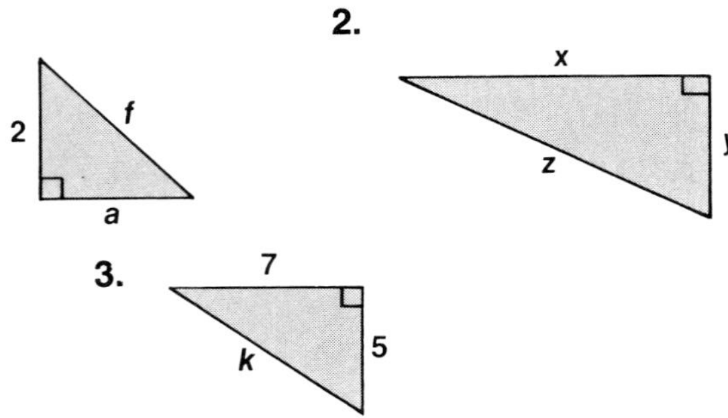

Figure 1: $\_\_\_\_^2 + \_\_\_\_^2 = \_\_\_\_^2$

Figure 2: $\_\_\_\_^2 + \_\_\_\_^2 = \_\_\_\_^2$

Figure 3: $\_\_\_\_^2 + \_\_\_\_^2 = \_\_\_\_^2$

Find the length of the missing side in each of these figures.

4.

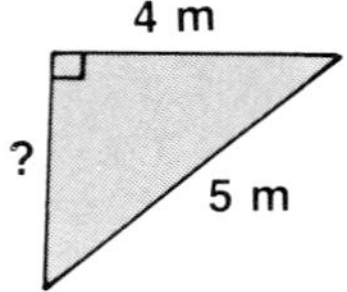

____

5.

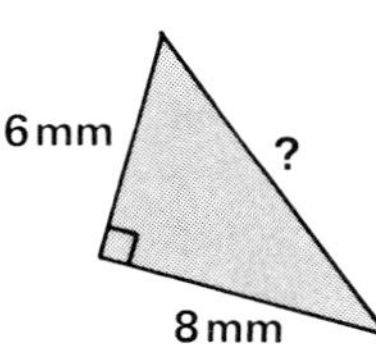

____

6.

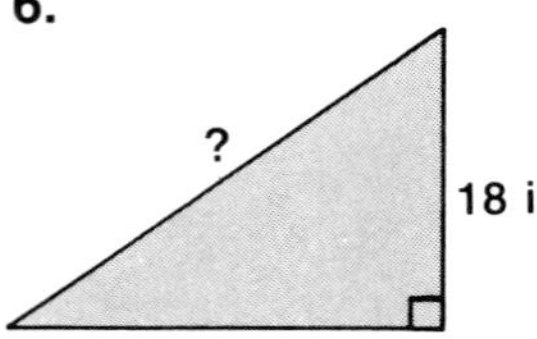

____

7.

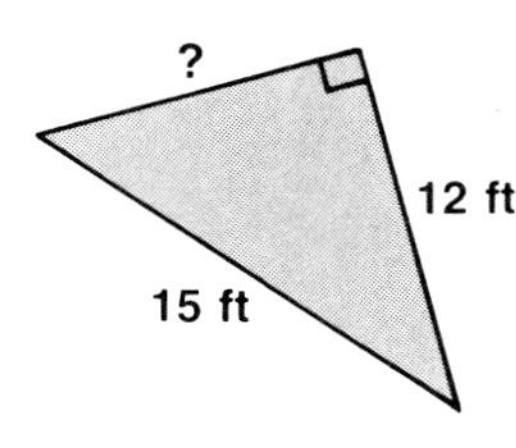

____

Warm-up Answers
1. $2^2 + a^2 = f^2$ **2.** $y^2 + x^2 = z^2$ **3.** $5^2 + 7^2 = k^2$
4. 3 m **5.** 10 mm **6.** 30 in **7.** 9 ft

On the Springboard

1. A right triangle has a hypotenuse of 25 inches and one leg 24 inches long. How long is the other leg in inches?

   (1) 7 (2) 17 (3) 19

   (1) (2) (3)

2. A man left his camp and walked 6 miles north and then 8 miles due west. How far was he from his camp in miles?

   (1) 14 (2) 10 (3) 9

   (1) (2) (3)

Check your Springboard answers on page 285. If you got these right, you have understood the Pythagorean Theorem well enough to go on to "The Real Thing." If you missed any, review and check your work as needed. When you can correct your answers and are confident that you are ready, go on to "The Real Thing."

The Real Thing

1. A jogger ran 1.5 miles east and 2.0 miles south. How many miles diagonally was she from her starting point?

   (1) 0.50 (2) 1.32 (3) 2.50
   (4) 3.50 (5) 6.25

   (1) (2) (3) (4) (5)

2. A 10-foot ladder is leaning against a wall. The bottom end of the ladder is 6 feet from the wall. How many feet above the ground does the ladder reach the wall?

   (1) 8 (2) 12 (3) 14
   (4) 16 (5) 64

   (1) (2) (3) (4) (5)

3. The guy wires of a radio tower are attached halfway up the tower. Each guy wire is 75 feet long and is attached to the ground 60 feet from the tower. How many feet tall is the tower?

(1) 45 (2) 80 (3) 90
(4) 270 (5) 5,625

① ② ③ ④ ⑤

4. A baseball diamond is 90 feet on each side. What is the approximate distance in feet from home plate to second base?

(1) 90 (2) 127 (3) 180
(4) 442 (5) 4,050

① ② ③ ④ ⑤

Items 5 and 6 refer to the following situation.

Phyllis built a display table that had the shape of a right triangle. The sides that form the right angle were 9 feet and 12 feet. She paid $2.39 *a yard* for material to put a skirt around the table.

5. How many feet long was the third side of Phyllis' table?

(1) 15 (2) 21 (3) 25
(4) 35 (5) 225

① ② ③ ④ ⑤

6. How much did Phyllis spend for the material to make the skirt around the table?

(1) $14.34 (2) $16.73 (3) $17.92
(4) $23.90 (5) $28.68

① ② ③ ④ ⑤

Check your answers and reread your score on page 292.

LESSON 38

Circles

Polygons are closed figures that have straight sides. Other closed figures have curved sides. The most common of these figures is the **circle.** Before working with circles, you must learn the words used to describe the parts of a circle. Here they are.

Coming to Terms

circle a closed curve in which all points are the same distance from the center

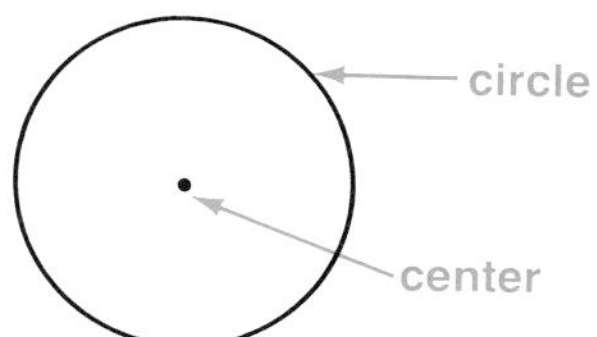

circumference the perimeter of, or distance around, a circle

radius a line segment that connects the center of a circle to any point on the circle. The *endpoints* of any radius are the center of the circle and a point on the circle.

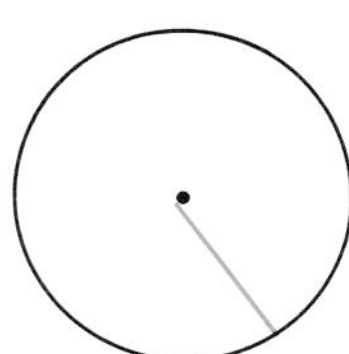

diameter a line segment that connects two points on a circle and passes through the center of the circle. A diameter is twice the length of a radius. A diameter is the longest segment connecting two points on a circle.

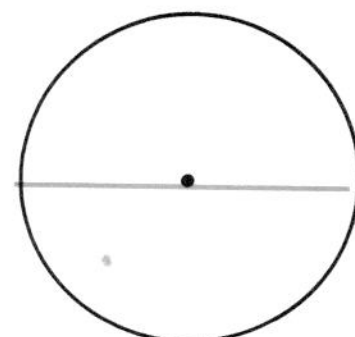

Circumference of a Circle

You can find the perimeter of a polygon by adding the lengths of its sides. For a circle, there are no sides to add. You need a different method for finding the distance around a circle.

The perimeter of a circle is usually called its **circumference.** To compute the circumference, just multiply the length of a **diameter** by a special number called *pi.* For questions on the GED Mathematics Test, you should use 3.14 as the value of pi unless you are told otherwise.

Pi is the name of a Greek letter that is written π. It is pronounced "pie." If you have to use a value of π other than the usual 3.14, you will be told to do so.

The circumference, C, of a circle is equal to π times the length of a diameter, d:

$$C = \pi d$$

A circle is usually referred to by naming its center.

Here's an Example

■ A circular pond is 14 meters across. What is the circumference of the pond in meters? (Use $\pi = 3.14$.)

To solve, use the formula $C = \pi d$.

The greatest distance across a circle is the diameter, so $d = 14$ meters.

Write the circumference formula, put in the numbers you know, and do the calculations. You can check the multiplication on a piece of scratch paper if you wish.)

$$C = \pi d$$
$$C = 3.14 \times 14$$
$$C = 43.96$$

The circumference is 43.96 meters. Let's see how to work backwards from the circumference.

■ Find the diameter in inches of a circular metal disk 88 inches around. (Use $\pi = 22/7$).

To solve, put the information into the circumference formula.

Step 1 Write down the formula and then fill in the information you have. Use $22/7$ for π.

$$C = \pi d$$
$$88 = \frac{22}{7} \times d$$

Step 2 Reason out what to do next. The formula says that $22/7$ multiplied by the diameter, d, is 88. To find d, you must divide 88 by $22/7$. When you do that you get

$$d = 88 \div \frac{22}{7}$$
$$d = \overset{4}{\cancel{88}} \times \frac{7}{\underset{1}{\cancel{22}}}$$
$$d = 28$$

The diameter of the metal disk is 28 inches. It may be a good idea to remember that to find the diameter of a circle you can divide the circumference by π.

Try It Yourself

Sometimes you are given the radius instead of the diameter. To get the diameter, simply double the radius.

■ Find the circumference of a circle with a radius of 3 inches. (Use $\pi = 3.14$.)

First, you should have doubled the radius to get the diameter as 6 inches. When you put the information into the formula, did you get $C = 3.14 \times 6$? Then, did you multiply and get 18.84 inches as your answer for the circumference?

If you are asked to find the radius of a circle given its circumference, the best way is to find the diameter first. To do this, divide the circumference by π. Once you know the diameter, divide it by 2 to get the radius. Here's one to try.

■ Find the radius of a circle with a circumference of 44 inches (Use $\pi = 22/7$.)

You should have divided the circumference, 44, by $22/7$. Since $44 \div 22/7$ equals $44 \times 7/22$, or 14, the diameter is 14 inches. For the radius, divide 14 inches by 2. You find that the radius is 7 inches.

Warm-up

Find the circumference of each of the circles described here. Use $\pi = 3.14$.

1. Diameter of 5 ft ______

2. Radius of 2.5 in ______

3. Diameter of 10 m ______

4. Radius of 6 cm ______

5. Diameter of 6 cm ______

6. Radius of 1.3 mm ______

On the Springboard

1. The circular mirror of a reflecting telescope is 15 feet across. How far is it around the rim of the telescope in feet? (Use 3.14 for π.)

   (1) 4.71 (2.) 23.55 (3) 47.1

   (1) (2) (3)

2. What is the distance in feet around a circular pool with a radius of 12 feet. (Use 3.14 for π.)

   (1) 75.36 (2) 81.46 (3) 87.56

   (1) (2) (3)

3. A circular coffee filter made of paper is 22 inches around. If the filter is folded in half, what will be the length in inches of the straight fold? (Use 22/7 for π.)

   (1) $3\frac{1}{2}$ (2) 7 (3) 14

   (1) (2) (3)

Check your Springboard answers on page 285, and, if you got them right, go on to "The Real Thing." If you missed any, check your work. Review if you think you need to do so. When you have fixed your mistakes, you can go on to "The Real Thing."

Warm-up Answers
1. 15.7 ft **2.** 15.7 in **3.** 31.4 m **4.** 37.68 cm
5. 18.84 cm **6.** 8.164 mm

The Real Thing

1. A circular track at a fair has a diameter of 77 feet. What is its circumference in feet? (Use 22/7 for π.)

   (1) $24\frac{1}{2}$ (2) 121 (3) 224
   (4) 242 (5) 484

   (1) (2) (3) (4) (5)

2. The circumference of a bicycle tire is $66\frac{6}{7}$ inches. What is the diameter of the tire to the nearest whole inch? (Use 22/7 for π.)

   (1) 20 (2) 21 (3) 24
   (4) 210 (5) 468

   (1) (2) (3) (4) (5)

3. A circular running track has a radius of 150 feet. If Rosita ran 6 laps around the track, how many feet did she run? (Use 3.14 for π.)

   (1) 942 (2) 1,413 (3) 2,826
   (4) 5,446 (5) 5,652

   (1) (2) (3) (4) (5)

4. Mr. Cass wants to put a circular edging around each of 8 trees. The edging comes in 30-foot rolls that cost $5.89 a roll. How much will it cost Mr. Cass to put edging around the 8 trees? (Use 3.14 for π.)

   (1) $19.98 (2) $24.14
   (3) $27.80 (4) $36.27
   (5) Insufficient data is given to solve the problem.

Check your answers on page 293.

Area of a Circle

You know how to find the area of a rectangle, a square, and a triangle. The area of a circle is the amount of surface space inside the circle. It is easy to calculate the area of a circle if you know its radius. Use this formula:

$$A = \pi r^2$$

Here's an Example

■ A circular cardboard disk has a radius of 10 inches. What is its area in square inches? (Use 3.14 for π.)

To solve, use the formula $A = \pi r^2$.

Step 1 Write the formula and put in the values you know.

$$A = \pi r^2$$
$$A = 3.14 \times 10^2$$

Step 2 Do the calculations. Since $10^2 = 10 \times 10 = 100$, you get

$$A = 3.14 \times 100$$
$$A = 314$$

The area of the cardboard disk is 314 square inches.

You can use the formula $A = \pi r^2$ to find the radius or the diameter of a circle if you know the area of the circle.

■ A circular lid had an area of 154 square inches. How wide is it in inches? (Use $\pi = 22/7$.)

To solve, use the formula $A = \pi r^2$ to find the radius. Then double the radius to find the total width of the lid.

Step 1 Write the formula and put in what you know.

$$A = \pi r^2$$
$$154 = \frac{22}{7} \times r^2$$

Step 2 Reason out the next step. Since you have to multiply r^2 by $22/7$ to get 154, it follows that r^2 must be equal to 154 divided by $22/7$.

$$r^2 = 154 \div \frac{22}{7}$$
$$r^2 = 154 \times \frac{7}{22}$$
$$r^2 = 49$$

Step 3 What is the square root of 49? In other words, what number squared equals 49? The answer is 7, so $r = 7$.

Step 4 You want the diameter, so double the radius. The diameter is $2 \times 7 = 14$.

The width of the lid is 14 inches.

Try It Yourself

Remember that the area formula uses the radius.

■ Five hoops, each measuring 3 feet in diameter, are to be covered with paper and used in a parade float. Approximately how many square feet of paper will be needed? (Use 3.14 for π.) Round the answer to the nearest square foot.

____

Did you multiply the area of one hoop by five? Did you remember, too, to divide the diameter by 2 to get the radius to use in the formula? If your answer was 35 square feet, you're right.

A Test-Taking Tip

You need not memorize the formulas for the circumference and area of a circle unless you want to do so. They will be on the GED Mathematics Test formula page.

Warm-up

Find the area of each circle described. Use 3.14 for π.

1. Radius of 10 ft ________________
2. Diameter of 10 ft ________________
3. Diameter of 4 m ________________
4. Radius of 20 mm ________________

Warm-up Answers
1. 314 sq ft **2.** 78.5 sq ft **3.** 12.56 m^2 **4.** 1,256 mm^2

On the Springboard

4. A circular table top has a diameter of 6 feet. What is the area of the table top in square feet?

   (1) 9.42 (2) 28.26 (3) 113.04

   1 2 3

5. Alan bought a giant pizza. He measured its radius as 10 inches. He split it evenly with a friend. How many square inches of pizza did each get?

   (1) 153 (2) 155 (3) 157

   1 2 3

Check your Springboard answers on page 286. If you got these right, go on to "The Real Thing." If you missed either, check your work. Review if you need to. When you're confident that you can answer both questions correctly, go ahead to "The Real Thing."

The Real Thing 

5. The diameter of a circular table top is 28 inches. How many square inches of felt does Casey need to cover this table top? (Use $22/7$ for π.)

   (1) 88 (2) 138 (3) 506
   (4) 616 (5) 17,248

   1 2 3 4 5

6. The area of a circular window is 616 square inches. How many inches long is the piece of wood that goes across the center of this window? (Use $22/7$ for π.)

   (1) 28 (2) 48 (3) 88
   (4) 92 (5) 382

   1 2 3 4 5

7. Swimming-pool regulations specify that there must be 12 square feet of space for each person in a wading pool. What is the maximum number of people who will be allowed in a circular wading pool that has a radius of 20 feet? (Use 3.14 for π.)

   (1) 14 (2) 52 (3) 104
   (4) 208 (5) 419

   1 2 3 4 5

8. A city planner decided to lay colored cement in circles in a mall. Each circle had a diameter of 14 feet. How many square feet of ground did 20 of these circles cover? (Use $22/7$ for π.)

   (1) 154 (2) 616 (3) 880
   (4) 3,080 (5) 12,320

   1 2 3 4 5

9. One quart of paint covers 100 square feet of concrete. About how many quarts of paint are needed to paint 120 circular concrete stepping stones, each with a diameter of 2 feet? (Use 3.14 for π.)

   (1) 1 (2) 3 (3) 4 (4) 8 (5) 15

   1 2 3 4 5

Check your answers and record all your scores on page 293.

LESSON 39

Composite Figures

You know how to find the perimeters and areas of triangles, quadrilaterals, and circles. Many figures do not fit exactly into any of these categories, but with just a little thinking you can usually find a way to see them as combinations of triangles, quadrilaterals, and circles. Figures that are combinations of simpler figures are called **composite figures.**

This is a combination of a square and half a circle.

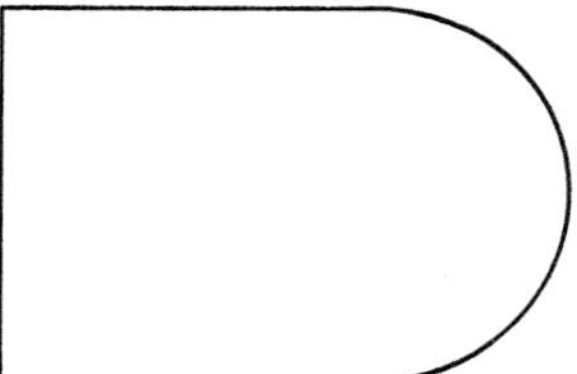

If you draw this dotted line, it is easy to see this fact.

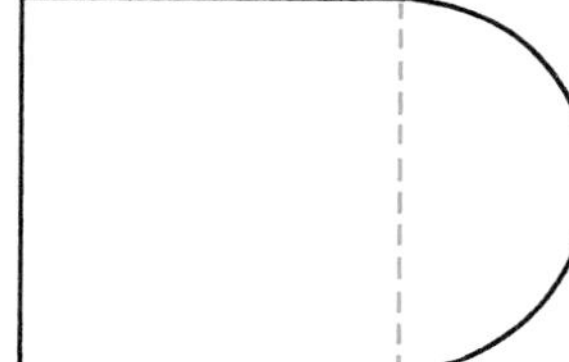

Since composite figures can be thought of as being built from simpler figures, it is often possible to find their perimeters and areas. You just use what you know about finding the perimeters and areas of the simpler figures.

Coming to Terms

composite figure a complex figure that can be thought of as being built of combinations of simpler figures

Finding Perimeters

When you take the GED Mathematics Test, you may be asked to find the perimeter of a complicated looking figure. Use your imagination to see if you can break the figure into parts that you know how to deal with.

Here's an Example

In the figures in this lesson you can assume that square-looking corners really are square, that is, that they measure 90°.

In many problems involving composite figures, you have to figure out some other measurements before you can answer the question. The following perimeter problem illustrates this.

■ Mr. Thomas wants to enclose an area as a computer work station. The area he wants to enclose is pictured in the diagram. How many feet of wallboard will he need?

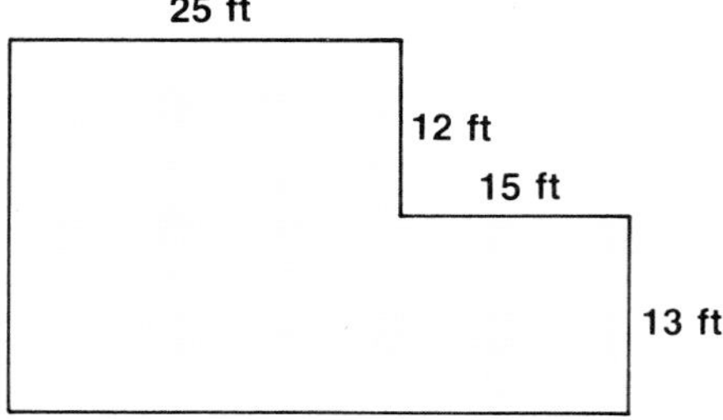

To solve, find the lengths of all the sides and add. You may find it useful to draw some dotted lines to help you figure out the lengths of the two sides whose length you were not given. You can write in measurements as you figure them out. You might wind up with a diagram that looks like this.

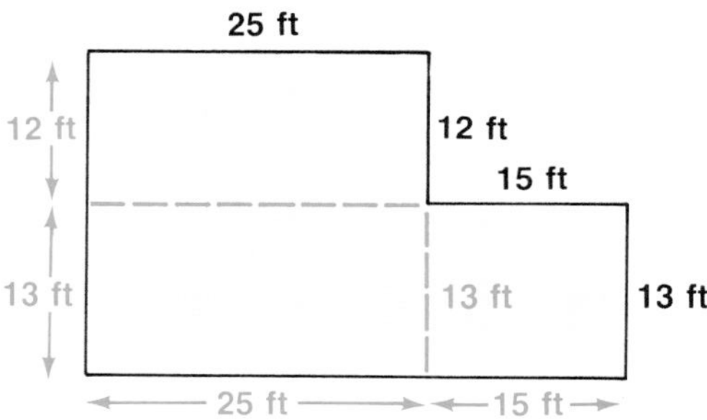

Now you can see that the length of the left side is 12 + 13, or 25 ft. The length of the bottom side is 25 + 15, or 40 ft. Since you now know the lengths of all sides, you can add them to find the perimeter.

$$25 + 12 + 15 + 13 + 40 + 25 = 130$$

Mr. Thomas needs 130 feet of wallboard.

■ For the holidays, a store is going to trim its roof with lights. The roof is flat and has the shape and dimensions shown in the diagram. How many feet of lights does the store need?

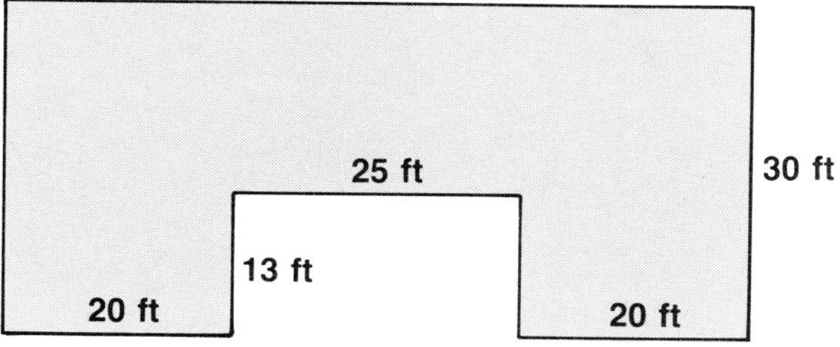

Did you figure out the lengths of the three sides not labeled in the diagram? Did you find that one of these measures 13 feet, another 30 feet, and the third 65 feet? When you added the lengths of all 8 sides, you should have gotten a total of 216 feet.

A Test-Taking Tip

A question on the GED Mathematics Test may ask you to find the perimeter of a polygon in which the sides may be given in different units. For example, one side may be labeled in inches and another in feet. Be sure that you change all measures to the same kind of unit before you add. You could also be given the measures of the sides in feet and inches or meters and centimeters. Be careful to add only like measures. If your answer can be changed to a better form, be sure to do so. For example, change 4 feet 16 inches to 5 feet 4 inches.

Warm-up

Find the perimeters of these figures. All measurements are in feet, and the figures are composed of squares, equilateral triangles, and rectangles. You may want to write in all the measures first before doing the addition.

1.

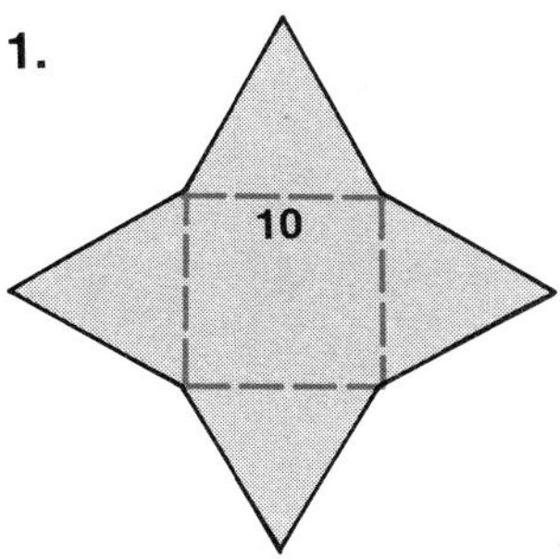

______

2.

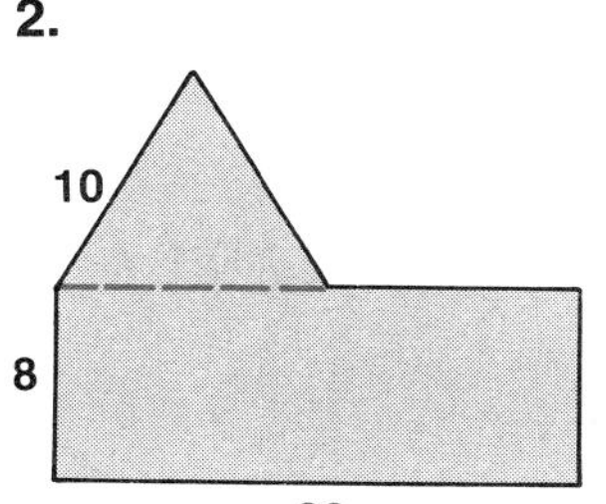

______

3.

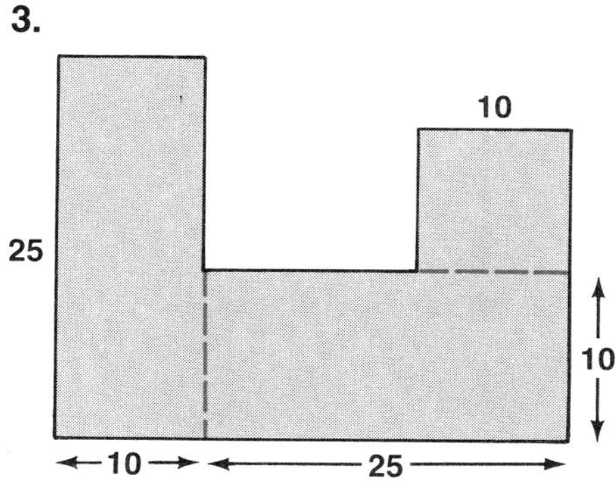

______

Warm-up Answers
1. 80 ft **2.** 66 ft **3.** 140 ft

On the Springboard

1. What is the perimeter of this figure? The curve at the left end is a half of circle. (Use $\pi = \frac{22}{7}$.)

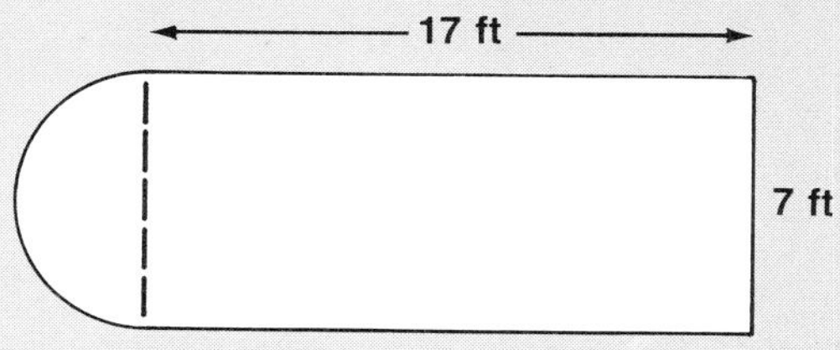

(1) 48 ft (2) 52 ft (3) 63 ft

① ② ③

2. What is the perimeter of this figure?

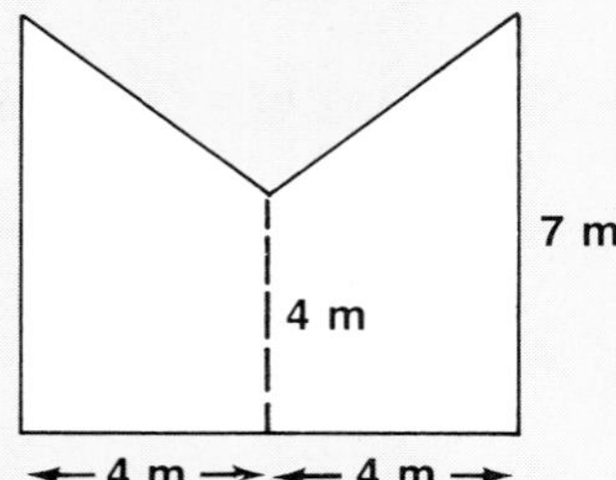

(1) 30 m (2) 32 m
(3) Insufficient data is given to solve the problem.

(1) (2) (3)

3. What is the perimeter of this figure?

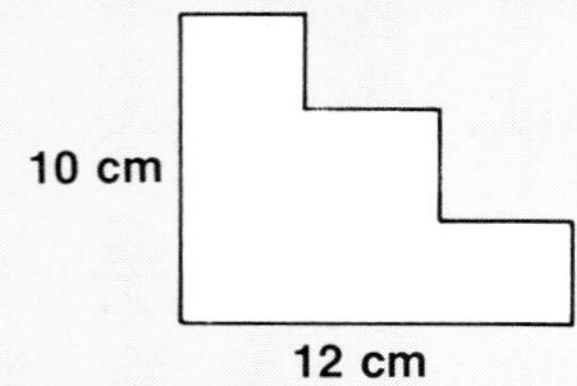

(1) 44 cm (2) 34 cm (3) 32 cm

(1) (2) (3)

Check your Springboard Answers on page 286. If you got these right, go on to "The Real Thing." If you made any errors, check and correct your work. When you are satisfied that you understand how to get the correct answers, go on.

"The Real Thing" 

1. Mrs. Prichard is building the table top shown below and is going to put a 2-inch edging completely around the table. How many feet of edging does she need?

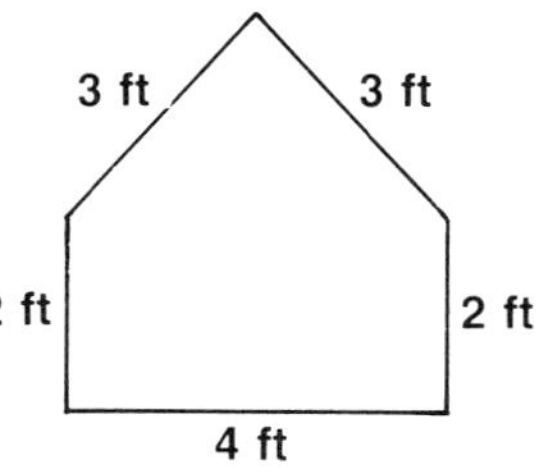

(1) 14 (2) 12 (3) 10
(4) 9 (5) 8

(1) (2) (3) (4) (5)

Questions 2 and 3 refer to the following information.

The Macgregors' vegetable garden is shown below. They bought low fencing at $1.03 a foot to put around the garden.

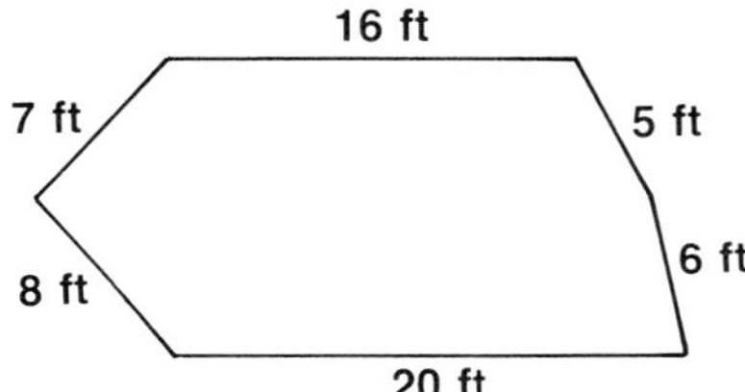

2. How many feet of fencing do the Macgregors need if they put it completely around their garden?

(1) 26 (2) 36 (3) 52
(4) 62 (5) 74

(1) (2) (3) (4) (5)

3. Mrs. Macgregor gave the clerk $100. How much change did she get back?

(1) $36.04 (2) $36.14
(3) $37.04 (4) $37.14
(5) Insufficient data is given to solve the problem.

(1) 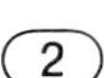  (4) (5)

4. The perimeter of the figure shown below is 90 meters. Assuming all angles in the figure are right angles, how many meters long is side C?

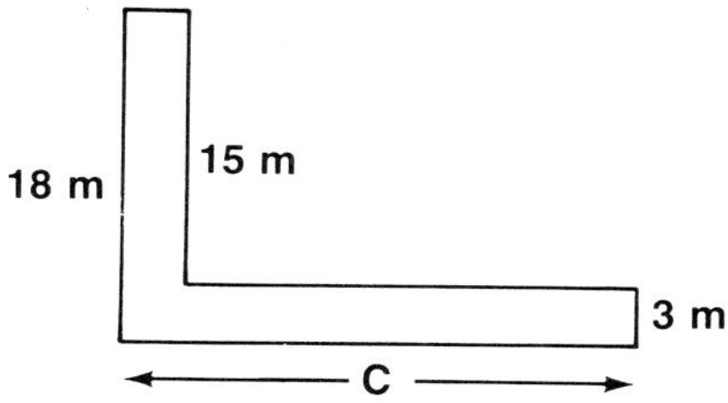

(1) 14 (2) 22 (3) 27 (4) 54
(5) Insufficient data is given to solve the problem.

Check your answers on pages 293–294.

Finding Areas

To find the area of a composite figure, you just find the areas of the simpler figures of which it is made up and add.

Here's an Example

Keep alert to notice these figures can be divided into triangles, squares, and rectangles.

■ This is a plan of a small corner lot. What is the area of the lot in yards?

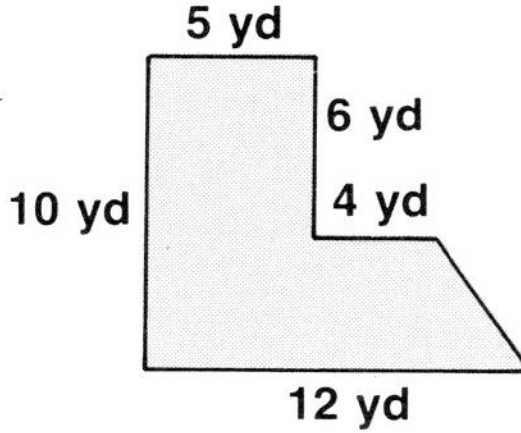

To solve, divide up the shape into simpler shapes, find their areas, and add. Here are two of the possible ways of dividing the composite figure into simple polygons.

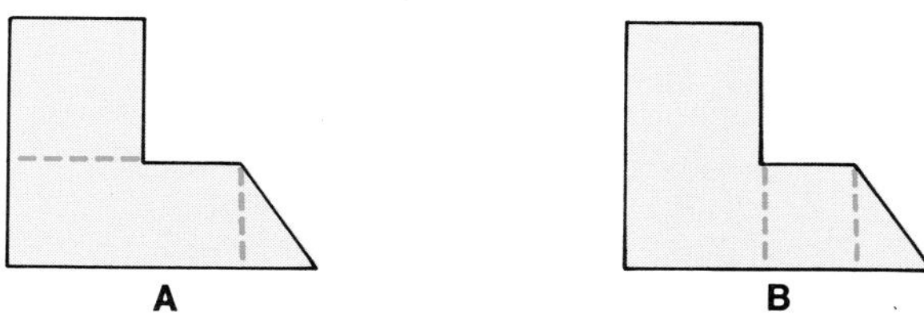

Figure A shows how the original figure can be divided into two rectangles and a triangle. Figure B shows how it can be divided into a rectangle, a square, and a triangle. First you can use Figure A, then use Figure B just to show that you get the same total area either way.

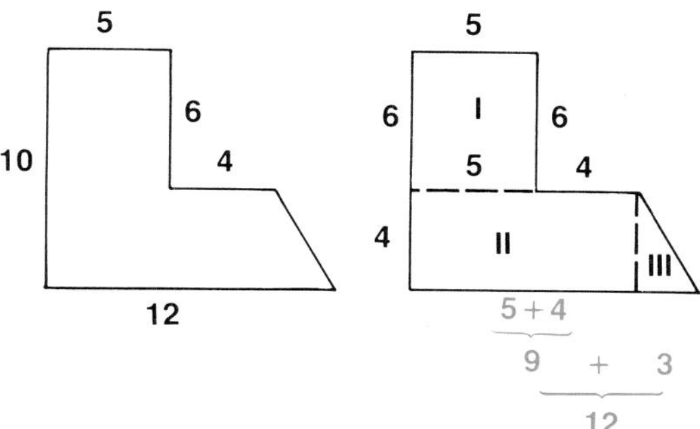

The bottom of the top rectangle must be 5 yards long, so the length of the top side and bottom side of the lower rectangle must be $5 + 4 = 9$ yards. Since the base of the figure is 12 yards, the part that is the base of the triangle must be $12 - 9 = 3$ yards. Now that you have the dimensions of all the simpler figures, you can find their areas and add.

Area I = Area of a Rectangle =
$\ell w = 5 \times 6 =$ 30
Area II = Area of a Rectangle =
$\ell w = 4 \times 9 =$ 36
Area III = Area of a Triangle =
$\frac{1}{2} bh = \frac{1}{2} \times 3 \times 4 =$ +6
72

Composite Figure A has an area of 72 square yards.

Look at how you would figure the area if you use the breakdown in Figure B. All the parts needed to find the total area are labeled. Can you see from where all the new measurements are taken? Find the area of each piece.

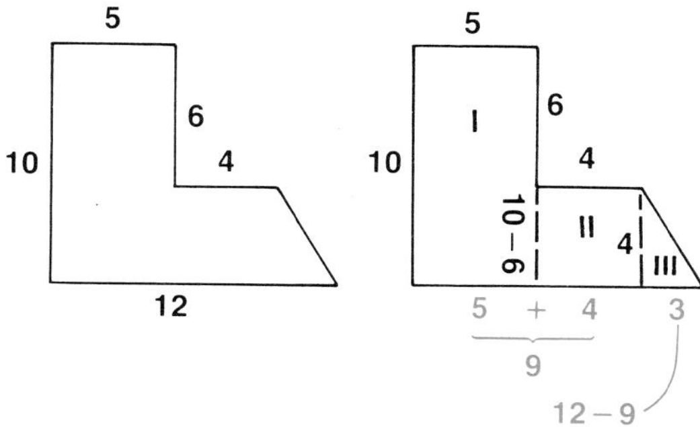

Area I = **Area of a Rectangle** = $\ell w = 5 \times 10 =$ 50
Area II = **Area of a Square** = $s^2 = 4^2 =$ 16
Area III = **Area of a Triangle** = $\frac{1}{2} bh = \frac{1}{2} \times 3 \times 4 =$ +6

72

Composite Figure B has an area of 72 square yards. No matter how you divide the composite figure, the area is the same.

Try It Yourself

The simplest way of dividing up a composite figure is usually the best.

■ The figure shows the end of a barn and its dimensions in feet. A painter needs to know its area in order to get the right amount of paint. What is its area in square feet?

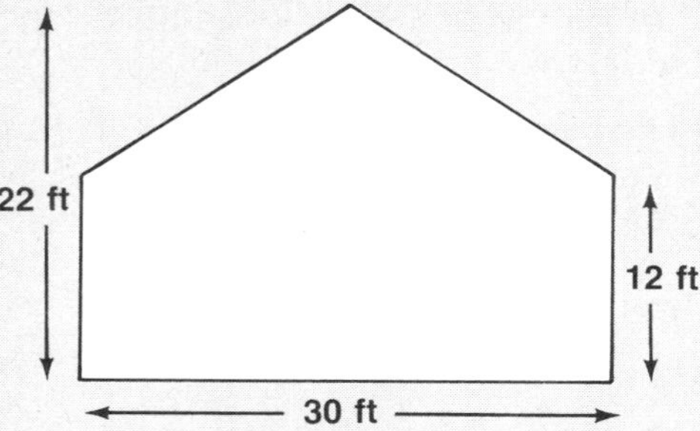

The most obvious way to divide the shape is probably into one triangle and one rectangle.

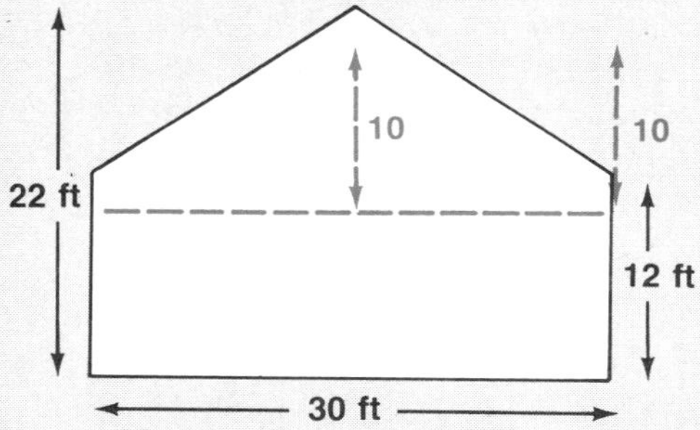

Here 22 − 12 = 10 gives the height of the triangle. The base of the triangle equals the width of the barn, 30. If you used this plan, you should have gotten the following:

Triangle	$\frac{1}{2} bh = \frac{1}{2} \times 30 \times 10$	= 150
Rectangle	$\ell w = 30 \times 12$	= 360
	Total	510

The area of the end of the barn is 510 square feet.

Warm-up

The height of an equilateral triangle, whose sides are 10 units long, is 8.7 units (to the nearest tenth of a unit). Use this fact to find how many square units there are in the areas of the Warm-up figures on page 261.

1. ________ **2.** ________ **3.** ________

On the Springboard

4. This is a diagram of the side view of a building. All dimensions are in feet. What is the area of the building in square feet?

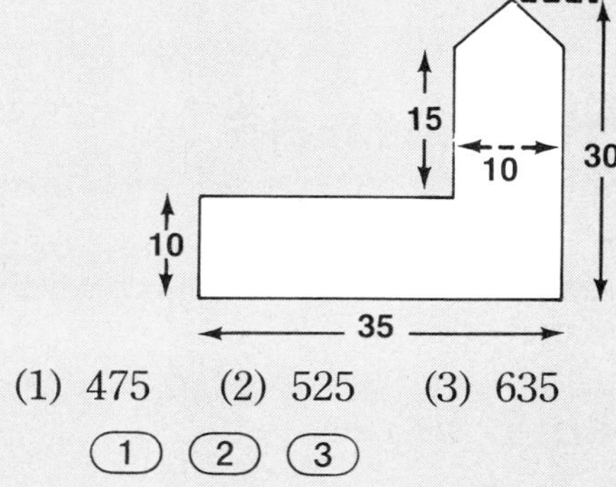

(1) 475 (2) 525 (3) 635

① ② ③

5. Some of the dimensions of a goldfish pond are given in the diagram in feet. What is the area of the pond in square feet? (Use $\pi = 3.14$.)

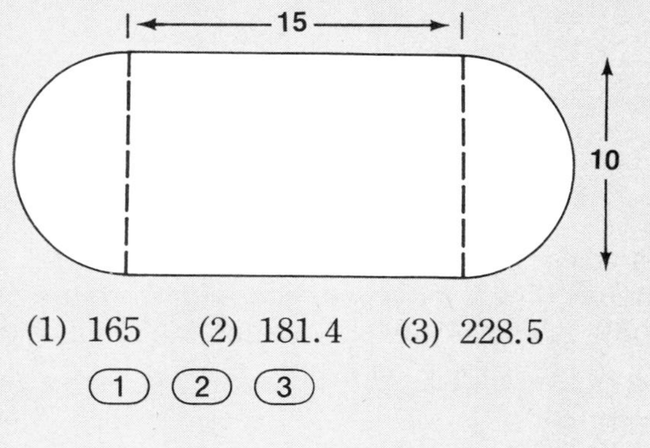

(1) 165 (2) 181.4 (3) 228.5

① ② ③

Check your Springboard answers on pages 286–287. If you got both of these right, go on to "The Real Thing." If you missed either one, find and correct your mistakes, then go on.

Warm-up Answers
1. 274 **2.** 203.5 **3.** 600

The Real Thing

Questions 5 and 6 refer to the following diagram, which represents the grounds in back of a school building.

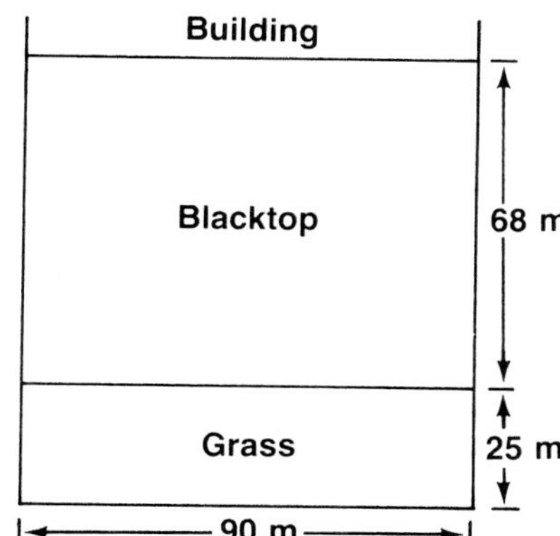

5. What is the area of the blacktop in square meters?

(1) 2,250 (2) 3,870 (3) 6,120
(4) 8,280 (5) 8,370

1 2 3 4 5

6. What is the total area of the blacktop and the grass in square meters?

(1) 8,370 (2) 8,280 (3) 6,120
(4) 3,870 (5) 2,250

1 2 3 4 5

7. The diagram below shows a desk top. What is the area of the desk top in square inches?

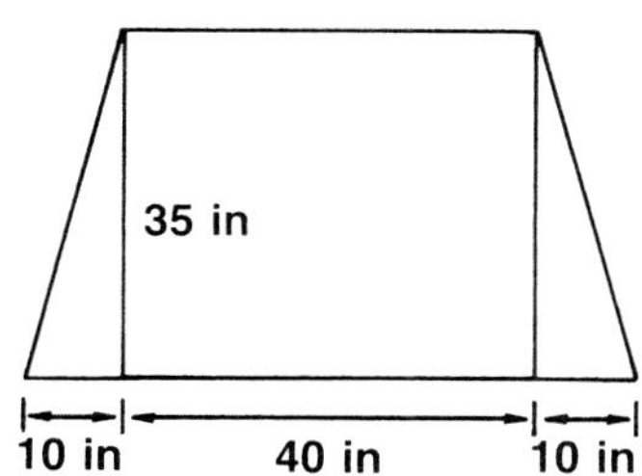

(1) 175 (2) 350 (3) 1,050
(4) 1,400 (5) 1,750

1 2 3 4 5

8. The Hildebrandts' driveway is shown below. What is its area in square feet?

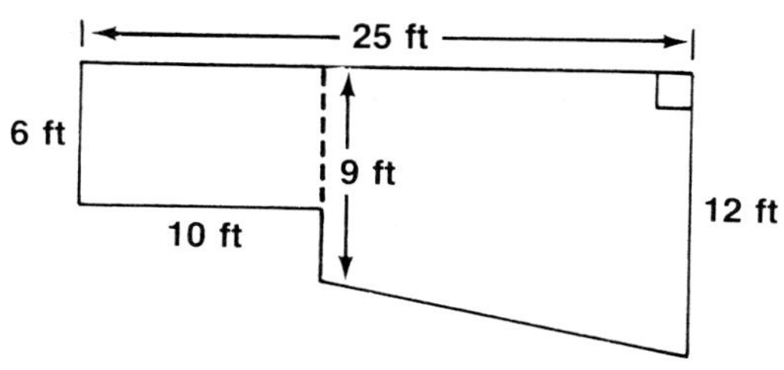

(1) 150.0 (2) 157.5 (3) 195.0
(4) 217.5 (5) 270.0

Questions 9 and 10 refer to the following information and diagram. Use 3.14 for π.

The park commissioners are going to build the infield and running track shown below. They received bids for edging, sod, and surfacing the track. The low bids were $9.99 for 30-foot rolls of edging and $1.75 for each square yard of sod. It will take approximately 20 days for the work crew to complete the project.

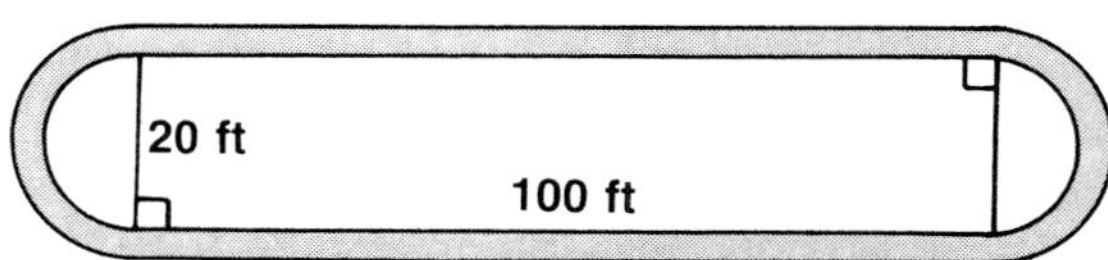

9. What is the perimeter of the infield in feet?

(1) 231.4 (2) 262.8
(3) 2,314.0 (4) 3,256.0
(5) Insufficient data is given to solve the problem.

1 2 3 4 5

10. Approximately how much will it cost to surface the running track?

(1) $1,358 (2) $1,259
(3) $875 (4) $741
(5) Insufficient data is given to solve the problem.

Check your answers and record all your scores on page 294.

LESSON 40

Solid Geometry

In Lesson 32, you learned that the volume of an object is the number of cubic units of space that it takes up. In the geometry lessons so far, you've been working with flat, plane figures. Figures that have **volume,** however, have not only length and width but also height (that is, depth or thickness). Such figures are part of solid geometry.

Finding the Volume of a Rectangular Solid

In Lesson 32, you learned how to find the volume of a box, which, in geometry, is called a **rectangular solid.** The sides of a rectangular solid are rectangles. From Lesson 32, you know that the cubic measure of a rectangular solid is found by multiplying length times width times height. Note that the units of volume in the metric system are represented by special symbols

Metric Unit	Symbol
cubic meter	m^3
cubic centimer	cm^3
cubic millimeter	mm^3

Coming to Terms

volume the number of cubic units of space a figure occupies (volume $= \ell \times w \times h$)

rectangular solid solid figure with rectangular sides, shaped like a box.

Here's an Example

You will find the volumes of rectangular solids just as you did the cubic measure of boxes: multiply length times width times height. Use the formula $V = \ell \times w \times h$, or $V = \ell wh$.

■ What is the volume in cubic inches of the waste basket shown in the figure?

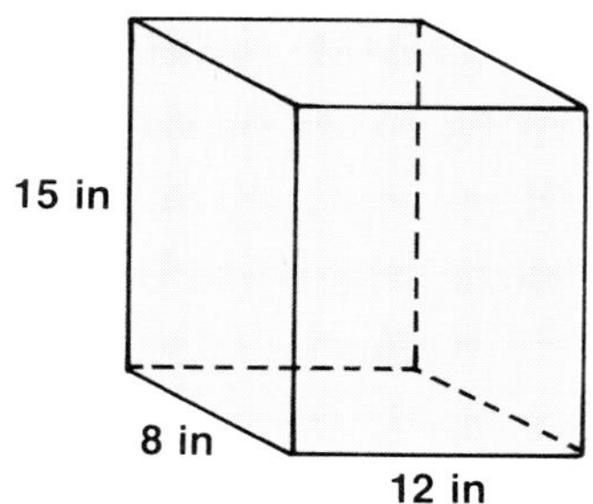

To solve, use the formula $V = \ell wh$. Fill in the information and multiply.

$$V = 12 \times 8 \times 15$$
$$V = 1{,}440$$

The volume of the waste basket is 1,440 cubic inches.

Try It Yourself

Handle this problem the same way as the previous example. Think of the room as a big box.

■ What is the volume of a room that is 10 feet long, 8 feet wide, and 8 feet high?

When you put the information in the formula you get $V = 10 \times 8 \times 8$ or $V = 640$. The volume of the room is 640 cubic feet.

Warm-up

Find the volumes in cubic units of these rectangular solids.

1.

8 ft
12 ft
10 ft

V = __________

2.

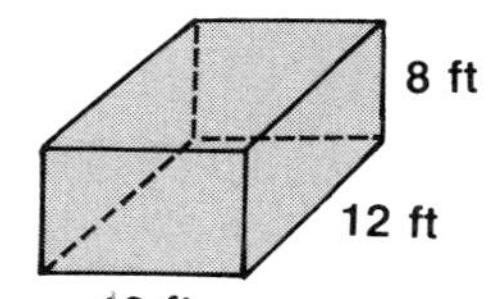

V = __________

3.

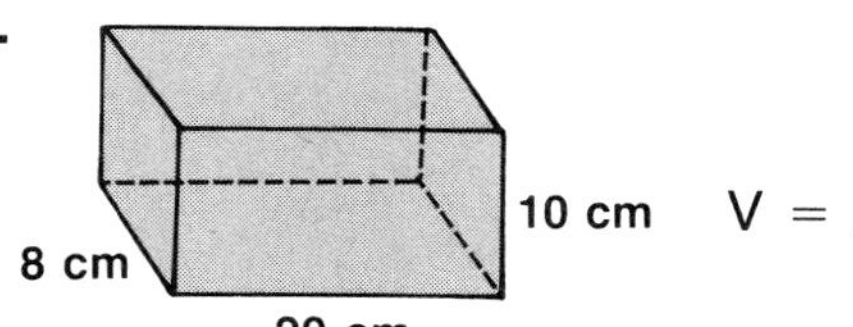

V = __________

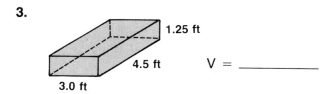

Find the volumes of the rectangular solids with these measurements.

	Length	Width	Height	Volume
5.	2 ft	9 ft	12 ft	________
6.	4 m	5 m	4 m	________
7.	6 in	2 ft	4 ft	________

On the Springboard

1. What is the volume in cubic feet of a moving van with measurements of 48 feet by 8 feet by 12 feet? The van is box-shaped.

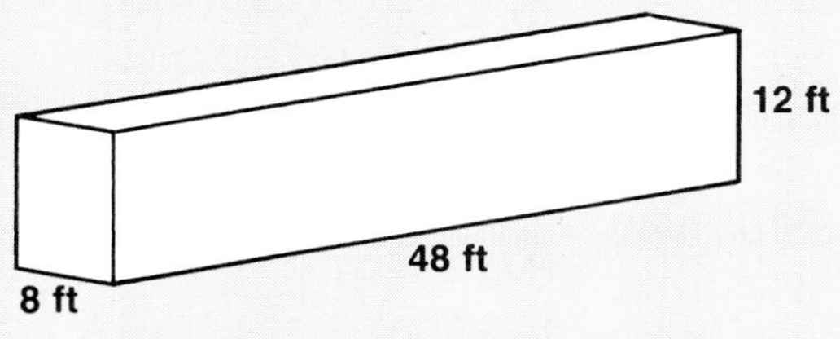

(1) 1,536 (2) 2,304 (3) 4,608

(1) (2) (3)

2. How many cubic feet of dirt will be removed for a basement 100 feet long, 40 feet wide, and 8 feet deep? Assume that the basement is box-shaped.

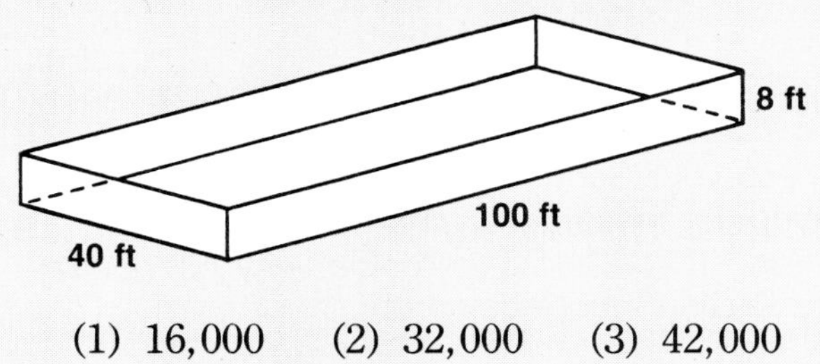

(1) 16,000 (2) 32,000 (3) 42,000

(1) (2) (3)

Check your Springboard answers on page 287. If you got both right, go on to "The Real Thing." If you made any errors, correct your work and then go on.

Warm-up Answers
1. 960 cu ft **2.** 1,600 cm^3 **3.** 16.875 cu ft **5.** 216 cu ft **6.** 80 m^3 **7.** 4 cu ft or 6,912 cu in Don't forget to first change all three numbers to the same unit, all inches or all feet.

The Real Thing

1. The rectangular storage space of a grain building is 90 feet long and 50 feet wide. It can be filled to a depth of 8 feet. How many cubic feet of grain can be stored in this space?

(1) 3,600 (2) 4,500 (3) 32,000
(4) 36,000 (5) 45,000

(1) (2) (3) (4) (5)

2. A rectangular air mattress is 4 feet wide, 6 feet long, and 6 inches deep. What is its volume in cubic feet?

(1) 12 (2) 18 (3) 24 (4) 144
(5) Insufficient data is given to solve the problem.

(1) (2) (3) (4) (5)

3. The trailer of a truck is 9 feet high. Its volume is 1,440 cubic feet. How many feet long is the trailer?

(1) 20 (2) 25 (3) 160 (4) 180
(5) Insufficient data is given to solve the problem.

(1) (2) (3) (4) (5)

Check your answers on page 294.

Finding the Volume of a Cube

A **cube** is a rectangular solid whose length, width, and height are all equal. All three are equal to the length of one side, *s*, of the figure. You can find the volume of a cube in the same way as that for any other rectangular solid, but since length, width, and height are all equal to *s*, the formula gives you $V = s \times s \times s$ or $V = s^3$.

Coming to Terms

cube a rectangular solid whose length, width, and height are all the same. The top, bottom, and all the side faces of a cube are squares.

Here's an Example

■ What is the volume of a microwave oven that measures 15 inches on a side?

To solve, use the formula for the volume of a cube. Put 15 in place of *s* in the formula and compute,

$$V = s^3$$
$$V = \quad ^3$$
$$V = 15 \times 15 \times 15$$
$$V = 3{,}375$$

The volume of the oven is 3,375 cubic inches.

Try It Yourself

■ A gift box measures 6 inches on a side. What is its volume in cubic inches?

The box is a cube. Did you use the formula $V = s^3$ to get $V = 6^3 = 6 \times 6 \times 6$? Since $6 \times 6 \times 6 = 216$, the answer is 216 cubic inches.

If you like fractions, you may have noticed that 6 inches is ½ foot, so the volume is also $(½)^3$, or $½ \times ½ \times ½, = ⅛$ cubic feet. Usually you will be told in what units the answer should be.

A Test-Taking Tip

Before you do any computing of volume on the GED Mathematics Test, be sure that all your measurements are expressed in the same unit. If the height is in inches and the width and length in feet, you must put them all in inches or all in feet before going ahead.

Warm-up

Find the volume of the solids described below. Give your answers in the units asked for.

1. 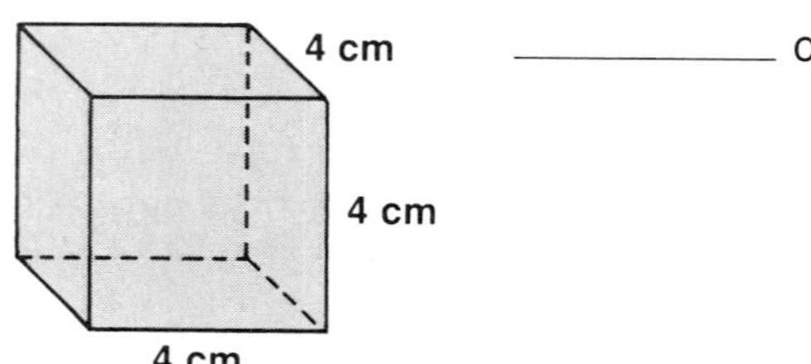

_________ cm^3

2. A cube measuring 0.05 m on each edge. _________ m^3

On the Springboard

3. A cube has edge measurements of 0.2 meters.

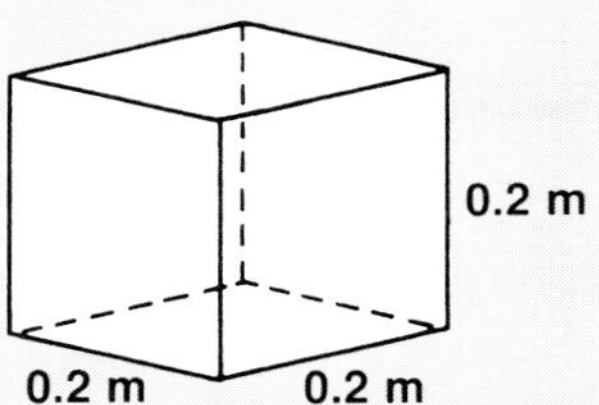

What is its volume in cubic meters?

(1) 0.8 (2) 0.08 (3) 0.008

① ② ③

4. What is the volume in cubic inches of the cube pictured?

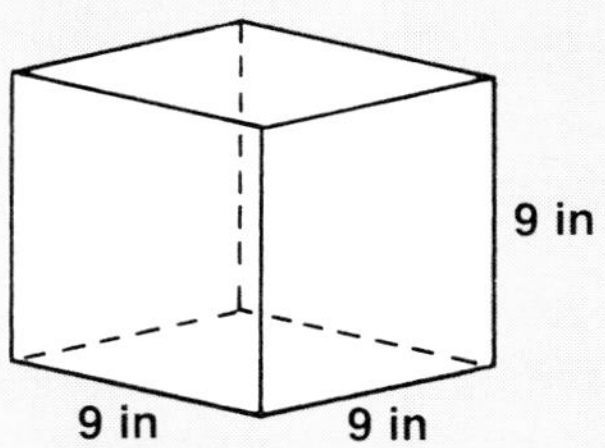

(1) 81 (2) 721 (3) 729

① ② ③

Check your Springboard answers on page 287. If you got these right, go on to "The Real Thing." If you had any trouble, why not review the section before going on.

The Real Thing 

4. Carl built a planter that measured 3 feet on each side. How many cubic feet of dirt does he need to fill this planter?

(1) 6 (2) 9 (3) 27 (4) 81
(5) Insufficient data is given to solve the problem.

① ② ③ ④ ⑤

Warm-up Answers
1. 64 cm^3 **2.** 0.000125 m^3

5. The volume of a box shaped like a cube is 64 cubic inches. What is the measure in inches of each side?

   (1) 5 (2) 4 (3) 3 (4) 2 (5) 1

   1 2 3 4 5

6. A livestock feeder built a storage shed that measured 14 feet by 14 feet by 14 feet. If one bushel of feed occupies 1.24 cubic feet of space, approximately how many bushels of feed can be stored in the shed?

   (1) 1,785 (2) 2,213 (3) 2,531
   (4) 2,744 (5) 3,403

Check your answers on page 294.

Finding the Volume of a Cylinder

A **cylinder** is shaped like a tin can—it has a circular top and bottom. To find the volume of a cylinder, you can multiply the area of one of the circular ends by the height of the cylinder. Each end of the cylinder is a circle, and its area is $A = \pi r^2$, where r is the radius of the end. So the formula for the volume of a cylinder is $V = \pi r^2 h$, where r is the radius and h is the height. Use 3.14 for π unless you are told to do otherwise.

Coming to Terms

cylinder a three-dimensional figure whose base and top are circles and whose sides are straight

Here's an Example

Remember, if you know the diameter of a cylinder and need to find the radius, you divide the diameter by 2.

■ A silo has a diameter of 15 feet and a height of 40 feet. How many cubic feet of grain can the silo hold?

To solve, use the formula $V = \pi r^2 h$. You may find it helpful to have a sketch.

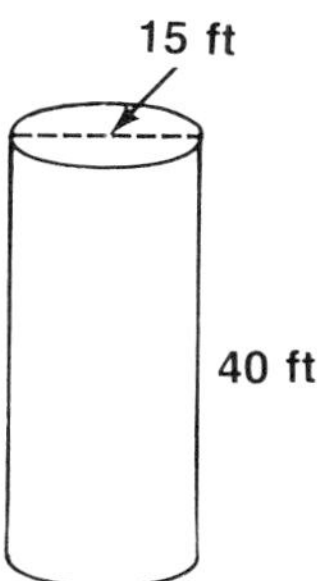

Step 1 Find the radius of the cylinder by dividing the diameter by 2.

$$r = 15 \div 2$$
$$r = 7.5$$

Step 2 You know that $r = 7.5$ and $h = 40$. Put this information in the formula and do the calculations. Use 3.14 for π.

$$V = \pi r^2 h$$
$$V = 3.14 \times 7.5^2 \times 40$$
$$V = 7,065$$

The silo holds 7,065 cubic feet of grain.

Try It Yourself

Divide the diameter by 2 to get the radius.

■ A soup can is 6 centimeters in diameter and 10 centimeters high. How many cubic centimeters of soup could the can hold?

The diameter of the can is 6, so the radius is 3. Since a value for π was not mentioned, you should have used 3.14. The formula gives you $V = 3.14 \times 3^2 \times 10$, or $V = 3.14 \times 9 \times 10$. When you did the calculation, did you get $V = 282.6$? The can holds 282.6 cm^3 of soup.

A Test-Taking Tip

Formulas for calculating the volumes of the figures will be on the formula page of the GED Mathematics Test. If a GED question requires a formula not on the formula page, you will be given the formula in the problem.

Warm-up

What are the volumes of these cylinders?

1. A cylinder with a radius of 3 feet and a height of 15 feet. ____________________

2. A cylinder with a diameter of 5 meters and a height of 10 meters. ______________________

3. A cylinder with a radius of 7 inches and a height of 4 inches. (Use $\pi = 22/7$.)

____

4. A cylinder with a diameter of 7 millimeters and a height of 12 millimeters. (Use $\pi = 22/7$.)

____

On the Springboard

5. What is the volume in cubic inches of a cylinder with a diameter of 3 inches and a height of 2 inches? (Use $\pi = 3.14$)

   (1) 7.065 (2) 9.42 (3) 14.13

   ① ② ③

6. Find the volume in cubic centimeters of a cylinder that is 7 centimeters high and has a radius of 2 centimeters. (Use $\pi = 3.14$.)

   (1) 87.92 (2) 96.72 (3) 102.06

   ① ② ③

Check your Springboard answers on page 287. If you got these right, go straight to "The Real Thing." If you had an error, check your work. Did you use the *radius* in the formula? Did you square the radius? Make sure that you find your error. Then you should be ready to go on.

"The Real Thing"

7. A cylindrical water heater has a diameter of 18 inches and a height of 42 inches. Approximately how many cubic inches of water does it hold? (Use 3.14 for π.)

   (1) 28 (2) 756 (3) 2,374
   (4) 10,682 (5) 42,729

   ① ② ③ ④ ⑤

8. A car cylinder has diameter of $3\frac{1}{2}$ inches and a depth of 4 inches. If the car has 4 cylinders, what is the volume in cubic inches of all 4 cylinders? Use $\frac{22}{7}$ for π.)

   (1) $38\frac{1}{2}$ (2) 128 (3) 154
   (4) 512 (5) 616

   ① ② ③ ④ ⑤

9. The diameter of a cylindrical vat is 42 feet. The volume of the vat is 16,632 cubic feet. How many feet deep is the vat? (Use $\frac{22}{7}$ for π.)

   (1) 12 (2) 18 (3) 22
   (4) 180 (5) 1,386

   ① ② ③ ④ ⑤

10. A rectangular box is 12 inches by 3 feet by 6 inches. What is the volume of the box in cubic feet?

    (1) 1.5 (2) 10 (3) 18
    (4) 216 (5) 2,592

    ① ② ③ ④ ⑤

11. The inside volume of a rectangular freezer is 9 cubic feet. If it is 2 feet wide and 1.5 feet deep, what is the length of the freezer in feet?

    (1) 2.5 (2) 3.0 (3) 5.5
    (4) 6.0 (5) 27.0

    ① ② ③ ④ ⑤

12. A rectangular steel block measures 6 inches by 8 inches by 12 inches. If steel weighs 0.283 pound per cubic inch, what is the approximate weight of the block in pounds?

    (1) 163 (2) 489 (3) 1,956
    (4) 2,035 (5) 24,424

    ① ② ③ ④ ⑤

Check your answers and record all your scores on page 295.

Warm-up Answers
1. 423.9 cu ft **2.** 196.25 m^3 **3.** 616 cu in **4.** 462 mm^3. If you got 2 or 4 wrong, maybe you forgot to divide the diameter by 2 to get the radius. Did you also remember to square the radius?

LESSON 41

Coordinate Geometry

This lesson contains basic information about plotting points on a grid. To use this lesson properly, you will need some graph paper with ¼-inch squares.

In coordinate geometry, you'll use what you know about locating positive and negative numbers on a number line to locate points on a plane, flat surface called a **grid**.

Coordinate Grids

In coordinate geometry, you **plot** (locate and mark) points on a grid. This is like placing pegs in certain positions on a pegboard.

Look at this grid used for plotting points. Notice the two heavy lines that cross in the center. The horizontal line is called the *horizontal* **axis,** or **x-axis.** The vertical line is called the *vertical axis,* or **y-axis.** The arrows on the ends of each axis indicate that the lines continue endlessly. These lines, *axes,* break the grid into four sections called **quadrants.**

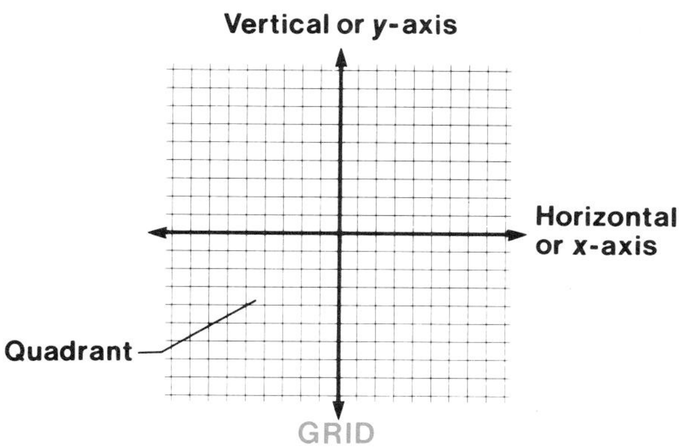

Coming to Terms

grid a network of horizontal and vertical lines that divide a flat surface into equal squares. Grids are used to help locate and mark points.

plot (or graph) to locate and mark points on a grid

axis one of the heavy lines drawn on a coordinate grid (plural is *axes*). The axes are the "home base" lines used in plotting points on a grid.

x-axis a horizontal number line on a coordinate grid

y-axis a vertical number line on a coordinate grid

quadrant one of the four sections formed when the two axes are drawn on a coordinate grid

Ordered Pairs

If you put your socks on first and then your shoes, the result is very different from first putting on your shoes and then your socks. The *order* in which you do these two things makes quite a difference.

Points on a grid are named by using **ordered pairs** of numbers. An *ordered pair* is two numbers written inside parentheses, with a comma between the two numbers. For example, (2, 3) or (−4, 0) or (−1, −5), and (0, 2) are all ordered pairs. The numbers in an ordered pair can be positive, negative, or zero. They are called *ordered pairs* because the order in which the numbers are written makes a difference: (2, −3) and (−3, 2) are two different ordered pairs, even though they contain the same numbers.

The numbers within each ordered pair are called *coordinates.* The first number in the ordered pair is called the **x-coordinate;** it marks the position of the point you are plotting along the *x*-axis. The second number is called the **y-coordinate;** it marks the position of the point along the *y*-axis.

Ordered pairs of numbers are used to pinpoint the locations of points on a grid. You start at the **origin** of the grid—the point where the horizontal and vertical axes cross.

This diagram shows a vertical axis and a horizontal axis. The lines of the grid are spaced 1 unit apart. The point where the axes cross is the origin.

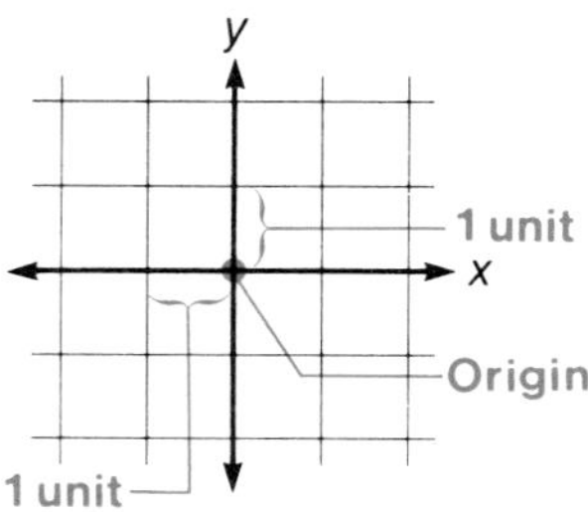

Coming to Terms

ordered pair two numbers written inside parentheses and separated by a comma

x-coordinate first number in an ordered pair. The *x*-coordinate tells how far left or right a point is using the *x*-axis as a reference.

y-coordinate second number in an ordered pair. The *y*-coordinate tells how far up or down a point is using the *y*-axis as a reference.

origin the point on a grid where the axes cross

Here's an Example

Now you'll see how ordered pairs can be used on the grid.

■ Plot the ordered pair (2, 3).

To solve, start at the origin (where the axes cross). Use the *x*-coordinate to decide how far left or right to go and the *y*-axis to decide how far up or down.

Step 1 Start at the origin. The *x*-coordinate is (positive) 2, so move 2 units to the right (that is, in the positive direction).

Step 2 The second number in the ordered pair is 3. So, from the point on the *x*-axis that you reached in Step 1, move up 3 units, parallel to the *y*-axis.

Step 3 You arrive at the point shown in the diagram at the top of the next column. Mark it with a dot and label it (2, 3).

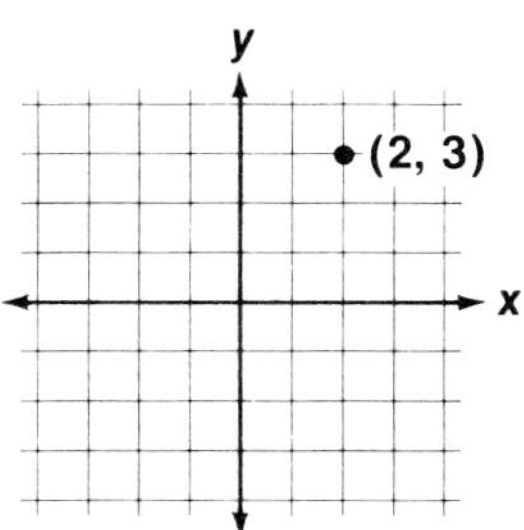

Notice that if you plot (0, 0), the point you mark will be the origin. The *x*-coordinate is 0, so you move neither up nor down. You stay at your starting place, which is the origin. If you were asked to plot (0, −5) you would just go down the *y*-axis 5 units because your *x*-axis movement is 0 units. If you had to plot (−2, 0) you would just go left 2 units along the *x*-axis. Since the *y*-coordinate is 0, you move neither up nor down.

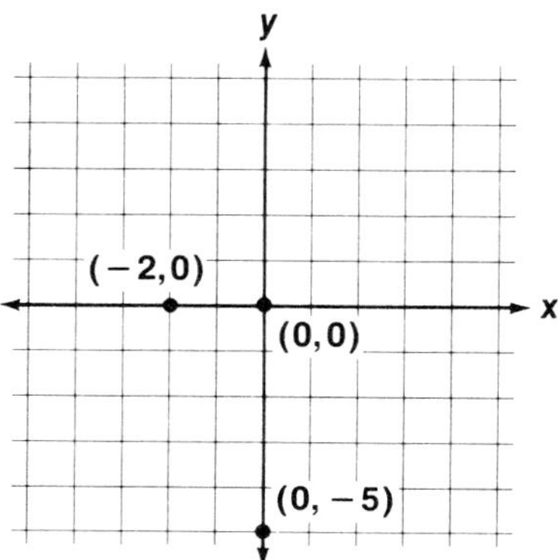

Try It Yourself

■ Plot these four points: (1, 2), (−1, 3), (−2, −2), and (3, −1). Use a piece of ¼-inch graph paper, or the following grid if this is your own book.

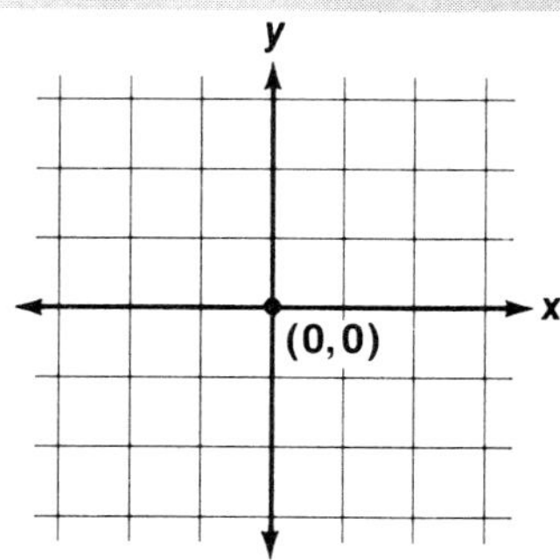

For the first ordered pair, (1, 2), did you start with the origin? Then did you move 1 to the *right* and 2 *up?* In the second ordered pair, (−1, 3), did you remember to first move 1 to the *left* because the first coordinate is a negative number? For the third ordered pair, (−2, −2), did you move 2 *left* and then 2 *down?* Finally, did you move 3 *right* and then 1 *down* for the ordered pair (3, −1)? You should have one dot in each quadrant. Did you label all four points with their ordered pairs?

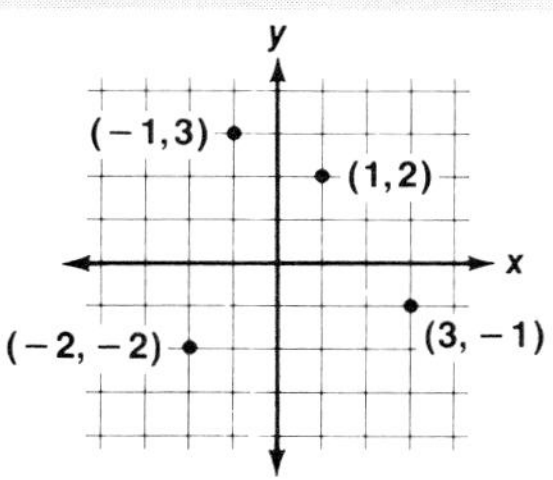

Warm-up

Plot each ordered pair. Use graph paper, or the grid below if this is your own book. Label each point with the letter given for that ordered pair.

1. (5, 4) A	**6.** (−2, −3) F	**11.** (−3,0) K
2. (0, −6) B	**7.** (1, 2) G	**12.** (−1, −3) L
3. (−2, 5) C	**8.** (0, 3) H	**13.** (1, −5) M
4. (3, −2) D	**9.** (−1, −5) I	**14.** (−3, 2) N
5. (4, 0) E	**10.** (0, 0) J	**15.** (−1, −1) O

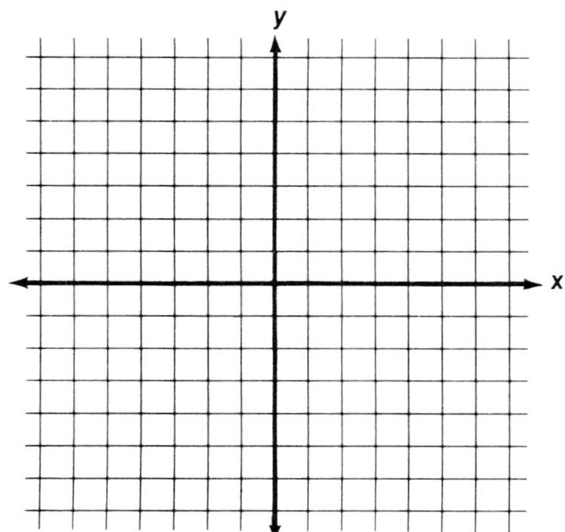

Warm-up Answers

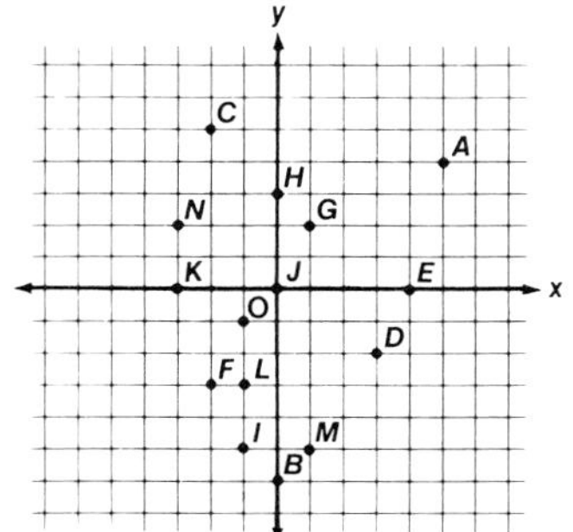

On the Springboard

1. In the ordered pair (2, −4), what is the *x*-coordinate?

   (1) −4 (2) −2 (3) 2

   (1) (2) (3)

2. What is the *y*-coordinate of the ordered pair (2, −4)?

   (1) −6 (2) −4 (3) 2

   (1) (2) (3)

<u>Questions 3 and 4</u> refer to this diagram.

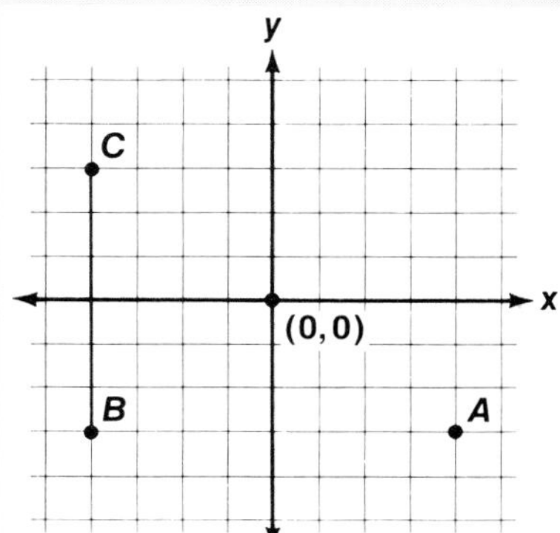

3. What ordered pair corresponds to the point where *BC* crosses the *x*-axis?

   (1) (−4, 0) (2) (−3, 0)
   (3) (0, −3)

   (1) (2) (3)

4. Which point is exactly 4 units from (0, 3)?

   (1) A (2) B (3) C

   (1) (2) (3)

Check your Springboard answers on page 287. If you got these right, go on to "The Real Thing." If you missed any, check your work. Review this section if you wish, and, when you are confident that you understand how to answer all these correctly, go on to "The Real Thing."

“The Real Thing”

Questions 1 and 2 refer to the ordered pair $(-1, 3)$, which locates a point on a coordinate grid.

1. Which number is the y-coordinate?

   (1) -3 (2) -1 (3) 1
   (4) 2 (5) 3

   (1) (2) (3) (4) (5)

2. Which number is the x-coordinate?

   (1) -3 (2) -1 (3) 1
   (4) 2 (5) 3

   (1) (2) (3) (4) (5)

Questions 3–5 refer to the following coordinate grid.

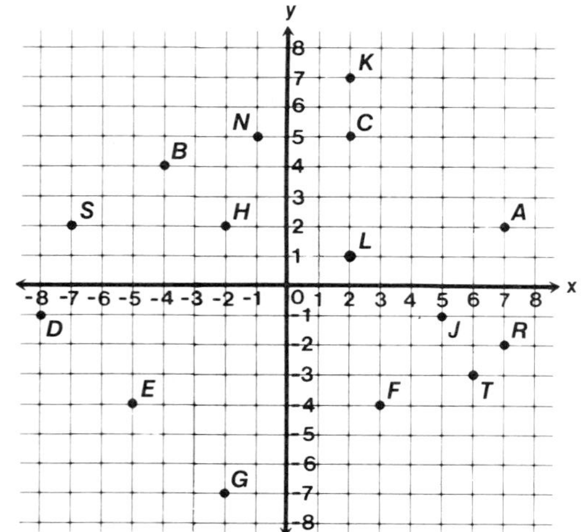

3. What are the coordinates of point C?

   (1) $(-2, -5)$ (2) $(-2, 5)$ (3) $(2, 5)$

   (4) $(5, -2)$ (5) $(5, 2)$

   (1) (2) (3) (4) (5)

4. What point corresponds to $(5, -1)$?

   (1) N (2) J (3) G (4) A (5) T

   (1) (2) (3) (4) (5)

5. What are the coordinates of point G?

   (1) $(-2, 7)$ (2) $(-2, -7)$
   (3) $(-7, 2)$ (4) $(-7, -2)$ (5) $(2, 7)$

   (1) (2) (3) (4) (5)

Check your answers and record your scores on page 295.

LESSON 42

Distance on a Coordinate Grid

Finding the distance between two cities on a map can be difficult, but finding the distance between two points on a coordinate grid is easy, because the distance is always the straight line from one point to the other. The distance from point A to point B is the length of the line segment from A to B.

Measuring Horizontal and Vertical Distances

You can find the horizontal and vertical distance on a grid by counting units.

Here's an Example

If the line segment from point A to point B is horizontal, you can often find the distance by simply counting units.

■ What is the distance between point A and point B in the figure below?

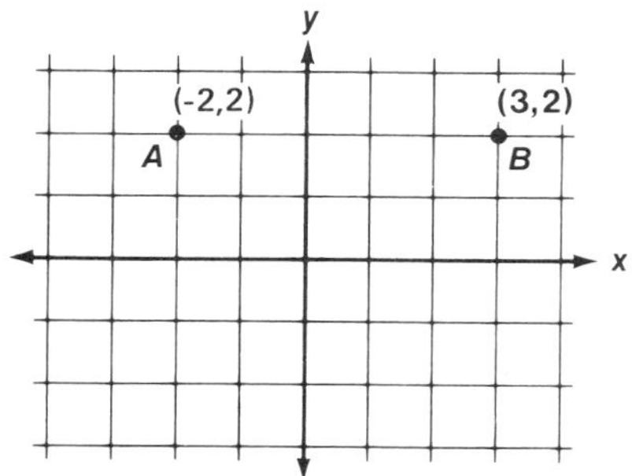

To solve, count the units between A and B. Look at the line segment connecting the two points. Both points are the same distance above the x-axis, so the line segment connecting the points is horizontal, that is, parallel to the x-axis.

Since the x-coordinates of the two points are not fractions, just count the units from one point to the other. The distance is 5 units.

Sometimes you may be given the ordered pairs for two points and asked to find the distance between them. For example, suppose the ordered pairs are (−6, 3) and (4, 3).

One easy way to get the distance is to plot the ordered pairs. You can see that the segment's horizontal. Again, you can simply count units to find the distance. You can also use logic. The second number in both pairs is the same (3). The points are the same distance above the *x*-axis. This means that the line joining the two points is parallel to the *x*-axis. The −6 in the first pair shows that the first point is 6 units to the left of the *y*-axis. The 4 in the other pair shows that the other point is 4 units to the right of the *y*-axis. The distance between them must be 6 units + 4 units or 10 units.

Try It Yourself

This time the line segment is vertical, but you can use the same kind of thinking as in the last example.

■ What is the distance between points *C* and *D* in this figure?

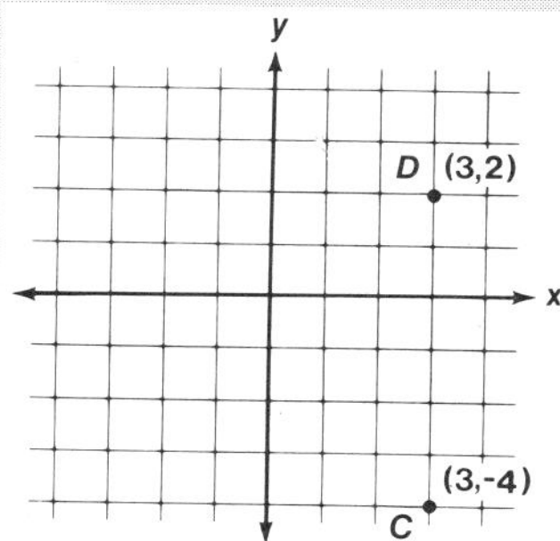

Did you notice that the line segment joining the two points is parallel to the *y*-axis? If you did, you realize that you could just count the units from one point to the other. The first point is 2 units *above* the *x*-axis and the second is 4 units *below* the *x*-axis. Altogether there are 6 units from one point to the other. The distance between the two points is 6 units.

If you had only been given the ordered pairs, you could have plotted them yourself or used logic. The first number in the two ordered pairs is the same, so the line segment joining the points must be parallel to the *y*-axis. The 2 in (3, 2) means that the point is 2 units above the *x*-axis. The −4 in (3, −4) means that the point is 4 units below the *x*-axis. So the distance between the points must be 2 + 4 or 6 units.

Warm-up

What is the distance between these points on the grid below. Assume the small squares in the grid are 1 unit on each side.

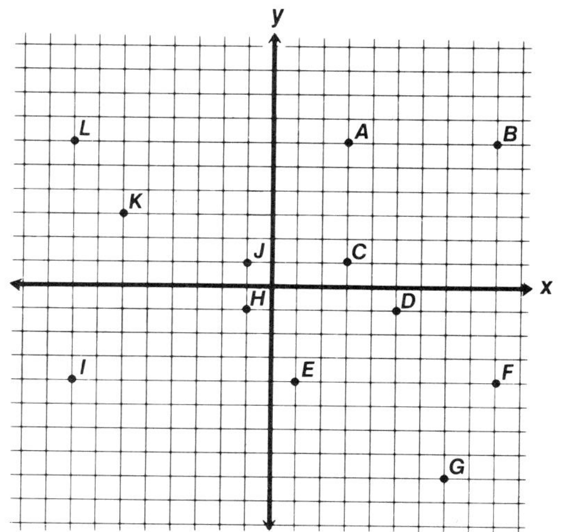

1. *H* and *D* ______
2. *A* and *L* ______
3. *J* and *H* ______
4. *E* and *F* ______
5. *A* and *C* ______
6. *I* and *L* ______
7. *B* and *F* ______
8. *J* and *C* ______

What is the distance between each pair of points?

9. (−3, 5) and (−3, −4) ______
10. (0, 7) and (0, − 1) ______
11. (−2, 5) and (3, 5) ______
12. (3, −10) and (3, 0) ______

Warm-up Answers
1. 6 **2.** 11 **3.** 2 **4.** 8 **5.** 5 **6.** 10
7. 10 **8.** 4 **9.** 9 **10.** 8 **11.** 5 **12.** 10

On the Springboard

Questions 1 and 2 refer to this grid.

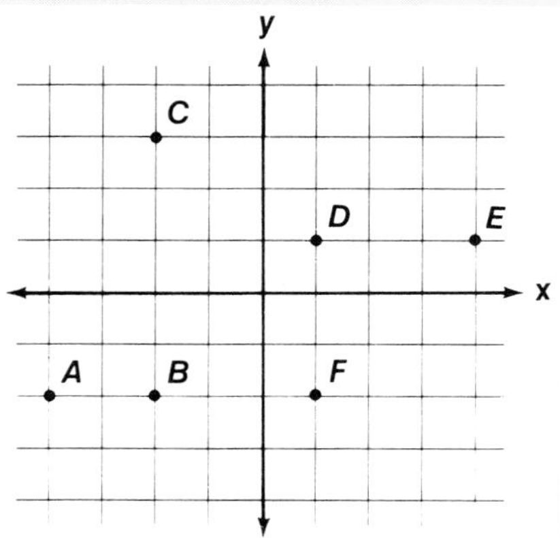

1. The distance between B and C is

   (1) 1 (2) 2 (3) 5

   ① ② ③

2. The distance between D and F is

   (1) 0 (2) 1 (3) 3

   ① ② ③

Check your Springboard answers on page 287. If you got both right, go on to "The Real Thing." If you missed either, you may want to review before going on.

The Real Thing

Questions 1–3 refer to the following grid.

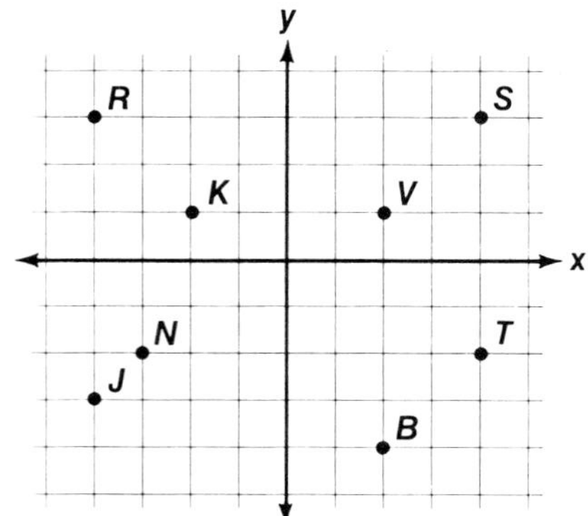

1. The distance between K and V is how many units?

   (1) 4 (2) 5 (3) 6 (4) 7 (5) 8

   ① ② ③ ④ ⑤

2. The distance between S and T is how many units?

   (1) 4 (2) 5 (3) 6 (4) 7 (5) 8

   ① ② ③ ④ ⑤

3. Which of the points marked on the grid are 7 units apart?

   (1) R and J (2) V and B (3) T and S
   (4) N and T (5) R and S

   ① ② ③ ④ ⑤

Check your answers on page 295.

Finding Perpendicular Distance

Perpendicular means meeting at right angles. Finding perpendicular distances is just as easy as counting units. In the following explanation, the symbol ⟷ over two letters means we are talking about a whole line. For example, $\overleftrightarrow{QR}$ means *line QR.*

Here's an Example

On this kind of problem, the easiest solution may entail your drawing a line on the grid.

■ What is the perpendicular distance on this grid from S to $\overleftrightarrow{QR}$?

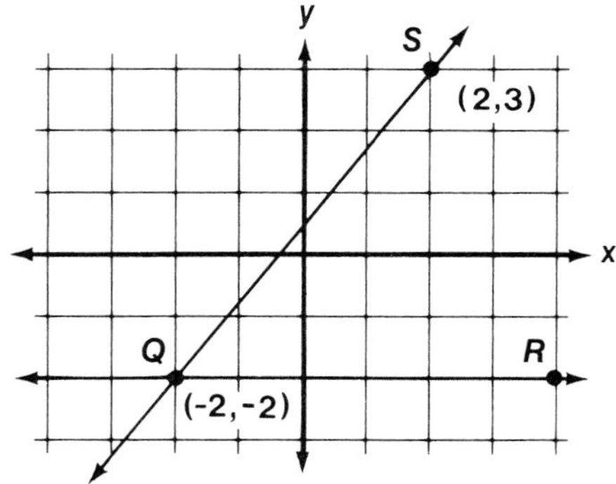

Solution: Draw a line segment perpendicular to $\overleftrightarrow{QR}$ from point S and find its length.

Step 1 Draw a line segment straight down from point S to meet $\overleftrightarrow{QR}$ at right angles (perpendicular). Label the point where it hits $\overleftrightarrow{QR}$ with the letter T.

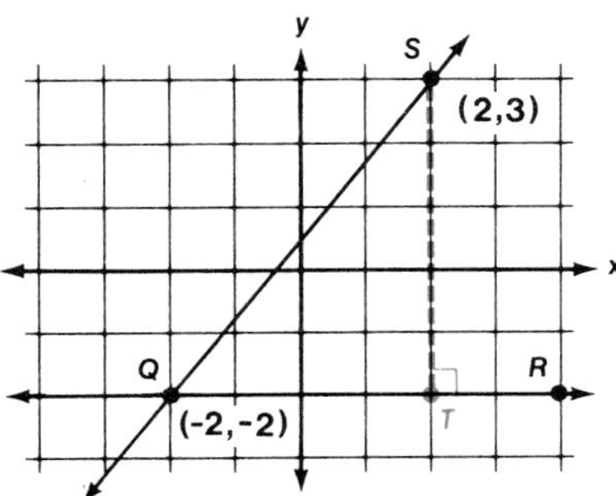

Step 2 Find the length of ST. Point S is 3 units above the x-axis. $\overleftrightarrow{QR}$ *is* 2 units below the x-axis. The total distance is 3 + 2, or 5 units.

The perpendicular distance from S to $\overleftrightarrow{QR}$ is 5 units.

Try It Yourself

It often takes a lot of words to state a simple problem. Don't be put off by the words.

■ In the figure below, line PR is parallel to the x-axis. Lines QR and PR intersect at R, which corresponds to the ordered pair (3, 2). What is the perpendicular distance from Q to $\overleftrightarrow{PR}$?

____

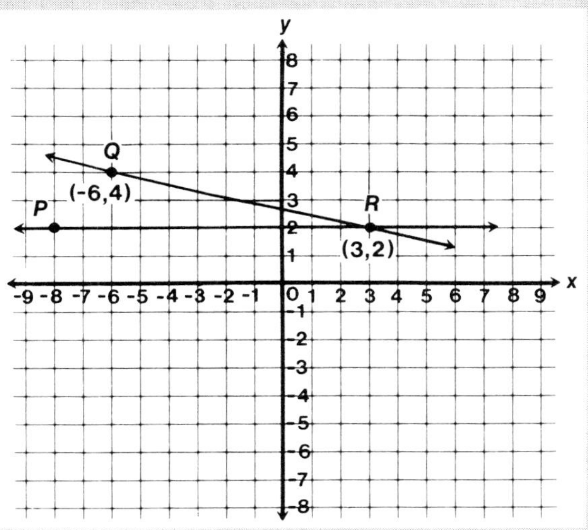

Did you draw a perpendicular from Q to $\overleftrightarrow{PR}$? $\overleftrightarrow{PR}$ is two units above the x-axis. Point Q is 4 units above the x-axis. So the length of the segment from Q perpendicular to $\overleftrightarrow{PR}$ is 2 units.

Warm-up

Find the perpendicular distance between the following points and lines on the grid below.

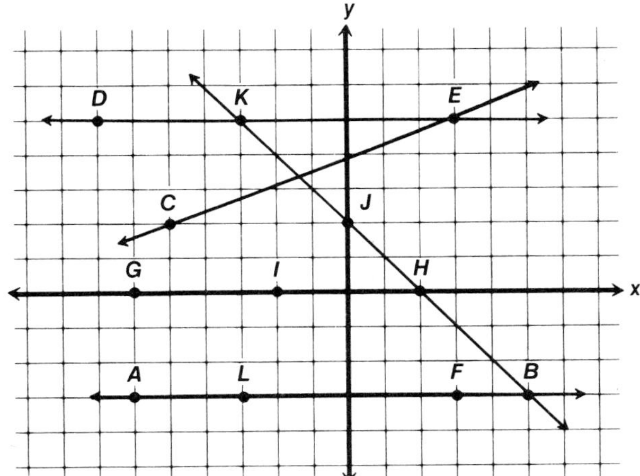

1. point C and line DE _______________
2. point I and line AB _______________
3. point F and line DE _______________
4. point J and line DE _______________
5. point K and line GH _______________

Warm-up Answers
1. 3. **2.** 3 **3.** 8 **4.** 3 **5.** 5

On the Springboard

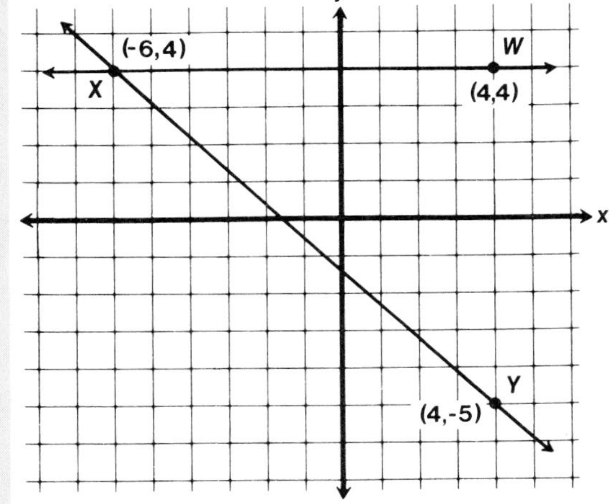

3. The perpendicular distance from Y to $\overline{XW}$ in the figure above is

(1) 4 (2) 5 (3) 9

① ② ③

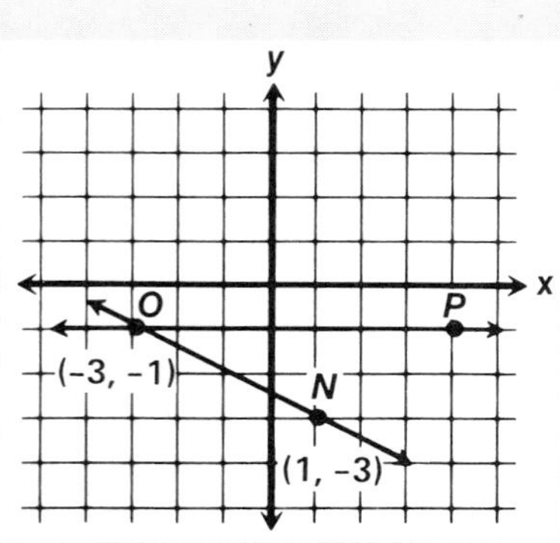

4. What is the perpendicular distance from point N to $\overline{OP}$ in the figure above?

   (1) 1 (2) 2 (3) 3

   (1) (2) (3)

Check your Springboard answers on page 287. If you got these right, go on to "The Real Thing." If you made an error, check your work before going on.

The Real Thing

Questions 4–6 refer to the following grid.

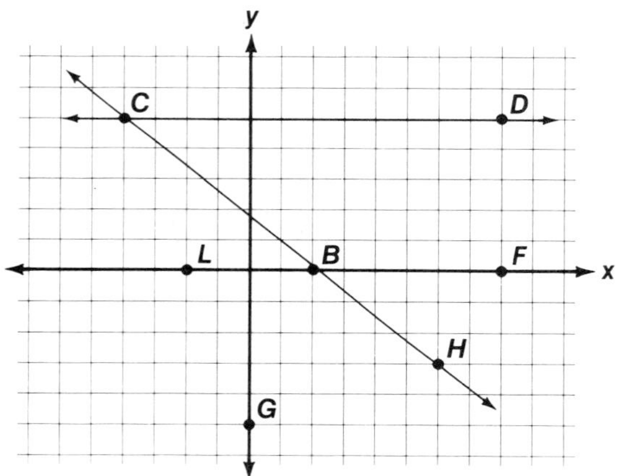

4. The perpendicular distance from point L to line CD is how many units?

   (1) 4 (2) 5 (3) 6 (4) 8 (5) 10

   (1) (2) (3) (4) (5)

5. The perpendicular distance from point H to $\overleftrightarrow{CD}$ is how many units?

   (1) 4 (2) 5 (3) 6 (4) 8 (5) 10

   (1) (2) (3) (4) (5)

6. From which point is the perpendicular distance to line CD 10 units?

   (1) B (2) F (3) G (4) H (5) L

   (1) (2) (3) (4) (5)

Check your answers on page 295.

Finding Diagonal Distance

You cannot just count units to find diagonal distances on a grid. But finding diagonal distances on a grid is related to something you have learned about right triangles. If two points are not on the same horizontal or vertical line, you find the distance between them by drawing a right triangle and using the Pythagorean Theorem.

You may want to refresh your memory about the Pythagorean Theorem before studying this section. If you do, take some time now to look back at Lesson 37 on pages 252–255. With the Pythagorean Theorem, finding diagonal distances is not very hard.

Here's an Example

■ What is the distance between points R and S on the grid below?

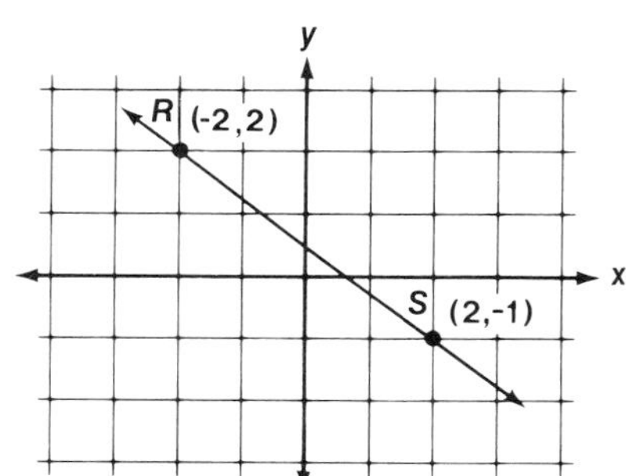

To solve, draw a horizontal line through one point and a vertical line through the other. Then, use the Pythagorean Theorem.

Do you remember how to use the Pythagorean Theorem

$$a^2 + b^2 = c^2$$

where a and b are legs of a right triangle and c is the hypotenuse? Here are the steps for finding the length of $\overline{RS}$, the diagonal distance between the points R and S.

Step 1 Draw a vertical line segment through point R and a horizontal segment through point S, as shown in the next diagram. Call the point where they meet T. (You could have called it by any letter you wanted.)

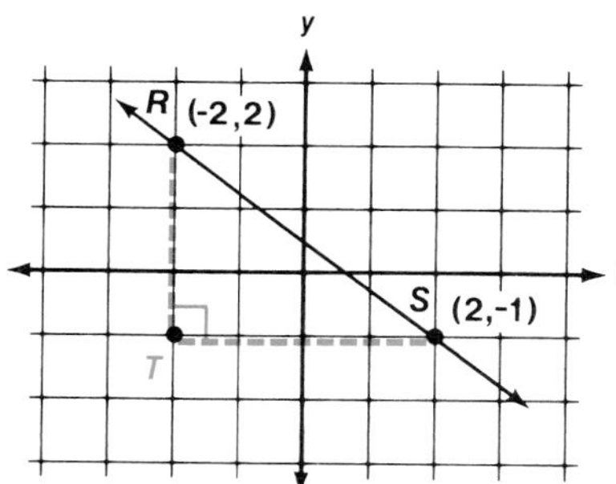

Step 2 Find the length of $\overline{RT}$ and $\overline{TS}$. You can do this by counting the units from R to T and from T to S. From R to T is 3 units and from T to S is 4 units.

Step 3 $\overline{RT}$ and $\overline{TS}$ are the legs of the right triangle you drew ($\triangle RTS$). Use the Pythagorean Theorem to find the hypotenuse ($\overline{RS}$) of this right triangle.

$$a^2 + b^2 = c^2$$
$$\overline{RT}^2 + \overline{TS}^2 = \overline{RS}^2$$
$$3^2 + 4^2 = \overline{RS}^2$$
$$9 + 16 = \overline{RS}^2$$
$$25 = \overline{RS}^2$$
$$5 = \overline{RS}$$

$\overline{RS}$, the diagonal distance from point R to point S, is 5 units. You might find it hard to believe, but there was nothing completely new to you in this example except the first grid you looked at. If you carefully follow the steps, each step is a repeat of something you learned in a previous lesson.

There is another way of doing this problem if you forget about drawing the perpendiculars. On the GED Mathematics Test formula page, you will see this distance formula:

$$d = \sqrt{(x_2 - x_1)^2 + (y_2 - y_1)^2}$$

The $\sqrt{}$ means square root. The x_1 and x_2 are the first numbers in the ordered pairs for two points on the coordinate grid, and y_1 and y_2 are the second numbers of the ordered pairs. (The first point is (x_1, y_1) and the second point is (x_2, y_2). If you just put the numbers into the formula and do the calculations, you will get the distance between the two points.

■ What is the distance from point (12, 15) to point (4, 9)?

Use $d = \sqrt{(x_2 - x_1)^2 + (y_2 - y_1)^2}$. The x-numbers are 12 and 4. The y-numbers are 15 and 9. Put them in the formula.

$$d = \sqrt{(12 - 4)^2 + (15 - 9)^2}$$

This becomes $d = \sqrt{8^2 + 6^2} = \sqrt{100}$. Since $10 \times 10 = 100$, $\sqrt{100}$ is 10. So, $d = 10$.

The distance between the two points is 10 units.

This distance formula will be on the formula page if you want to use it. It works for every pair of points on the coordinate grid.

Try It Yourself

With a grid in front of you it's easy to use the triangle method.

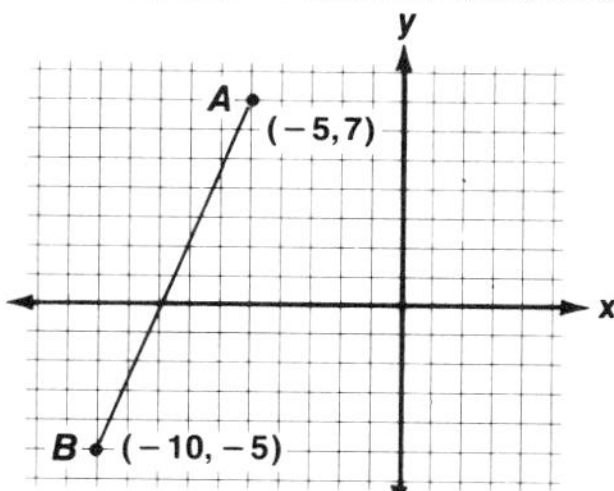

■ What is the distance between points A and B?

____

Were you able to set up two perpendicular line segments? One could have gone horizontally to the right from B. The other could have run vertically down from A. Then did you count the units in each? The vertical segment is 12 units. The horizontal segment is 5 units. Did you then remember the formula ($a^2 + b^2 = c^2$)? Then $12^2 + 5^2 = c^2$. You find that $c = 13$, so $\overline{AB}$ is 13 units.

If you did the problem by the distance formula, you had to remember how to deal with signed numbers. The x-numbers were -10 and -5 and the y-numbers were 7 and -5. So $x_2 - x_1 = -10 - (-5) = -10 + 5 = -5$ and $y_2 - y_1 = 7 - (-5) = 7 + 5 = 12$. If you put that into the formula, you got $d = \sqrt{(-5)^2 + 12^2} = \sqrt{25 + 144}$. So $d = \sqrt{169}$, or $d = 13$, as in the other method.

Warm-up

Write down the steps you would use to find the distance from (0, 8) to (−6, 0) using the triangle method. You can use this grid if you are free to write in the book.

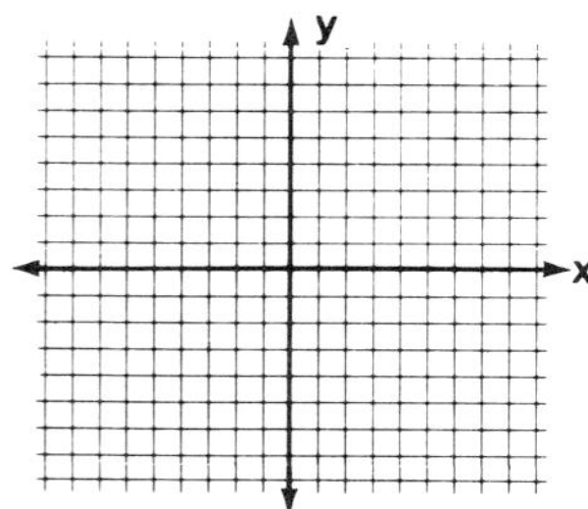

Warm-up Answer
Join (0, 8) to (−6, 0). Draw a right triangle. Find the lengths of the perpendicular sides. These lengths are 8 units and 6 units. Use the Pythagorean Theorem to get $6^2 + 8^2 = c^2$. So $100 = c^2$, and $c = 10$.

On the Springboard

Items 6–9 refer to the following grid.

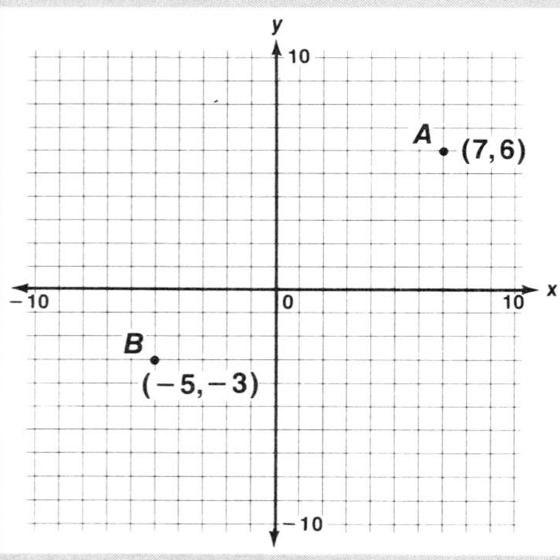

5. What is the perpendicular distance from point A to the y-axis?

   (1) 6 (2) 7 (3) 8

   ① ② ③

6. What is the perpendicular distance from point B to the x-axis?

   (1) −3 (2) −2 (3) 3

   ① ② ③

7. Use the Pythagorean Theorem or the distance formula to find the diagonal distance from point A to point B.

   (1) 12 (2) 15 (3) 18

   ① ② ③

Check your Springboard answers on pages 287–288. If you got them all right, go on to "The Real Thing." If you got any wrong, particularly number 7, you should probably review this section and try again the ones you missed. When you feel confident that you're ready, go on.

"The Real Thing"

Questions 7–10 refer to the following grid.

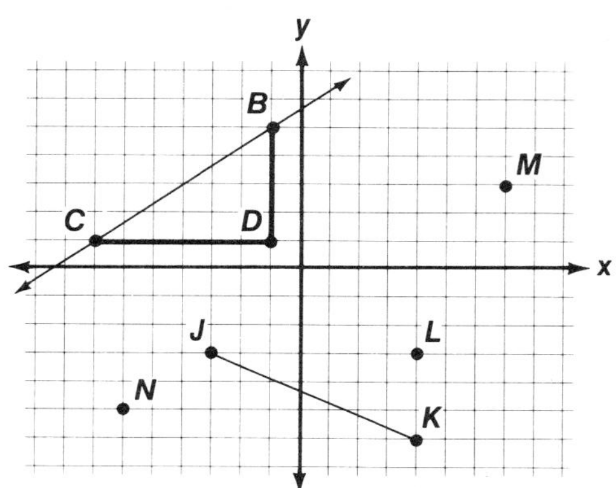

7. How many units long is $\overline{CD}$?

   (1) 1 (2) 4 (3) 5 (4) 6 (5) 7

   ① ② ③ ④ ⑤

8. How many units long is segment BD?

   (1) 6 (2) 5 (3) 4 (4) 3 (5) 2

   ① ② ③ ④ ⑤

9. Approximately how many units long is segment BC?

   (1) 6.8 (2) 7.2 (3) 8.4
   (4) 20.0 (5) 52.0

   ① ② ③ ④ ⑤

10. Approximately how many units long is segment J?

(1) 7.6 (2) 8.1 (3) 8.3
(4) 40.0 (5) 58.0

(1) (2) (3) (4) (5)

Check your answers on page 296.

Finding Midpoints

If you have any two numbers, you can calculate their average by adding them and dividing by 2. The average of the two numbers is exactly halfway between them. For example, take 4 and 12. Their average is $(4 + 12) \div 2$, or 8, and 8 is exactly halfway between 4 and 12. Or try -6 and 8. Their average is $(-6 + 8) \div 2$. Since $-6 + 8 = 2$, and $2 \div 2 = 1$, the average of -6 and 8 is 1. Is 1 exactly halfway between -6 and 8? Yes it is, and you can easily convince yourself by drawing a number line. The distance from -6 to 1 is 7 units, and the distance from 1 to 8 is also 7 units.

You can use this fact about the average of two numbers to find **midpoints** of line segments on a coordinate grid. If you know the ordered pairs for the points at the ends of a line segment, the midpoint (the point halfway between the endpoints) can be found by finding the average of the x-coordinates and the average of the y-coordinates of the end points.

Coming to Terms

midpoint the point on a line segment that is exactly halfway between the endpoints of the segment.

Here's an Example

To find the midpoint of a line segment on a coordinate grid, use the ordered pairs for the endpoints. Average their x-coordinates, then their y-coordinates.

■ Find the midpoint of the line segment joining (2, 8) to (6, 10).

To solve, find the average of the x-coordinates and the average of the y-coordinates.

Step 1 Add the x-coordinates (the first number in each pair) and divide the answer by 2.

$(2 + 6) \div 2 = 8 \div 2 = 4$

The x-coordinate of the midpoint is 4.

Step 2 Do the same with the y-coordinates.

$(8 + 10) \div 2 = 18 \div 2 = 9$

The y-coordinate of the midpoint is 9.

Step 3 Write down the averages in an ordered pair and you get (4, 9).

The midpoint is the point (4, 9).

You can plot the points on a grid to see that this is really the midpoint. It makes no difference if the line segment is parallel to one of the axes or at a diagonal. The rule works for all circumstances. Let's quickly find the midpoint of the line segment joining $(-4, -5)$ to (6, 17).

x-coordinate: $-4 + 6 = 2$, and $2 \div 2 = 1$
y-coordinate: $-5 + 17 = 12$, and $12 \div 2 = 6$

The midpoint is (1, 6). Make a sketch if you want to check.

Try It Yourself

Here you get a picture, but you do the work exactly the same way as you did before.

■ Find the ordered pair for the midpoint of line segment AB.

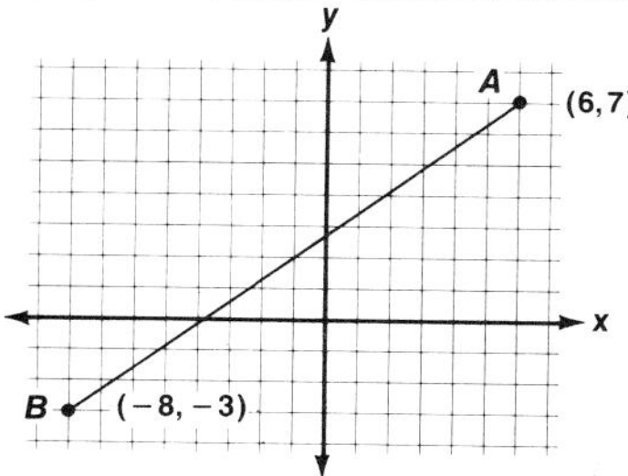

If you added x-coordinates and divided by 2, you should have gotten $6 + (-8) = -2$, and $-2 \div 2 = -1$ for the x-coordinate. For the y-coordinate did you get $7 + (-3) = 4$ and $4 \div 2 = 2$? The midpoint is $(-1, 2)$.

So far all the answers have been whole numbers, but you can end up with points like (2½, 9) or (3¼, 1½) and they are perfectly good points.

Warm-up

What are the midpoints of the line segments joining these pairs of points?

1. (2, 5) and (8, 11) Midpoint: (_____, _____)

2. (3, 5) and (2, 7) Midpoint: (_____, _____)

3. (−3, 5) and (−2, −7) Midpoint: (_____, _____)

4. (½, 8) and (3½, −8) Midpoint: (_____, _____)

Warm-up Answers
1. (5, 8) **2.** (2½, 6) **3.** (−2½, −1) **4.** (2,0)

On the Springboard

Questions 8–10 refer to the parallelogram below.

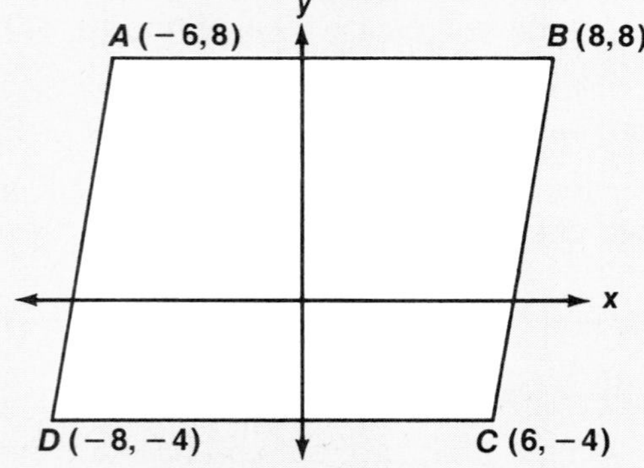

What are the coordinates of the points named?

8. The midpoint of the diagonal AC.

(1) (0, 2) (2) (0, 3) (3) 6, 4)

① ② ③

9. The midpoint of diagonal BD.

(1) (−1, −2) (2) (0, 2) (3) (0, 3)

① ② ③

10. The midpoint of AD.

(1) (−9, −4) (2) (−7, 2)
(3) (7, −2)

① ② ③

Check your Springboard answers on page 288. If you got all these right, go on to "The Real Thing." If you missed any, check your work.

The Real Thing

11. The endpoints of a line are (−2, −3) and (6, 11). What is the midpoint of this line?

(1) (2, 4) (2) (3, 9) (3) (4, 7)
(4) (4, 8) (5) (8, 14)

① ② ③ ④ ⑤

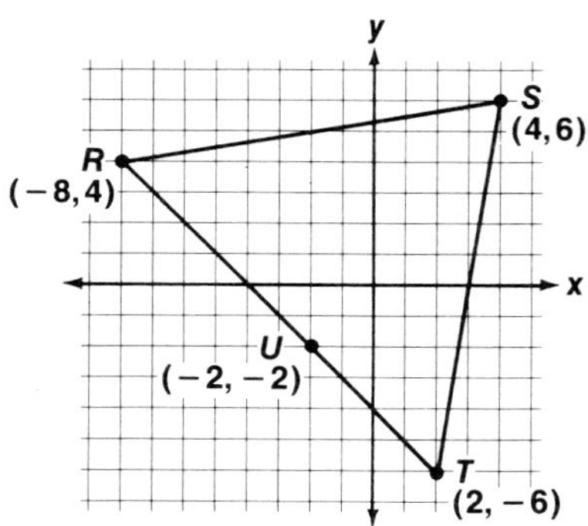

12. What is the midpoint of segment RU on the grid above?

(1) (−1, 1) (2) (−5, 1)
(3) (−6, −4) (4) (−6, 6)
(5) (−10, 2)

① ② ③ ④ ⑤

Questions 13 and 14 refer to the following grid.

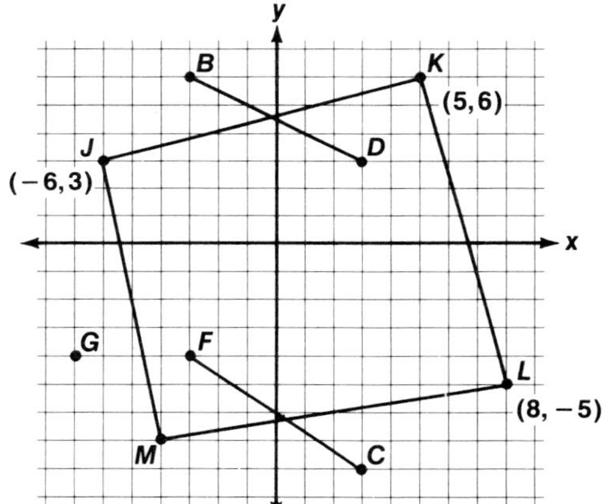

13. The perpendicular distance from point J to the x-axis is

(1) 3 (2) 4 (3) 5 (4) 6 (5) 7

① ② ③ ④ ⑤

14. The perpendicular distance between which point and the y-axis is 6 units?

(1) G (2) J (3) K (4) L (5) M

① ② ③ ④ ⑤

Check your answers and record all your scores on page 296.

Answers: On the Springboard

33 Angles

(page 226)

1. (2) 145°
Angle x and the 35° angle have one side in common. The other two sides point in opposite directions and form a straight angle. The straight angle has a measure of 180°.

$35° + m\angle x = 180°$

Therefore, $m\angle x = 180° - 35° = 145°$

2. (1) 20°
The measures of the angles marked x and 70° add up to the measure of the large angle marked with the right angle symbol (⌉). Since all right angles have a measure of 90°, $m\angle x + 70° = 90°$. To find $m\angle x$, subtract.

$m\angle x = 90° - 70° = 20°$

3. (1) 45°
The angles marked x and 45° are vertical angles. Since vertical angles have equal measures, $m\angle x = 45°$.

4. (3) 130°
Two of the sides of the angles marked x and 50° form a straight angle. Since the measure of a straight angle is 180°, $m\angle x + 50° = 180°$. Therefore, $m\angle x = 180° - 50° = 130°$.

5. (3) 60°
The measures of the angles marked 20°, 10°, and x add up to 90°, the measure of the right angle (marked with the ⌉ symbol).

$20° + 10° + m\angle x = 90°$
$30° + m\angle x = 90°$

To find $m\angle x$, subtract.

$m\angle x = 90° - 30° = 60°$

(page 229)

6. (2) Each pair is always equal in measure.

Option (1) is not correct. Except for vertical angles, the pairs of angles listed cannot be formed without three lines.

Option (3) is not correct. When two paralled lines are cut by a transversal, 4 pairs of vertical angles are formed, not 2 pairs.

7. (2) 30°
$\angle g$ and $\angle f$ are vertical angles. Since vertical angles have the same measure, $m\angle f = 30°$.

Angles f and b are corresponding angles formed by the transversal that cuts lines x and y. Since lines x and y are parallel, these corresponding angles ($\angle f$ and $\angle b$) have the same measure; therefore $m\angle b = 30°$.

8. (3) 90°
$\angle 2$ is marked with the right-angle symbol, so its measure is 90°. But $\angle 3$ and $\angle 2$ are vertical angles and therefore have equal measures. This means that the measure of $\angle 3$ must also be 90°.

34 Quadrilaterals

(page 232)

1. (1) rectangle
The angles are all right angles, so opposite sides are parallel and equal in length. The figure is, therefore, a rectangle.

Certainly the figure could not be a square, because all sides of a square must be the same length.

2. (1) parallelogram
The diagram makes it appear that opposite sides are equal as well as parallel. Certainly *ABCD* could not be either a rectangle or a square, since $\angle B$ has a measure of 80°. In rectangles and squares the angles all measure 90°.

3. (1) 2 in
All sides of a square are equal, so side *XY* must be just as long as side *ZY,* which the diagram shows to be 2 inches long.

4. (3) 165°
The sum of the angles of a quadrilateral is always 360°.

$\underbrace{110° + 30° + 55°} + m\angle R = 360°$

$195° \qquad + m\angle R = 360°$

To find what $m\angle R$ is, you can subtract: $m\angle R = 360° - 195° = 165°$.

(page 234)

5. (3) 2(2) + 2(9)
The question gives you the information that the figure is a parallelogram. Therefore the opposite sides are equal. This means that, when you add the sides to find the perimeter, you will have 2 sides of length 2 feet and 2 sides of length 9 feet.

The sides of length 2 feet will have a total length of 2(2) feet, and the sides of length 9 feet will add a total of 2(9) feet. The perimeter will be the sum of 2(2) and 2(9). $P = 2(2) + 2(9)$

6. (2) 16
The formula for the perimeter of a square is $P = 4s$. If you put 64 in place of P, you get $64 = 4s$. So ask yourself: 64 equals 4 times what number? To find out, divide 64 by 4. $64 \div 4 = 16$ The sides of the square are 16 feet long.

7. (3) 16.00
To find the perimeter of the trapezoid, add the lengths of all the sides.

$P = 3.17 + 5.50 + 3.33 + 4.00 = 16.00$ (in)

8. (2) 26
Use the formula for the perimeter of a rectangle. Use 10 in place of ℓ (length) and 3 in places of w (width). $P = 2\ell + 2w$.

$P = 2(10) + 2(3)$
$P = 20 + 6$
$P = 26$ (inches)

(page 237)

9. (2) B
Find the area of each space by multiplying length by width.

A = 50 × 25 = 1,250 (sq ft)
B = 40 × 30 = 1,200 (sq ft)
C = 35 × 35 = 1,225 (sq ft)

The area that is closest to 1,100 square feet is 1,200 square feet. So space B is the best choice.

10. (2) 1,575
Find the area of the entire rectangular plot (including the pool).

$A = \ell w$
$A = 60 \times 30$
$A = 1{,}800$ (sq ft)

Now find the area of the square pool

$A = s^2$
$A = 15^2$
$A = 225$ (sq ft)

To find how many square feet will be covered with gravel, subtract the area of the pool (225) from the area of the entire plot (1,800).

$1{,}800 - 225 = 1{,}575$ (sq ft)

11. (3) 13,500
The height of the building is extra information—you really don't need it. To answer the question, find the area of the rectangular roof.

$A = \ell w$
$A = 50 \times 30$
$A = 1{,}500$ (sq yd)

Now multiply by 9 because you were asked for the number of *square feet,* and the area you just found was in square yards. (There are 9 square feet in 1 square yard.)

$1{,}500 \times 9 = 13{,}500$ (sq ft)

35 Triangles

(page 241)

1. (2) 45°
One approach is to try each choice. You know that all three angles should add up to 180°. Would option (1) do? If $\angle B$ were 30°, then $\angle A$ would also be 30°. In this case the sum of all three angles would be $90° + 30° + 30° = 150°$. So (1) cannot be correct.

What about (2)? If $\angle B$ is 45°, then $\angle C$ is also 45°. Would all the angles add up to 180°? Yes, since $90° + 45° + 45° = 180°$. So the answer is option (2).

Another approach uses reasoning. All the angles must add up to 180°.

$m\angle A + m\angle B + m\angle C = 180°$

But you could put $m\angle A$ in place of $m\angle B$ (the angles have equal measures) and 90° in place of $m\angle C$. You get

$$\underbrace{m\angle A + m\angle A}_{} + 90° = 180°$$
$$2\,(m\angle A) \quad + 90° = 180°$$

2 ($m\angle A$) must equal $180° - 90°$ or 90°, therefore $m\angle A$ is ½ of 90° or 45°. So $m\angle B$ is 45°.

2. (2) 30°, 110°, 40°
Option (2) is the correct answer because it is the only option for which all the angles add up to 180°. The angles of a triangle *always* add up to 180°.

Option (1): $30° + 90° + 61° = 181°$
Option (2): $30° + 110° + 40° = 180°$
Option (3): $55° + 45° + 90° = 190°$

3. (2) 52°
The angles must add up to 180°.

$$\underbrace{42° + 86°}_{} + \text{(third angle)} = 180°$$
$$128° \quad + \text{(third angle)} = 180°$$

To find what you'd add to 128° to get 180°, subtract.

third angle $= 180° - 128°$
$= 52°$

(page 243)

4. (3) C
In each triangle there are two sides 5 inches long. The larger the angle formed by these two sides, the longer the side opposite (the third side) will be. Therefore the side opposite the 55° angle in triangle *C* is longer than each of the bottom sides of the other two triangles. From this you can conclude that when you add the lengths of the sides of the triangles to get their perimeters, the greatest perimeter will be for triangle *C.*

5. (3) C
The perimeter for triangle *C* is 17 inches, since $3 + 8 + 6 = 17$.

You can figure out the perimeter for triangle *B*. The angles opposite the 5-inch sides are equal. (In a triangle, angles opposite equal sides are equal.) From this you can conclude that each of these angles is 60°. (This is necessary if all the angles are to add up to 180°.) Therefore triangle *B* is equilateral, which tells you the bottom side also measures 5 inches. Now you know the measures of all three sides and can calculate the perimeter.

$5 + 5 + 5 = 15$ (in)

You can't be quite sure what the length of the bottom side of triangle *A* is, but it's less than 5 inches. (The angle formed by the 5-inch sides in triangle *A* is smaller than the angle formed by the 5-inch sides in triangle *B*.) Since in triangle *A* the bottom side is less than 5 inches, its perimeter will be less than that of triangle *B*.

From all of this you can conclude that triangle *C* (with a perimeter of 17 inches) has the greatest perimeter of all three triangles.

(page 245)

6. (2)
Use the formula $A = ½\, bh.$ Put 12 in place of *b* and 12 in place of *h*.

$A = \frac{1}{2}bh$

$A = \frac{1}{2} \times 12 \times 12$

$A = \frac{1}{2} \times 144$

$A = 72$

7. (2) 4
Use the formula $A = ½\, bh$. You know the area and height, so put that information in the formula. Put 32 in place of *A* and 16 in place of *h.*

$A = \frac{1}{2}bh$

$32 = \frac{1}{2} \times b \times 16$

$32 = \frac{1}{\cancel{2}} \times \frac{b}{1} \times \frac{\overset{8}{\cancel{16}}}{1}$

$32 = b \times 8$

To find b (the number you'd multiply by 8 to get 32), divide: $b = 32 \div 8 = 4$.

36 Congruent and Similar Triangles

(page 248)

1. (3) 54°
You know that the measure of $\angle A$ is 36°, you also know that $\angle C$ is 90°, since it's marked with the right-angle symbol in the diagram. The angles of triangle ABC must add up to 180°, so $m\angle B = 54°$. $\angle B$ and $\angle E$ are corresponding parts of the two congruent triangles. This means that $m\angle E$ must also be 54°.

2. (2) 100°
The sum of the angles of triangle ABD is equal to 180°. From this and the information in the diagram, you can conclude that $m\angle A = 100°$. Now use the fact that opposite angles of a parallelogram are equal. $\angle A$ and $\angle C$ are opposite angles of parallelogram $ABCD$. So $m\angle C$ is also equal to 100°.

(page 250)

3. (3) DEF
Look carefully at the diagram. $\angle ABC$ is the largest angle of triangle ABC. Since the triangles are similar, it corresponds to the largest angle of the other triangle, and that angle is angle DEF.

4. (2) AC
DF is the longest side of triangle DEF. So it must correspond to the longest side of the other triangle. The longest side of triangle ABC is side AC, so AC is the correct answer.

5. (3) 12 ft
Use the fact that the triangles are similar. Corresponding sides of similar triangles are proportional. AB corresponds to DE and BC corresponds to EF. Write a proportion involving these sides.

$$AB : DE = BC : EF$$

Now put in the information you are given about AB, DE, and BC. The proportion becomes

$$4 : 8 = 6 : ? \text{ or } \frac{4}{8} = \frac{6}{?}$$

Here a question mark has been used for EF, since that is the length you are trying to find. Solve $4/8 = 6/?$ by using a grid.

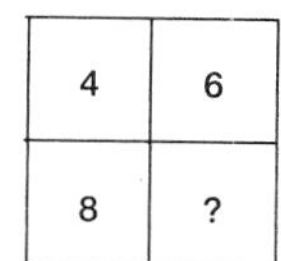

$$6 \times 8 = 48$$
$$48 \div 4 = 12$$

The length of EF is 12 feet.

37 Pythagorean Theorem

(page 254)

1. (1) 7
Use the Pythagorean Theorem. Use 24 for the value of a (one of the legs) and 25 for c. The value of b is the length of the other leg, which is what you're trying to find.

$$a^2 + b^2 = c^2$$
$$24^2 + b^2 = 25^2$$
$$576 + b^2 = 625$$

Subtract to find b^2.

$$b^2 = 625 - 576 = 49$$

You now know that $b^2 = 49$. Ask yourself what number you could square to get 49. The answer is 7 ($7^2 = 7 \times 7 = 49$). So the length, b, of the other leg is 7 inches.

2. (2) 10
You'll probably find it helpful to sketch a diagram to picture the situation.

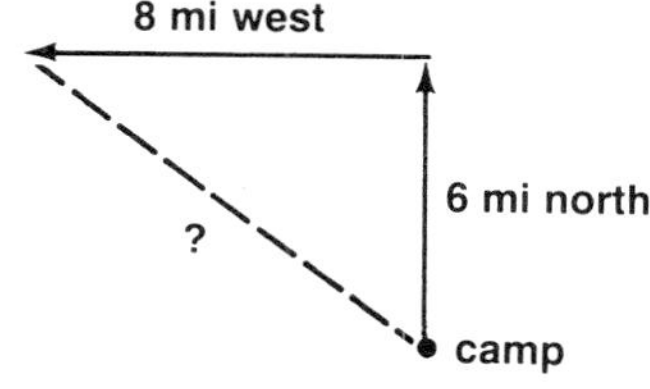

To solve the problem, you need to find the length of the dotted line segment. This line segment is the hypotenuse of a right triangle whose legs are 6 miles long and 8 miles long. Use the Pythagorean Theorem: $a^2 + b^2 = c^2$.

Put 6 in place of a and 8 in place of b. You get

$$6^2 + 8^2 = c^2$$
$$36 + 64 = c^2$$
$$100 = c^2$$

To find c, ask what number you could square to get 100. The answer is 10 ($10^2 = 10 \times 10 = 100$). So the man was 10 miles from his camp.

38 Circles

(page 257)

1. (3) 47.1
Use the formula $C = \pi d$ to calculate how far it is around the rim of the mirror. Use 3.14 for π and 15 for d.

$$C = \pi d$$
$$C = 3.14 \times 15$$
$$C = 47.10, \text{ or } 47.1$$

The distance around the rim is 47.1 feet.

2. (1) 75.36
Double the radius to find the diameter. Since $2 \times 12 = 24$, the diameter of the garden is 24 feet. Now you can use 24 for d in the formula $C = \pi d$ to calculate the circumference of (distance around) the garden.

$$C = \pi d$$
$$C = 3.14 \times 24$$
$$C = 75.36$$

The distance around the garden is 75.36 feet.

3. (2) 7
When the coffee filter is folded in half, the crease line of the fold will be a diameter of the filter. You can use the formula $C = \pi d$ to find the diameter. Use 22 for C and $22/7$ for π.

$$C = \pi d$$
$$22 = \frac{22}{7} \times d$$

To find d, divide 22 by $22/7$.

$$d = 22 \div \frac{22}{7}$$
$$= \frac{22}{1} \div \frac{22}{7}$$
$$= \frac{\overset{1}{\cancel{22}}}{1} \times \frac{7}{\underset{1}{\cancel{22}}} = \frac{7}{1} = 7$$

The fold line is 7 inches long.

(page 259)

4. (2) 28.26
You can calculate the area of the table top by using the formula $A = \pi r^2$. First, however, divide the diameter by 2 to find the radius, *r*, of the table top.

$r = 6 \div 2 = 3$

Use $A = \pi r^2$ to calculate the area. Put 3 in place of *r* and 3.14 in place of π.

$A = \pi r^2$
$A = 3.14 \times 3^2$
$A = 3.14 \times 3 \times 3$
$A = 3.14 \times 9$
$A = 28.26$

The area of the table top is 28.26 square feet.

5. (3) 157
First use $A = \pi r^2$ to find the area of the pizza. Use 3.14 for π and 10 for *r*.

$A = \pi r^2$
$A = 3.14 \times 10^2$
$A = 3.14 \times 10 \times 10$
$A = 3.14 \times 100$
$A = 314$ (sq in)

Alan and Sue are going to split the pizza evenly, so divide by 2 to find how many square inches each will get.

$314 \div 2 = 157$

39 Composite Figures

(pages 261–262)

1. (2) 52 ft
The figure can be thought of as 3 sides of a rectangle plus half of a circle. To find the perimeter, first find the circumference and divide by 2.

$C = \pi d$

$C = \frac{22}{7} \times \frac{7}{1}$

$C = \frac{22}{1}$ or 22 (ft)

Half of this is 11 feet, which is the curved part on the left. Add this and the lengths of the 3 straight sides to find the perimeter.

$11 + 17 + 7 + 17 = 52$ (ft)

If your answer was 63 feet, you probably forgot to use just half the circumference.

2. (2) 32 m
Here is the figure with a horizontal dotted line drawn. The lengths that have been added to the figure are easy to figure out by remembering that opposite sides of a rectangle are equal.

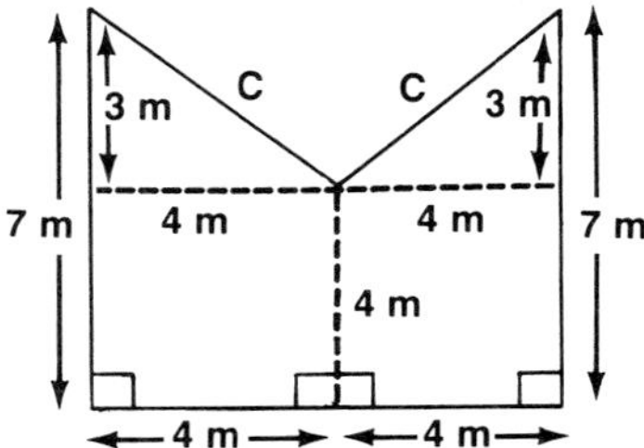

The two triangles at the top are congruent right triangles. You know that in each the legs have lengths 3 and 4. You can use the Pythagorean Theorem to find the lengths of their hypotenuses.

$3^2 + 4^2 = C^2$
$9 + 16 = C^2$
$25 = C^2$

Since $5^2 = 25$, $C = 5$.

You can now add to find the perimeter of the figure.

$5 + 5 + 7 + 4 +$
$4 + 7 = 32$

3. (1) 44 cm
Look at this copy of the figure you were given. Some letters have been added for sides that you were not told the lengths of.

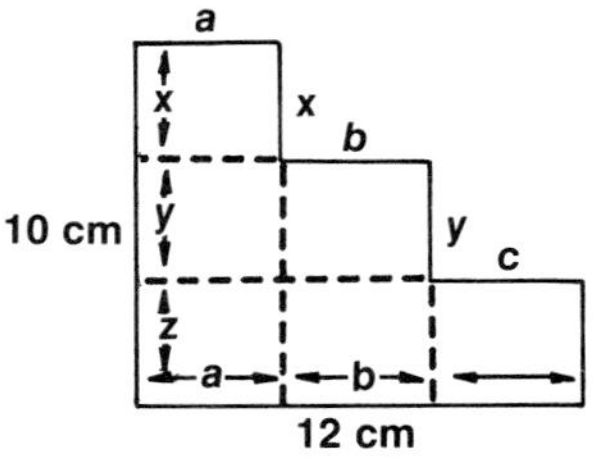

You do not know the lengths *a*, *b*, and *c*, but you *do* know that their *sum* is equal to 12 cm. Similarly, you do not know *x*, *y*, and *z*, but their *sum* is equal to 10 cm. So the perimeter of the figure is

$10 + 12 +$
$\underbrace{(a + b + c)}_{12} + \underbrace{(x + y + z)}_{10} =$

$10 + 12 + 12 + 10 = 44$

(page 264)

4. (2) 525
Here is a copy of the figure given in the question with some dotted lines added.

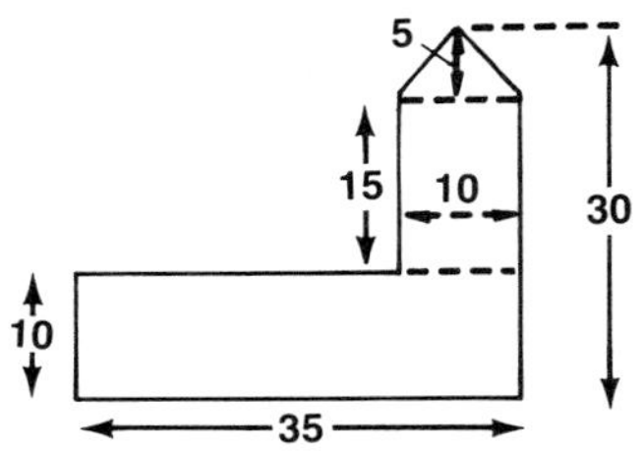

As you can see, you can think of the original figure as made up of two rectangles and a triangle.

To find the area, find the areas of the parts and add. The area of the bottom rectangle is easy. The rectangle is 35 feet long and 10 feet wide.

$A = \ell w$
$A = 35 \times 10$
$A = 350$ (sq ft)

The top rectangle is 15 feet long and 10 feet wide. Use the formula to calculate its area.

$A = \ell w$
$A = 15 \times 10$
$A = 150$ (sq ft)

The triangle has a base of 10 feet and its height is 5 feet (30 minus the sum of 10 and 15).

$A = \frac{1}{2} bh$

$A = \frac{1}{2} \times 10 \times 5$
$A = 25$ (sq ft)

Add to find the area of the whole figure.

$350 + 150 + 25 = 525$ (sq ft)

5. (3) 228.5
The figure can be thought of as made up of a rectangle and two half-circles. If you put the two half-circles together, you'd have a whole circle of diameter 10, that is, of radius 5. Calculate its area by using $A = \pi r^2$.

$A = \pi r^2$
$A = 3.14 \times 5^2$
$A = 3.14 \times 25$
$A = 78.5$ (sq ft)

The rectangular part of the figure is 15 feet long and 10 feet wide. Calculate its area.

$A = \ell w$
$A = 15 \times 10$
$A = 150$ (sq ft)

Add to find the area of the entire figure.

$78.5 + 150 = 228.5$ (sq ft)

40 Solid Geometry

(page 267)
1. (3) 4,608
To find the volume of the van, use the formula $V = \ell wh$. Replace ℓ with 48, w with 8, and h with 12.

$V = \ell wh$
$= 48 \times 8 \times 12$
$= 4{,}608$ (cu ft)

2. (2) 32,000
Use the formula $V = \ell wh$. Replace ℓ with 100, w with 40, and h with 8.

$V = \ell wh$
$= 100 \times 40 \times 8$
$= 32{,}000$ (cu ft)

(page 268)
3. (3) 0.008
Use the formula $V = s^3$. Replace s with 0.2 and then compute.

$V = s^3$
$= (0.2)^3$
$= 0.2 \times 0.2 \times 0.2$
$= 0.008$ (cu m)

4. (3) 729
Use the formula $V = s^3$. Replace s with 9 and compute.

$V = s^3$
$= 9^3$
$= 9 \times 9 \times 9$
$= 729$

(page 270)
5. (3) 14.13
First divide the diameter (3) by 2 to find the radius, r.

$r = 3 \div 2 = 1.5$

Next use the formula $V = \pi r^2h$. Use 1.5 for r, 2 for h, and 3.14 for π.

$V = \pi r^2h$
$= 3.14 \times (1.5)^2 \times 2$
$= 3.14 \times 1.5 \times 1.5 \times 2$
$= 14.13$ (cu in)

6. (1) 87.92
Use the formula $V = \pi r^2h$. Use 2 for r, 7 for h, and 3.14 for π.

$V = \pi r^2h$
$= 3.14 \times 2^2 \times 7$
$= 3.14 \times 4 \times 7$
$= 87.92$ (cu cm)

41 Coordinate Geometry

(page 273)
1. (3) 2
The x-coordinate of an ordered pair is the first number in the ordered pair, so the x-coordinate of $(2, -4)$ is 2.

2. (2) -4
The y-coordinate of an ordered pair is the second number in the ordered pair, so the y-coordinate of $(2, -4)$ is -4.

3. (1) $(-4, 0)$
Start at the origin and go straight left to the point where BC crosses the x-axis. You have traveled 4 units. So the x-coordinate is -4. The point is *on* the x-axis, so you go 0 units (neither up nor down) parallel to the y-axis. The y-coordinate is 0. Therefore the ordered pair is $(-4, 0)$.

4. (3) C
Locate the point (0, 3). It is on the y-axis, exactly 3 units above the origin. You can see once you've located this point that it lies exactly 4 units to the right of point C, so the answer is point C.

42 Distance on a Coordinate Grid

(page 276)
1. (3) 5
Both points B and C are 2 units to the left of the y-axis, so $\overline{BC}$ is vertical. Point B is 2 units *below* the x-axis and point C is 3 units *above* the x-axis. Therefore the distance between B and C is 5 units.

2. (3) 3
DF is parallel to the y-axis, since both points are 1 unit to the right of the y-axis. Point D is 1 unit *above* the x-axis, and point F is 2 units *below* the x-axis. Therefore the distance between D and F is 3 units.

(pages 277–278)
3. (3) 9
Study the diagram. Line XW is parallel to the x-axis, and both Y and W are 4 units to the right of the y-axis. So segment YW is perpendicular to line XW. Since W is 4 units *above* the x-axis and Y is 5 units *below* the x-axis, the distance from Y to W (and therefore to XW) is 9 units.

4. (2) 2
If you were to draw a perpendicular from point N to line OP, it would cross OP at $(1, -1)$. Since $(1, -1)$ is 1 unit below the x-axis and N is 3 units below the x-axis, the distance from N to line OP is 2 units.

(page 280)
5. (2) 7
The perpendicular distance from point A to the y-axis is 7, since the first coordinate of any point indicates how far to the left or right of the y-axis the point is.

6. (3) 3
The perpendicular distance from point B to the x-axis is 3, since the second coordinate of $(-5, -3)$ indicates that B is 3 units directly *below* the x-axis.

7. (2) 15
To use the Pythagorean Theorem, draw a horizontal line through B and a vertical line through A. Call the point where these lines intersect point C.

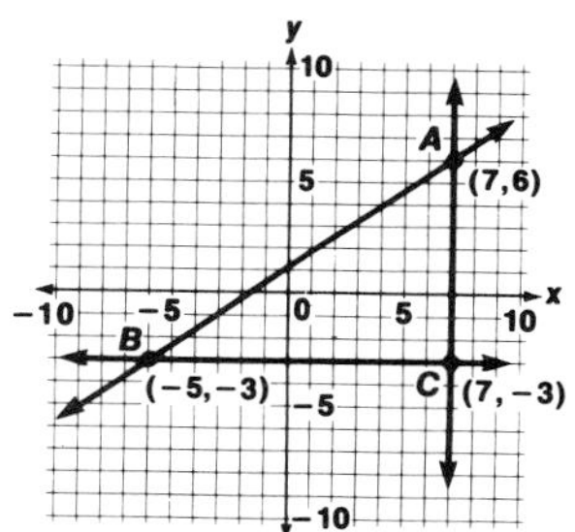

Count to find the lengths of segments *BC and AC*; you find that *BC* is 12 units long and *AC* is 9 units long. Segments *BC* and *AC* are the legs of the right triangle *ABC.* The length of the hypotenuse *AB* is the distance from *A* to *B.* Use the Pythagorean Theorem.

$$\begin{aligned} a^2 + b^2 &= c^2 \\ 12^2 + 9^2 &= \overline{AB}^2 \\ 144 + 81 &= \overline{AB}^2 \\ 225 &= \overline{AB}^2 \end{aligned}$$

Since $15^2 = 225$, *AB* equals 15 units.

You can also find the distance from *A* to *B* by using the distance formula. Let *A* be the point (x_1, y_1), so that x_1 is 7 and y_1 is 6. Let *B* be (x_2, y_2), so that x_2 is -5 and y_2 is -3.

$$\begin{aligned} d &= \sqrt{(x_2 - x_1)^2 + (y_2 - y_1)^2} \\ &= \sqrt{(-5 - 7)^2 + (-3 - 6)^2} \\ &= \sqrt{(-12)^2 + (-9)^2} \\ &= \sqrt{144 + 81} \\ &= \sqrt{225} \\ &= 15 \end{aligned}$$

(page 282)

8. (1) (0, 2)
Average the *x*-coordinates of *A* and *C*: $-6 + 6 = 0$ and $0 \div 2 = 0$. The *x*-coordinate of the midpoint is 0.

Average the *y*-coordinates of *A* and *C*: $8 + (-4) = 4$ and $4 \div 2 = 2$. The *y*-coordinate of the midpoint is 2.

The midpoint is the point for (0, 2).

9. (2) (0,2)
Average the *x*-coordinates of *B* and *D*: $-8 + 8 = 0$ and $0 \div 2 = 0$. The *x*-coordinate of the midpoint is 0.

Average the *y*-coordinates of *B* and *D*: $-4 + 8 = 4$ and $4 \div 2 = 2$. The *y*-coordinate of the midpoint is 2.

The midpoint has coordinates (0, 2).

10. (2) (−7, 2)
Average the *x*-coordinates of *A* and *D*: $-6 + -8 = -14$ and $-14 \div 2 = -7$. The *x*-coordinate of the midpoint is −7.

Average the *y*-coordinates of *A* and *D*: $8 + (-4) = 4$ and $4 \div 2 = 2$. The *y*-coordinate of the midpoint is 2.

The coordinates of the midpoint are (− 7, 2).

Answers: "The Real Thing"

33. Angles

(pages 226–227)

1. (3) ∠*FDE*
Point *D* is the vertex of the angle. The vertex is always named between the points on the sides.

2. (4) supplementary
The sum of angles *G* and *H* is 180 (60 + 120 = 180) degrees. Supplementary angles are a pair of angles whose measures have a sum of 180.

3. (4) ∠*d* and ∠*a*
Look for pairs of "opposite" angles formed by the two straight lines. There are two such pairs of angles, but the pair made up of ∠*a* and ∠*d* is the only one you can name with the labels in the diagram. Choice number 4 is the only one that contains both of these.

4. (4) ∠*c* and ∠*d*
The information that ∠*RXT* measures 90° tells you that ∠*d* is a complement of ∠*a*. You can see from the diagram that m∠*a* + m∠*d* = m∠*RXT* = 90°.

Study the diagram. The right angle symbol tells you that m∠*PXQ* = 90°. Since ∠*PXS* is a straight angle, m∠*PXS* = 180°. Also

$$\underbrace{m\angle PXQ}_{90^\circ} + m\angle QXS = \underbrace{\angle PXS}_{180^\circ}$$

So m∠*QXS* must equal 90°. Since m∠*c* + m∠*a* = m∠*QXS* = 90°, you can conclude that ∠*c* is a complement of ∠*a*.

(pages 229–230)

5. (4) 140°
Angles *b* and *c* have the same measure because they are vertical angles. Since the measure of angle *c* is 40°, the measure of angle *b* is 40°.

Angles *a* and *b* are supplementary, so the sum of their measures is 180°. The measure of angle *b* is 40°. The m∠*a* is 180° − 40°, or 140°.

6. (1) alternate exterior
Lines *RS* and *TU* are parallel. Angles *b* and *g* are outside the parallel lines and are on opposite sides of the transversal.

7. (2) 55°
Angles *e* and *g* are vertical angles, which means that their measures are the same. Since the measure of angle *e* is 55°, the measure of angle *g* is also 55°.

8. (5) 360°
From number 5 you know that the measure of angle *e* and of angle *g* is 55°.

Angles *e* and *f* are supplementary angles, so the measure of angle *f* is 180° − 55°, or 125°.

Angles f and h are vertical angles, so the measure of angle h is also 125°.

Add to find the sum of the measures of angles e, f, g, and h.

$55° + 125° + 55° + 125° = 360°$

9. (3) 80
The measure of $\angle YTS$ is 130° because $\angle YTS$ and $\angle XST$ are alternate interior angles. The measure of $\angle STR$ is 50°, because vertical angles have the same measure. The measure of $\angle YTR$ equals $m\angle YTS - m\angle STR$ or $130° - 50°$ or 80°. But $\angle YTR$ and $\angle SRT$ are alternate interior angles and so have the same measure, 80°.

KEEPING TRACK
Perfect score = 9

Your score ☐

34 Quadrilaterals

(pages 234–235)

1. (5) $73\frac{1}{6}$
Use the formula $P = \ell + \ell + w + w$ to find the perimeter of the room.

$$\begin{aligned} 24\tfrac{1}{4} &= 24\tfrac{3}{12} \\ 24\tfrac{1}{4} &= 24\tfrac{3}{12} \\ 12\tfrac{1}{3} &= 12\tfrac{4}{12} \\ +12\tfrac{1}{3} &= 12\tfrac{4}{12} \\ \hline & 72\tfrac{14}{12} = 73\tfrac{2}{12} = 73\tfrac{1}{6} \end{aligned}$$

2. (4) $36\frac{5}{6}$
To find the perimeter, add the lengths of the four sides.

$$\begin{aligned} 10\tfrac{1}{4} &= 10\tfrac{3}{12} \\ 6 &= 6 \\ 15\tfrac{3}{4} &= 15\tfrac{9}{12} \\ +\ 4\tfrac{5}{6} &= 4\tfrac{10}{12} \\ \hline & 35\tfrac{22}{12} = 36\tfrac{10}{12} = 36\tfrac{5}{6} \end{aligned}$$

3. (4) 756.5
If you multiple the length of a side by 4, you get the perimeter of the square (3,026). Divide the perimeter (3,026) by 4 to find the length of each side.

$3{,}026 \div 4 = 756.5$

4. (2) 36
First use $P = 2\ell + 2w$ to find the perimeter.

$$\begin{aligned} P &= (2 \times 12) + (2 \times 9) \\ &= 24 + 18 \\ &= 42 \end{aligned}$$

You do not put baseboard where there are doorways, so subtract.

42	total distance around
− 6	doorways
36	

5. (3) \$25.92
First find the perimeter.

$$\begin{aligned} P &= (2 \times 24) + (2 \times 12) \\ &= 48 + 24 \\ &= 72 \end{aligned}$$

There is 72 feet of fencing, which costs \$0.36 per running foot. Multiply \$0.36 by 72.

$\$0.36 \times 72 = \$25.92$

(pages 237–238)

6. (3) 15.0
A dollar bill is a rectangle, so use $A = \ell w$ to find the area.

$$\begin{aligned} A &= 6 \times 2\tfrac{1}{2} \\ &= \frac{\overset{3}{\cancel{6}}}{1} \times \frac{5}{\underset{1}{\cancel{2}}} = 15 = 15.0 \end{aligned}$$

7. (1) $4\frac{1}{2}$
First change 15 feet to yards. Since 3 feet = 1 yard, divide.

$15 \div 3 = 5$ (yards)

Use the formula for the area of a parallelogram and put in the values.

$$\begin{aligned} A &= bh \\ 22\tfrac{1}{2} &= b \times 5 \end{aligned}$$

Ask yourself, "What number must I multiply 5 by to get 22½?" To find this number divide 22½ by 5.

$$\begin{aligned} b &= 22\tfrac{1}{2} \div 5 \\ &= \tfrac{45}{2} \div \tfrac{5}{1} \\ &= \frac{\overset{9}{\cancel{45}}}{2} \times \frac{1}{\underset{1}{\cancel{5}}} \\ &= \tfrac{9}{2} = 4\tfrac{1}{2} \end{aligned}$$

8. (3) 156
Find the area of the wall, window included.

$$\begin{aligned} A &= \ell w \\ &= 22 \times 8 \\ &= 176 \end{aligned}$$

Find the area of the window.

$$\begin{aligned} A &= \ell w \\ &= 5 \times 4 \\ &= 20 \end{aligned}$$

Subtract to find the difference.

176	area with window
− 20	area of window
156	area without window

9. (1) $1\frac{1}{3}$
Find the perimeter of the table.

$$\begin{aligned} P &= 2\ell + 2w \\ &= (2 \times 32) + (2 \times 18) \\ &= 64 + 36 \\ &= 100 \end{aligned}$$

The perimeter is 100 inches. Change 100 inches to feet by dividing by 12.

$100 \div 12 = 8\frac{1}{3}$

Mr. Leadbetter needs 8⅓ feet of edging. Since he has 7 feet, he needs 8⅓ − 7, or 1⅓ feet more.

10. (4) 6
To find the perimeter of the park, add the lengths of the sides.

$0.28 + 0.21 + 0.19 + 0.17 = 0.85$

You would run 0.85 of a mile if you went around the park once. To find how many times you would go around it to run 5 miles, divide 5 by 0.85.

```
          5.8
0.85.)5.00.0
       4 25
         75 0
         68 0
          7 0
```

Since the answers are whole numbers, round to the nearest whole number, 6.

11. (4) \$176.80
First find the perimeter.

$$\begin{aligned} P &= 2\ell + 2w \\ &= (2 \times 22) + (2 \times 12) \\ &= 44 + 24 \\ &= 68 \end{aligned}$$

The Hansens need 68 feet of curbing at \$2.60 per running foot. Multiply to find the cost.

$\$2.60 \times 68 = \$176.80$

12. (3) 100
Find the area of the garden.

$A = \ell w$
$A = 22 \times 12$
$A = 264$

The area is 264 square feet. Each plant needs 2.4 square feet. To find how many plants can be put in, divide the area by 2.4.

$264 \div 2.4 = 110$

The answers are rounded off. The choice closest to 110 is 100.

13. (5) Insufficient data is given to solve the problem.

The problem does not tell you the length of the eighth side.

14. (3) $399
You are asked to find the cost of square yards, so first find the area of the room in square yards. Since the measurements are given in feet you'll have to divide by 3 ($12 \div 3 = 4$, $18 \div 3 = 6$, $9 \div 3 = 3$). Then find the area of each rectangle.

$A = 4 \times 6 = 24$
$= 3 \times 6 = 18$

Add the two to get the area of the entire living/dining room.

$24 + 18 = 42$

There are 42 square yards. Multiply times the cost per yard.

$42 \times 9.50 = 399$

It will cost the Garcias $399.

15. (4) 625
The flower bed is square, so the 4 sides are equal. Divide.

$100 \div 4 = 25$

Each side is 25 feet long. To find the area, use the formula.

$A = s^2$
$= 25^2 = 625$

The area of the flower bed is 625 square feet.

KEEPING TRACK
Perfect score = 15
Your score ☐

35 Triangles

(pages 241–242)

1. (4) 48°
The right-angle symbol in angle *A* means that its measure is 90°. Add the angle measures you know.

42°	∠C
+ 90°	∠A
132°	sum of two angles

The sum of the angles of a triangle must equal 180°. Subtract to find the size of the third angle.

180	total sum of angles
− 132	sum of two angles
48	third angle (*B*)

2. (2) 70°
Draw the triangle described. Label ∠*S* as 40°. The measures of angles opposite the equal sides are also equal. If m∠*S* = 40°, then 140° (180° − 40°) is left for the other two angles. Divide 140° by 2 to find m ∠*T*.

m∠*T* = 140° ÷ 2 = 70°

3. (3) 60
See the additional angles labeled in the figure below:

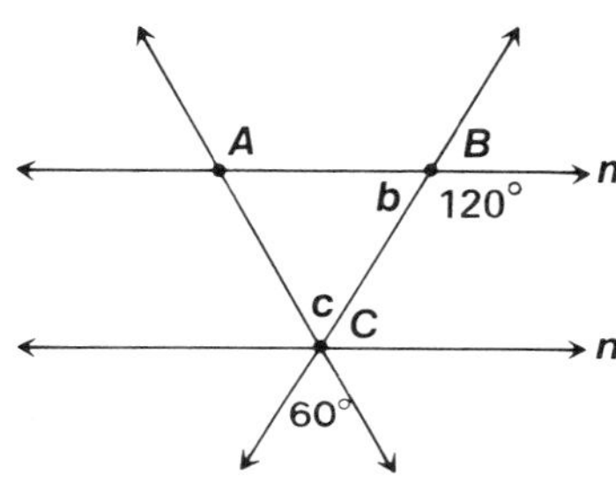

∠*B* + 120° = 180° because these two angles are supplementary. Thus, ∠*b* = 60°. ∠*c* = 60° because it forms vertical angles with another 60° angle. Thus, two angles of △*ABC* measure 60°. The third angle, ∠*BAC*, must also measure 60° for the three angles of the triangle to total 180°.

4. (4) 90°
If m∠*X* is twice the measure of *each* of the other two angles, then *each* of the other two angles must be exactly ½ the measure of ∠*X*, whatever that is.

Try the choices. What about (1)? If m∠*X* = 22.5°, the other two angles would each have to be 11.25°. The three angles don't add up to 180°, so (1) is not correct.

Continue in this way. Only choice (4) will work. If m∠*X* = 90°, each of the other two angles would be ½ of 90° or 45°. The angles will then be 90°, 45°, and 45°, and these do add up to 180°.

(page 243)

5. (4) $13\frac{1}{2}$
Add the lengths of the three sides to find the perimeter.

$3\frac{1}{2} + 4\frac{3}{4} + 5\frac{1}{4}$

$3\frac{2}{4} + 4\frac{3}{4} + 5\frac{1}{4} = 12\frac{6}{4}$

$= 13\frac{2}{4} = 13\frac{1}{2}$

6. (5) 13.80
The sides of an equilateral triangle are the same length. Multiply 4.6 by 3 to find the perimeter.

$3 \times 4.6 = 13.8$ or 13.80

7. (2) 23.2
Subtract to find the length of the two equal sides.

78.4	perimeter
− 32.0	longest side
46.4	two equal sides

To find the length of each equal side, divide 46.4 by 2.

$46.4 \div 2 = 23.2$

8. (3) 70
The sections of road form a triangle with sides *AB, BC,* and *CA.* The perimeter (length of whole trip) is 167, *AB* is 38, and *BC* is 59. You have to find the length of *CA.* Add to find the combined lengths of *AB* and *BC*.

$38 + 59 = 97$

Subtract to find the length of *CA.*

167	perimeter
− 97	length of *AB* and *BC*
70	length of *CA*

(pages 245–246)

9. (4) 1,121.0
Use the formula *A* = ½ *bh* to find the area of the sail.

$A = \frac{1}{2} \times 38 \times 59$
$= 19 \times 59$
$= 1{,}121 = 1{,}121.0$

10. (5) Insufficient data is given to solve the problem.

The problem does not give you the measures of the base or height of the triangle.

11. (3) $2,512.08
First find the area of the room.

$A = \frac{1}{2}bh$

$= \frac{1}{2} \times 24 \times 18 = 216$

The area of the room is 216 square yards and carpeting costs $11.63 a square yard. Multiply to find the total cost.

$\$11.63 \times 216 = \$2,512.08$

12. (3) $37.50
First find the area of the ceiling.

$A = \frac{1}{2}bh$

$= \frac{1}{2} \times 20 \times 15 = 150$

The area is 150 square feet. Each roll of paper covers 30 square feet. Divide to find how many rolls are needed.

$150 \div 30 = 5$

To find the cost, multiply $7.50, the cost per roll, by 5.

$\$7.50 \times 5 = \$37.50$

13. (5) 7.5
Use the formula $A = \frac{1}{2}bh$ to find the area.

$A = \frac{1}{2} \times 5 \times 3$

$= \frac{1}{2} \times 15 = 7\frac{1}{2}$ or 7.5

14. (3) 112 ft
Find the perimeter of the flower bed by using the formula for the area of a rectangle. Use 36 for ℓ and 20 for w.

$P = 2\ell + 2w$
$= (2 \times 36) + (2 \times 20)$
$= 72 + 40 = 112$

15. (5) 5
First find the area of one wall.

$A = \ell w$
$= 15 \times 8 = 120$

Three walls have this area. Multiply 120 by 3 to find the total wall area.

$3 \times 120 = 360$

Next find the area of the ceiling.

$A = \frac{1}{2}bh$

$= \frac{1}{2} \times 15 \times 12 = 90$

Add to find the area of the room.

360	area of walls
+ 90	area of ceiling
450	total area of room

A quart of paint covers 100 square feet. Divide 450, which is the total room area, by 100.

$450 \div 100 = 4.5$

Since 4.5 quarts of paint are needed, the Kershaws will have to buy 5 quarts.

16. (2) $60.00
First change the dimensions to feet by multiplying each by 3. The base is 6 feet and the height is 11 feet. Next find the area of the patio.

$A = \frac{1}{2}bh$

$= \frac{1}{2} \times 6 \times 11 = 33$

To find the cost of installing the patio, multiply $1.85 by 33 square feet, which is the area of the patio.

$\$1.85 \times 33 = \$61.05$

The answer choices are rounded to the nearest $10. The best choice is $60.00.

17. (3) $16.00
First find the perimeter.

$P = 3\frac{2}{3} + 2 + 4\frac{1}{6}$

$= 3\frac{4}{6} + 2 + 4\frac{1}{6} = 9\frac{5}{6}$

Next multiply $1.68, the cost of the edging, by 9⅚.

$\$1.68 \times 9\frac{5}{6}$

$\frac{\overset{.28}{\cancel{\$1.68}}}{1} \times \frac{59}{\underset{1}{\cancel{6}}} = \$16.52$

The best choice is $16.00.

18. (1) 36
An isosceles triangle has two equal sides. Multiply by 2 to find the length of these sides.

$42 \times 2 = 84$

Subtract to find the length of the third side.

120	perimeter
− 84	two equal sides
36	third side

19. (2) 2,002
The dotted line in the diagram shows that the lot can be thought of as having two pieces, a rectangular piece and a triangular piece.

Use the formula to find the area of the rectangular piece.

$A = \ell w$
$= 68 \times 26 = 1,768$

To find the area of the triangular piece, you need to know the base and height of the triangle. The 86 foot side of the lot is made up of the length of the rectangle (68 feet) and the base of the triangle. Subtract to get the base (b) of the triangle.

$b = 86 - 68 = 18$

The height of the triangle is 26 feet. Use the formula to find the area of the triangle.

$A = \frac{1}{2}bh$

$= \frac{1}{2} \times 18 \times 26 = 234$

Add the area of the rectangular piece and the area of the triangular piece to find the area of the whole lot.

$1,768 + 234 = 2,002$

KEEPING TRACK
Perfect score = 19

Your score ☐

36 Congruent and Similar Triangles

(page 248)

1. (1) 50°
The measures of angles J, L, R, and T are equal, and each is 65°. Add to find the combined measure of $\angle R$ and $\angle T$.

$65° + 65° = 130°$

The sum of the angles of a triangle is 180°. Subtract to find $\angle S$.

180°	all angles of the triangle
− 130°	angles R and T
50°	$\angle S$

2. (3) 25.6
Sides JK and KL are the same length (9 feet) because they are opposite equal angles.

Sides JL and RT have the same length (7.6 feet) because they are corresponding sides of congruent triangles. Add to find the perimeter.

$9 + 9 + 7.6 = 25.6$

(page 251)

3. (1) 30.80
Set up a grid with the information on Margaret and the tree. Use the grid to solve.

	Margaret	Tree	
actual height	5.6	?	actual height
shadow length	12	66	shadow length

$5.6 \times 66 = 369.6$
$369.6 \div 12 = 30.8$

4. (3) 30.0
Set up a grid with the information on a scale drawing and the actual garden. Use the grid to solve the problem.

	scale drawing	flower bed	
height of scale	12	?	actual height
base of scale	22	55	actual base

$12 \times 55 = 660$
$660 \div 22 = 30$

5. (5) Insufficient data is given to solve the problem.

The problem does not tell you Darwin's actual height.

6. (2) 840.0
The triangles are similar, so you can set up a proportion.

720 is to 1,120 as
540 is to *CD*

Put this proportion into a grid so that it can be solved.

is to	720	540	is to
	1,120	?	
	as		

$1{,}120 \times 540 = 604{,}800$
$604{,}800 \div 720 = 840$, or 840.0

7. (1) 5
The triangles are similar, so you can set up a proportion.

ML is to *JL* as *MN* is to *JK*
24 is to 12 as 10 is to *JK*

Use a grid to solve.

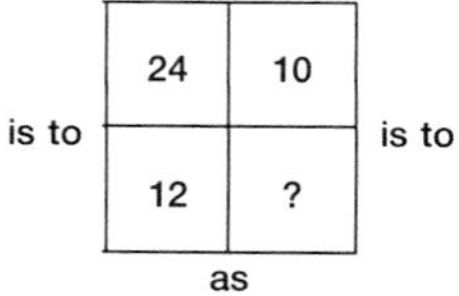

$12 \times 10 = 120$
$120 \div 24 = 5$

8. (2) 16
The triangles are similar triangles, so you can set up a proportion.

BC is to *CE* as *AB* is to *ED*
10 is to 20 as 8 is to *ED*

Put this proportion into a grid so that it can be solved.

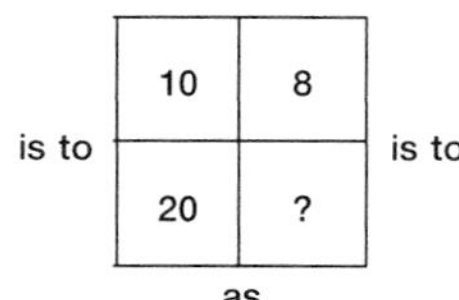

$20 \times 8 = 160$
$160 \div 10 = 16$

9. (2) 62°
Since $ED = FD$, angles *E* and *F* are equal.

In $\triangle ABC$, the sum of all the angles is 180°. But the sum of angles *A* and *C* is 118°, which means that $m\angle B = 180° - 118° = 62°$.

Since corresponding parts of congruent triangles are equal, $\angle D$ must have a measure of 62°.

KEEPING TRACK
Perfect score = 9
Your score ☐

37 Pythagorean Theorem

(pages 254–255)

1. (3) 2.50
The right triangle formed has legs that are 1.5 and 2.0 miles long. Find the hypotenuse (diagonal distance) of this triangle.

$$a^2 + b^2 = c^2$$
$$(1.5)^2 + 2^2 = c^2$$
$$2.25 + 4 = c^2$$
$$6.25 = c^2$$
$$2.5 = c$$

The diagonal distance is 2.50 miles.

2. (1) 8
The ladder, the wall, and the ground form a right triangle. The ground and the wall are the legs. The ladder is the hypotenuse. Solve for the distance from the ground to the top of the ladder.

$$a^2 + b^2 = c^2$$
$$6^2 + b^2 = 10^2$$
$$36 + b^2 = 100$$

Subtract to find b^2.

$$b^2 = 100 - 36 = 64$$

Since $8^2 = 64$, $b = 8$.

3. (3) 90
Two right triangles are formed. Which ever one you use, it will have one leg that is 60 feet and a hypotenuse that is 75 feet. Use the Pythagorean theroem to find the length of the other leg.

$$a^2 + b^2 = c^2$$
$$60^2 + b^2 = 75^2$$
$$3{,}600 + b^2 = 5{,}625$$

Subtract to find b^2.

$$b^2 = 5{,}625 - 3{,}600 = 2{,}025$$
$$b = 45$$

The wires are attached halfway up the tower. Multiply 45 by 2 to find the full height of the tower.

$$45 \times 2 = 90$$

4. (2) 127
The baseball diamond forms a square. The diagonal from home plate to second base forms the hypotenuse of a right triangle whose legs are both 90 feet long. Use the Pythagorean theorem to find the hypotenuse.

$$a^2 + b^2 = c^2$$
$$(90)^2 + (90)^2 = c^2$$
$$8100 + 8100 = c^2$$
$$16{,}200 = c^2$$

16,200 is between 14,400 (120×120) and 16,900 (130×130). The number whose square is 16,200 must be between 120 and 130. The only answer choice that does this is 127.

5. (1) 15
The two legs of the right triangle are 9 feet and 12 feet. Use the Pythagorean theorem to find the length of the hypotenuse.

$$a^2 + b^2 = c^2$$
$$9^2 + 12^2 = c^2$$
$$81 + 144 = c^2$$
$$225 = c^2$$
$$15 = c$$

6. (5) $28.68
First find the perimeter.

9 + 12 + 15 = 36 feet

Next change 36 feet to yards by dividing 36 by 3. (3 ft = 1 yd)

36 ÷ 3 = 12 yards

To find the cost of the material, multiply $2.39 by 12.

$2.39 × 12 = $28.68

KEEPING TRACK
Perfect score = 6
Your score ☐

38 Circles

(page 257)

1. (4) 242
Use the formula $C = \pi d$.

$C = \pi d$

$C = \frac{22}{7} \times 77 = 242$

2. (2) 21
The circumference is given and you have to find the diameter. Use $C = \pi d$.

$C = \pi d$

$66\frac{6}{7} = \frac{22}{7} \times d$

To find d, divide $66\frac{6}{7}$ by $\frac{22}{7}$.

$d = 66\frac{6}{7} \div \frac{22}{7}$

$= \frac{\overset{234}{\cancel{468}}}{\underset{1}{\cancel{7}}} \times \frac{\overset{1}{\cancel{7}}}{\underset{11}{\cancel{22}}}$

$= \frac{234}{11} = 21\frac{3}{11}$

Round $21\frac{3}{11}$ inches to the nearest whole inch, 21 inches.

3. (5) 5,652
First multiply the radius by 2 to get the diameter. You find that d is 300 feet. Then use $C = \pi d$ find the circumference.

$C = \pi d$
$= 3.14 \times 300 = 942$

The distance around the track is 942 feet. Rosita ran around the track 6 times. Multiply 942 by 6 to find how far she ran.

942 × 6 = 5,652

4. (5) Insufficient data is given to solve the problem.
The problem does not give you information about how large the circles of edging will be.

(page 259)

5. (4) 616
First find the radius of the table top by dividing the diameter by 2.

$r = 28 \div 2 = 14.$

Next use the formula $A = \pi r^2$ to find the area.

$A = \pi r^2$

$= \frac{22}{\underset{1}{\cancel{7}}} \times \overset{2}{\cancel{14}} \times 14$

$= 22 \times 28$
$= 616$

6. (1) 28
The area is given and you must find the radius.

$A = \pi r^2$

$616 = \frac{22}{7} \times r^2$

To find r^2, divide 616 by $\frac{22}{7}$.

$r^2 = 616 \div \frac{22}{7}$

$= \frac{\overset{308}{\cancel{616}}}{1} \times \frac{7}{\underset{11}{\cancel{22}}}$

$= \frac{2156}{11} = 196$

You now know that $r^2 = 196$. To find r, try numbers to see if you can find one whose square is 196. You find that $14^2 = 14 \times 14 = 196$. So $r = 14$. To find the diameter, multiply the radius by 2. You get 14 × 2 or 28.

7. (3) 104
First find the area.

$A = \pi r^2$
$= 3.14 \times 20 \times 20 = 1,256$

The area of the wading pool is 1,256 square feet. There must be 12 square feet for each person. To find the maximum number of people, divide 1,256 by 12.

$1,256 \div 12 = 104\frac{2}{3}$

You need a whole-number answer, so round mixed number to the nearest whole number, 104.

8. (4) 3,080
The diameter of each circular block is 14 feet. The radius is half this length, or 7 feet. Use the formula to find the area of 1 circular block.

$A = \pi r^2$

$= \frac{22}{\underset{1}{\cancel{7}}} \times \overset{1}{\cancel{7}} \times 7 = 154$

Multiply by 20 to find the area of 20 of these circular blocks.

20 × 154 = 3,080

9. (3) 4
The diameter of each block is 2 feet. The radius is half this length, or 1 foot. Use the formula to find the area of 1 circular block.

$A = \pi r^2$
$= 3.14 \times 1 \times 1 = 3.14$

There are 120 of these stones. Multiply by 3.14 by 120 to find the total area.

120 × 3.14 = 376.8 (sq ft)

One quart of paint will cover 100 square feet. To find how many quarts of paint are needed, divide 376.8 by 100.

376.8 ÷ 100 = 3.768 (quarts)

Round 3.768 up to the nearest whole number, 4.

KEEPING TRACK
Perfect score = 9
Your score ☐

39 Composite Figures

(pages 262–263)

1. (1) 14
Add the lengths of the sides to find the distance around the table.

3 + 3 + 2 + 4 + 2 = 14

2. (4) 62
Add the lengths of the sides to find the distance around the garden.

16 + 5 + 6 + 20 +
8 + 7 = 62

3. (2) $36.14
First find the cost of the fencing. In number 2, you found that 62 feet were needed. Multiply 62 by $1.03, the cost of fencing.

$1.03 × 62 = $63.86

Subtract to find the amount of change.

	$100.00	amount given to clerk
−	63.86	cost of item
	$ 36.14	change received

4. (3) 27

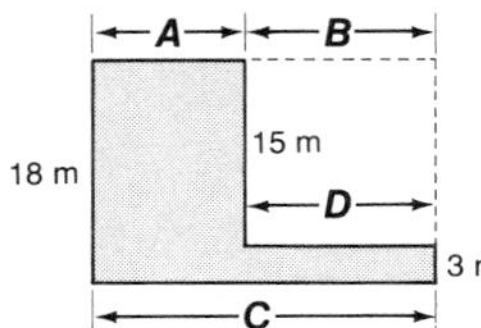

Add the lengths of the sides whose measures you know.

$18 + 15 + 3 = 36$

Subtract from the perimeter to find what part of the perimeter is made up of sides *A, B,* and *C.*

$90 - 36 = 54$

This is twice the length of side *C,* since *A* and *B* add up to as much as *C.* So divide 54 by 2 to find *C:* $54 \div 2 = 27$.

(page 265)

5. (3) 6,120
The blacktop section is a rectangle that is 90 meters long and 68 meters wide. Use $A = \ell w$ to find the area.

$A = \ell w$
$= 90 \times 68$
$= 6{,}120$ (sq m)

6. (1) 8,370
The blacktop and the grass form a rectangle that is 90 meters long and 68 + 25, or 93, meters wide. Use $A = \ell w$ and put in the values for ℓ and w.

$A = \ell w$
$= 90 \times 93$
$= 8{,}370$ sq m

You could also find the area of the grass (2,250 square meters) and add it to the area of the blacktop (6,120 square meters).

7. (5) 1,750
Think of the top as a rectangle and 2 congruent triangles. The rectangle is 40 inches long and 35 inches wide. $A = \ell w$ to find the area.

$A = \ell w$
$= 40 \times 35$
$= 1{,}400$ (sq in)

Each triangle has a base of 10 inches and a height of 35 inches. To find the area of one triangle, use $A = \frac{1}{2}bh.$

$A = \frac{1}{2}bh$
$= \frac{1}{2} \times 10 \times 35$
$= 175$ sq in

This is the area of one triangle. To find the area of both triangles, multiply by 2.

$2 \times 175 = 350$ (sq in)

Add to find the total area.

1,400	area of rectangle
+ 350	area of triangles
1,750	area of desk top

8. (4) 217.5
The part of the driveway on the right can be thought of as a rectangle plus a triangle. The height of the triangle is 12 − 9, or 3 feet, and the base is 25 − 10, or 15 feet.

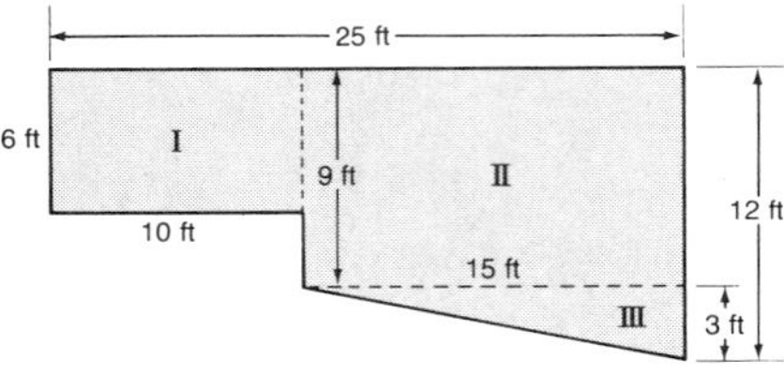

Find the area of rectangular section I.

$A = \ell w$
$= 10 \times 6 = 60$ (sq ft)

Find the area of rectangular section II.

$A = \ell w$
$= 15 \times 9 = 135$

Find the area of triangular section III.

$A = \frac{1}{2}bh$
$= \frac{1}{2} \times 15 \times 3 =$
$\frac{45}{2} = 22\frac{1}{2}$

Add all three areas: $60 + 135 + 22\frac{1}{2} = 217\frac{1}{2}$, or 217.5.

9. (2) 262.8
You can see that the figure is made up of a rectangle and 2 half circles. First, figure out the perimeter of the circle. You know it has a diameter of 20 feet, so use $C = \pi d$.

$C = 3.14 \times 20 = 62.8$

For the length, use 2×100 or 200. Add.

$62.8 + 200 = 262.8$

10. (5) Insufficient data is given to solve the problem.

You were not given any cost information on surfacing the track.

KEEPING TRACK
Perfect score = 9
Your score ☐

40 Solid Geometry

(page 267)

1. (4) 36,000
The building is a rectangular solid. Use the formula $V = \ell wh$ to find the volume. Substitute the values that are known. Find the unknown value.

$V = \ell wh$
$= 90 \times 50 \times 8$
$= 36{,}000$ (cu ft)

2. (1) 12
Change 6 inches to feet by dividing by 12: $6 \div 12 = 0.5$. Then use the formula to find the volume.

$V = \ell wh$
$= 6 \times 4 \times 0.5 = 12$ (cu ft)

3. (5) Insufficient data is given to solve the problem.

The problem does not tell you how wide the trailer is.

(pages 268–269)

4. (3) 27
The planter is a cube. Use the formula $V = s^3$ to find the volume.

$V = s^3$
$= 3 \times 3 \times 3 = 27$ (cu ft)

5. (2) 4
Use $V = s^3$, and put 64 in place of V.

$V = s^3$
$64 = s^3$

What number used as a factor three times equals 64? From number 4, you know that $3 \times 3 \times 3 = 27$. Next try 4: $4 \times 4 \times 4 = 64$. Each side is 4 feet long.

6. (2) 2,213
First find the volume of the storage shed.

$V = s^3$
$= 14 \times 14 \times 14$
$= 2{,}744$ cu ft.

One bushel occupies 1.24 cubic feet, so divide 2,744 by 1.24 to find approximately how many bushels can be stored in the shed.

$2{,}744 \div 1.24 = 2{,}212.90$

Since the answer choices are all whole numbers, round the answer to the nearest whole number: 2,213.

(page 270)

7. (4) 10,682
Use the formula $V = \pi r^2 h$ to find the volume of the water heater. The formula uses the length of the radius, so first divide the diameter by 2 to get the radius: $18 \div 2 = 9$.

$V = \pi r^2 h$
$= 3.14 \times 9 \times 9 \times 42$
$= 10{,}682.28$ (cu in)

Round the answer to the nearest whole number: 10,682.

8. (3) 154
First find the volume of 1 cylinder. The diameter is $3\frac{1}{2}$ inches so the radius is $1\frac{3}{4}$, or $\frac{7}{4}$ inches.

$V = \pi r^2 h$

$$= \frac{\overset{11}{\cancel{22}}}{\underset{1}{\cancel{7}}} \times \frac{\overset{1}{\cancel{7}}}{\underset{2}{\cancel{4}}} \times \frac{7}{\underset{1}{\cancel{4}}} \times \frac{\overset{1}{\cancel{4}}}{1} = \frac{77}{2}$$

$= 38\frac{1}{2}$ (cu in)

There are 4 cylinders. Multiply the volume of 1 cylinder ($38\frac{1}{2}$) by 4.

$4 \times 38\frac{1}{2} = 154$ (cu in)

9. (1) 12
The diameter is 42 feet, so the radius is 21 feet.

$V = \pi r^2 h$

$$16{,}632 = \frac{22}{\underset{1}{\cancel{7}}} \times \frac{\overset{3}{\cancel{21}}}{1} \times \frac{21}{1} \times h$$

$16{,}632 = 1{,}386 \times h$

To find h, divide 16,632 by 1,386.

$h = 16{,}632 \div 1{,}386$
$= 12$ (ft)

10. (1) 1.5
Change inches to feet by dividing by 12. ($12 \div 12 = 1$ and $6 \div 12 = 0.5$)

$V = \ell wh$
$= 1 \times 3 \times 0.5 = 1.5$ (cu ft)

11. (2) 3.0
Use $V = \ell wh$ and put in the known values.

$V = \ell wh$
$9 = \ell \times 2 \times 1.5$
$9 = \ell \times 3$

To find ℓ, divide 9 by 3.

$\ell = 9 \div 3 = 3$ (ft)

12. (1) 163
First find the volume of the bar.

$V = \ell wh$
$= 12 \times 6 \times 8 = 576$ (cu in)

Steel weighs 0.283 pounds per cubic inch. Multiply 576 by 0.283 to find the total weight of the bar.

$0.283 \times 576 = 163.008$ lb

Round to the nearest whole number: 163.

KEEPING TRACK
Perfect score = 12
Your score ☐

41 Coordinate Geometry

(page 274)

1. (5) 3
In an ordered pair, the second number always refers to the y-coordinate.

2. (2) -1
The first number in an ordered pair always refers to the x-coordinate.

3. (3) (2, 5)
Begin at the origin. To get to point C, move 2 units to the right and 5 units up. The ordered pair for C is (2, 5).

4. (2) J
Start at the origin. Since the first number (5) is positive, move 5 units to the right. The second number (-1) is negative, so move down 1 unit. Point J is at this location.

5. (2) $(-2, -7)$
Begin at the origin. To get to point G, move 2 units to the left and 7 units down. The ordered pair for G is $(-2, -7)$.

KEEPING TRACK
Perfect score = 5
Your score ☐

42 Distance on a Coordinate Grid

(page 276)

1. (1) 4
The segment KV is parallel to the x-axis. K is 2 units to the left of the y-axis and V is 2 units to the right of the y-axis. The distance between K and V is $2 + 2$, or 4 units.

2. (2) 5
The segment ST is parallel to the y-axis. S is 3 units to above the the x-axis and T is 2 units below the x-axis. The distance between points S and T is $2 + 3$, or 5 units.

3. (4) N and T
N is 3 units to the left of the y-axis and T is 4 units to the right of the y-axis. The line segment that joins N and T is parallel to the x-axis. The distance between N and T is $3 + 4$, or 7 units.

(page 278)

4. (2) 5
Line CD is parallel to the x-axis. Draw a line segment from L to line CD so that the segment is at right angles to the line. Count the units from L to the point where the segment meets the line. The segment is 5 units long.

5. (4) 8
Draw a line segment from H to line CD so that the segment is at right angles to the line. Count the units from H to the point where the segment meets the line. The total length is 8 units.

6. (3) *G*
The *y*-axis contains point *G* and is perpendicular to line *CD*. If you count the units from point *G* to line *CD* as you move along the *y*-axis, you find that the distance from *G* to line *CD* is 10 units.

(pages 280–281)
7. (4) 6
Segment *CD* is parallel to the *x*-axis, so you just count the units. The distance from *C* to *D* is 6 units.

8. (3) 4
Segment *BD* is parallel to the *y*-axis, so you just count the units. The distance from *B* to *D* is 4 units.

9. (2) 7.2
Segment *BC* is the hypotenuse of the right triangle *BCD.* In numbers 7 and 8 you found the lengths of the legs to be 4 and 6. Use the Pythagorean Theorem to find the length of hypotenuse *BC.*

$$4^2 + 6^2 = BC^2$$
$$16 + 36 = BC^2$$
$$52 = BC^2$$

Since $7^2 = 49$ and $8^2 = 64$ and 52 is between 49 and 64, the square root of 52 must be between 7 and 8. Therefore the best answer choice is 7.2.

10. (1) 7.6
Use the formula

$$d = \sqrt{(x_2 - x_1)^2 + (y_2 - y_1)^2}$$

For (x_1, y_1) use the coordinates of point *K:* (4, −6). For (x_2, y_2) use the coordinates of point *J:* (−3, −3).

$$d = \sqrt{(-3 - 4)^2 + (-3 - (-6))^2}$$
$$= \sqrt{(-7)^2 + (3)^2}$$
$$= \sqrt{49 + 9}$$
$$= \sqrt{58}$$

Since $7^2 = 49$ and $8^2 = 64$ and 58 is between 49 and 64, $\sqrt{58}$ is between 7 and 8. So 7.6 is the best answer choice.

(page 282)
11. (1) (2, 4)
Add the *x*-coordinates (−2 + 6 = 4) and divide by 2 (4 ÷ 2 = 2). The *x*-coordinate of the midpoint is 2. Add the *y*-coordinates (−3 + 11 = 8) and divide by 2 (8 ÷ 2 = 4). The *y*-coordinate of the midpoint is 4. The midpoint is (2, 4).

12. (2) (−5, 1)
To find the coordinates of the midpoint, average the *x*- and then the *y*-coordinates of the endpoints. The sum of the *x*-coordinates is −8 + (−2) = −10 and −10 ÷ 2 = −5. The sum of the *y*-coordinates is 4 + (−2) = 2 and 2 ÷ 2 = 1. The midpoint of segment *RU* is (−5, 1).

13. (1) 3
Draw a line straight down from *J* to the *x*-axis and count the units. The perpendicular distance from *J* to the *x*-axis is 3 units.

14. (2) *J*
Think of drawing a perpendicular line from each point to the *y*-axis. Point *J* is the only point that is 6 units from the *y*-axis.

KEEPING TRACK
Perfect score = 14
Your score ☐

Keeping Track

Now enter all your scores from the Keeping Track boxes on the lines below.

Lesson	Perfect Score	Your Score
Angles	9	______
Quadrilaterals	15	______
Triangles	19	______
Congruent and Similar Triangles	9	______
Pythagorean Theorem	6	______
Circles	9	______
Composite Figures	9	______
Solid Geometry	12	______
Coordinate Geometry	6	______
Distance on a Coordinate Grid	14	______
TOTAL	98	______

Now look at how you did in the *Geometry* section. Be sure to review those lessons in which you could use more practice. That will help you when you complete the *Extra Practice* section that follows. After that you'll be ready to take the first Posttest.

Extra Practice: Geometry

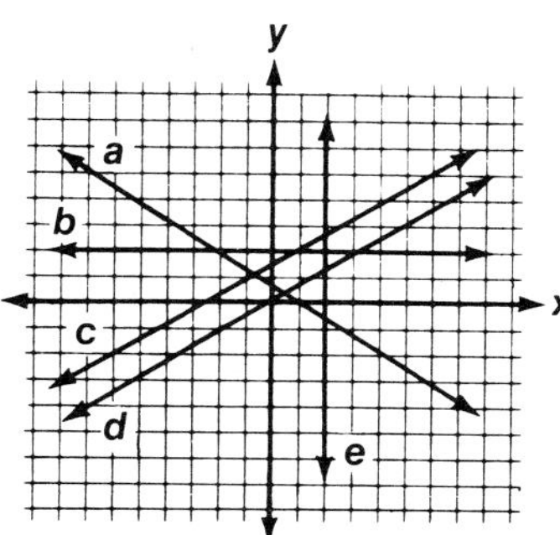

1. Which line in the graph is perpendicular to the x-axis?

   (1) a (2) b (3) c (4) d (5) e

   ① ② ③ ④ ⑤

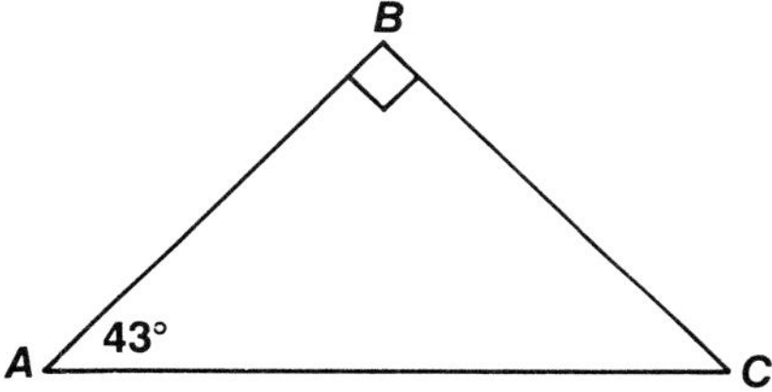

2. What is the measure of $\angle C$ in the triangle?

   (1) 137° (2) 133° (3) 47°
   (4) 43° (5) 37°

   ① ② ③ ④ ⑤

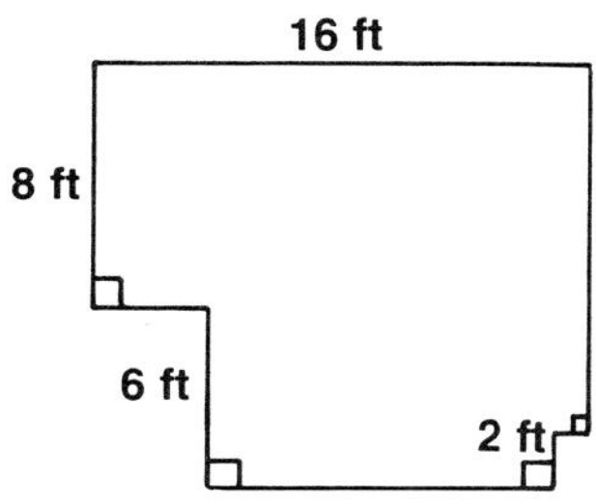

3. How many feet of baseboard are needed for the room in the figure?

   (1) 32 (2) 44 (3) 56
   (4) 60 (5) 224

   ① ② ③ ④ ⑤

4. What is the approximate surface area in square feet of a round mirror that has a radius of 3 feet? (Use $\frac{22}{7}$ for π).

   (1) 4 (2) 7 (3) 8 (4) 9 (5) 28

   ① ② ③ ④ ⑤

5. A cylinder with a radius of 1 inch and a height of 5.5 inches has a capacity of how many inches?

   (1) 6.44 (2) 9.465 (3) 15.465
   (4) 17.27 (5) 23.71

   ① ② ③ ④ ⑤

6. For the points $M(-2, 3)$ and $N(6, -5)$, find the midpoint of line segment MN.

   (1) (4, 4) (2) (1, 2) (3) (2, -1)
   (4) (-8, -8) (5) $(\frac{1}{2}, \frac{1}{2})$

   ① ② ③ ④ ⑤

7. Give the circumference in inches of a compact disc that has a radius of 6 inches.

   (1) 9.14 (2) 15.14 (3) 18.84
   (4) 37.68 (5) 113.04

   ① ② ③ ④ ⑤

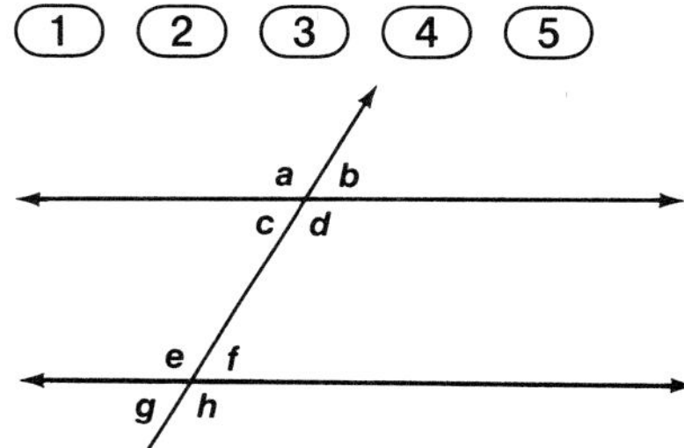

8. If $\angle a$ is twice as large as $\angle b$, what is the measurement of $\angle h$?

   (1) 30° (2) 45° (3) 60°
   (4) 90° (5) 120°

   ① ② ③ ④ ⑤

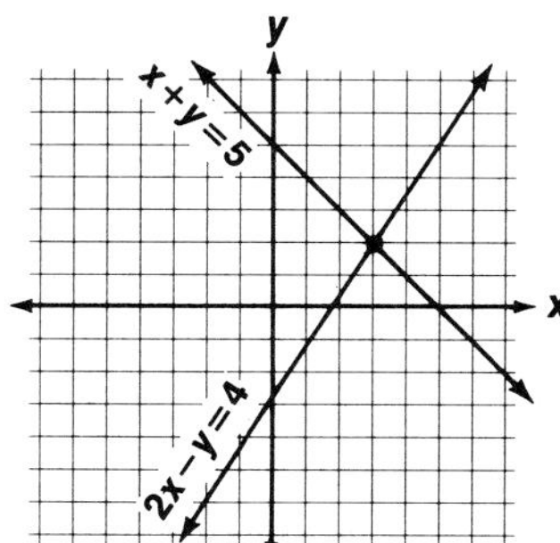

9. What are the coordinates of the intersection of the lines on the graph?

   (1) (-1, 6) (2) (2, 3) (3) (3, 2)
   (4) (-1, 6) (5) (5, 6)

   ① ② ③ ④ ⑤

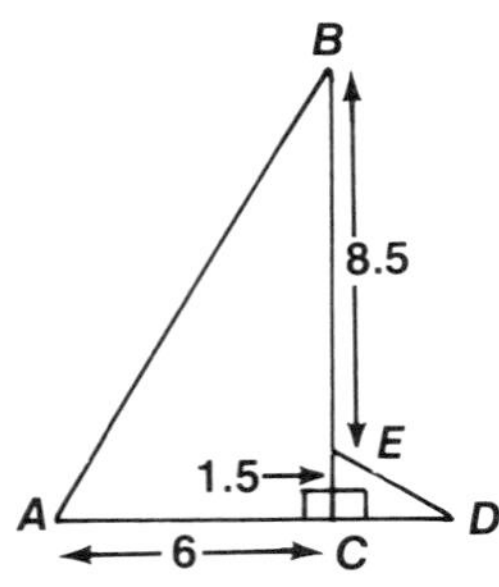

10. In the figure above, triangles ABC and EDC are similar. What is the length of CD?

(1) 3.4 (2) 2.5 (3) 2.125
(4) 1.06 (5) .9

(1) (2) (3) (4) (5)

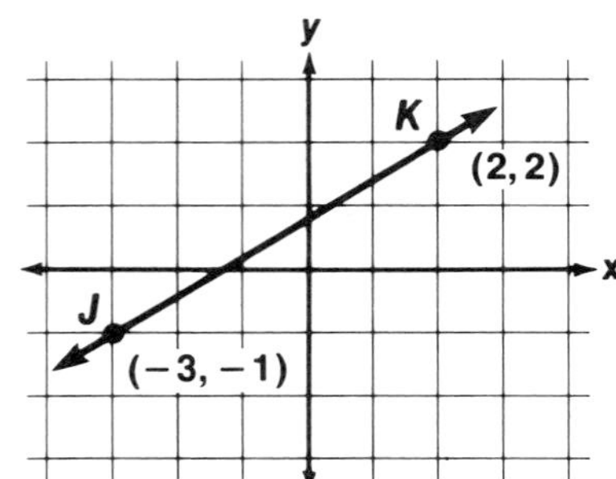

11. Find the ordered pair for the midpoint of line segment JK.

(1) $(-1, 0)$ (2) $(-1, 1)$ (3) $(-\frac{1}{2}, 1)$
(4) $(-\frac{1}{2}, \frac{1}{2})$ (5) $(1, \frac{1}{2})$

(1) (2) (3) (4) (5)

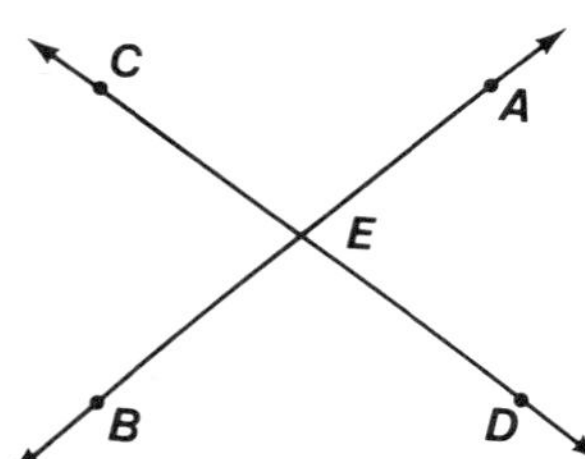

12. In the figure above, find the measurement of $\angle AEC$ if $\angle AED$ equals 78°.

(1) 12° (2) 22° (3) 60° (4) 102°
(5) Insufficient data is given to solve the problem.

(1) (2) (3) (4) (5)

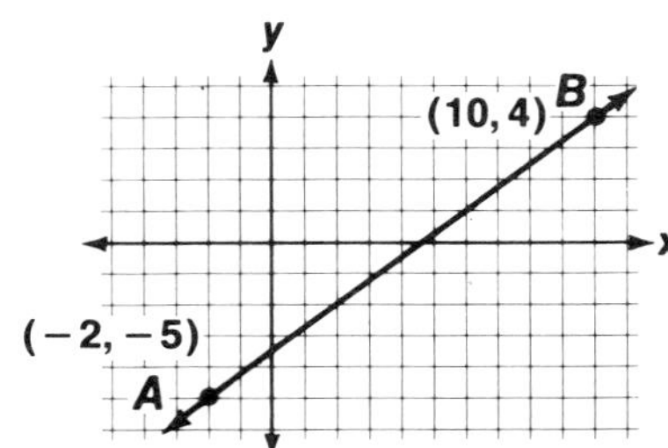

13. What is the distance between points A and B in the figure above?

(1) $\sqrt{21}$ (2) $\sqrt{65}$ (3) 15
(4) 21 (5) 225

(1) (2) (3) (4) (5)

Items 14 and 15 refer to the graph.

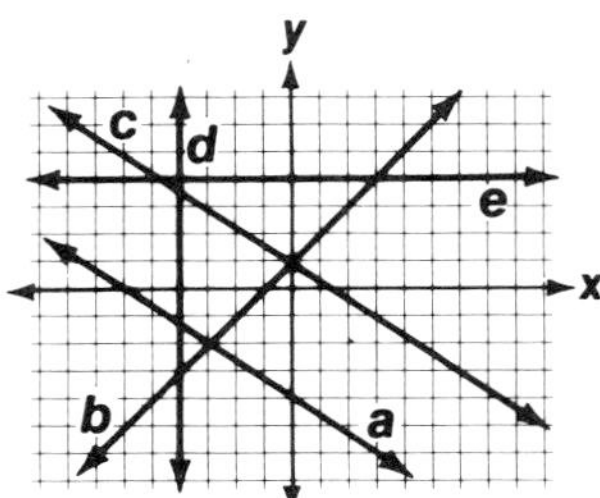

14. Lines a and b intersect at the ordered pair $(-3, -2)$. What is the perpendicular distance from the point of intersection to the x-axis?

(1) 2 (2) 4 (3) 5 (4) 6 (5) 8

(1) (2) (3) (4) (5)

15. Lines c and d intersect at the ordered pair $(-4, 4)$ and lines b and c intersect at the ordered pair $(0, 1)$. What is the diagonal distance between these two points of intersection?

(1) 3 (2) 4 (3) 5 (4) 6 (5) 7

(1) (2) (3) (4) (5)

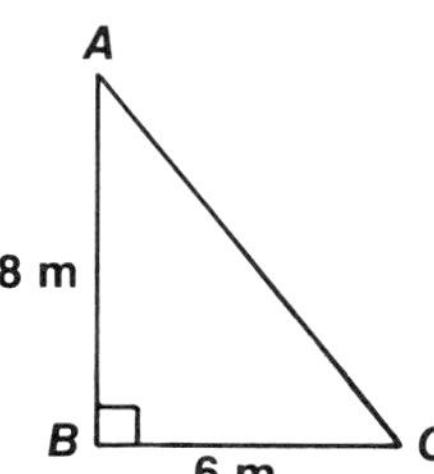

16. What is the perimeter of the triangular garden shown in the figure above?

(1) 14 m (2) 24 m (3) 48 m
(4) 100 m (5) 114 m

(1) (2) (3) (4) (5)

Answers to Extra Practice begin on page 360. Record your score on the Progress Chart on the Inside Back Cover.

Algebra

To many people algebra looks strange. It often uses letters like *a*'s, *b*'s, and *c*'s or *x*'s, *y*'s, and *z*'s instead of numbers. What these people don't realize is that such letters simply stand for certain unknown amounts. The unknown amounts can be people's ages, how long it takes to finish a job, or any other number. Algebra is really not as strange as it seems.

Although the GED Mathematics Test measures your ability to solve algebra problems, only about 17 of the 56 questions are algebra questions. You can use your arithmetic skills to figure out that this is about 30 percent—or not quite one out of every three questions. Some of the algebra questions may require a special knowledge of algebra, but you can solve others just by relying on your practical experience and common sense.

You may be happy to know that the GED Mathematics Test counts as algebra questions some problems that you think of as ordinary arithmetic. Look at this problem, for example.

1. If 3 cinnamon rolls cost $1.39, what is the cost of a dozen rolls?

   (1) $4.17 (2) $4.46 (3) $5.26
   (4) $5.46 (5) $5.56

   ① ② ③ ④ ⑤

Did you choose (5) as the correct answer? If so, you probably used your arithmetic skills and your own practical experience to do a type of algebra problem known as a proportion. Did you divide $1.39 by 3 to find the cost of one roll? Then did you multiply that number by 12 to find the cost of a dozen? In this *Algebra* section you will learn how to solve proportion problems quickly and easily. Remember, on the GED Test you can use whatever method you like for solving a problem. What is important is getting the right answer.

Now look at an algebra problem that uses a formula.

2. If $b = 10$ and $a = 0$, evaluate $7a^2 + 5b^2$.

   (1) 100 (2) 250 (3) 500
   (4) 1,000 (5) 5,000

   ① ② ③ ④ ⑤

If you thought (3) was correct, you substituted correctly in the equation. This problem is simply a substitution problem. You substitute the number for the letter and then use basic arithmetic skills to compute your answer.

If someone asked you a moment ago if you knew any algebra, you probably would have said, "Forget it!" But if you got this problem right, you already have the idea. In this *Algebra* section you'll learn how to work with formulas. Remember, though, the GED Test will not expect you to memorize formulas. If a formula is needed, it will be given on the formula page of the GED Test. All the formulas you will be given on that page are printed on the inside front cover of this book. When you solve the problems in *"The Real Thing"* sections, the *Extra Practice* and the *Posttests*, don't look back in the lessons to find algebraic formulas. Instead, use the formula page on the inside front cover. That way, you'll be working with formulas the way you'll have to when you take the actual GED Test.

Here's one last algebra problem. Like many algebra problems, it uses a letter to stand for an unknown number.

3. S stands for Sergio's age. His father's age is 7 years more than 3 times Sergio's age. Which of the following expressions represents his father's age?

   (1) $3S$ (2) $3S + 7$
   (3) $3(S + 7)$ (4) $\frac{S + 7}{3}$
   (5) $3S - 7$

   ① ② ③ ④ ⑤

Did you notice that this problem doesn't even require you to add, subtract, multiply, or divide? All you have to do is use logic. Did you pick (2)? If so, you understand the basics of writing algebraic expressions. You may have thought you didn't know anything about algebraic expressions before you answered this question.

A complete study of algebra would take much more space than the pages of this book allow. However, this book does teach you enough of the basics of algebra for you to make good choices on the GED Test. It also provides a good foundation if you decide to go on for some other training where you need algebra. What's more, you may very well find that you enjoy working through this *Algebra* section of your book.

LESSON 43

Numerical Expressions

You are now entering the world of algebra. Does that make you nervous? Don't be. Anyone who knows algebra learned it *step by step.* You can too.

Before you can use algebra to solve problems, you have to learn the rules of the game. This lesson will introduce you to those rules.

What Is an Expression?

Numbers can be combined in many different ways. They can be written with positive and negative signs; parentheses; addition, subtraction, division, and multiplication signs; and exponents. In mathematics, such a combination is called a **numerical expression.**

Coming to Terms

numerical expression a group of numbers combined with signs of operation (addition, subtraction, multiplication, division) or with exponents. A numerical expression may include parentheses.

Here's an Example

■ Look at this expression: $5^2 - 2(3 + 4)$

There are many things going on in this expression. The number 5 is being squared. 3 is being added to 4. Something is being subtracted.

Did you notice the 2 outside and next to the parentheses? This is a special way to write multiplication. You can use parentheses and leave out the multiplication sign. This means that when you see a number written next to a set of parentheses, you must multiply that number by whatever is in parentheses. No other sign ($+$, $-$, or $\div$) can *ever* be left out like this; only the multiplication sign can be left out and replaced by parentheses. In this expression, $2(3 + 4)$ means the same as $2 \times (3 + 4)$.

Order of Operations

When you have many numbers with plus and minus signs, multiplication and division signs, parentheses, and exponents, there has to be some order to doing the calculations. Mathematicians have agreed on what this order should be so that everyone can get the same answer.

In most algebra problems, you will have to do more than one operation. And you must do the operations in a certain order. Which operation do you do first—add, multiply, or what? The order of operations below shows you what to do first, second, and so on. Follow these steps.

Step 1 Do what is in parentheses first.

Step 2 Do work involving exponents.

Step 3 Multiply and divide from left to right, just as you read the expression.

Step 4 Add and subtract, also from left to right.

The key words in order are parentheses, exponents, multiply, divide, add, and subtract. Take the first letter of each word, and you get the acronym PEMDAS. Some people use a memory trick to remember the correct order of the letters. They use *Perfectly easy, my dear Aunt Sally.*

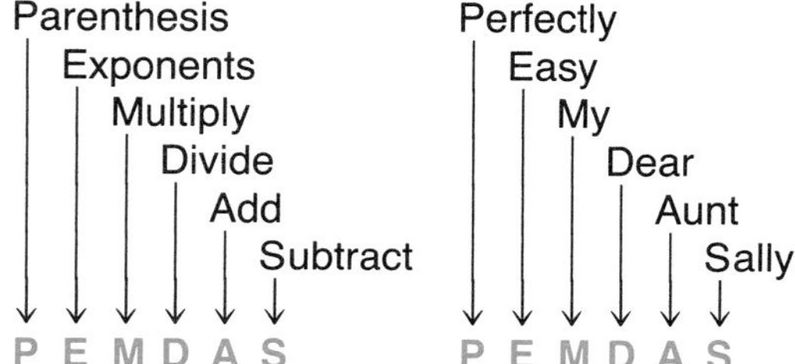

However you choose to remember it, be sure to learn the PEMDAS rule for doing the calculations for numerical expressions.

Coming to Terms

order of operations an agreed-upon system for simplifying numerical expressions. The rule is sometimes called the PEMDAS rule: parentheses, exponents, multiply, divide, add, and subtract.

Here's an Example

When you simplify an expression, you find what it is equal to after you've done everything. You follow the order of operations rule until you can go no farther.

To simplify $5^2 - 2(3 + 4)$, you follow the order of operations rule using PEMDAS.

Begin with $5^2 - 2(3 + 4)$

		Think	Write
Step 1	Parentheses.	$3 + 4 = 7$	$5^2 - 2(7)$
Step 2	Exponents.	$5^2 = 25$	$25 - 2(7)$
Step 3	Multiply/divide.	$2(7) = 14$	$25 - 14$
Step 4	Add/subtract.	$25 - 14 = 11$	11

The answer is 11.

Did you remember that 2(7) means 2×7?

Try It Yourself

■ Simplify the expression: $6(8 - 11) + 2^3$.

____

Write down each step as you go along. Don't try to do more than one step at a time.

Did you follow these steps?

Step 1	Parentheses.	$6(-3) + 2^3$
	(Did you remember that to find $8 - 11$, you can *add* 8 and -11?)	
Step 2	Exponents.	$6(-3) + 8$
Step 3	Multiply.	$-18 + 8$
Step 4	Add.	-10

If you did not take these steps in *exactly* the same order, go back over the order of operations.

☑ A Test-Taking Tip

You *must* know the order of operations rule (PEMDAS) from memory. This rule is not written down for you anywhere on the GED Mathematics Test. You will be expected to know it.

Warm-up

Simplify these expressions using PEMDAS. Write down the steps as in "Here's an Example." For problem 5, remember that any number (other than zero) with a zero exponent is equal to 1.

1. $(13 - 7) + 2^3 =$ __________
2. $-5 + 17 =$ __________
3. $6^2 - 7(3 - 4) =$ __________
4. $3 + 14 - 5(6) + 7 =$ __________
5. $4^0 - 2(6 - 2) =$ __________

On the Springboard

1. Simplify the expression $8 + (5 - 2)^2$.

   (1) 9 (2) 17 (3) 121

   ① ② ③

2. Which of the following could *not* be a step in the simplifying of $7^2 - 4(18 - 6) + 9$?

   (1) $49 - 4(18 - 6) + 9$
   (2) $1 + 9$ (3) $49 - 48 + 9$
   (4) $49 - 4(12) + 9$ (5) $45(12) + 9$

   ① ② ③

Check your Springboard answers on page 344. If you got both right, go on to "The Real Thing." If you did not, it may be a good idea to check everything with PEMDAS before going on.

Warm-up Answers

1.	Parentheses	$13 - 7 = 6$	$6 + 2^3$
	Exponents	$2^3 = 8$	$6 + 8$
	Add	$6 + 8 = 14$	14
2.	Add	$-5 + 17 = 12$	12
3.	Parentheses	$3 - 4 = -1$	$6^2 - 7(-1)$
	Exponents	$6^2 = 36$	$36 - 7(-1)$
	Multiply	$7 \times -1 = -7$	$36 - (-7)$
	Subtract	$36 - (-7) = 43$	43
4.	Multiply	$5(6) = 30$	$3 + 14 - 30 + 7$
	Add/Subtract	$3 + 14 - 30 + 7$	
		$17 - 30 + 7$	
		$-13 + 7 = -6$	-6
5.	Parentheses	$6 - 2 = 4$	$4^0 - 2(4)$
	Exponents	$4^0 = 1$	$1 - 2(4)$
	Multiply	$2 \times 4 = 8$	$1 - 8$
	Subtract	$1 - 8 = -7$	-7

The Real Thing

1. What is the value of $32 - 28 + 41 - 19 + 56$?

   (1) 25 (2) 47 (3) 63
   (4) 82 (5) 106

   ① ② ③ ④ ⑤

2. What is the value of $93 - (48 + 23) + (47 - 25)$?

   (1) 23 (2) 44 (3) 67
   (4) 90 (5) 106

   ① ② ③ ④ ⑤

3. What is the value of $6^2 - (25 - 2^3) + (16 \div 4)$?

   (1) 23 (2) 35 (3) 48
   (4) 56 (5) 64

   ① ② ③ ④ ⑤

Check your answers on page 348.

Laws for Rearranging Numerical Expressions

There are three laws that make it easier to simplify expressions. If you know these laws, you can save time in simplifying expressions on the GED Mathematics Test.

Order Law (Commutative Law)

The **order law** says that you can reverse the order of numbers in an addition problem or a multiplication problem and still get the same answer.

Here's an Example

■ Look at these two expressions.

$2 + 3$
$3 + 2$

Notice that they both equal 5.

If you reverse the order of the numbers in $2 + 3$, you get $3 + 2$. The answer for the addition is the same either way.

$$2 + 3 = 3 + 2$$

Although you can reverse any addition or multiplication problem, you cannot reverse subtraction or division. You know that $3 - 5$ does *not* equal $5 - 3$, nor does $30 \div 5$ equal $5 \div 30$.

Coming to Terms

order law (commutative law) the rule that says numbers in an addition problem or multiplication problem may be reversed without changing the answer. For example, $2 + 3 = 3 + 2$ and $2 \times 3 = 3 \times 2$.

Try It Yourself

■ Which of these shows the order law? Write *yes* or *no.*

$3 + 4 = 4 + 3$ ____________

$6 + 5 = 11$ ____________

$0 (5) = 5 (0)$ ____________

Did you say *yes* for the first and third?

■ Now fill in the blanks. Use the order law.

$8 + 9 =$ ________ $+$ ________

$5 (6 + 7) = ($________ $+ 7) 5$

Did you fill in the blanks like this? $8 + 9 = 9 + 8$, and $5 (6 + 7) = (6 + 7) 5$.

Grouping Law (Associative Law)

The **grouping law** says that when you are adding three or more numbers or multiplying three or more numbers, you can group the numbers in different ways and still get the same answer.

Here's an Example

■ Look at this expression.

$$(4 + 5) + 3 = 4 + (5 + 3)$$

The numbers are in the same order on both sides of the equal sign. Only the parentheses have been moved.

Simplify each side separately.

Left Side		Right Side
$(4 + 5) + 3$	$=$	$4 + (5 + 3)$
$9 + 3$	$=$	$4 + 8$
12	$=$	12

Can you see that no matter how the three numbers being added are grouped, the two sides are equal?

The same thing is true of multiplication.

■ Look at this expression. Simplify each side separately.

$$(4 \times 5) \times 3 = 4 \times (5 \times 3)$$

Left Side		Right Side
$(4 \times 5) \times 3$	$=$	$4 \times (5 \times 3)$
20×3	$=$	4×15
60	$=$	60

You can regroup numbers that are multiplied or added. You cannot do this when you subtract or divide. For example $(7 - 4) - 1$ is *not equal to* $7 - (4 - 1)$ because $(7 - 4) - 1 = 2$, but $7 - (4 - 1) = 4$.

Try It Yourself

■ Which of the following shows the grouping law? Write *yes* or *no.*

$7 + (5 + 2) = (7 + 5) + 2$ ____________

$8 \times (6 \times 3) = 8 \times (3 \times 6)$ ____________

$[(9)(5)] (2) = (9) [(5)(2)]$ ____________

$7 + 0 + 4 = 11$ ____________

Did you say *yes* for the first and third?

■ Now fill in the blanks using the grouping law.

$9 \times (6 \times$ ______$) = ($______ $\times$ ______$) \times 4$

Does your answer look like this?

$9 \times (6 \times 4) = (9 \times 6) \times 4$

Coming to Terms

grouping law (associative law) the rule stating that three or more numbers to be added or multiplied can be grouped in different ways without changing the answer. For example, $(2 + 3) + 4 = 2 + (3 + 4)$, and $(2 \times 3) \times 4 = 2 \times (3 \times 4)$.

Law for Spreading Multiplication over Addition (Distributive Law)

The **law for spreading multiplication over addition** says that if you are to multiply one number by the sum of two or more other numbers, you can multiply the first number by *each* of the other numbers and then add.

Here's an Example

■ Look at this example to see what the spreading law means.

$$2(3 + 4) = 2(3) + 2(4)$$

Do you see how the 2 gets spread to the other two numbers (3 and 4)?

$$2(3 + 4) = 2(3) + 2(4)$$

Notice that the + sign stays between the 2(3) and the 2(4). If you simplify each side separately, you see that the expressions $2(3 + 4)$ and $2(3) + 2(4)$ really are equal.

$2(3 + 4)$	$=$	$2(3) + 2(4)$
$2(7)$	$=$	$6 + 8$
14	$=$	14

Try It Yourself

■ Which of these shows the law of spreading multiplication over addition? Write *yes* or *no.*

$3 + 4 = 4 + 3$ ____________

$6 \times (4 \times 7) = (6 \times 4) \times 7$ ____________

$9(1 + 7) = 9(1) + 9(7)$ ____________

If you understand the spreading law, you should have chosen only the third.

■ Fill in the blanks using the spreading law.

$3(4 + 9) = 3($______$) +$ ______(9)

______$(12 + 4) =$ ______$(12) + 6($______$)$

Did you fill in the blanks like this?

$3(4 + 9) = 3(4) + 3(9)$
$6(12 + 4) = 6(12) + 6(4)$

Coming to Terms

law for spreading multiplication over addition (distributive law) the rule that states that multiplying a number by a sum is the same as multiplying the first number by each number making up the sum and then adding. For example, $2\,(3 + 4) = 2(3) + 2(4)$. You can also spread multiplication over subtraction.

Warm-up

If the statement is true write *T* and if it is false write *F*.

1. $27(4 + 16) = 27(4) + 27(16)$ ______

2. $4 \div 98 = 98 \div 4$ ______

3. $6 + (87 + (-2)) = (6 + 87) + (-2)$ ______

4. $4(36) - 3(36) = 3(36) - 4(36)$ ______

5. $-17 + (4 \times 5) = (4 \times 5) + (-17)$ ______

On the Springboard

3. Which of the following is equal to $89 \times 3 + 89 \times 7$?

   (1) $3 + 89 \times 7$
   (2) $89(3 + 7)$
   (3) $7 \times (89 + 3)$

   (1) (2) (3)

4. Which of the following equals $3 + (4 + 9)$?

   (1) $(3 + 4) + 9$
   (2) $3(4) + 3(9)$
   (3) $3 \times (4 + 3) \times 9$

   (1) (2) (3)

Warm-up Answers
1. *T* **2.** *F* **3.** *T* **4.** *F* **5.** *T*

Check your Springboard answers on page 344. If you got them both right, go on to the next section about writing numerical expressions. If you got either wrong, find your mistakes. Review if you wish. When you're sure you're ready, go on.

Writing Numerical Expressions

Sometimes you have to write a numerical expression for a practical situation.

Here's an Example

How you set up the numbers is more important than getting a final answer in this section.

■ A part-time typist worked 2 days one week and 4 days the next. Each day he earned \$50. Write an expression that tells how much he earned in the two weeks.

In the Arithmetic section of this book you worked with many problems of this type. You may have already calculated in your head that the typist earned \$300. In this Algebra section you are not being asked for a final number as an answer but for an expression that tells how to find the answer. Do you recall the system for solving word problems that was introduced in Lesson 1?

Step 1 Read the problem carefully.

Step 2 Decide what it is you are trying to find.

Step 3 Decide what operations to use.

Step 4 Set up the problem.

Step 5 Compute the answer.

Step 6 Check the computation.

Step 7 See if the answer makes sense.

Here you only have to go as far as Step 4. The typist worked $2 + 4$ days and earned \$50 for each day, so the expression for the total amount earned is $50(2 + 4)$.

Try It Yourself

Write a mathematical expression for this situation.

■ In a discount store Alice and her sister bought 2 bottles of shampoo for \$2.29 each and a paperback novel for \$2.95. They split the cost of these items. Give an expression that shows what Alice paid in dollars and cents.

____

If you began by showing the total cost of the items, you were off to a good start. The cost of the two bottles of shampoo can be shown as 2(\$2.29). This must be added to the cost of the novel, \$2.95. You should have written all this as 2(\$2.29) + \$2.95. To show how much Alice paid, did you indicate that this total cost must be divided by 2? Your final expression should have been $\frac{2(\$2.29) + \$2.95}{2}$ or (2(\$2.29) + 2.95) ÷ 2. If you left out the \$ signs, that was all right. If you got either of these as your answer, you're further along in algebra than you may think.

Warm-up

Write an expression for each situation.

1. The price of a jar of baby food if 3 jars cost \$0.98 ______________________________

2. The cost of 3 photocopies of each of 15 pages if copies cost \$.10 each

____

3. The average monthly electric bill if the annual number of kilowatt-hours used is 4,632 and the cost is \$.062 per kilowatt-hour

____

Warm-up Answers

1. $\frac{\$0.98}{3}$ or \$0.98 ÷ 3

2. \$.10(3)(15) or \$.10(3 × 15) or \$0.10 × 3 × 15

3. $\frac{4{,}632 \times \$0.062}{12}$ or $\left(\frac{4{,}632}{12}\right)$ \$0.062 or \$0.062(4,632 ÷ 12)

On the Springboard

5. Four students in an algebra class got final exam scores of 90, 85, 80, and 75. Which expression shows their average score?

(1) $\frac{4}{90 + 85 + 80 + 75}$

(2) $4(90 + 85 + 80 + 75)$

(3) $\frac{90 + 85 + 80 + 75}{4}$

(1) (2) (3)

6. Jennifer bought 3 avocados for \$0.50 each and an apple for \$0.20. Which of the following expressions shows the amount she spent?

(1) 3(\$0.50) + \$0.20

(2) $\frac{\$1.50}{3}$ + \$0.20

(3) 3(\$1.50 + \$0.20)

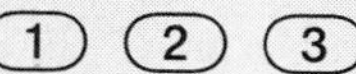

Check your Springboard answers on page 344. If you got both correct, go on to "The Real Thing." If you missed any, check carefully to make sure you know how to correct your work. Then go on to "The Real Thing."

The Real Thing

4. A machinist made 85 parts each day for 3 days and 75 parts for each of 2 days. Which expression tells how many parts the machinist made in these 5 days?

(1) (3 × 85) − 75
(2) (3 × 2) × (85 + 75)
(3) (3 × 85) + (2 × 75)
(4) 85 + 75 + 2 + 3
(5) 5 × 85 + 75)

(1) (2) (3) (4) (5)

5. Six Brand A razor blades cost \$1.29. Nine Brand B razor blades cost \$2.54. Which expression below tells how much more a Brand B blade costs than a Brand A blade?

(1) $3.83 - 2.54 - 15$
(2) $(2.54 \div 9) - (1.29 \div 6)$
(3) $\frac{2.54}{6} - \frac{1.29}{9}$
(4) $(6 + 9) - (1.29 + 2.54)$
(5) $1.29(6 + 9) - 2.54(6 + 9)$

(1) (2) (3) (4) (5)

6. Jane earned \$8 per hour for 7 hours and \$6.25 per hour for 5 hours. Which expression below tells how much Jane earned for the 12 hours?

(1) $12(8 + 6.25)$
(2) $(7 \times 8) + 6.25$
(3) $(5 + 7) \times (8 + 6.25)$
(4) $(7 \times 8) + (5 \times 6.25)$
(5) $(8 \times 5) \times (7 \times 6.25)$

(1) (2) (3) (4) (5)

7. On sale, 4 bars of soap cost \$0.88. The regular price of the 4 bars is \$0.97. Which expression tells how much Stefan saved by buying 20 bars on sale?

(1) $4(0.97 - 0.88) + 20$
(2) $5 + (0.97 - 0.88)$
(3) $(5 + 0.88) - 0.97$
(4) $(20 \times 0.88) - (20 \times 0.97)$
(5) $\frac{20}{4}(0.97 - 0.88)$

(1) (2) (3) (4) (5)

8. The Ryans bought a refrigerator for \$980. They paid \$200 down and arranged to pay the balance in 9 equal monthly payments. Which expression tells how much each monthly payment will be?

(1) $\frac{980 - 200}{9}$
(2) $9(980 - 200)$
(3) $\frac{980 + 200}{9}$
(4) $9(980 + 200)$
(5) $980 - \frac{200}{9}$

(1) (2) (3) (4) (5)

Check your answers and record all your scores on pages 348–349.

LESSON 44

Algebraic Expressions

Algebraic expressions are like numerical expressions except that you can use letters as well as numbers in such expressions.

Using Letters to Express Unknown Numbers

Remember that an algebraic expression is like a numerical expression except that it has one or more letters in it. To write an algebraic expression for a problem, you do just what you did to write a numerical expression. Instead of having all numbers, you will have one or more letters.

Here's an Example

This example will involve an addition expression.

■ If a person is y years old now, what expression shows how old he will be 8 years from now?

To solve, write an addition expression that uses what you are told.

Here "8 years from now" shows that you have to add 8 to however many years old the person is now. If he is 16 years old now, he will be 16 + 8, or 24 years old, 8 years from now. Since you don't know exactly how old this person is, the age is an open question. That's why a letter (y) is used instead of one specific number. If y stands for how many years old the person is, then the expression $y + 8$ shows what the person's age will be 8 years from now, *however* old the person is now.

Now try one with subtraction.

■ Ethel weighs x pounds. What expression shows how much she would weigh if she lost 10 pounds?

To solve, write an expression using x and 10.

Here the word *lost* is the key to deciding what operation to use. You will have to *subtract* 10. You aren't told exactly what she weighs now. That's why the letter x is used in stating the question. She would weigh $x - 10$ pounds.

The next one has multiplication in it.

■ Peggy's shopping bill was d dollars. Sally's bill was four times as much. What expression tells how much Sally's bill was in dollars?

To solve, you will write a multiplication expression, since Sally spent four times as much as Peggy. So if Peggy's bill was d dollars, Sally's bill must have been $4 \times d$, or $4d$. (You learned that leaving out the $\times$ and using parentheses is a way to write multiplication.

Finally, here's an example with division.

■ Prairie High School had p students. Lake High School had only a third as many students. What expression tells how many students are in Lake High School?

Do you see that, in this problem, the words *a third as many students* are your key? The number of students in Lake High School is $p \div 3$. This could also be written as

$$p/3 \text{ or } \frac{p}{3}.$$

Sometimes the expressions are more complex than those in these examples. In the last example, if Lake had 45 students *less* than a third of Prairie, the expression would have been $p/3 - 45$. Writing algebraic expressions for problems is really no more difficult than writing numerical expressions for problems for which all the numbers are known.

Try It Yourself

■ There are p pounds of honey to be shared equally among 9 people. Write an expression telling how many pounds of honey each person will get.

If you had been given a number instead of p, you would have divided this number by 9, right? If you did the same with p, you have the idea. The answer is $\frac{p}{9}$, or $p \div 9$, pounds. (The form $9\overline{)p}$ is not used for answers in algebra.)

■ A square piece of linoleum measures s feet on each side. A piece 8 square feet in area is removed so the linoleum will fit around a refrigerator. Write an expression telling how many square feet of linoleum are left?

The area before a piece is removed is s^2. To show how much is left after 8 square feet are removed, you need an expression that shows 8 being subtracted from s^2. Did you write $s^2 - 8$?

Warm-up

On your own paper, write algebraic expressions for the following:

1. What number is 6 more than half the number m?
2. What number is 14 less than 3 times the number m?
3. Mark got b bricks in the first load delivered for the wall he is building. He needs 5 times that number. How many bricks does he need?
4. John had 5 times as much money as Paul, but he has to divide it evenly among 6 people. If Paul had d dollars, how many dollars did each of the 6 people receive from John?

On the Springboard

1. Leon went on a diet at a weight of e pounds. How many pounds did he weigh after losing 35 pounds?

   (1) $35(e - 35)$ (2) $e - 35$ (3) $e/35$

   ① ② ③

2. If you paid c dollars per year on your car loan, how many dollars per month would that be?

   (1) $c/12$ (2) $12/c$ (3) $12(1 + c)$

   ① ② ③

Check your Springboard answers on pages 344–345. If you got these right, go on to the next section. If you missed any, review before you go on.

Warm-up Answers
1. $\frac{1}{2}n + 6$, or $\frac{n}{2} + 6$ **2.** $3m - 14$ **3.** $5b$ **4.** $\frac{5d}{6}$, or $5d \div 6$

Combining Like Terms

Terms are algebraic expressions that are added or subtracted. Sometimes you find two **like terms** in an expression. When you find like terms, you can combine them to make the original expression simpler. In the expression $3x + 4x + 2y$, the terms are $3x$, $4x$, and $2y$. The terms $3x$ and $4x$ are like terms. You can combine them by adding. You get $7x$. So, the original expression $3x + 4x + 2y$ can be simplified to $7x + 2y$. You cannot combine $7x + 2y$ into a single simpler term, because $7x$ and $2y$ are not like terms. One contains x but not y; the other contains y but not x.

Coming to Terms

like terms those terms using the same letter or letters and in the same way. Only like terms can be added or subtracted to get one single term.

Here's an Example

Here is an algebraic expression that can be simplified by combining like terms.

■ $3x + 5y + 6xy + 2x + 4xy + x^2 + y + 3x^2$

There are 8 different terms in the expression, but there are only 4 *kinds* of terms. Look for the letters that the terms contain. Also keep an eye out for letters that have exponents. The terms that are circled and joined by curved lines show the 4 kinds of like terms.

$$3x + 5y + 6xy + 2x + 4xy + x^2 + y + 3x^2$$

Here the terms are all being added.

	Like Terms	Combined
There are the x terms:	$3x$ and $2x$	$5x$
There are the y terms:	$5y$ and y(or $1y$)	$6y$
There are the xy terms:	$6xy$ and $4xy$	$10xy$
There are the x^2 terms:	x^2 (or $1x^2$ and $3x^2$)	$4x^2$

When the like terms are combined, the original expression can be simplified to this expression with just 4 terms.

$$5x + 6y + 10xy + 4x^2$$

Instead of circling like terms, you could have rearranged the original expression to put like terms next to one another and then combined.

$$3x + 2x + 5y + y + 6xy + 4xy + x^2 + 3x^2$$
$$5x \quad + \quad 6y \quad + \quad 10xy \quad + \quad 4x^2$$

■ Look at this algebraic expression whose terms are being added and subtracted.

$$3x + 5xy + 6y + 8 - 6x + 4y - xy - 10$$

Before you combine like terms in an expression, change subtraction to addition.

$$3x + 5xy + 6y + 8 - 6x + 4y - xy - 10$$
$$3x + 5xy + 6y + 8 + (-6x) + 4y + (-xy) + (-10)$$

Now you can simplify by combining all terms with the same labels. Notice that there are three different labels: x, xy, and y. There are two terms with no letter labels at all—they are simply numbers with no letters. These two terms can be combined. Now combine the x, the xy, and the y terms. Did you notice the term $-xy$? It is a short way of writing $-1xy$. When you combine $-xy$ and $5xy$, you get $4xy$ ($-1 + 5 = 4$). Your final answer is $-3x + 4xy + 10y - 2$.

When you combine like terms, always read carefully to find terms with *exactly* the same labels. Keep in mind that you can't combine a term with another term containing that letter and an exponent.

Try It Yourself

■ Simplify, $4m + (-5n) - (2m^2) - (-m) + n$

____

Did you change signs and combine like terms in this way?

$$4m + (-5n) - (2m^2) - (-m) + n$$
$$4m + (-5n) + (-2m^2) + (+m) + n$$
$$(4m + m) + (-5n + n) + (-2m^2)$$
$$5m + (-4n) + (-2m^2)$$

Then did you change to subtraction: $5m - 4n - 2m^2$? Did you get $5m - 4n - 2m^2$? If so, you just simplified a rather complex expression. If not, check to see if you changed subtraction to addition before combining like terms. Also check to make sure that the terms you combined had either a letter part that was exactly the same or no letter part at all. If you wrote the terms in a different order, that's all right. For example, $5m - 2m^2 - 4n$ is just as correct as $5m - 4n - 2m^2$.

A Test-Taking Tip

If you are asked to combine like terms on the GED Mathematics Test, it's important to recognize a correct answer that looks a little different. For example, you might get $-6x^2 + 5x$, but the answer might be listed as $5x - 6x^2$. Could you still see that your answer is right?

Warm-up

Combine like terms to simplify these expressions.

1. $5a + 8a =$ ______

2. $6x + -3x =$ ______

3. $-8y - y =$ ______

4. $4b - (-4b) + 6b =$ ______

5. $-10c - 3d + 8c =$ ______

6. $8t - 5 + 10 + t =$ ______

7. $15 - 7n - (-9n) =$ ______

8. $3xy + 6x + -4xy =$ ______

9. $-5d + 6 + 5d =$ ______

10. $8x - 6x + -3x =$ ______

Circle the letters of the two expressions that are equivalent.

11. **a.** $-4d + 2d^2 + 9$
b. $9 + 2d^2 - 4d$
c. $-4d - 9 + 2d^2$

12. **a.** $-5x + 1xy$
b. $1xy - (-5x)$
c. $xy + 5x$

13. **a.** $3a^2b + 2ab^2 + b$
b. $5a^2b + b$
c. $b + 2ab^2 + 3a^2b$

14. **a.** $6(9rs) + 0$
b. $69rs + 0$
c. $54rs$

Warm-up Answers
1. 13*a* **2.** 3*x* **3.** −9*y* **4.** 14*b* **5.** −2*c* − 3*d*
6. 9*t* + 5 **7.** 15 + 2*n* **8.** 6*x* − *xy* **9.** 6 **10.** −*x*
11. a,b **12.** b,c **13.** a,c **14.** a,c

On the Springboard

3. Simplify this expression: $6a + a - 5a$.

(1) a (2) $2a$ (3) $11a$

① ② ③

4. Combine like terms in this expression: $2x + 3y + 5x - 5y + x + 6y$.

(1) $4x + 8y$ (2) $8x + 4y$
(3) $2x - y$

① ② ③

Check your Springboard answers on page 344. If you got both right, go on to "The Real Thing." If you missed either, go over both until you can correct your work. Then go on to "The Real Thing."

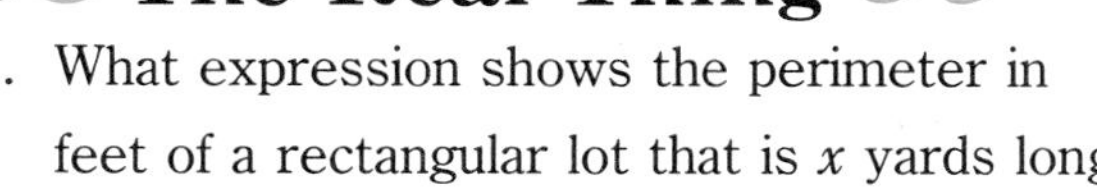

The Real Thing

1. What expression shows the perimeter in feet of a rectangular lot that is x yards long and $\frac{1}{3}x$ yards wide?

(1) $\frac{8}{9}x$ (2) $\frac{4}{3}x$ (3) $\frac{8}{3}x$
(4) $4x$ (5) $8x$

① ② ③ ④ ⑤

2. Jerome bought x apples at \$0.15 each and y apples at \$0.12 each. He sold all of the apples at a ball game for \$0.20 each. Which expression tells how much profit he made (in dollars and cents) from the sale of these apples?

(1) $0.05x + 0.08y$ (2) $0.08y - 0.05x$
(3) $0.20y - 0.20x$ (4) $0.35x + 0.32y$
(5) $0.40x + 0.27y$

① ② ③ ④ ⑤

3. The Johnsons bought a rectangular lot that was m feet by n feet. The Howards bought a rectangular lot that was $(m - 3)$ feet by $(2n + 7)$ feet. Which expression tells how much less the perimeter of the Johnsons' lot is than the perimeter of the Howards' lot?

(1) $2n + 8$ (2) $2n + 20$
(3) $2m + 2n$ (4) $2m + 4n + 8$
(5) $4m + 6n + 8$

① ② ③ ④ ⑤

4. One week, Theodore sold b photographs, c oil paintings, and d water colors. The following week, he sold 3 fewer photographs, 4 times as many oil paintings, and $\frac{1}{3}$ as many water colors. Which expression tells how many pictures he sold in the 2 weeks?

(1) $b - 3c + \left(\frac{2}{3}\right)d + 3$

(2) $2b + 4c + \left(1\frac{1}{3}\right)d - 3$

(3) $2b + 4c + \left(\frac{2}{3}\right)d - 3$

(4) $2b + 5c + \left(1\frac{1}{3}\right)d - 3$

(5) $5b + \left(1\frac{1}{3}\right)c + 2d - 3$

(1) (2) (3) (4) (5)

Check your answers on page 349.

Evaluating Algebraic Expressions

In some problems on the GED Mathematics Test you will be given an algebraic expression such as $2a - 3c + 4ac - c$ and *values* of each letter, for example, $a = 5$ and $c = -1$. You will have to evaluate, or find the value of, the expression. **Evaluating expressions** is important. If you substitute the numbers for the letters, you can find a value for the expression.

Coming to Terms

evaluating expressions substituting number values for the letters in algebraic expressions, and then using the order of operations to get a single value for each resulting numerical expression

Here's an Example

■ Evaluate $2a - 3c + 4ac$ using $a = 3$ and $c = 2$.

To solve, replace a with 3 and c with 2 throughout the expression. You may need to add some parentheses before calculating.

Step 1 Replace a with 3 and c with 2.

$2a - 3c + 4ac$
$2(3) - 3(2) + 4(3)(2)$

Step 2 Calculate term by term, then simplify.

$2(3) - 3(2) + 4(3)(2)$
$6 - 6 + 24$
$0 + 24$
24

The result is 24.

■ Evaluate $2a - 3c + 4ac$ using $a = 5$ and $c = -2$

Step 1 Replace a with 5 and c with -2. Use parentheses to keep things straight.

$2a - 3c + 4ac$
$2(5) - 3(-2) + 4(5)(-2)$

Step 2 Calculate term by term, then simplify.

$2(5) - 3(-2) + 4(5)(-2)$
$10 - (-6) + (-40)$

Subtracting -6 is the same as adding $+6$

$10 + (+6) + (-40)$
$16 + (-40)$
-24

The result is -24.

Try It Yourself

■ Use $x = 4$, $y = 3$, $z = 1$ and evaluate the expression $x^2 - 2xy + z$.

Did you replace the letters with numbers to get $4^2 - 2(4)(3) + 1$? Did you find that this numerical expression equals $16 - 24 + 1$? If you simplified this numerical expression correctly, you should have gotten -7 as your final answer.

A Test-Taking Tip

If you are asked to evaluate an expression on the GED Mathematics Test, you will always be told the value of each letter. There is no way you could evaluate an expression without this information.

Warm-up

Evaluate the first six expressions using $a = 4$, $b = 2$, and $c = 0$. It may help to use a separate piece of paper for your steps.

1. $a^2 - 4c =$ ____________________

2. $a - b - c =$ ____________________

3. $bc^2 - a =$ ____________________

4. $b - a^3 =$ ____________________

5. $3ac - 7ab =$ ____________________

6. $5(a + b) - c =$ ____________________

Now evaluate these expressions using $a = -4$, $b = 2$, and $c = 0$.

7. $\frac{a - c}{b - a} =$ ____________________

8. $c - b - a =$ ____________________

On the Springboard

5. Find the value of $e + 5f$ if $e = 2$ and $f = 3$.

   (1) 10 (2) 13 (3) 17

   ① ② ③

6. If $p = \frac{1}{2}$ and $q = 1$, find the value of pq.

   (1) $\frac{1}{2}$ (2) $1\frac{1}{2}$ (3) 2

   ① ② ③

7. Find the value of $2x - (y^2)$, if $x = -2$ and $y = 2$.

   (1) -12 (2) -8 (3) -4

   ① ② ③

Check your Springboard answers on page 345. If you got them all right, you did very well. Go on to "The Real Thing." If you had some errors, check your work carefully until you can correct your mistakes. When you're confident that you're ready, go on to "The Real Thing."

Warm-up Answers
1. 16 **2.** 2 **3.** -4 **4.** -62 **5.** -56 **6.** 30
7. $\frac{-4}{6}$ or $\frac{-2}{3}$ **8.** 2

“The Real Thing”

Questions 5–7 refer to the information below.

This year, twice Enrico's age (b) plus 12 is equal to his father's age (f). Enrico's sister's age (s) is $\frac{1}{3}$ of Enrico's age. Enrico's uncle's age (u) is equal to Enrico's father's age minus his sister's age.

5. If Enrico is 18, how old is his father? Use $2b + 12 = f$.

   (1) 30 (2) 42 (3) 46
   (4) 48 (5) 56

   ① ② ③ ④ ⑤

6. If Enrico is 18, how old is his sister? Use $\frac{1}{3}b = s$.

   (1) 6 (2) 9 (3) 10
   (4) 12 (5) 16

   ① ② ③ ④ ⑤

7. If Enrico is 18, how old is his uncle? Use $(2b + 12) - \left(\frac{1}{3}\right)b = u$.

   (1) 28 (2) 30 (3) 32
   (4) 36 (5) 42

   ① ② ③ ④ ⑤

8. Mr. O'Bara had x quarters, $2y$ dimes, and $3z$ nickels. Let $x = 7$, $y = 2$, and $z = 4$. How many coins (t) did Mr. O'Bara have in all? Use $x + 2y + 3z = t$.

   (1) 21 (2) 22 (3) 23
   (4) 28 (5) 29

   ① ② ③ ④ ⑤

9. Let $x = 3$, $y = 7$, and $z = 10$. In dollars and cents how much money (c) did Mr. O'Bara have in all? Use $0.25x + 0.10(2y) + 0.05(3z) = c$.

   (1) \$0.75 (2) \$1.40 (3) \$1.50
   (4) \$2.90 (5) \$3.65

   ① ② ③ ④ ⑤

Questions 10 and 11 refer to the following information.

A child's ticket costs \$1.50 and an adult's ticket costs \$2.75. Let x be the number of children's tickets sold. Let y be the number of adult tickets sold. Then the total value of the tickets sold (in dollars and cents) is $1.50x + 2.75y$.

10. Debbie and Chuck sold 15 children's tickets and 20 adult tickets. How much did they collect for these tickets?

(1) \$22.50 (2) \$55.00 (3) \$68.25
(4) \$71.25 (5) \$77.50

① ② ③ ④ ⑤

11. Chris sold 13 children's tickets and 18 adult tickets. Kim sold 15 children's tickets and 12 adult tickets. Chris collected how much more money than Kim?

(1) \$13.50 (2) \$16.50 (3) \$59.25
(4) \$72.25 (5) \$131.50

① ② ③ ④ ⑤

12. Ellen practiced n minutes on each of 2 days, $(n + 15)$ minutes on each of 3 days, and $2n$ minutes on each of 4 days. Which expression tells how many minutes of practice she averaged each day?

(1) $(5n + 17)/9$ (2) $9(13n + 45)$
(3) $13n + 45$ (4) $(13n + 45)/9$
(5) $13n/9$

① ② ③ ④ ⑤

13. Bill has f nickels, g dimes, and h dollar bills. Pam has 5 more nickels than Bill has, 3 times as many dimes, and half as many dollar bills. Which expression tells how many *coins* they have together?

(1) $2f + 4g + 5$
(2) $5f + 2g + 0.5h$
(3) $6f + 3g + 0.5h$
(4) $7f + 1.5g + 4h$
(5) $2f + 4g + 1.5h + 5$

① ② ③ ④ ⑤

Check your answers on pages 349–350 and record all your scores on page 350.

LESSON 45

Algebraic Equations

In Lesson 44, you learned about algebraic expressions. An **equation** is a statement that two expressions are equal. Look at these equations:

$$3 + 4(5) = 27 - 4$$
$$3x + 5 = 4x - 2$$
$$17 = 10 - y$$

You know that these are equations because each has two expressions with an equal sign between them. The first is a **numerical equation** because both the expressions (that is, $3 + 4(4)$ and $27 - 4$) are numerical expressions. If either or both of the expressions in an equation is an *algebraic* expression, the equation is called an **algebraic equation.** $3x + 5 = 4x - 2$ is an algebraic equation with two algebraic expressions. The third equation, $17 = 10 - y$, is also an algebraic equation even though it contains just one algebraic expression.

Coming to Terms

equation a statement that two expressions are equal. An equation consists of two expressions with an equal sign between them.

numerical equation an equation in which both the expressions (before and after the equal sign) are numerical expressions. A numerical equation contains no letters.

algebraic equation an equation in which one or both of the expressions is an algebraic equation. Algebraic equations contain one or more letters.

Here's an Example

■ Tell whether these are or are not equations. If they are equations, tell whether they are numerical or algebraic equations.

a. $x^2 + 7x = 5x - 3$
b. $x + 6y$
c. $4(3) - 8 = 16 \div 2$
d. $2y - 3 =$

$x^2 + 7x = 5x - 3$ is an algebraic equation, because it contains two algebraic expressions with an equal sign between them.

$x + 6y$ is not an equation, since it does not contain an equal sign.

$4(3) - 8 = 16 \div 2$ is a numerical equation. It has two numerical expressions, $4(3) - 8$ and $16 \div 2$, with an equal sign between them. It is not an algebraic equation, because the question contains no letters.

$2y - 3 =$ is not an equation. It has an equal sign, but there is no expression after the equal sign. An equation must have *two* expressions, one before the equal sign and one after.

Try It Yourself

■ Tell whether each of these is or is not an algebraic equation. Answer *yes* or *no*.

$5 = 3y + 2$ ______

$8x - 2y + 3$ ______

You should have indicated that the first one is an algebraic equation. It has two expressions with an equal sign between them. The fact that one of the expressions contains a letter tells you that the equation is an *algebraic* equation. The second is just an expression, so you should have answered *no*.

Warm-up

Are the following algebraic equations? Answer *yes* or *no*.

1. $90 = 8 - 3x$ ______

2. $2a - 6b + 15$ ______

3. $3c - 6 = 2c$ ______

4. $6x = 0$ ______

5. $10 - 3b$ ______

6. $5^2 - 3^2 = 4^2$ ______

Warm-up Answers
1. yes **2.** no **3.** yes **4.** yes **5.** no **6.** no (It is a *numerical* equation but not an *algebraic* equation.)

Roots of Equations

Suppose an algebraic equation contains just one letter. If you put a number in place of the letter everywhere in the equation (the *same* number throughout), you'll get a numerical equation. If the numerical equation is *true,* that is, if it is a solution to the equation, then the number that you put in is called a **root** of the equation.

Sometimes the word *solution* is used instead of *root.* You can see if any number is a root of an equation by replacing the letter with the number. If the expression on the left of the equal sign has the same value as the one on the right-hand side of the equal sign, then you know that the number you put in was a root of the equation.

Coming to Terms

root any number or numbers that, when put in place of a letter in an algebraic equation, makes both sides of the equation equal

Here's an Example

■ Find out if $x = 6$ is a root of the equation $3x + 5 = 4x - 2$.

To see if $x = 6$ is a root of the equation, replace x with 6 and simplify.

Step 1 Write the equation and replace x with 6.

$$3x + 5 = 4x - 2$$
$$3(6) + 5 = 4(6) - 2$$

Step 2 Simplify both sides of the numerical equation.

$$3(6) + 5 = 4(6) - 2$$
$$18 + 5 \qquad 24 - 2$$
$$23 \qquad 22$$

Step 3 Compare the values you get when you simplified. Here the left hand side is 23 and the right hand side is 22. Since these are not equal, the numerical equation $3(6) + 5 = 4(6) - 2$ is *false.*

$x = 6$ is not a root.

■ Find out whether $x = 7$ is a root of the equation $3x + 5 = 4x - 2$.

To see if $x = 7$ is a root, put 7 in place of x and simplify.

Step 1 Write the equation and then replace x with 7.

$$3x + 5 = 4x - 2$$
$$3(7) + 5 = 4(7) - 2$$

Step 2 Simplify both sides of the numerical equation.

$$\begin{array}{ccc} 3(7) + 5 & = & 4(7) - 2 \\ 21 + 5 & & 28 - 2 \\ 26 & & 26 \end{array}$$

Step 3 Compare the values on both sides. You got 26 both times. So the numerical equation is *true.*

$x = 7$ is a root of the equation $3x + 5 = 4x - 2$.

Try It Yourself

■ See if $y = 1$ is a root of the equation $12 - 5y = 2y + 5$.

Did you put 1 in place of y and get $12 - 5(1) = 2(1) + 5$? Did you simplify both sides and get 7 in both places? You should have found that the numerical equation is *true,* which means that $y = 1$ is a root of the equation.

Warm-up

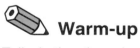

Tell whether the value given is a root of the equation. Write *yes* or *no.* (Use scrap paper for calculations.)

1. $x - 5 = 10$ for $x = -5$ ______
2. $-7 = y + 3$ for $y = -10$ ______
3. $10 - 3x = 2x + 5$ for $x = 1$ ______
4. $0 = 10y - 2$ for $y = 50$ ______
5. $\frac{1}{2}x = 16$ for $x = 32$ ______
6. $\frac{1}{4} = \frac{8}{3}y$ for $y = \frac{3}{32}$ ______
7. $2x - 5x + 3 = -6 - x$ for $x = -2$ ______
8. $0.5y = 20$ for $y = 40$ ______
9. $x^2 - 5x + 6 = 0$ for $x = 2$ ______
10. $x^2 - 5x + 6 = 0$ for $x = 3$ ______

Solving Equations

When you solve an algebraic equation, your aim is to find the values that make the equation true. The letter in the equation is often called the *unknown.* The way that usually works best is to concentrate on getting the unknown alone on one side of the equation and the numerical values on the other.

Here's an Example

■ Solve the equation $3 + x = 15$. To solve the equation, add the same number to each side of it to isolate the unknown.

If you add -3 to the left side of the equation, then x will stand alone. But you are not finished. The important rule to remember is *what you do to one side of the equation you MUST do to the other side.*

In this case, you must add -3 to *both* sides.

$$\begin{array}{rcr} 3 + x & = & 15 \\ -3 & & -3 \\ \hline 0 + x & = & 12 \end{array} \quad \text{or } x = 12$$

Did you notice that in order to eliminate a number, you added its opposite? If the problem was $-3 + x = 15$, you could eliminate -3 by adding $+3$.

$$\begin{array}{rcr} -3 + x & = & 15 \\ +3 & & +3 \\ \hline 0 + x & = & 18 \end{array} \quad \text{or } x = 18$$

Are you beginning to see a pattern here? To isolate the unknown in a *multiplication* problem, you *divide* both sides of the equation by the same number.

$$3x = 15$$
$$\frac{3x}{3} = \frac{15}{3}$$
$$x = 5$$

Warm-up Answers
1. no **2.** yes **3.** yes **4.** no **5.** yes **6.** yes **7.** no **8.** yes **9.** yes **10.** yes

To isolate the unknown in a *division* problem, you *multiply* both sides of the equation by the same number.

$$3 = \frac{x}{5}$$
$$3(5) = \frac{x}{5}(5)$$
$$15 = x$$

Solving equations often requires more than just one step. There are equations in which you must add the opposite and then multiply or divide. Always add first. Then multiply or divide. It's a good idea to check your answer by putting the number you got in place of the letter in the original equation.

Look at this equation that involves more than one step.

■ If $9y + 2 = 20$, then $y =$

To solve the equation, first do addition; then do multiplication or division.

Step 1 Add -2 to both sides.

$$9y = 18$$

Step 2 Divide both sides by 9.

$$y = 2$$

Step 3 Check your answer by substituting 2 for y in the original equation.

$$9(2) + 2 = 20$$
$$18 + 2 = 20$$
$$20 = 20$$

If $9y + 2 = 20$, then $y = 2$.

Try It Yourself

■ Find r when $\frac{r}{2} = 5$. ____________

Did you see you had to multiply to undo the division? Did you multiply *both* sides by 2 to get $r = 10$?

Now try a problem with two steps.

■ If $7p - 4 = 24$, $p =$ ____________

Did you first change subtraction to addition? Then did you add 4 to both sides to get $7p = 28$? Did you divide both sides by 7? You should have $p = 4$.

Warm-up

What number should you add to both sides to isolate the unknown?

1. $x + 3 = 5$ ________

2. $x - 3 = 5$ ________

3. $4 + x = 9$ ________

4. $-4 + x = 8$ ________

What number should you divide both sides by to isolate the unknown?

5. $4x = 24$ ________

6. $68 = 34x$ ________

7. $\frac{1}{2}x = 12$ ________

What number would you multiply both sides by to isolate the unknown?

8. $\frac{x}{3} = 9$ ____ **9.** $\frac{y}{5} = 2$ ____ **10.** $\frac{r}{6} = \frac{1}{2}$ ____

Warm-up Answers

1. -3 **2.** $+3$ **3.** -4 **4.** $+4$ **5.** 4 **6.** 34 **7.** $\frac{1}{2}$
8. 3 **9.** 5 **10.** 6

A Test-Taking Tip

On the GED Mathematics Test you may be asked to solve an equation just when you are running short of time. You can substitute the choices one by one in most cases until you find the one that works and then you have the answer.

On the Springboard

1. $x - 5 = 10$ $x =$ ______

(1) 10 (2) 15 (3) 50

① ② ③

2. $\frac{1}{2}x = 16$ $x =$ ______

(1) 16 (2) 18 (3) 32

① ② ③

3. $0 = \frac{1}{4}y - 2$ $y =$ ______

(1) -10 (2) -8 (3) 8

(1) (2) (3)

4. $0.5y = 20$ $y =$ ______

(1) 40   (2) 80   (3) 100

(1) (2) (3)

Check your Springboard answers on pages 344–345. If you got these right, go on to "The Real Thing." If you missed any, check your work. Review the section if you wish. When you're confident that you're ready, go on to "The Real Thing."

The Real Thing

1. What is the root of $x - 7 = 18$?

   (1) 7   (2) 11   (3) 18
   (4) 25   (5) 28

   (1) (2) (3) (4) (5)

2. What is the root of $13 = x + 6$?

   (1) 3   (2) 7   (3) 13
   (4) 19   (5) 78

   (1) (2) (3) (4) (5)

3. What is the root of $3x - 4 = 5$?

   (1) 3   (2) 6   (3) 9   (4) 17   (5) 60

   (1) (2) (3) (4) (5)

4. What is the root of $(4/5)x + 7 = 15$?

   (1) 4   (2) 5   (3) 10   (4) 37   (5) 54

   (1) (2) (3) (4) (5)

5. If the number of quarters in Andy's bank is increased by 7, he will have 23 quarters. Which of the following equations can be used to solve this problem?

   (1) $x + 23 = 7$ (2) $7x = 2(3)$
   (3) $x = 23 + 7$ (4) $x - 7 = 23$
   (5) $x + 7 = 23$

   (1) (2) (3) (4) (5)

Check your answers on page 350.

An Unknown on Both Sides of the Equal Sign

Sometimes an equation will have letters, or unknowns, on both sides of the equation. In order to solve them, work to get all the unknowns to one side of the equation. It doesn't matter which side; use whichever is easiest. Then get all the numbers on the other side. Remember, *what you do to one side of the equation, you must do to the other.*

Here's an Example

■ Find x if $12 - 5x = 2x + 5$.

To solve, isolate the terms containing x on one side, then solve as usual.

Step 1 Get all the unknowns on one side. One way to do this is to add $5x$ to each side.

$$\begin{array}{rcr} 12 + -5x & = & 2x + 5 \\ +5x & & +5x \\ \hline 12 \quad -0 & = & 7x + 5 \end{array}$$

Step 2 Isolate the unknown by adding -5 to each side.

$$\begin{array}{rcr} 12 & = & 7x + 5 \\ -5 & & -5 \\ \hline 7 & = & 7x \end{array}$$

Step 3 To find x, divide both sides by 7.

$$\frac{7}{7} = \frac{7x}{7}$$

$$1 = x \quad \text{or} \quad x = 1$$

Remember that you can always check your work by putting your answer into the original equation to see if you get a true numerical equation.

Try It Yourself

■ Solve for f: $8f + 5 = 14 + 11f$

$f =$ ______

Did you add $-8f$ to each side first? That gave you $5 = 14 + 3f$. Then did you add -14 to each side to isolate the $3f$? If you did, you got $-9 = 3f$. Finally, did you divide each side by 3 to undo the multiplication? Your answer should have been $-3 = f$.

If one side of an equation has more than one unknown, combine like terms first and then solve. For example, $4r - 5 - r = 13 - 2r - 4r$ should first be changed to addition $4r + -5 + -r = 13 + -2r + -4r$. Then combine like terms to get $3r + -5 = 13 + -6r$. Work as usual.

Warm-up

What would you add to each side of these equations to get all the unknowns on the left-hand sides?

1. $3x + 5 = x + 2$ ________

2. $3y + 5 = 2 - y$ ________

3. $6 + 4x = 8 - \frac{1}{2}x$ ________

4. $7c + 2 = c + 2c + 6$ ________

Solve these equations and check by substituting your answer in the original equation.

5. $b + 5 = 5b - 15$ $b =$ ________

6. $3f + 15 = 6f + 4f - 13$

$f =$ ________

7. $5x + 75 = 2x + 51$ $x =$ ________

8. $y - 6 - 4y = 12 + 3y$ $y =$ ________

Warm-up Answers
1. add $-x$ (or add $-3x$) **2.** add y (or add $-3y$) **3.** add $\frac{1}{2}x$ (or add $-4x$) **4.** add $-3c$ (or add $-7c$) **5.** $b = 5$ **6.** $f = 4$ **7.** $x = -8$ **8.** $y = -3$

On the Springboard

5. $x - 3 = 10x + 3$ $x =$ ______

(1) $-\frac{3}{2}$ (2) $-\frac{6}{7}$ (3) $-\frac{2}{3}$

(1) (2) (3)

6. $5 - 3a = a + 3$ $a =$ ______

(1) -2 (2) $\frac{1}{2}$ (3) 1

(1) (2) (3)

7. $8 = 4b - 16$ $b =$ ______

(1) $\frac{1}{2}$ (2) 2 (3) 6

(1) (2) (3)

8. $6 - 2x = 6 + 3x$ $x =$ ______

(1) -12 (2) 0 (3) $\frac{5}{12}$

(1) (2) (3)

9. $\frac{2}{3}y = \frac{8}{15}$ $y =$ ______

(1) $\frac{4}{5}$ (2) $\frac{16}{15}$ (3) $\frac{6}{5}$

(1) (2) (3)

Check your Springboard answers on page 345. If you got these right, go on to "The Real Thing." If you got any wrong, go over your work and correct your mistakes. When you can solve all the equations correctly, you are ready to go on.

The Real Thing

6. Becky says that if the number of pencils on her desk is increased by 9, she will have 4 more than twice the number she has now. Which of the following equations can be used to solve this problem?

(1) $2x + 9 = 4 - x$
(2) $x + 9 = 2x + 4$
(3) $x + 4 = 2x - 9$
(4) $x + 2x = x + 4$
(5) $2x + 9 = x + 4$

(1) (2) (3) (4) (5)

7. If a number is subtracted from 25, the difference is 8 less than twice the number. What is the number?

(1) 2 (2) 7 (3) 11
(4) 18 (5) 33

(1) (2) (3) (4) (5)

8. What is the root of $4x - 8.275 = 3(x + 5.487)$?

(1) 14.975 (2) 18.492 (3) 21.153
(4) 24.736 (5) 29.328

(1) (2) (3) (4) (5)

9. What is the root of $9.42 - 2x = 3x + 2.57$?

(1) 0.64 (2) 1.37 (3) 2.40
(4) 6.00 (5) 11.99

① ② ③ ④ ⑤

10. What is the root of $\left(\frac{2}{3}\right)x - \frac{1}{2} = \frac{3}{4} + \left(\frac{1}{6}\right)x$?

(1) $\frac{1}{3}$ (2) $\frac{5}{6}$ (3) $1\frac{1}{4}$
(4) $1\frac{7}{12}$ (5) $2\frac{1}{2}$

① ② ③ ④ ⑤

Check your answers on pages 350–351.

Using Equations to Solve Word Problems

Word problems can often be translated into equations. You then solve the equation to find the unknown. First make sure that you know what the problem is asking you to find. Call that amount *x* or *y* or whatever letter you like. Then translate the problem into an equation containing the letter you selected for the unknown.

Here's an Example

■ The average score for a series of tests is found by dividing the total number of points for all the tests by the number of tests taken. After 8 tests Eamon found that his average was 85. What was his total number of points for all the tests?

Step 1 Find what is asked for. Call it *x*. Here you are asked for the total number of points on all the tests. So *x* stands for the total number of points on all the tests.

Step 2 Write down the problem as an equation. Here the problem says

$$\frac{\text{total points}}{\text{number of tests}} = \text{average}$$

Put numbers and *x* in place of the words.

$$\frac{x}{8} = 85$$

Step 3 Solve the equation you get. Here you undo division by multiplying both sides by 8 and you get

$$x = 85 \times 8 = 680$$

The total number of points on all the tests was 680. Look at how all the words in the original problem got boiled down to the short equation in step 2. The algebraic equation is much simpler.

Try It Yourself

■ At a class reunion there were twice as many women present as there were men. In all, there were 186 class members at the reunion. How many were men?

____

If you let the number of men present equal *x*, then the number of women was twice that, or $2x$. When you added men and women, you got $x + 2x = 186$. To solve, you add x and $2x$ to get $3x$, so that the equation becomes $3x = 186$. If you divide both sides by 3, you get $x = 62$. The answer is 62 men.

Warm-up

Write equations for these problems. You do not need to solve the equations.

1. Lynn bought 3 blouses that were on sale. Each cost the same. The total tax was \$2. In all she paid \$47. How much did each blouse cost? (Let *b* be the cost of each blouse.)

____

2. Jeff types ¾ as many pages in one hour as Linda. In the last hour of the afternoon Jeff and Linda typed a total of 14 pages. How many pages an hour does Linda type? (Let *p* be the number of pages an hour that Linda types.)

____

3. Marco has an average score of 76 on his first three exams in history. After the next test, his average score was 78. What was his score on the fourth test? (Let *x* be his score on the fourth test.) ____________________

Warm-up Answers
1. $3b + 2 = 47$ **2.** $p + \frac{3}{4}p = 14$ **3.** $\frac{3(76) + x}{4} = 78$

On the Springboard

10. The steel mill employed five times as many men as women. The total work force was 1,272. How many of the workers were women?

    (1) 156 (2) 212 (3) 1,260

    (1) (2) (3)

11. Edith started on her first full-time job last year. This year she is earning 1.1 times as much as she did last year. Last year and this year together she earned \$21,000. How much did she earn her first year?

    (1) \$5,000 (2) \$10,000 (3) \$11,000

    (1) (2) (3)

12. In a certain brand of mixed nuts, there are twice as many ounces of pecans as of cashews and three times as many ounces of peanuts as of cashews. How many ounces of cashews are there in a bag that weighs 48 ounces?

    (1) 4 (2) 6 (3) 8

    (1) (2) (3)

Check your Springboard answers on page 346. If you got these all right, you did very well. Go on to "The Real Thing." If you missed any, read the problem again very carefully and see if you can find your mistakes. When you are confident that you're ready, go on.

The Real Thing

11. Which of the following equations has 6 as a root?

    (1) $6x = 36$ (2) $x - 4 = 3$
    (3) $5 + x = 13$ (4) $\frac{x}{4} = 2$
    (5) $21 = 3x$

    (1) (2) (3) (4) (5)

12. If a number is added to 18, the result is 25. Which of the following equations can be used to find the number?

    (1) $18 - x = 25$ (2) $x + 25 = 18$
    (3) $18 + x = 25$ (4) $x - 18 = 25$
    (5) $x - 25 = 18$

    (1) (2) (3) (4) (5)

13. What is the root of $3x - 2.87 = 3.46$?

    (1) 1.77 (2) 2.11 (3) 3.46
    (4) 6.33 (5) 18.99

    (1) (2) (3) (4) (5)

14. What is the root of $\frac{3}{4} + 2x = 1\frac{2}{3}$?

    (1) $\frac{1}{2}$ (2) $\frac{2}{3}$ (3) $\frac{5}{12}$ (4) $\frac{11}{24}$ (5) $\frac{5}{6}$

    (1) (2) (3) (4) (5)

15. Juan bought 75 sheets of graph paper. He sold it at a price of 5 sheets for \$0.22. After he had sold all of the paper, he had \$0.18 less than twice the cost of the paper. Which of the following equations can be used to find how much Juan paid for the paper?

    (1) $75 - 0.18x = 5(x - 2)$
    (2) $5(75 \times 0.22) = 2(x - 0.18)$
    (3) $0.22\left(\frac{75}{5}\right) = 2x - 0.18$
    (4) $\frac{x}{2} - 0.22 = 5(75 - 0.18)$
    (5) $0.18(x - 0.22) = 22(75 \times 5)$

    (1) (2) (3) (4) (5)

16. What is the root of $4.271 - 5x = 2(x - 3.468)$?

    (1) 0.765 (2) 0.983 (3) 1.213
    (4) 1.572 (5) 1.601

    (1) (2) (3) (4) (5)

17. What is the root of $6x - \frac{2}{3} = 9x + \frac{5}{6}$?

    (1) $-\frac{1}{3}$ (2) $-\frac{1}{2}$ (3) $+\frac{1}{4}$
    (4) $+\frac{2}{3}$ (5) $+\frac{3}{4}$

    (1) (2) (3) (4) (5)

18. Ling earned \$46.85, which is \$32.15 less than twice the amount that Tina earned. How much did Tina earn?

(1) \$21.82 (2) \$25.18 (3) \$27.75
(4) \$32.40 (5) \$39.50

① ② ③ ④ ⑤

19. Roger's sister is 15 years old. If her age is 3 years more than $\frac{1}{4}$ of Roger's age, how old is Roger?

(1) 32 (2) 36 (3) 40
(4) 48 (5) 56

① ② ③ ④ ⑤

Check your answers on pages 351 and 352 and record all your scores on page 352.

LESSON 46

Inequalities

In mathematics and real life, there are many situations in which two quantities may not be equal. In these situations, you may need one of the following *inequality* signs instead of an *equal* sign.

Inequality

Sign	How to Read
$<$	"is less than"
$>$	"is greater than"
$\leq$	"is less than or equal to"
$\geq$	"is greater than or equal to"

Notice that the last two signs allow the *possibility* of equality. An algebraic statement that uses one of these four signs between two expressions is called an **inequality.**

Coming to Terms

inequality an algebraic statement that has two expressions with an inequality sign between them. The inequality signs are $<$ (is less than), $>$ (is greater than), $\leq$ (is less than or equal to) and $\geq$ (is greater than or equal to).

Here's an Example

■ Let's see what the inequality $x < -3$ means using the number line.

(any x on this side)

The inequality $x < -3$ is read "x is less than -3."

This inequality says that x is to the *left* of -3 on the number line. Numbers such as -4, $-5\frac{1}{2}$, and -7.6 are values of x that make $x < -3$ true. *Every* number to the left of -3 will make the inequality true. However, -3 will not, and neither will any number to the right of -3.

It is easy to remember what the inequality signs mean. Just remember that in a true inequality, the signs $<$ and $>$ *point toward the smaller number.*

A Test-Taking Tip

If you are given a question about inequalities on the GED Mathematics Test, it will be a simple one. Do not worry about knowing all the ins and outs. Become familiar with inequalities and then, if one does appear on the test, you will be able to make a good choice.

Try It Yourself

■ Can you read this inequality and tell in your own words what it means? $x \geq 2$ __________

____

Did you read "x is greater than or equal to 2"? This means that x can be any number to the right of 2 on the number line or even 2 itself.

Warm-up

Write these inequalities in words, just as you would read them.

1. $y \leq -1$ ____________________

____

2. $x \geq -6$ ____________________

____

3. $x > 0$ ____________________

____

4. $y < 3$ ____________________

____

Warm-up Answers
1. y is less than or equal to -1 **2.** x is greater than or equal to -6 **3.** x is greater than 0 **4.** y is less than 3

Solving Inequalities

Solving inequalities is much like solving equations. There is, however, one important difference. If you multiply or divide both sides of an inequality by a negative number, you must change the direction in which the inequality points.

Here's an Example

You can add or subtract a number from both sides of an inequality so long as you add or subtract the *same number from both sides.*

■ Solve this inequality: $x + 4 < 6$.

To solve, isolate the unknown by adding or subtracting.

Step 1 Add -4 to both sides of the inequality.

$$x + 4 - 4 < 6 - 4$$

Step 2 Simplify both sides. You get $x < 2$.

The answer $x < 2$ means that if x is any number less than 2, it will make the original inequality true. In other words, if you add any number less than 2 to 4 in the original inequality your answer will be a number less than 6.

You would deal with a negative number in the same way.

■ Solve $x - 4 < 9$.

Step 1 Add $+4$ to both sides and you get

$$x - 4 + 4 < 9 + 4$$

Step 2 Simplify both sides. You get $x < 13$.

This is the answer. It means that if you subtract 4 from any number less than 13, you will get a number less than 9 as the result. You can also multiply or divide both sides by any *positive* number.

■ Solve $\frac{1}{2}x - 3 < 5$

Step 1 Isolate the unknown by adding 3 to each side and simplifying.

$$\tfrac{1}{2}x - 3 + 3 < 5 + 3$$

$$\tfrac{1}{2}x < 8$$

Step 2 Multiply both sides by 2 and simplify.

You get $2 \times \frac{1}{2}x < 2 \times 8$ or $x < 16$.

The answer is $x < 16$.

The only difference between solving inequalities and solving equations occurs if you multiply or divide both sides of the inequality by a *negative* number. *If you multiply or divide both sides by a negative number you must change the direction in which the inequality sign points.*

Look at $x > 4$. Any number greater than 4 makes this a true statement. Let's multiply both sides by -1. If you wrote the result as $-x > -4$ and left the inequality sign pointing to the right, you would be wrong. In the case of $x > 4$, 5 is a value of x that makes $x > 4$ true. But -5 is *not* greater than -4. The signs of the numbers changed from positive to negative when you multiplied by -1, so you have to reverse the direction of the inequality sign. That is, you must change $>$ to $<$. *Any time you multiply or divide both sides of an inequality by a negative number, you must reverse the direction of the inequality sign.*

Try It Yourself

■ Solve $4x \geq -12$.

Did you divide both sides by 4? If you did this and then simplified correctly, you got the correct answer, which is $x \geq -3$.

■ Solve $3x + 5 < 20$.

Did you first add -5 to each side to get $3x < 15$? Did you then divide both sides by 3? Your final answer should have been $x < 5$.

■ Solve $-2x > 8$.

If you divided both sides by -2, you were on the right track. Did you remember to reverse the inequality sign? Your answer should be $x < -4$.

Warm-up

Do what is indicated to the right of each inequality. Write the correct result on your own paper.

1. $4x - 7 > 10$; add $+7$ to both sides.
2. $3 < 8 - 2x$; add -8 to both sides.
3. $-5b > 10$; divide both sides by -5.
4. $6 - 2k \leq 1$; add -6 to both sides.
5. $-\frac{1}{4}x \geq 9$; multiply both sides by -4.
6. $5 + \frac{x}{2} > 0$; add -5 to both sides.

On the Springboard

1. Solve the inequality $3x \leq 24$.

   (1) $3x \leq 8$ (2) $x \leq 8$
   (3) $x \geq 8$

   ① ② ③

2. Solve $-6x > 18$.

   (1) $x > -3$ (2) $x < 3$ (3) $x < -3$

   ① ② ③

3. Solve the inequality $4x - 6 \geq 30$.

   (1) $x \geq 9$ (2) $x \leq 9$ (3) $x < 9$

   ① ② ③

Check your Springboard answers on page 346. If you got these right, you're ready to go on to "The Real Thing." If you missed any, check your work before going on.

The Real Thing

1. Which of the following lists contains only numbers that make $-3x < 12$ true?

   (1) $-10, 0, 4$ (2) $-4, -1, 8$
   (3) $-2, 2, 10$ (4) $-6, 3, -18$
   (5) $-3, -6, -8$

   ① ② ③ ④ ⑤

2. Which is the solution of $5 + \frac{1}{3}x \geq 17$?

   (1) $x > 22$ (2) $x \leq 51$
   (3) $x \geq 8$ (4) $x \geq 36$
   (5) Insufficient data is given to solve the problem.

   ① ② ③ ④ ⑤

Warm-up Answers
1. $4x > 17$ **2.** $-5 < -2x$ **3.** $b < -2$ **4.** $-2k \leq -5$
5. $x \geq -36$ **6.** $\frac{x}{2} > -5$

3. Which of the following numbers will make both $2x - 11 \leq 10$ and $16 - 3x \leq 4$ true?

(1) 3 (2) 7 (3) 12
(4) 16 (5) 19

(1) (2) (3) (4) (5)

4. Which of the following inequalities is true when x is replaced by 5?

(1) $2x + 3 \leq 15$ (2) $5 - 3x > 26$
(3) $5x - 7 < 13$ (4) $7x - 22 > 20$
(5) $3x + 8 \geq 29$

(1) (2) (3) (4) (5)

5. What is the solution of $3 - 2x < \frac{3}{4}$?

(1) $x < \frac{3}{4}$ (2) $x > 1\frac{1}{8}$
(3) $x < -\frac{2}{3}$ (4) $x < 0$
(5) $x < -1\frac{3}{4}$

(1) (2) (3) (4) (5)

6. Which is the solution of $10.1 > 3x + 2.75$?

(1) $x > 2.75$ (2) $x < 7.35$
(3) $x > 10.1$ (4) $x < 2.45$
(5) $x > -3.37$

(1) (2) (3) (4) (5)

7. Which is the solution of $5.438 - 4x \geq 62.95$?

(1) $x \leq 62.95$ (2) $x > 51.512$
(3) $x < 21.752$ (4) $x > -8.756$
(5) $x \leq -14.378$

(1) (2) (3) (4) (5)

8. Which is the solution of $x - 2\frac{5}{6} \geq \frac{1}{2} - 4x$?

(1) $x \geq \frac{2}{3}$ (2) $x \leq \frac{1}{2}$ (3) $x \geq -\frac{1}{4}$
(4) $x < -2\frac{5}{6}$ (5) $x > -4\frac{1}{2}$

(1) (2) (3) (4) (5)

9. Which of the following numbers makes both $x \geq -\frac{1}{2}$ and $x < 5$ true?

(1) 0 (2) 5 (3) 10 (4) 15 (5) 20

(1) (2) (3) (4) (5)

Check your answers and record your score on page 352.

LESSON 47

Algebraic Equations with Two Unknowns

Suppose you and a friend have invested some money in a folk musician to help her get started. A year later, when she begins to make a profit, she tells the two of you that she can pay back \$450. You have invested twice as much as your friend has. To be fair, how much should the musician pay each of you at this time?

Since you and your friend did not contribute the same amount, it makes sense that you should not just divide the \$450 by two and each get half.

An easy way to solve problems like this is to use two unknowns. If you use two unknowns you must have two equations to be able to find the values of those unknowns.

Solving by Substitution

It takes two equations to solve for two unknown numbers. If you are told that $x + y = 450$, there are hundreds of values for x and y that make this a true statement. For example, $x = 10$ and $y = 440$ will work. But so will $x = 5$ and $y = 445$, and so on; if you are given another equation that is true for these numbers such as $y = 2x$, there is a good chance you can identify exactly one value for x and one value for y that will make *both* equations true at the same time.

Here's an Example

■ If $x + y = 450$ and $y = 2x$, what are the values of x and y?

To solve, put $2x$ in place of y in $x + y = 450$ and solve for x.

Step 1 Since the second equation tells you that y equals $2x$, put $2x$ in place of y in the first equation to get an equation in just one unknown. You get $x + 2x = 450$.

Step 2 Solve $x + 2x = 450$ as usual. You get $3x = 450$

Divide by 3 on both sides. You get $x = 150$.

Step 3 The second equation told you that $y = 2x$, so put the value you have found for x into this equation to get a value for y.

$$y = 2x$$
$$y = 2(150)$$
$$y = 300$$

If you replace x by 150 and y by 300, you can easily check that these numbers make *both* equations ($x + y = 450$ *and* $y = 2x$) true. These are, in fact, the only numbers for x and y that work in *both* equations at once.

Try It Yourself

■ If $y + x = 12$ and $y = 3x$, find y. (Notice that you are asked for y, not for both x and y.)

Since $y = 3x$, you can put $3x$ in place of y in the first equation. If you do this, you get $3x + x = 12$. If you solve for x, you should get $4x = 12$, so that $x = 3$. The problem asked you for y. You should have gone a step further and put the value of x into the equation $y = 3x$ to get $y = 3 \times 3 = 9$. The answer is $y = 9$.

Up to here all the answers have been positive numbers, but you could get negative numbers as well.

Warm-up

1. If $x - y = 8$ and $x = 3y$, then $x =$ ______

2. If $x + y = 24$ and $y = 5x$, then $y =$ ______

3. If $3x + y = 2$ and $y = 2x$, then $y =$ ______

4. If $x + 2y = 8$ and $y = \frac{1}{2}x$, then $x =$ ______

Warm-up Answers
1. 12 **2.** 20 **3.** $\frac{4}{5}$ **4.** 4

☑ A Test-Taking Tip

On the GED Mathematics Test itself, you may be given two equations and then asked either what x is or what y is. Be sure to find *exactly* what the question asks for. If you are asked for y, be sure to find y, not x. You'll possibly find the value of x as well. *But remember to answer with the value you were asked for.* Read carefully.

On the Springboard

1. $3x + 2y = 17$ and $y = x + 1$ What is y?

   (1) 3 (2) 4 (3) 5

   ① ② ③

2. $2x + y = 10$ and $y = x + 1$ What is x?

   (1) -5 (2) 3 (3) 4

   ① ② ③

3. $2x + y = -2$ and $y = x + 7$ What is x?

   (1) -3 (2) -2 (3) 4

   ① ② ③

Check your Springboard answers on page 346. If you got these right, go on to "The Real Thing." If you missed any, check and correct your work before you go on.

"The Real Thing"

1. If $y = 2x$ and $3x + y = 25$, what are the values of x and y?

   (1) $x = -10, y = 20$
   (2) $x = -5, y = -10$
   (3) $x = -5, y = 10$
   (4) $x = 5, y = 10$
   (5) $x = 5, y = -10$

   ① ② ③ ④ ⑤

2. If $y + 1 = x$ and $y = 2x - 3$, what are the values of x and y?

(1) $x = 1, y = 2$
(2) $x = 2, y = 1$
(3) $x = 2, y = 3$
(4) $x = 4, y = 5$
(5) $y = 5, x = 4$

(1) (2) (3) (4) (5)

3. If $y = -x$ and $x + 2y = 3$, what are the values of x and y?

(1) $x = -3, y = 3$
(2) $x = -1, y = 1$
(3) $x = 1, y = -1$
(4) $x = 3, y = -3$
(5) $x = 4, y = -4$

(1) (2) (3) (4) (5)

Check your answers on page 353.

Solving Word Problems

You can now use your ability to solve equations in two unknowns to solve word problems. Look again at the example on page 323.

Here's an Example

■ You and a friend have invested in a folk musician. After a year, she begins to make a profit and says she can pay back \$450. You have invested twice as much as your friend has. To be fair, how much should the musician pay each of you at this time?

If you let x equal what your friend should receive and y equal what you should receive, you can use the information in the problem to set up two equations.

The total paid back is \$450, so

$$x + y = 450$$

You invested twice as much as your friend, so you should get back twice as much of the money as your friend.

$$y = 2x$$

The first example in the lesson showed that when you solve $x + y = 450$ and $y = 2x$, you get $x = 150$ and $y = 300$. Your friend should get \$150 and you should get \$300.

Try It Yourself

■ A truck delivered 620 pounds of bananas to 2 food stores one day. One store got 3 times what the other store got. What were the weights of the 2 loads?

____

Were your equations $x + y = 620$ and $y = 3x$? If you had $x + y = 620$ and $x = 3y$, that's just as good. Did you get 465 pounds for one and 155 pounds for the other?

Warm-up

You and a friend were counting some books. When the counting was completed, he had x of them and you had y of them.

Write equations for these different statements.

1. You counted three times as many as he did.

____

2. He counted twice as many as you.

____

3. There were 410 books altogether.

____

4. You counted 7 fewer than he did.

____

5. He counted 19 more than you did.

____

6. If he gave you 3 books from his pile, you would then each have the same number of books.

____

Warm-up Answers
1. $y = 3x$ **2.** $x = 2y$ **3.** $x + y = 410$ **4.** $y + 7 = x$ or $y = x - 7$ **5.** $y + 19 = x$ or $y = x - 19$ **6.** $y + 3 = x - 3$ (When he gave you 3 books, he ended up with 3 less than he had at first.)

☑ A Test-Taking Tip

If you have trouble with equation problems, let these problems go until the end. Then go back and try out each answer in the equations. Do not leave any answers blank on the test.

On the Springboard

4. A freight car contained 36 tons of wood and steel products. The steel products weighed three times as much as the wood products. Which of these pairs of equations would you use to find the weights of each kind of product?

   (1) $\begin{cases} x + y = 36 \\ y = x - 3 \end{cases}$ (2) $\begin{cases} x + y = 36 \\ y = x + 3 \end{cases}$

   (3) $\begin{cases} x + y = 36 \\ y = 3x \end{cases}$

   ① ② ③

5. Betty's long distance phone bill was four times her water bill for the month of June. The total amount of the bills was $120. How much was the phone bill in dollars?

   (1) $48 (2) $76 (3) $96

   ① ② ③

Check your Springboard answers on page 346. If you got these right, go on to "The Real Thing." If you made an error, check your work before going on.

The Real Thing

4. During the month of April, Mrs. Johnson purchased 25 daily papers and 5 Sunday papers at a total cost of $12.25. In May, she bought 27 daily papers and 1 Sunday paper for a total of $7.95. What was the cost of a daily paper?

   (1) $0.10 (2) $0.15 (3) $0.20
   (4) $0.25 (5) $0.30

   ① ② ③ ④ ⑤

5. In one hour, Sam earned $0.66 less than Judy. Together they earned $29 an hour. How much did Sam earn per hour?

   (1) $14.17 (2) $15.66 (3) $16.34
   (4) $17.32 (5) $18.91

   ① ② ③ ④ ⑤

6. Dennis bought some pencils and notebooks. He could buy 12 pencils and 5 notebooks for $4.08 or 1 pencil and 8 notebooks for $5.80. What was the cost of each notebook?

   (1) $0.36 (2) $0.48 (3) $0.72
   (4) $0.84 (5) $0.96

   ① ② ③ ④ ⑤

7. A rectangle is twice as long as it is wide. Its perimeter is 42 inches. What is the length of the rectangle in inches?

   (1) 7 (2) 14 (3) 22
   (4) 44 (5) 50

   ① ② ③ ④ ⑤

8. Clarice raises rabbits and pigeons. When asked how many of each she had, she replied that she could count 78 feet but only 27 heads. How many rabbits did she have?

   (1) 10 (2) 12 (3) 15
   (4) 20 (5) 27

   ① ② ③ ④ ⑤

9. There are 92 members in the club this year. If the number of men is 19 less than twice the number of women, how many men belong to the club?

   (1) 24 (2) 28 (3) 37
   (4) 48 (5) 55

   ① ② ③ ④ ⑤

10. In 3 more years, Wanda's grandfather will be 6 times as old as Wanda was last year. When Wanda's present age is added to her grandfather's age, the sum is 68. How old is Wanda now?

    (1) 8 (2) 9 (3) 10
    (4) 11 (5) 12

    ① ② ③ ④ ⑤

Check your answers and record all your scores on pages 353–354.

LESSON 48

Factoring

If you multiply 2 and 3, you get 6. The numbers 2 and 3 are called **factors** of 6 because they *make* 6 when you multiply them together. A word that is similar to *factor* is *factory,* a place where things are made. The number 6, which you get when you multiply 2 and 3, is called the **product** of 2 and 3. The same name is given to what comes out of a factory. So 2 and 3 are factors and their product is 6. Likewise, the product of the factors 1, 5, and 7 is 35.

Coming to Terms

factors numbers that are multiplied together

product the result of multiplying factors

Multiplying Unknowns

In Lesson 43, you learned that multiplication can be spread over addition. For example,

$$4(5 + 7) = 4(5) + 4(7)$$

No matter what numbers you use, the law for spreading multiplication over addition works. Since the letters in algebraic expressions are really just unspecified numbers, you can use the spreading law with algebraic expressions. No matter what values the letters may have, the following equations are true because of the spreading law:

$$a(b + c) = ab + ac$$
$$2(x + y) = 2x + 2y$$
$$x(2x + 1) = x(2x) + x(1)$$

Multiplication can also be spread over subtraction. To see this with numbers, look at this example.

$$5(10 - 2) = 5(10) - 5(2)$$
$$5(8) = 50 - 10$$
$$40 = 40$$

You can spread multiplication over subtraction when the things you multiply are algebraic expressions.

$$a(b - c) = ab - ac$$
$$2(x - y) = 2x - 2y$$
$$x(2x - 1) = x(2x) - x(1)$$

When you are multiplying algebraic expressions, it is sometimes possible to rearrange factors to write your answers in a simpler form.

Here's an Example

■ Multiply $2x$ by $3x + 7$ and simplify. To solve, use the spreading law and then rearrange factors in each term to simplify.

Step 1 Use the spreading law.

$$2x(3x + 7) = 2x(3x) + 2x(7)$$

Step 2 Rearrange factors in each term and simplify.

$$2x(3x) + 2x(7)$$
$$2(3)xx + 2(7)x$$
$$6x^2 + 14x$$

Here's an example where you can spread multiplication over subtraction.

■ Multiply x and $2y - 5x$ and simplify.

Step 1 Use the spreading law.

$$x(2y - 5x) = x(2y) - x(5x)$$

Step 2 Simplify individual terms.

$$x(2y) - x(5x)$$
$$2xy - 5xx$$
$$2xy - 5x^2$$

The answer is $2xy - 5x^2$. Notice that although it would not be incorrect to leave $x(2y)$ as it is nor to write it as $x2y$, it is customary to rearrange the factors to put the number first and the letters last, that is, $2xy$.

Try It Yourself

■ Multiply $3x$ and $2y + 3x$ and simplify.

____

Did you first use the spreading law to get $3x(2y) + 3x(3x)$? Did you rearrange factors in $3x(2y)$ to get $(3)(2)xy$ or $6xy$? Then, did you rearrange factors in $3x(3x)$ to get $(3)(3)xx$ or $9x^2$. Was your final answer $6xy + 9x^2$?

■ Multiply $-3x$ and $(2x + 8)$ and simplify.

When you used the spreading law, did you get $(-3x)(2x) + (-3x)8$? When you rearrange factors in $(-3x)(2x)$, you get $(-3)(2)xx$ or $-6x^2$. When you rearrange factors in $(-3x)8$, you get $(-3x)(8)x$ or $-24x$. The final answer is $-6x^2 + (-24x)$. It is more common to get rid of the parentheses as follows:

$$-6x^2 + (-24x)$$
$$-6x^2 - 24x$$

You can do this because *subtracting* $24x$ is the same as *adding* its opposite, $-24x$.

Warm-up

Use the spreading law to fill in the blanks.

1. $x(b + 2) = xb + x(\_\_\_\_\_)$
2. $4m(3a + c) = (4m)(\_\_\_\_\_) + 4m(c)$
3. $2y(x - y) = 2y(\_\_\_\_\_) - 2y(y)$
4. $-5x(y + 2) = (-5x)y + (-5x)(\_\_\_\_\_)$
5. $-3y(2y - x) = (-3y)(2y) - (-3y)(\_\_\_\_\_)$

On the Springboard

1. Multiply $2x$ and $10x + 10y$ and simplify.

   (1) $20x + 20xy$ (2) $20x^2 + 20xy$
   (3) $200x^2y$

   ① ② ③

2. Multiply $-x$ and $2y - x$ and simplify.

   (1) $2xy - x^2$ (2) $-2xy + x$
   (3) $-2xy + x^2$

   ① ② ③

3. Multiply $6x$ and $\frac{1}{2}y + 6x$ and simplify.

   (1) $12xy + 6x^2$ (2) $3xy + 6x$
   (3) $3xy + 36x^2$

   ① ② ③

Warm-up Answers
1. 2 **2.** $3a$ **3.** x **4.** 2 **5.** x

Check your Springboard answers on page 347. If you got them all right, go on to the next section. If you missed any, go back and review your work before you continue.

Factoring

Here's an example of the spreading law. But the law has been used in reverse.

$$3x + 3y = 3(x + y)$$

On the left side, you see the factor 3 in $3x$ and in $3y$. On the right side, you see the multiplication that you would have done with the spreading law to get $3x + 3y$. Using the spreading law in reverse is called **factoring.** When you *multiply,* you start with factors and find the product. When you *factor,* you start with the product and find the factors.

Coming to Terms

factoring the process of finding the factors that can be multiplied to obtain a given expression as a product

Here's an Example

■ Factor $mx + my$.

To solve, change $mx + my$ into a product by using the spreading law.

Step 1 Look for a factor present in each term of $mx + my$ and write it outside a pair of parentheses.

$$m(\quad)$$

Step 2 Extract this factor from each term and write what is left from those terms in the parentheses.

$$mx + my$$
$$m(x + y)$$

The answer is $m(x + y)$.

You can check by multiplying $m(x + y)$. As you can see, you get back the original $mx + my$. This system works even if the expression has exponents.

■ Factor $3x^2 + 3y^2$.

Here 3 is the common factor, so start with 3(   ). Extract the factor 3 from $3x^2$ and from $3y^2$ and write the result in the parentheses. You get $3(x^2 + y^2)$ as the answer.

The method also works when the common factor is a letter or when exponents are present.

■ Factor $9x + ax^2$.

Step 1 Find the common factor and write it outside a pair of parentheses. Here the common factor is x.

$$x(\quad)$$

Step 2 Extract the factor x from each of the terms: $9x$ and ax.[2] Write the result in the parentheses.

$$x(9 + ax)$$

The answer is $x(9 + ax)$.

Try It Yourself

■ Factor $5a + 5x$. ____________________

If you saw that 5 was the common factor and followed the system, you should have gotten $5(a + x)$. It doesn't matter how many terms there are, as long as the factor you choose to pull out is in *all* of them.

■ Factor $3x + 3m + 3y + 3$.

____

If you put 3 outside the parentheses and extracted it from each of the terms in the original expression, including the 3, you should have gotten $3(x + m + y + 1)$.

Warm-up

Factor these expressions. You can check your answers by multiplying out.

1. $ax + bx$ ____________________

2. $ax^2 + bx$ ____________________

3. $x^2a + 3a$ ____________________

4. $ax + 3x^2$ ____________________

5. $4a + 4b + 4$ ____________________

On the Springboard

4. Factor $3ab + 2ac + ad$.

(1) $a(3b + 2c + d)$
(2) $3a(b + c + d)$
(3) $3b(a + b + c)$

① ② ③

5. Which of these is a true statement?

(1) $ab + ab^2 = ab(1 + b)$
(2) $x + xy = x(x + y)$
(3) $3x + 2y = xy(3 + 2)$

① ② ③

Check your Springboard answers on page 347. If you got all of these right, go on to "The Real Thing." If you made any errors, you may find it helpful to review. Then try again on the ones you missed. When you feel that you're ready, go on to "The Real Thing."

Warm-up Answers
1. $x(a + b)$ **2.** $x(ax + b)$ **3.** $a(x^2 + 3)$ **4.** $x(a + 3x)$
5. $4(a + b + 1)$

The Real Thing

1. Which of the following expressions correctly factors $(2c + 2d)$?

   (1) $2(c + d)$ (2) $\frac{1}{2}(2c + 2d)$

   (3) $1(c + 2d)$ (4) $-2(c + d)$

   (5) $\frac{c + d}{2}$

   (1) (2) (3) (4) (5)

2. Which of the following expressions is correct?

   (1) $r^2 + 2rs = 2rs(r + 1)$
   (2) $3n + 6n^2 = 3n(3n + 2)$
   (3) $4m + 4 = 4(m + 4)$
   (4) $2v + 2w = 4(v + w)$
   (5) $3j^2 + 12j = 3j(j + 4)$

   (1) (2) (3) (4) (5)

3. Which of the following expressions correctly factors $5jk + 5k^2$?

   (1) $5k(j + k)$
   (2) $5(kj + k)$
   (3) $k(5j + 1)$
   (4) $5k^2(j + 1)$
   (5) $5jk(1 + k)$

   (1) (2) (3) (4) (5)

4. Which of the following is a factor of 10?

   (1) 3 (2) 4 (3) 5 (4) 6 (5) 7

   (1) (2) (3) (4) (5)

5. Mr. and Mrs. McDougall paid $(2r^2 - rs)$ cents for a bag of $(2r - s)$ oranges. How many cents did they pay for each orange?

   (1) 2 (2) $-s$ (3) rs
   (4) r (5) $2r^2$

   (1) (2) (3) (4) (5)

6. Which of the following correctly factors $16d^2e^2 - 24de$?

   (1) $8(2d^2e^2 + 3e^2)$
   (2) $8de(2de - 3)$
   (3) $de(16de - 24)$
   (4) $4(4d^2e^2 - 6de)$
   (5) $2de(8de - 3e)$

   (1) (2) (3) (4) (5)

Check your answers and record your scores on page 354.

LESSON 49

Quadratic Equations

Multiplying Two Expressions with Two Terms

Look at how the law of spreading multiplication over addition can be used to multiply two expressions with two terms.

Here's an Example

■ $(x + 2)(x + 5)$

Step 1 First think of this as spreading $(x + 2)$ over $(x + 5)$. You will multiply $(x + 2)$ by x, next $(x + 2)$ by 5, and then add.

$$(x + 2)(x + 5) = (x + 2)x + (x + 2)5$$

Step 2 Use the spreading law on $(x + 2)x$ and on $(x + 2)5$.

$$\underbrace{(x + 2)x}_{x^2 + 2x} + \underbrace{(x + 2)5}_{5x + 10}$$

$$x^2 + 2x + 5x + 10$$

Step 3 Simplify by combining like terms. The only like terms in the last step are $2x$ and $5x$. Since $2x + 5x = 7x$, you get

$$x^2 + 7x + 10$$

The final answer is: $(x + 2)(x + 5) = x^2 + 7x + 10$.

The following pattern gives the same result. Multiply the first term in the first parentheses by each of the terms in the second parentheses. Write the result.

$$(x + 2)(x + 5) \qquad x^2 + 5x$$

Then multiply the second term in the first parentheses by each term in the second parentheses and write the result.

$$(x + 2)(x + 5) = x^2 + 5x + 2x + 10$$

(You get the same result as in Step 2, but with $5x$ and $2x$ written in a different order.) Then, by combining terms you get $x^2 + 7x + 10$.

To make sure you have the pattern clear in your mind, look at several more examples.

■ Multiply (x + 2)(x + 3).

The method works even if there are minus signs in the parentheses. Just remember to use the exact signs.

■ Multiply $(x - 3)(x + 2)$.

Step 1 Multiply the terms in the second parentheses by the x in the first parentheses.

$(x - 3)(x + 2) \qquad x^2 + 2x$

Step 2 Now multiply by the -3.

$(x - 3)(x + 2) = x^2 + 2x - 3x - 6$

Notice that the equal sign was not written until the second step because the lefthand side did *not* equal the right-hand side until that step.

Step 3 Combine like terms.

$x^2 + 2x - 3x - 6 = x^2 - x - 6$

So far in your multiplication of two expressions with two terms you have always had *three* terms in your final answer. It is possible to finish up with only *two* terms in your final answer. Notice how this can happen.

■ Multiply $(x - 4)(x + 4)$.

Combining the first two steps you would get

$$(x - 4)(x + 4) = x^2 + 4x - 4x - 16$$

This simplifies to $x^2 - 16$ because the two middle terms add up to zero.

Try It Yourself

■ Multiply $(x + 7)(x + 4)$.

____

For the first step you should have had $x^2 + 4x$. The second step should have given you $7x + 28$. The combined terms add to $x^2 + 11x + 28$.

If you didn't get it right, stop now and look through the section again to see what you are missing. The rest of the lesson depends on your being able to do this kind of multiplication.

If you understand everything so far, try these two.

■ $(x + 7)(x - 4)$
$(x - 7)(x - 4)$

____

____

The first one gives you $x^2 - 4x + 7x - 28$. When you combine like terms, you get $x^2 + 3x - 28$. The second one gives $x^2 - 7x - 4x + 28$, and when you combine like terms you get $x^2 - 11x + 28$. If you made any slips, check the in-between stages carefully.

■ Multiply: $(x + 5)(x - 5)$

____

Did you get $x^2 - 5x + 5x - 25$? When you combined like terms, you should have gotten $x^2 - 25$. Whenever you notice that the two expressions you are going to multiply are the same except that one has a + sign and the other − sign, the answer will have only two terms.

Warm-up

Multiply these expressions. (Two of them will end up with only two terms in the answer.

1. $(x + 1)(x + 2) =$ ______________
2. $(x + 2)(x + 3) =$ ______________
3. $(x + 3)(x - 3) =$ ______________
4. $(x + 8)(x - 8) =$ ______________
5. $(x - 9)(x - 4) =$ ______________

Warm-up Answers
1. $x^2 + 3x + 2$ **2.** $x^2 + 5x + 6$ **3.** $x^2 - 9$ **4.** $x^2 - 64$
5. $x^2 - 13x + 36$

On the Springboard

1. Multiply $(x + 9)(x - 10)$.

(1) $x^2 - 19x - 90$
(2) $x^2 + 19x - 90$
(3) $x^2 - x - 90$

① ② ③

2. Multiply $(y + 7)(y - 7)$.

(1) $y^2 - 14y - 49$ (2) $y^2 - 49$
(3) $y^2 + 49$

(1) (2) (3)

3. Multiply $(a + 6)(a - 1)$.

(1) $a^2 + 7a - 6$ (2) $a^2 - 5a - 6$
(3) $a^2 + 5a - 6$

(1) (2) (3)

Check your Springboard answers on page 347. If you got them all right, you're doing very well. Go on to the next section. If you missed any, check carefully to see where your error was. When you are confident of your work, go on to the next section.

Factoring Expressions with Three Terms

In mathematics, for every operation you can *do*, there is usually some way of *undoing* it. You have just multiplied factors to get expressions and now you will see how to change certain expressions back into factors. The best way to start is to look for patterns in what you already know.

You have already worked these three.

1. $(x + 7)(x + 4) = x^2 + 11x + 28$

2. $(x + 7)(x - 4) = x^2 + 3x - 28$

3. $(x - 7)(x - 4) = x^2 - 11x + 28$

All three sets of parentheses have the same numbers in them. Only the signs are different. Look carefully at the answers and you will notice these things.

a. If the sign of the last term of the answer is $+$, the signs in both parentheses are the same as the sign of the second term of the answer. Look at 1 and 3. The final term is $+28$ in both cases (and the sign of the second term, $+11x$ in 1 and $-11x$ in 3) is the sign in both parentheses.

b. If the sign of the final term is $-$, as in 2, the signs in the parentheses are different and the larger number of the two numbers in the parentheses has the sign of the second term.

These two things are always true if the expression has factors. With these two rules it is easy to know what the signs were in the original parentheses.

Here's an Example

■ Fill in the missing signs in the parentheses to show what factors were multiplied to get the three-term expressions.

1. $(x \_\_\_ 2)(x \_\_\_ 4) = x^2 - 6x + 8$

2. $(x \_\_\_ 4)(x \_\_\_ 6) = x^2 + 10x + 24$

3. $(x \_\_\_ 4)(x \_\_\_ 6) = x^2 - 2x - 24$

These three examples include all the possibilities, so read this carefully and you will be able to do any problem like these.

For number 1,

Step 1 Look at the final sign. Here it is $+$, so you know that both signs in the parentheses should be the same. Also, both should be $-$, since the second term of the product has a $-$ sign.

Step 2 Write in the signs according to the rules: $(x - 2)(x - 4)$ is the answer. You can check by multiplying to see if you get $x^2 - 6x + 8$.

For number 2,

Step 1 Check the final sign. Again it is $+$. Since the second term in the expression you're factoring has a $+$ sign, a $+$ sign should go in both parentheses.

Step 2 Write the sign of the second term in the parentheses. $(x + 4)(x + 6)$ is the answer. You can check by multiplying. You should get $x^2 + 10x + 24$.

For number 3,

Step 1 Check the sign of the last term in the three-term expression. Here it is $-$, so the signs in the parentheses are different. The larger number has the same sign as the second term of the expression you're factoring, which here is a minus sign.

Step 2 Put in the signs according to step 1. $(x + 4)(x - 6)$ is the answer. You can check by multiplying.

Now look at these examples.

■ Given $x^2 + 11x + 28$, find its factors.

Step 1 Write down two pairs of parentheses and put an x as the first term in each.

$$(x \quad)(x \quad)$$

Step 2 Work out what the signs are going to be. Here the final term in what you're factoring is + and the second term sign is +, so + signs go in both parentheses.

$$(x + \quad)(x + \quad)$$

Step 3 Find two numbers that when multiplied make +28 and when added make +11. Several pairs can be multiplied to make 28: (1,28), (2,14), and (4,7). But only the last pair adds to make 11, so the numbers needed are 4 and 7.

Step 4 Write them in the parentheses.

$$(x + 4)(x + 7)$$

The numbers in the final expressions are $(x + 4)(x + 7)$.

■ Find the factors of $x^2 - 11x + 28$.

Step 1 Start off the parentheses with the first terms.

$$(x \quad)(x \quad)$$

Step 2 Find what the signs are going to be and put them in. Here you can see that the last term of what you're factoring is +28, so both signs are the same. The second term of what you're factoring is $-11x$, so both signs should be − signs.

$$(x - \quad)(x - \quad)$$

Step 3 Find a pair of numbers that when multiplied make +28 and when added make −11. The signs in the parentheses tell you that both numbers are negative. The pairs that multiply to +28 are the same as before except for the signs: (−1, −28), (−2, −14), and (−4, −7). The last pair not only multiplies to 28 but also adds to −11, so those are the numbers. Write 4 and 7 in the parentheses.

$$(x - 4)(x - 7)$$

$(x - 4)(x - 7)$ is the answer.

■ Find the factors of $x^2 + 3x - 28$.

Step 1 Write down the parentheses and the first terms.

$$(x \quad)(x \quad)$$

Step 2 Find what the signs must be. Here the −28 tells you that the signs in the parentheses will be different. Put them in the parentheses and remember that the larger number of the pair you are going to find is the positive number because the second term is positive.

$$(x - \quad)(x + \quad)$$

Step 3 Find two numbers that multiply to −28 and add to +3. You just found pairs of numbers that multiply to give −28; (−1, +28), (+1, −28), (2, −14), (−14,2), (+4, −7), and (−4, +7). The only pair that adds to +3 is (−4, +7) so put in those values.

$$(x - 4)(x + 7)$$

The final answer is $(x - 4)(x + 7)$.

When working backwards from the expression to the factors, you first get the signs. Then use the number clues given by the last term and the middle term.

Try It Yourself

■ Fill in the correct signs in the parentheses.

1. $(x$ ______ $4)(x$ ______ $3) = x^2 - x - 12$
2. $(x$ ______ $5)(x$ ______ $2) = x^2 + 7x + 10$
3. $(x$ ______ $6)(x$ ______ $4) = x^2 - 10x + 24$

If you understood the rules and inspected the signs carefully, you should have thought along the following lines.

The final sign of the expression you're factoring is −, so use rule 2. The larger number is 4, so that has the minus sign (the same sign as the second term of what you're factoring). The answer is $(x - 4)(x + 3)$.

The final sign of what you're factoring is +, so use rule 1. The second term of the expression you're factoring is +, so both signs are +. The answer is $(x + 5)(x + 2)$.

The last sign of what you're factoring is +, so use rule 1 and put the second term sign in both. The answer is $(x - 6)(x - 4)$.

Now that you can find what signs should go in the factors of an expression, you need to be able to determine what numbers should go in the factors. Look at these problems and see if there is a pattern that will help.

■ **1.** $(x + 7)(x + 4) = x^2 + 11x + 28$
2. $(x - 7)(x - 4) = x^2 - 11x + 28$
3. $(x + 7)(x - 4) = x^2 + 3x - 28$

____

____

____

Since all the products contain x^2, it is no surprise that each pair of parentheses has an x. Now look carefully at the numbers in the parentheses (the factors) and in the expressions (the products). There is one thing true in every case so you can use it as a rule to help you. It is this:

The last number in the expression is the product of the two numbers in the parentheses, and the number you see in the *middle* term of the product expression is the *sum* of the numbers in the parentheses.

We'll use a^2 instead of x^2 just for a change.

■ Factor these expressions.

1. $a^2 - 10a + 21$ ________________

2. $a^2 + 10a + 21$ ________________

3. $a^2 + 4a - 21$ ________________

These are the lines of thought you might have followed.

1. Last sign is +, so both signs are −, the same as the sign of the second term in the expression that is being factored. You need two negative numbers that multiply to give +21 and add to give −10. Since $(-3) \times (-7) = +21$ and $(-3) + (-7) = -10$, the numbers are −3 and −7. The answer is $(a - 3)(a - 7)$.

2. The last sign in what you're factoring is positive, so both signs in the factors will be + because of $+10x$. Two numbers that multiply to +21 and add to +10 are +3 and +7. The answer is $(a + 3)(a + 7)$.

3. The last term is negative, so the signs in the parentheses for the factors will be + and −. The larger number of the pair is positive because of the $+4a$. The pair that multiply to 21 are 3 and 7, and +7 and −3 add to 4, so that is the pair. The answer is $(a + 7)(a - 3)$.

Warm-up

Factor each of these expressions.

1. $x^2 + 5x + 6$ ________________

2. $a^2 - 5a - 14$ ________________

3. $a^2 + 5a - 14$ ________________

4. $x^2 + 20x + 100$ ________________

5. $x^2 - 16x + 55$ ________________

On the Springboard

What are the factors of these expressions?

4. $x^2 + 11x + 28$

(1) $(x - 7)(x + 4)$
(2) $(x + 7)(x + 4)$
(3) $(x + 9)(x + 3)$

(1) (2) (3)

5. $y^2 - 10y + 21$

(1) $(y - 3)(y - 7)$
(2) $(y + 3)(y + 7)$
(3) $(y + 7)(y - 10)$

(1) (2) (3)

6. $b^2 + 8b + 16$

(1) $(b + 2)(b + 8)$
(2) $(b + 4)(b + 4)$
(3) $(b + 12)(b - 4)$

(1) (2) (3)

Check your Springboard answers on page 347. If you got these right, well done! Go on to the next section. If you had any errors, find and correct your mistakes and then go on to the next section.

Warm-up Answers
1. $(x + 2)(x + 3)$ **2.** $(a - 7)(a + 2)$ **3.** $(a + 7)(a - 2)$ **4.** $(x + 10)(x + 10)$ (Yes, you do sometimes find both factors to be the same.) **5.** $(x - 5)(x - 11)$

Difference of Two Squares

You dealt with some cases involving only two terms when you learned how to multiply across parentheses.

$$(x + 3)(x - 3) = x^2 - 9$$
$$(x + 5)(x - 5) = x^2 - 25$$

Let's work one out to review why this happens.

$$(x + 6)(x - 6) = x^2 + 6x - 6x - 36.$$

This comes down to $x^2 - 36$ because the middle terms add to zero. You always get only two terms when the two factors are the same except for the signs. Notice too that x^2 is a square and so are the numbers 9, 25, and 36. In all of the above the righthand expression is the *difference of two squares.*

To factor a two-term expression that is the difference of two squares, just put the square root of each term in both pairs of parentheses and write opposite signs.

Here's an Example

■ What are the factors of $n^2 - 100$?

$n^2 - 100$ is the difference of two squares, $n \times n$, and 10×10, so write down the square roots in the parentheses with opposite signs $(n + 10)(n - 10)$ The answer is $(n + 10)(n - 10)$. If you wish, you can check by multiplying.

A special case that fools some people is $x^2 - 1$. This is actually a difference of two squares because the terms are $x \times x$ and 1×1. The factors of $x^2 - 1$ are therefore $(x + 1)(x - 1)$. Remember this one. It's the only one that isn't quite obvious.

Try It Yourself

■ Factor these expressions.

1. $a^2 - 81 =$ ________________

2. $c^2 - d^2 =$ ________________

You should have thought that $81 = 9 \times 9$, so the first answer is $(x + 9)(x - 9)$. The second one is easiest because you don't have any numbers. Here $c^2 = c \times c$, and $d^2 = d \times d$, so the answer is $(c + d)(c - d)$.

Solving an Equation with Squares

You know that $3 \times 0 = 0$, $5 \times 0 = 0$, and so on. Any number multiplied by zero, equals zero. Look at this algebraic equation.

$$ab = 0$$

If $a = 0$ or $b = 0$, the statement must be true. In fact, that's the only way ab could equal zero. *If the product of two factors is 0, then one or both of the factors must also be zero.* Using this reasoning, consider these two equations:

$$x(x + 3) = 0$$
$$(x + 4)(x - 7) = 0$$

In the first, two factors, x and $(x + 3)$ multiplied together to make 0. If either is equal to zero, the statement is true. So there are two choices: $x = 0$ or $x + 3 = 0$, in which case $x = -3$. The equation has *two roots,* $x = 0$ and $x = -3$. Either makes the equation $x(x + 3) = 0$ true.

In the second equation, there are also two factors, $(x + 4)$ and $(x - 7)$, that are multiplied to make 0. If either factor equals 0, the equation will be true. If $x + 4 = 0$, then $x = -4$. So $x = -4$ is one root. If $x - 7 = 0$, then $x = +7$. So $x = +7$ is another root.

Again, there are two possible roots or solutions. If you put either -4 or $+7$ in place of x in $(x + 4)(x - 7) = 0$, the resulting numerical equation is true. Now you know that, if $(x + 4)(x - 7) = 0$, $x = -4$ or $x = +7$. You can multiply out the parentheses as you have done earlier in this lesson.

$$(x + 4)(x - 7) = x^2 - 3x - 28$$

So you have also solved the equation $x^2 - 3x - 28 = 0$ by looking at its factors and letting each one equal zero in turn. To solve equations that contain x^2 or some other unknown squared, you need only find the factors, let each factor equal zero, and then find what value or values of the unknown makes the equation true.

Here's an Example

■ Find the roots of the equation $x^2 - 8x + 15 = 0$.

To solve, factor the left-hand side, put each factor equal to zero, and solve for x.

Step 1 Write out the parentheses with the unknown in each.

$$(x \quad)(x \quad) = 0$$

Step 2 Determine what the signs will be in the parentheses. Here the final term is +15, so both signs will be the same. Since the second term is $-8x$, both signs will be $-$ signs. Put them in.

$$(x - \quad)(x - \quad) = 0$$

Step 3 Determine the numbers. You'll need two numbers that multiply to give +15 and add to give −8. Trial and error shows that these are −5 and −3. You now have this equation.

$$(x - 3)(x - 5) = 0$$

Step 4 Put each factor equal to zero and solve the equations.

If $x - 3 = 0$, then $x = +3$.
If $x - 5 = 0$, then $x = +5$.

The answers are $x = +3$ and $x = +5$.

Try It Yourself

■ Find the roots of $x^2 + 12x + 27 = 0$.

____

Is your answer $(x + 3)(x + 9) = 0$? When you put each factor equal to zero, you should have found that $x = -3$ and $x = -9$ are the two roots.

Some equations with a squared unknown have only one root. Consider $(x - 4)(x - 4) = 0$. When you put the factors equal to 0 and solve for x, you get $x = 4$ for both factors. The equation has just one root, $x = 4$.

Warm-up

Find the roots of these equations.

1. $(x - 3)(x + 4) = 0$ ______________

2. $(y + 2)(y - 7) = 0$ ______________

3. $x^2 - 7x + 12 = 0$ ______________

4. $x^2 + 9x + 20 = 0$ ______________

5. $p^2 + 6p + 9 = 0$ ______________

Warm-up Answers
1. $x = -4$ and $x = 3$ **2.** $y = 7$ and $y = -2$ **3.** $x = 3$ and $x = 4$ **4.** $x = -5$ and $x = -4$ **5.** $p = -3$ (This equation has just one root)

A Test-Taking Tip

On the GED Mathematics Test there will probably be only one quadratic equation to solve. If you are pushed for time and the two roots are in the choices, you can substitute in the equation to see which choices make the equation true. If you are not given the roots and you don't have time to solve the equation, just guess.

On the Springboard

7. What are the roots of equation $n^2 + 3n + 2 = 0$?

   (1) $n = -2, n = 4$
   (2) $n = -2, n = -1$
   (3) $n = -2, n = 1$

   ① ② ③

8. What is the result when you add together the roots of the equation $x^2 - 13x + 42 = 0$?

   (1) − 13 (2) − 12 (3) 13

   ① ② ③

9. What is the result of multiplying the roots of this equation together? $(x - 6)(x + 4) = 0$

   (1) −24 (2) −10 (3) 24

   ① ② ③

Check your Springboard answers on page 347. If you got these right, well done! Go on to "The Real Thing." If you missed any, check your work before going on.

The Real Thing

1. One root of $b^2 - 4 = 0$ is −2. What is the other root?

   (1) −4 (2) −1 (3) 0
   (4) +2 (5) +3

   ① ② ③ ④ ⑤

2. What are the factors of $x^2 + x - 6$?

(1) $(x + 2)(x + 3)$ (2) $(x - 3)(x - 2)$
(3) $(x + 3)(x - 2)$ (4) $(x - 3)(x + 2)$
(5) $(x - 3)(x + 3)$

(1) (2) (3) (4) (5)

3. One root of $a^2 + a - 20 = 0$ is $+4$. What is the other root?

(1) -16 (2) -5 (3) $+2$
(4) $+4$ (5) $+10$

(1) (2) (3) (4) (5)

4. What is a root of $y^2 - 6y + 9 = 0$?

(1) -9 (2) -6 (3) -1
(4) $+2$ (5) $+3$

(1) (2) (3) (4) (5)

5. What is a root of $c^2 + 12c + 36 = 0$?

(1) -6 (2) -4 (3) $+2$
(4) $+8$ (5) $+12$

(1) (2) (3) (4) (5)

6. If $x^2 - 11x + 30 = 0$, which of the following are the roots?

(1) -2 or -15 (2) -1 or -30
(3) $+5$ or $+6$ (4) $+4$ or $+10$
(5) $+30$ or $+1$

(1) (2) (3) (4) (5)

7. Which of the following equations has factors of $(m + 4)(m - 2)$

(1) $m^2 + m - 12 = 0$
(2) $m^2 + 2m - 8 = 0$
(3) $m^2 - 5m + 6 = 0$
(4) $m^2 - 3m - 28 = 0$
(5) $m^2 + 3m - 10 = 0$

(1) (2) (3) (4) (5)

8. Which of the following equations has roots of $+7$ and -4?

(1) $a^2 - 5a - 14 = 0$
(2) $a^2 + 7a + 12 = 0$
(3) $a^2 - 9a + 14 = 0$
(4) $a^2 - 3a - 28 = 0$
(5) $a^2 + 9a + 20 = 0$

(1) (2) (3) (4) (5)

Check your answers and record your score on pages 354–355.

LESSON 50

Line Graphs and Slopes

Now you are in the last lap of the marathon and in a very short time you'll be ready to take the GED Mathematics Test—with much more knowledge and experience than you had when you began this book.

This last lesson looks again at equations with two variables, like $y = x + 7$ or $x - y = 6$. If you have the time, it will help you to review briefly what was dealt with in Lesson 48. This time the graphs of the equations will be drawn on a coordinate grid. The graphs will be straight lines. You will find out what it means when the lines for two equations cross. You will also learn how to find the slope of a straight line on a coordinate grid. It will help you to get some squared paper and to make graphs of your own as you study the lesson.

Point of Intersection

If an equation contains two simple variables, it can be drawn on a grid as a straight line.

Here's an Example

■ Draw the equation $y = x + 2$ on a grid.

If you can find two values of x and y that make this equation true, you can write the values as ordered pairs, plot two points on a grid, and join the points with a straight line to get the equation in picture form.

Step 1 In the equation $y = x + 2$, give x any value you like and find out what y would equal in that case. Let $x = 3$; then $y = 3 + 2$, so $y = 5$.
This gives you the ordered pair (3, 5).

Step 2 Do it again for another value of x.
Let $x = 0$; then $y = 0 + 2$, so $y = 2$.
This gives you the ordered pair (0, 2).

Step 3 Plot these points on a grid and draw a straight line through them.

You have drawn the graph of $y = x + 2$. Any point on the line corresponds to an ordered pair that will fit back into the equation and make it true. Notice, for example, that the line crosses the x-axis where $x = -2$ and $y = 0$. Put those values into the equation. Now $y = x + 2$ becomes $0 = -2 + 2$, which is a true equation.

Now, using the same grid, draw the graph of $y = 6 - x$.

Step 1 Choose a value for x and find the corresponding value for y. Let $x = 0$; then $y = 6 - 0 = 6$.
The ordered pair is (0, 6).

Step 2 Do it again for another value of x.
Let $x = 3$; then $y = 6 - 3 = 3$.
The ordered pair is (3, 3).

Step 3 Put these pairs on the same grid and draw a straight line through them. The grid showing both lines should look like this.

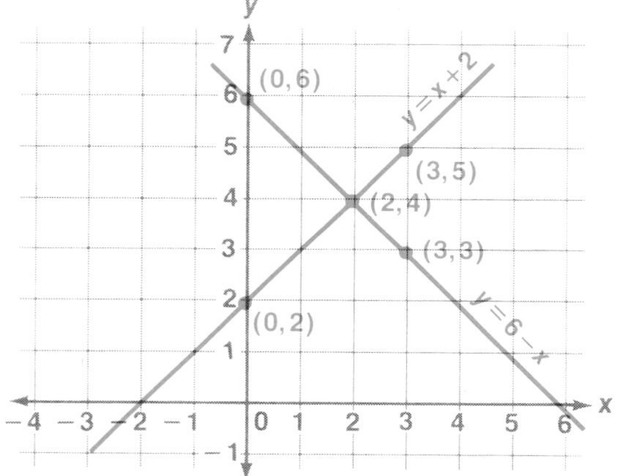

Notice that the two lines cross. If you did the drawing accurately, the ordered pair for the point where they cross is (2, 4). This is the only ordered pair in the world that makes both equations true at the same time. If you put $x = 2$ and $y = 4$ into both equations you get

$y = x + 2 \rightarrow 4 = 2 + 2$, which is true
and $y = 6 - x \rightarrow 4 = 6 - 2$, which is true.

Sometimes the lines cross at a place that is not easy to read, but you can still find the exact ordered pair there. Recall how you solved two equations with two unknowns. You just substituted the value of one variable into the other equation. When you did that, you were actually finding the ordered pair that made both equations true at the same time.

Let's do it for the equations we have just drawn. Solve to find the values of x and y that make both true at the same time.

■ $y = x + 2$ and $y = 6 - x$

Step 1 Write down one of the equations

$$y = x + 2$$

Step 2 Replace one of the variables with the value from the other equation. Here it is easiest to replace y with $6 - x$ in the equation $y = x + 2$. You get

$$6 - x = x + 2$$

Step 3 Solve this last equation for x to find one answer. Then put that value back into one of the original equations to find the second answer.

	$6 - x = x + 2$
Add x to each side	$6 + 0 = 2x + 2$
Add -2 to each side	$6 - 2 = 2x$
So	$4 = 2x$
Divide both sides by 2	$2 = x$

Put $x = 2$ back into $y = 6 - x$ and you get $y = 6 - 2 = 4$. The ordered pair that makes both equations true is (2, 4).

You can find the intersection point in two ways, either by examining a drawing or by calculating.

Try It Yourself

■ Find the intersection of the lines $\overleftrightarrow{RS}$ and $\overleftrightarrow{TU}$ by inspection of the grid, and your answer by substituting it back into the equations and solving the two equations.

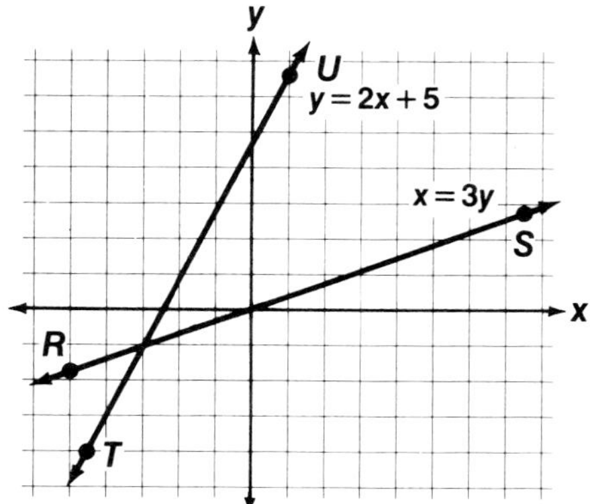

If you looked along the x-axis to -3 and then down the y-axis to -1 and wrote $(-3, -1)$ as the answer, you were correct. When you put the values $x = -3$ and $y = -1$ into the equations, you should have had for $y = 2x + 5$ the numerical equation $-1 = 2(-3) + 5$ or $-1 = -6 + 5$, which is true. The other should have given you $-3 = 3(-1)$, which is also true.

To solve the equations without a diagram, you can use $x = 3y$ and put $3y$ in place of x in the other equation. If you did that, you should have gotten $y = 2(3y) + 5$. When you solve for y, you get. $y = 6y + 5$. Therefore $-5y = 5$ and so $y = -1$. Putting $y = -1$ back into $x = 3y$ gives $x = -3$. All the results agree and you can always support one method with another.

A Test-Taking Tip

In the GED Mathematics Test you will have five options to a question about a point of intersection of two lines. Even if you can't see exactly where the lines meet, you can tell just by looking if both coordinates are positive or negative or mixed. Using that data, you can often tell which options to ignore and then substitute the others back into the equation to see which pair fits.

Warm-up

What are the coordinates for the point of intersection of the lines $\overleftrightarrow{NO}$ and $\overleftrightarrow{UV}$ in the figure? Use inspection or computation to find out.

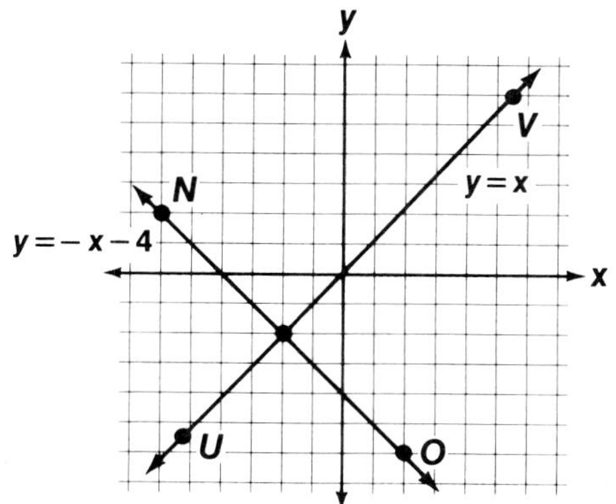

Warm-up Answer
(−2, −2)

On the Springboard

Use the labeled points of intersection on the grid below to answer these questions.

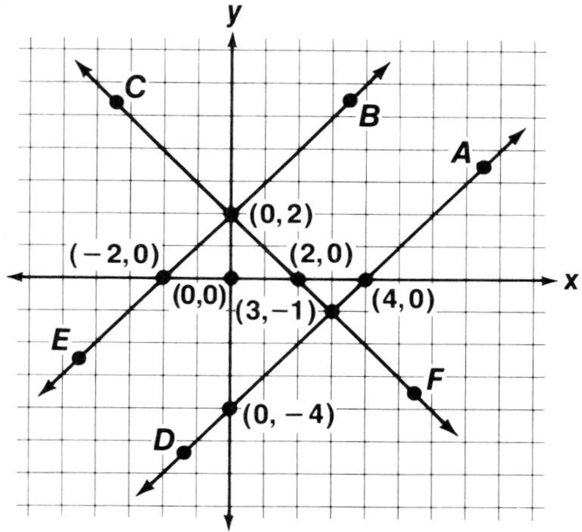

1. What ordered pair makes the equations of lines BE and CF true at the same time?

   (1) (−2, 0) (2) (2, 0) (3) (0, 2)

   ① ② ③

2. Which lines represent equations that cannot have an ordered pair in common?

   (1) $\overleftrightarrow{AD}$ and $\overleftrightarrow{BE}$ (2) $\overleftrightarrow{BE}$ and $\overleftrightarrow{CF}$
   (3) $\overleftrightarrow{CF}$ and $\overleftrightarrow{AD}$

   ① ② ③

3. What ordered pair makes the equations of lines AD and CF true at the same time?

   (1) (2, 0) (2) (4, 0) (3) (3, −1)

   ① ② ③

Check your Springboard answers on pages 347–348. If you got these right, go on to "The Real Thing." If you missed any, just make sure you were not doing something like reversing the x and y in the ordered pairs. Correct your work and then go on.

“The Real Thing”

1. The graph below shows lines for $2x - y = 3$ and $3x + y = 2$. What is their point of intersection?

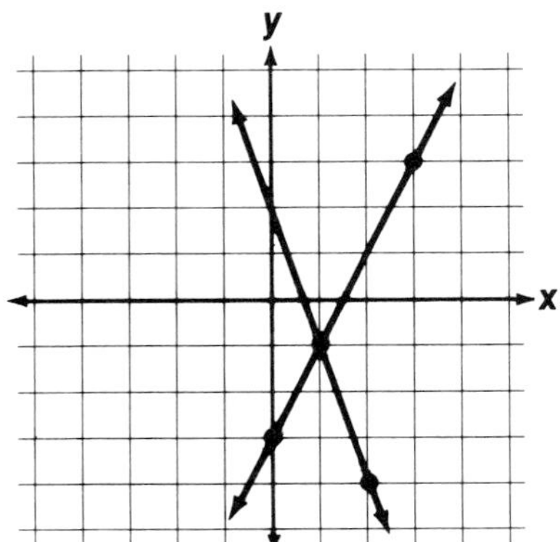

(1) (3, 3) (2) (1, −1) (3) (0, 2)
(4) (2, −4) (5) (0, −3)

(1) (2) (3) (4) (5)

2. Which is the point of intersection of the lines $3x - y = -1$ and $2x + y = -9$?

(1) (1, 9) (2) (3, −2) (3) (−1, 9)
(4) (−2, −5) (5) (−5, 3)

(1) (2) (3) (4) (5)

3. What is the equation of line AB?

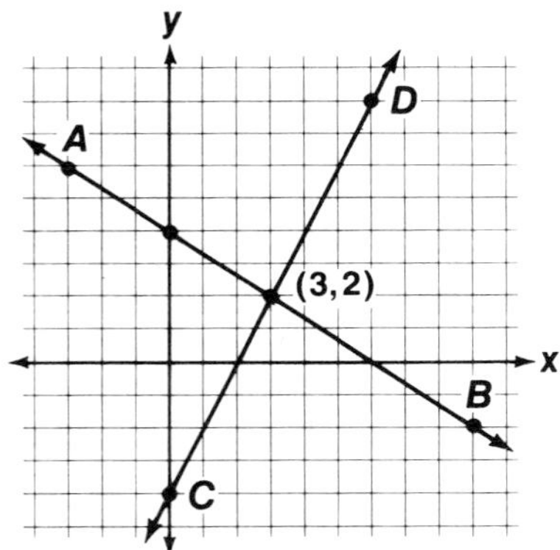

(1) $y = \left(-\frac{2}{3}\right)x + 4$

(2) $y = 2x - 4$

(3) $y = \left(\frac{3}{4}\right)x + 6$

(4) $y = \left(-\frac{1}{2}\right)x - 6$

(5) $y = x + 3$

(1) (2) (3) (4) (5)

Check your answers on page 355.

Finding the Slope of a Line

This is the last topic in the book, so hang on. The **slope** of a line is a number that tells how steeply a line rises on a grid and also which way it is slanting.

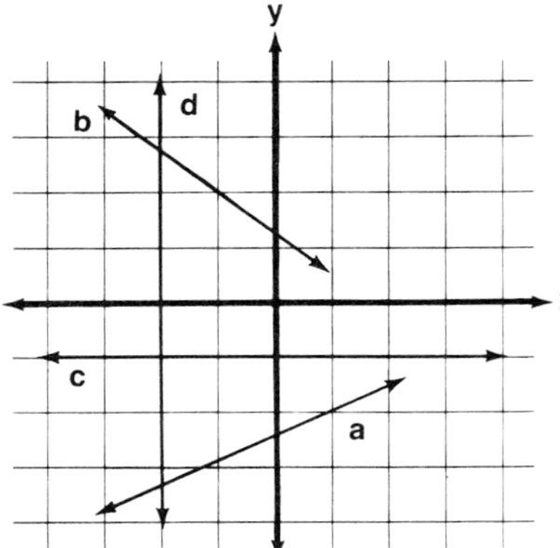

A line that slopes upward from left to right has a *positive slope.* Line *a* has a positive slope.

A line that slopes downward from left to right has a *negative slope.* Line *b* has a negative slope.

A line that is horizontal (straight across) has *0 (zero) slope.* The line neither rises nor falls. It stays parallel to the *x*-axis. Line *c* has 0 slope.

A line that is vertical (straight up and down) has *no slope.* It stays parallel to the *y*-axis at all points. Line *d* has no slope. *No slope and 0 slope are not the same.*

Coming to Terms

slope a number that tells how steep a line on a coordinate grid is. A line falling to the right has a negative slope. A line rising to the right has a positive slope. A horizontal line has 0 slope. A vertical line has no slope.

The slope of a line can be found in two ways, and you can usually choose the one you prefer.

Here’s an Example

The first way you can find the slope is by using the drawing of the grid. The second way uses a formula that will be given on the GED Mathematics Test formula sheet.

■ Find the slope of line AB in the grid below.

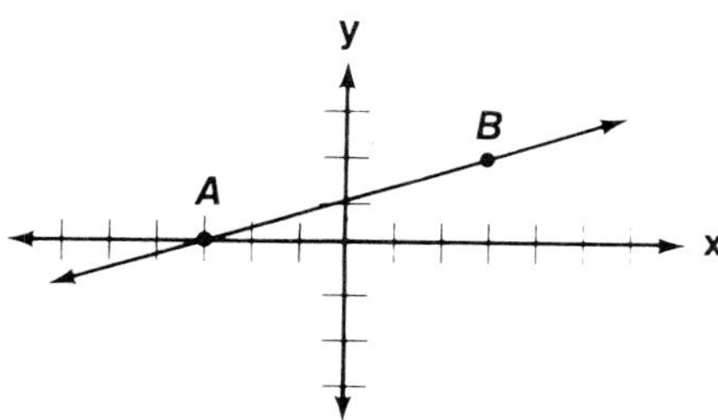

To solve, decide if the slope is positive or negative. Construct a right triangle from two points on the line to find the number that tells how steep the line is.

Step 1 Decide if the slope is positive or negative. Look back at the explanation just before the last *Coming to Terms.* Line AB is positive since it goes up as you look across the page from left to right.

Step 2 Choose two points on the line, and construct a right triangle as shown here. It doesn't matter on which side of the line the triangle is constructed. The answer will be the same.

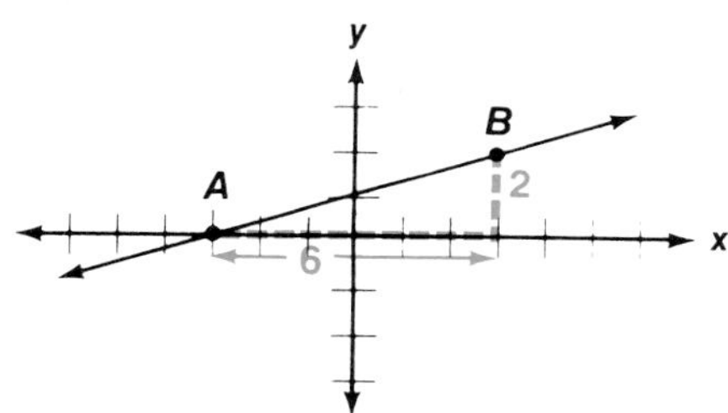

Step 3 This right triangle has two legs. One leg is vertical, or up-and-down. The other is horizontal, running across. Find the length of each leg by counting. The vertical leg is 2 units long, and the horizontal leg is 6 units long.

Step 4 Write a fraction with the lengths of the legs.

$$\frac{\text{length of vertical leg}}{\text{length of horizontal leg}} = \frac{2}{6}$$

Step 5 Reduce the fraction to lowest terms. Use a positive sign for a positive slope or a negative sign for a negative slope.

$$\frac{2}{6} = \frac{1}{3}$$

The slope of line AB is $+\frac{1}{3}$. Remember that the positive sign can be understood, so you could state that the slope of line AB is $\frac{1}{3}$.

The second way you can find slope is by using a formula that gives you the fraction

$$\frac{\text{length of vertical leg}}{\text{length of horizontal leg}}$$

directly, along with the correct *sign* for the slope.

Step 1 Choose any two convenient points on the line.

Step 2 Write down the coordinates of those points as (x_1, y_1) and (x_2, y_2). The small 1s tell you you're talking about the coordinates of the first point, and the 2s mean you're talking about the coordinates of the second point. Here (x_1, y_1) is $(-3, 0)$ and (x_2, y_2) is $(3, 2)$.

Step 3 Put these values into the formula.

$$\text{slope of line} = \frac{y_2 - y_1}{x_2 - x_1}$$

This gives you $\frac{2 - 0}{3 - (-3)} = \frac{2}{6} = \frac{1}{3}$.

You can take any point you like to be (x_1, y_1) and any other as (x_2, y_2), as long as you keep the order straight in the formula.

Try It Yourself

■ Find the slope of the line in this grid. Use either method.

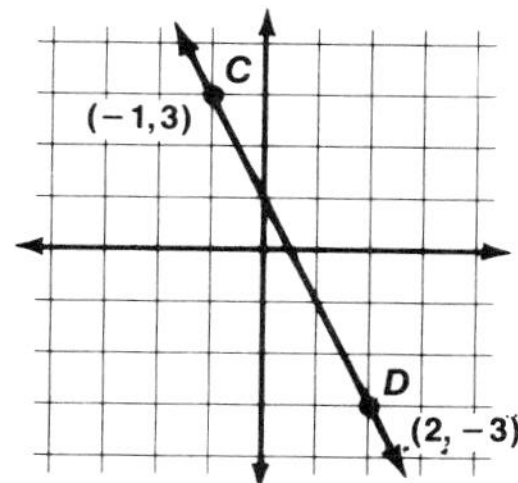

If you used the counting method, you would have

$$\frac{\text{vertical leg}}{\text{horizontal leg}} = \frac{6}{3} = \frac{2}{1} \text{ or } 2$$

But the slope is negative, so the answer is -2. If you used the formula for the points in the grid you should have $\frac{3 - (-3)}{-1 - 2} = \frac{6}{-3} = -2$. The formula method always gives you the correct sign for the slope.

A Test-Taking Tip

You might be asked to give the slope of a vertical or horizontal line on the GED Mathematics Test. A *vertical line* (up-and-down) has *no slope* at all. A *horizontal line* has a slope of *zero.* It might be helpful to remember these two facts.

Warm-up

Find the slopes of these lines by either method.

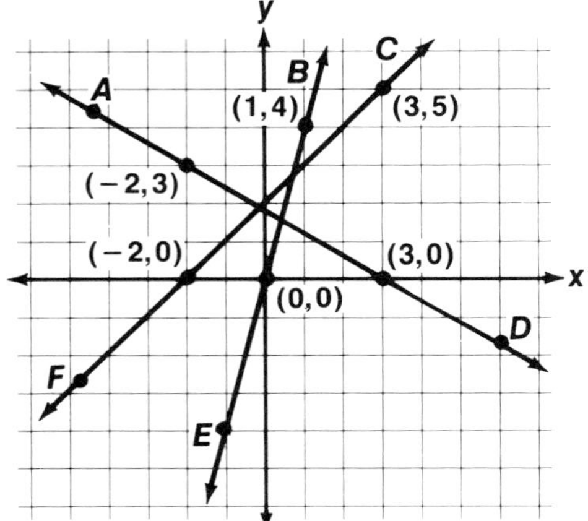

1. slope of line *AD* = ______________

2. slope of line *BE* = ______________

3. slope of line *CF* = ______________

On the Springboard

4. What is the slope of a straight line that passes through the points (5, 6) and (10, 11)?

   (1) -1 (2) $\frac{1}{2}$ (3) 1

   ① ② ③

5. What point do the lines $y = 2x + 3$ and $y = 3x + 1$ have in common? Write the ordered pair for the point.

   (1) (2, 7) (2) (−2, 7) (3) (2, −7)

   ① ② ③

Warm-up Answers

1. $\frac{-3}{5}$ (the slope is negative) **2.** 4 **3.** 1

Check your Springboard answers on page 348. If you got these right, you have grasped this last topic very well and you are ready for the last set of GED-level questions in "The Real Thing." If you got either wrong, check your work carefully and, when you are confident you're correct, go on to "The Real Thing."

The Real Thing

4. In the graph below, what is the slope of line *MN?*

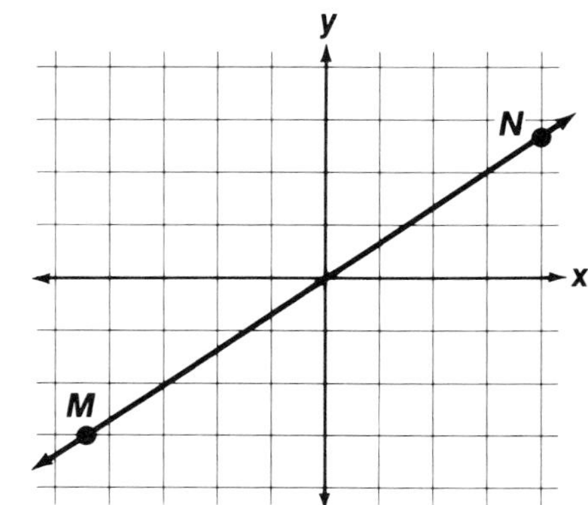

   (1) $-\frac{4}{3}$ (2) $-\frac{2}{3}$ (3) $\frac{2}{3}$

   (4) $\frac{3}{2}$ (5) 3

   ① ② ③ ④ ⑤

5. What is the slope of a line containing the points (−2, 5) and (4, 1)?

   (1) −6 (2) −4 (3) $-\frac{3}{2}$

   (4) $-\frac{2}{3}$ (5) $\frac{1}{4}$

   ① ② ③ ④ ⑤

6. What is the slope of a ramp that is 25 feet long horizontally and rises 6 feet?

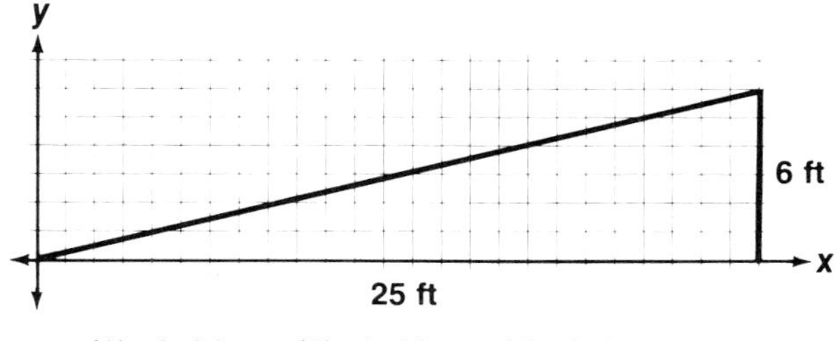

   (1) 0.06 (2) 0.12 (3) 0.24
   (4) 0.40 (5) 0.60

   ① ② ③ ④ ⑤

7. Line KL has no slope and the ordered pair for point K is (4, 6). Which of the following ordered pairs also gives a point on line KL?

(1) (−7, 6) (2) (4, −5) (3) (6, 1)
(4) (3, 4) (5) (−6, −2)

(1) (2) (3) (4) (5)

8. In the figure below, what is the intersection of lines d and c?

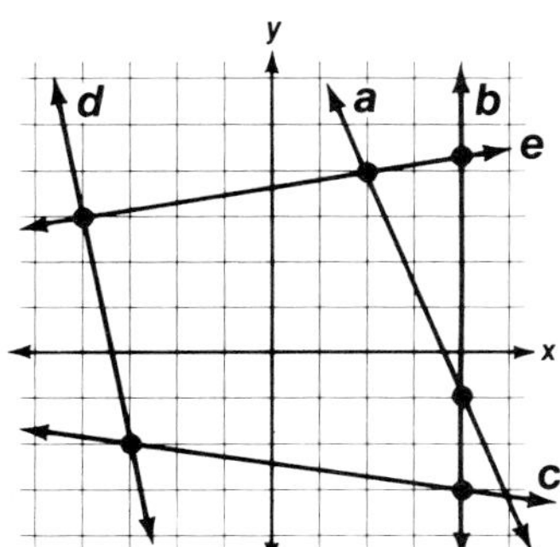

(1) (2, 4) (2) (4, −1) (3) (4, −3)
(4) (−3, −2) (5) (−4, 3)

(1) (2) (3) (4) (5)

9. What is the slope of a line containing the points (2, −4) and (7, 6)

(1) 0.67 (2) 1.00 (3) 1.25
(4) 1.50 (5) 2.00

(1) (2) (3) (4) (5)

10. The grade of a road is how far up it goes compared to how far it goes horizontally. A road has a grade of 15%. How many feet does it go up in a 200-foot horizontal distance?

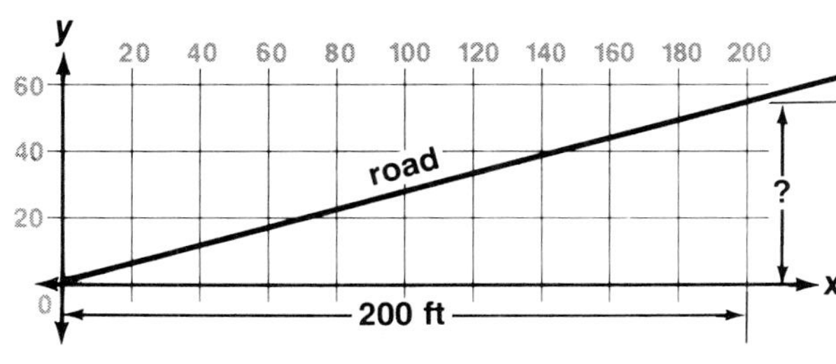

(1) 0.15 (2) 0.30 (3) 1.50
(4) 30.0 (5) 40.0

(1) (2) (3) (4) (5)

11. In the graph below, which two lines have *equal* slopes?

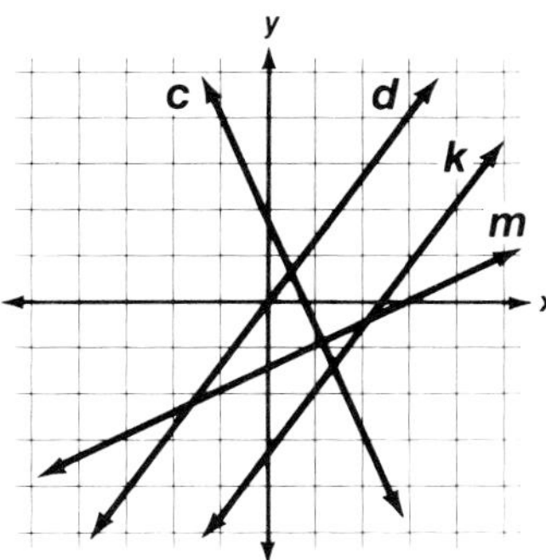

(1) d and k (2) c and m (3) k and c
(4) m and d (5) c and d

(1) (2) (3) (4) (5)

12. The graph shows the line for the equation $y = x$. What is the measure of angle a?

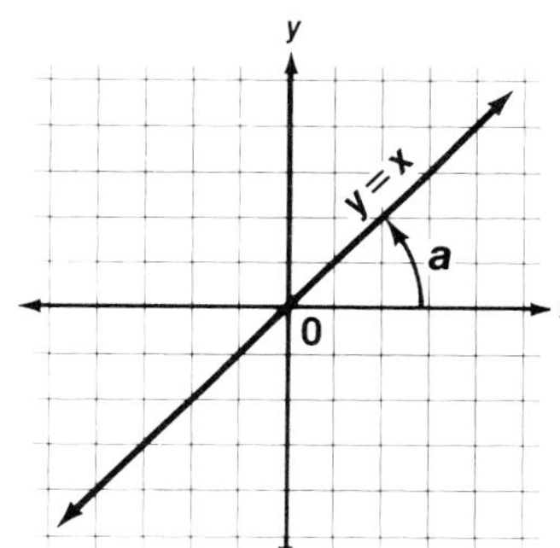

(1) 30° (2) 45° (3) 60° (4) 75°
(5) Insufficient data is given to solve the problem.

(1) (2) (3) (4) (5)

Check your answers and record all your scores on pages 355–356.

Answers: On the Springboard

43 Numerical Expressions

(page 301)

1. (2) 17
Use the PEMDAS rule. First do what's indicated inside the parentheses. Then handle the exponent portion, then the addition.

	$8 + (5 - 2)^2$
Parentheses:	$8 + (3)^2$
Exponent:	$8 + 9$
Add:	17

2. (5) 45 (12) + 9
Use the PEMDAS rule.

	$7^2 - 4\,(18 - 6) + 9$
Parenthesis:	$7^2 - 4\,(12) + 9$
Exponent:	$49 - 4\,(12) + 9$
Multiply:	$49 - 48 + 9$
Add/Subtract:	$1 + 9$
	10

(page 304)

3. (2) 89 (3 + 7)
You could approach this problem in two ways. Perhaps the best way (best because it's fastest) is to notice that if you spread multiplication over addition in (2) you get $89\,(3 + 7) = (89 \times 3) + (89 \times 7)$, and the expression after the equal sign is equal to $89 \times 3 + 89 \times 7$. (You don't *have* to have the parentheses in $89 \times 3 + 89 \times 7$, because the PEMDAS rule means that you would first do the multiplications and then the addition.)

Another approach is to simplify $89 \times 3 + 89 \times 7$. You will get 890. Find which of the answer choices can be simplified to get 890. Only 89 (3 + 7) works.

4. (1) (3 + 4) + 9
Notice that the expression 3 + (4 + 9) could be rearranged to get (3 + 4) + 9. This is so because of the grouping law. So the correct option is (1).

Another approach is to simplify 3 + (4 + 9) to get 15. Then simplify the expressions for each option. Only option (1) can be so simplified.

(page 305)

5. (3) $\frac{90 + 85 + 80 + 75}{4}$
To find the average score, you would first add all the scores.

$90 + 85 + 80 + 75$

Next you would divide by 4 (the number of students).

$(90 + 85 + 80 + 75) \div 4$

If you remember that this last expression can be written as a fraction, you get the expression in option (3).

$$\frac{90 + 85 + 80 + 75}{4}$$

6. (1) 3 ($1.50) + $0.20
First write the expression that tells what she spent for avocados. Multiply the number of avocados by the price for each.

3 × $1.50, or 3 ($1.50)

Add $0.20 (the amount for the apple. You get option (1).

3 ($1.50) + $0.20

44 Algebraic Expressions

(page 307)

1. (2) $e - 35$
Since he lost 35 pounds, you should *subtract* 35 from what he weighed when he started the diet. You get $e - 35$.

2. (1) $c/12$
You know that there are 12 months in a year, so *divide* the total amount he paid for the whole year by 12.

$c \div 12$ (dollars per month)

Since $c \div 12$ can also be written as $c/12$, the correct choice is (1).

(page 309)

3. (2) $2a$
Remember that a can be thought of as $1a$. Change subtraction to addition. Then combine like terms.

$6a + a - 5a$
$6a + 1a + (-5a)$

Since $6 + 1 + (-5) = 2$, your final expression (when you combine like terms) is $2a$.

4. (2) $8x + 4y$
Change subtraction to addition and rewrite x as $1x$.

$2x + 3y + 5x - 5y + x + 6y$
$2x + 3y + 5x + (-5y) + 1x + 6y$

Rearrange the terms to group like terms together. Then combine like terms.

$$\underbrace{2x + 5x + 1x}_{8x} + \underbrace{3y + (-5y) + 6y}_{4y}$$

(page 311)

5. (3) 17
Replace e with 2 and f with 3. Then simplify the numerical expression.

$e + 5f$
$2 + 5(3) = 2 + 15 = 17$

6. (1) $\frac{1}{2}$
Replace p with $\frac{1}{2}$ and q with 1.

Then simplify the numerical expression.

pq
$(\frac{1}{2})(1) = \frac{1}{2} \times \frac{1}{1} =$

$\frac{1 \times 1}{2 \times 1} = \frac{1}{2}$

7. (2) -8
Replace x with -2 and y with 2 and then simplify the resulting numerical expression.

$2x - y^2$
$2(-2) - 2^2 =$
$-4 - 4 = -8$

45 Algebraic Equations

(pages 315–316)

1. (2) 15
To solve $x - 5 = 10$, isolate the unknown. Remember that what you do to one side of an equation, you must do to the other.

	$x - 5 = 10$
Change − to +:	$x + (-5) = 10$
Add 5:	$+5 \quad +5$
	$x + 0 = 15$
Simplify:	$x = 15$

Check	$x - 5 = 10$
15 for x:	$15 - 5 = 10$
Subtract:	$10 = 10$

You could also get the answer by substituting answer choices in $x - 5 = 10$ to see which will result in a true numerical equation. You would find that 15 is the number that works.

2. (3) 32
Solve as usual.

$$\tfrac{1}{2}x = 16$$

Divide by ½, that is, multiply by $^2/_1$:

$$\tfrac{1}{2}x \times \tfrac{2}{1} = 16 \times \tfrac{2}{1}$$

Regroup the numbers being multiplied: $\tfrac{1}{2} \times \tfrac{2}{1} \times x = 16 \times \tfrac{2}{1}$

Simplify: $1 \times x = \tfrac{32}{1}$

$x = 32$

Check: $\tfrac{1}{2}x = 16$

32 for x: $\tfrac{1}{2} \times 32 = 16$

Multiply: $16 = 16$

You can also get the answer by substituting answer choices to see which one makes $\tfrac{1}{2}x = 16$ true.

3. (3) 8

$$0 = \tfrac{1}{4}y - 2$$

Change $-$ to $+$: $0 = \tfrac{1}{4}y + (-2)$

Add 2: $+2 \quad +2$

$$2 = \tfrac{1}{4}y$$

Divide by ¼, that is, multiply by $^4/_1$:

$$2 \times \tfrac{4}{1} = \tfrac{1}{4}y \times \tfrac{4}{1}$$

Simplify: $8 = y$
or $y = 8$

Check: $0 = \tfrac{1}{4}y - 2$

8 for y: $0 = \tfrac{1}{4} \times 8 - 2$

Multiply: $0 = 2 - 2$

Subtract: $0 = 0$

You could also get the answer by substituting answer choices to see which one makes $0 = \tfrac{1}{4}y - 2$ true.

4. (1) 40

$$0.5y = 20$$

Divide by 0.5: $0.5y/0.5 = 20/0.5$

Simplify: $y = 40$

Check: $0.5y = 20$

40 for y: $0.5 \times 40 = 20$

Multiply: $20 = 20$

You could also get the answer by substituting answer choices in $0.5y = 20$ to see which results in a true numerical equation.

(page 317)

5. (3) $-\tfrac{2}{3}$

$$x - 3 = 10x + 3$$

Add $-x$: $-x \quad -x$

$$0 - 3 = 9x + 3$$

Simplify: $-3 = 9x + 3$

Add -3: $-3 \quad -3$

$$-6 = 9x$$

Divide by 9: $\tfrac{-6}{9} = \tfrac{9x}{9}$

Simplify: $-\tfrac{2}{3} = x$

or $x = -\tfrac{2}{3}$

Check: $x - 3 = 10x + 3$

$-\tfrac{2}{3}$ for x: $(-\tfrac{2}{3}) - 3 = 10(-\tfrac{2}{3}) + 3$

Simplify: $-3\tfrac{2}{3} = -\tfrac{20}{3} + 3$

Change $-^{20}/_3$ to $-6^2/_3$:

$$-3\tfrac{2}{3} = -6\tfrac{2}{3} + 3$$

Simplify: $-3\tfrac{2}{3} = -3\tfrac{2}{3}$

You could also get the answer by substituting answer choices for x in $x - 3 = 10x + 3$ to see which one results in a true equation.

6. (2) $\tfrac{1}{2}$

$$5 - 3a = a + 3$$

Add $3a$: $+3a \quad +3a$

$$5 = 4a + 3$$

Add -3: $-3 \quad -3$

$$2 = 4a$$

Divide by 4: $\tfrac{2}{4} = \tfrac{4a}{4}$

Simplify: $\tfrac{1}{2} = a$

or $a = \tfrac{1}{2}$

Check: $5 - 3a = a + 3$

$\tfrac{1}{2}$ for a: $5 - 3(\tfrac{1}{2}) = \tfrac{1}{2} + 3$

Multiply: $5 - \tfrac{3}{2} = \tfrac{1}{2} + 3$

Subtract/Add: $3\tfrac{1}{2} = 3\tfrac{1}{2}$

You could also get the answer by substituting answer choices for a to see which one results in a true numerical equation.

7. (3) 6

$$8 = 4b - 16$$

Add 16: $+16 \quad +16$

$$24 = 4b$$

Divide by 4: $6 = b$
or $b = 6$

Check:	$8 = 4b - 16$
6 for b:	$8 = 4(6) - 16$
Multiply:	$8 = 24 - 16$
Subtract:	$8 = 8$

You could also get the answer by substituting answer choices for b in $8 = 4b - 16$ to see which one results in a true equation.

8. (2) 0

$$6 - 2x = 6 + 3x$$

Add $-3x$: $-3x \quad -3x$

$$-5x = 0$$

Divide by -5: $x = 0$

Check:	$6 - 2x = 6 + 3x$
0 for x:	$6 - 2(0) = 6 + 3(0)$
Multiply:	$6 - 0 = 6 + 0$
Subtract/Add:	$6 = 6$

You could also get the answer by substituting answer choices for x to see which one results in a true equation.

9. (1) $\tfrac{4}{5}$

$$\tfrac{2}{3}y = \tfrac{8}{15}$$

Divide by ⅔: that is, multiply by $^3/_2$:

$$\frac{\overset{1}{\cancel{2}}}{\underset{1}{\cancel{3}}}y \times \frac{\overset{1}{\cancel{3}}}{\underset{1}{\cancel{2}}} = \frac{\overset{4}{\cancel{8}}}{\underset{5}{\cancel{15}}} \times \frac{\overset{1}{\cancel{3}}}{2}$$

$$y = \tfrac{4}{5}$$

Check: $\tfrac{2}{3}y = \tfrac{8}{15}$

$\tfrac{4}{5}$ for y: $\tfrac{2}{3}(\tfrac{4}{5}) = \tfrac{8}{15}$

Multiply: $\tfrac{8}{15} = \tfrac{8}{15}$

You could also get the answer by substituting answer choices for y in $^2/_3 y = {}^8/_{15}$ to see which one results in a true equation.

(page 319)

10. (2) 212
Let w stand for the number of women. Since there are 5 times as many men as women, the number of men is $5w$. The sum of w and $5w$ equals the total work force (1,272). So use the equation $w + 5w = 1{,}272$.

Solve.

$$\begin{aligned} & w + 5w = 1{,}272 \\ \text{Combine like terms: } & 6w = 1{,}272 \\ \text{Divide by 6: } & w = 212 \\ \textit{Check: } & w + 5w = 1{,}272 \\ 212 \text{ for } w\text{: } & 212 + 5(212) = 1{,}272 \\ \text{Multiply: } & 212 + 1{,}060 = 1{,}272 \\ \text{Add: } & 1{,}272 = 1{,}272 \end{aligned}$$

There were 212 women in the work force.

11. (2) \$10,000
Let x stand for what Edith earned her first year. This year she earned $1.1x$ (that is, 1.1 times what she earned the first year). The sum of x and $1.1x$ equals her combined income for both years. (\$21,000). So solve the equation $x + 1.1x = 21{,}000$.

$$\begin{aligned} & x + 1.1x = 21{,}000 \\ \text{Combine like terms: } & \\ & 2.1x = 21{,}000 \\ \text{Divide by 2.1 } & x = 10{,}000 \\ \textit{Check: } & x + 1.1x = 21{,}000 \\ 10{,}000 \text{ for } x\text{: } & \\ & 10{,}000 + 1.1(10{,}000) = 21{,}000 \\ \text{Multiply: } & \\ & 10{,}000 + 11{,}000 = 21{,}000 \\ \text{Add: } & 21{,}000 = 21{,}000 \end{aligned}$$

Edith earned \$10,000 her first year.

12. (3) 8
Let c stand for the number of ounces of cashews. Then $2c$ represents the number of ounces of pecans, and $3c$ represents the peanuts. The sum of c, $2c$, and $3c$ equals the total number of ounces (48). So solve the equation $c + 2c + 3c = 48$.

$$\begin{aligned} & c + 2c + 3c = 48 \\ \text{Combine like terms: } & 6c = 48 \\ \text{Divide by 6: } & c = 8 \\ \textit{Check: } & c + 2c + 3c = 48 \\ 8 \text{ for } c\text{: } & 8 + 2(8) + 3(8) = 48 \\ \text{Multiply: } & 8 + 16 + 24 = 48 \\ \text{Add: } & 48 = 48 \end{aligned}$$

There were 8 ounces of cashews.

46 Inequalities

(page 322)

1. (2) $x \le 8$
To solve the inequality $3x \le 24$, divide both sides by 3. Since $3x \div 3 = x$ and $24 \div 3 = 8$, you get $x \le 8$.

2. (3) $x < -3$
To solve $-6x > 18$, divide both sides by -6. Remember that when you multiply or divide both sides of an inequality by a *negative* number, you must *reverse* the inequality sign.

$$-6x > 18$$

$\frac{-6x}{-6} < \frac{18}{-6}$ and $x < -3$

3. (1) $x \ge 9$
To solve $4x - 6 \ge 30$, first add 6 to both sides.

$$\begin{aligned} 4x - 6 + 6 &\ge 30 + 6 \\ 4x &\ge 36 \end{aligned}$$

Next divide both sides by 4.

$\frac{4x}{4} \ge \frac{36}{4}$ and $x \ge 9$

47 Algebraic Equations with Two Unknowns

(page 324)

1. (2) 4
First put $x + 1$ in place of y in $3x + 2y = 17$. Then solve for x.

$$\begin{aligned} 3x + 2(x + 1) &= 17 \\ 3x + 2x + 2 &= 17 \\ 5x + 2 &= 17 \\ 5x &= 15 \\ x &= 3 \end{aligned}$$

You are asked for the value of y, so put 3 in place of x in the equation $y = x + 1$.

$$\begin{aligned} y &= x + 1 \\ y &= 3 + 1 = 4 \end{aligned}$$

The value of y is 4.

2. (2) 3
Put $x + 1$ in place of y in $2x + y = 10$. Then solve for x.

$$\begin{aligned} 2x + (x + 1) &= 10 \\ 3x + 1 &= 10 \\ 3x &= 9 \\ x &= 3 \end{aligned}$$

The value of x is 3.

3. (1) -3
Put $x + 7$ in place of y in $2x + y = -2$. Then solve for x.

$$\begin{aligned} 2x + (x + 7) &= -2 \\ 3x + 7 &= -2 \\ 3x &= -9 \\ x &= -3 \end{aligned}$$

The value of x is -3.

(page 326)

4. (3) $\begin{cases} x + y = 36 \\ y = 3x \end{cases}$
Let x be the weight of the wood products and y be the weight of the steel products. The paragraph tells you that the total weight was 36 tons. This gives you the equation $x + y = 36$.

The paragraph also tells you that the steel products weighed 3 times as much as the wood products. This translates into the equation $y = 3x$.

You might have selected other letters when you were figuring out the equations, but the equations you got should have been similar to $x + y = 36$ and $y = 3x$ except for the letters.

5. (3) \$96
Let x be the amount of the water bill and y the amount of the phone bill. The problem tells you that the total amount of the bills was \$120, so this translates into the equation

$$x + y = 120$$

The problem also tells you that the phone bill was 4 times the water bill. This translates into

$$y = 4x$$

Solve. First replace y with $4x$ in $x + y = 120$, then solve for x.

$$\begin{aligned} x + y &= 120 \\ x + 4x &= 120 \\ 5x &= 120 \\ x &= 24 \end{aligned}$$

Next replace x with 24 in $y = 4x$ to find y (the amount of the telephone bill).

$$\begin{aligned} y &= 4x \\ y &= 4(24) \\ y &= 96 \end{aligned}$$

The phone bill was \$96.

48 Factoring

(page 328)

1. (2) $20x^2 + 20xy$
Use the spreading law to spread the multiplication over addition.

$2x(10x + 10y) =$
$2x(10x) + 2x(10y) =$
$20x^2 + 20xy$

2. (3) $-2xy + x^2$
Use the spreading law.

$(-x)(2y - x) =$
$(-1x)(2y - 1x) =$
$(-1x)(2y) - (-1x)(1x) =$
$-2xy - (-1x^2)$

Change subtraction to addition of the opposite.

$-2xy - (-1x^2) =$
$-2xy + (+1x^2) =$
$-2xy + x^2$

In arriving at the final expression you use the fact that $+1x^2 = x^2$.

3. (3) $3xy + 36x^2$
Use the spreading law.

$6x\,(\frac{1}{2}y + 6x) = 6x(\frac{1}{2}y) + 6x(6x)$
$= 3xy + 36x^2$

(page 329)

4. (1) $a(3b + 2c + d)$
The factor that is common to all three terms is a. So start with a ( ). Extract the factor a from $3ab$, $2ac$, and ad and write the result in the parentheses.

$3ab + 2ac + ad =$
$a(3b + 2c + d)$

5. (1) $ab + ab^2 = ab(1 + b)$
Probably the fastest way to get the answer is to look at each choice and see if the expression to the *right* of the = sign gives the expression to the left of the = sign when you use the spreading law.

In option (1), the spreading law can be used on $ab(1 + b)$ to get $ab(1) + ab(b)$, or $ab + ab^2$

Since $ab + ab^2$ *is* the expression to the left of the = sign, (1) must be the correct choice.

49 Quadratic Equations

(pages 331–332)

1. (3) $x^2 - x - 90$
Multiply as usual.

$(x + 9)(x - 10) =$
$x^2 - 10x + 9x - 90 =$
$x^2 - 1x - 90 =$
$x^2 - x - 90$

2. (2) $y^2 - 49$
Multiply as usual.

$(y + 7)(y - 7) =$
$y^2 - 7y + 7y - 49 =$
$y^2 - 49$

Did you remember that $-7y + 7y$ equals 0?

3. (3) $a^2 + 5a - 6$
Multiply as usual.

$(a + 6)(a - 1) =$
$a^2 - 1a + 6a - 6 =$
$a^2 + 5a - 6$

(page 334)

4. (2) $(x + 7)(x + 4)$
Study the expression $x^2 + 11x + 28$.
The last term has a + sign and so does the middle term. So both factors will have + signs.

$(x + \quad)(x + \quad)$

What numbers multiply to give 28 and add to give 11? The number pairs that have a product of 28 are (28, 1), (14, 2), and (7, 4). Of these, only the last pair has a sum of 11. The answer is

$(x + 7)(x + 4)$.

5. (1) $(y - 3)(y - 7)$
In $y^2 - 10y + 21$, the sign before the last term is + and the sign of the middle term is −, so both factors will have a − sign.

$(y - \quad)(y - \quad)$

What negative numbers multiply to give +21 and add to give −10? The numbers that will give +21 are (−1, −21) and (−3, −7). Of these, the second pair has a sum of −10, so write 3 and 7 after the minus signs in the parentheses.

$(y - 3)(y - 7)$

6. (2) $(b + 4)(b + 4)$
The sign before the term 16 is + and the sign before $8b$ is +. So the factors will both have + signs in them.

$(b + \quad)(b + \quad)$

The positive numbers that have a product of 16 are (1, 16), (2, 8), and (4, 4). Of these, only the last pair adds to give 8. Fill in 4s in both parentheses to get the final result.

$(b + 4)(b + 4)$

(page 336)

7. (2) $n = -2, n = -1$
To solve the equation, factor the left side. You need to find two numbers that multiply to give 2 and that add to give 3. The numbers 1 and 2 will work.

$n^2 + 3n + 2 = 0$
$(n + 1)(n + 2) = 0$

If $n + 1 = 0$, then $n = -1$.
If $n + 2 = 0$, then $n = -2$.

Option (2) is therefore correct.

8. (3) 13
First find the roots of $x^2 - 13x + 42 = 0$.
Factor the left side.

$x^2 - 13x + 42 = 0$
$(x - 6)(x - 7) = 0$

If $x - 6 = 0$, then $x = 6$.
If $x - 7 = 0$, then $x = 7$.

The question asks for the *sum* of the roots, so add: $6 + 7 = 13$.

9. (1) −24
To find the roots, see what numbers will make one of the factors in $(x - 6)(x + 4) = 0$ equal to 0.
If $x - 6 = 0$, then $x = 6$.
If $x + 4 = 0$, then $x = -4$.
The question asks what you would get if you multiplied the roots, so multiply: $6 \times (-4) = -24$.

50 Line Graphs and Slopes

(page 339)

1. (3) (0, 2)
Look at the graphs to see where lines BE and CF intersect. Even though you are not told in the graph what the equations for the lines are, you can be sure the pair of numbers for the point of intersection would make both

equations true. The grid shows that these lines intersect at the point (0, 2), so this pair of numbers is the correct one.

2. (1) *AD* and *BE*
You can see from the grid that *CF* has a point of intersection with *BE* and a point of intersection with *AD*. So the answer must be choice (1).

3. (3) (3, −1)
Look for the point on the grid where lines *AD* and *CF* intersect. This point is labeled (3, −1), so this ordered pair is the correct one.

(page 342)

4. (3) 1
Since you are not given a diagram, it's easiest to use the formula for calculating the slope. Let (x_1, y_1) be (5, 6) and let (x_2, y_2) be (10, 11). Then

$x_1 = 5$ and $y_1 = 6$
$x_2 = 10$ and $y_2 = 11$

Substitute in the formula.

$$\text{Slope of line} = \frac{y_2 - y_1}{x_2 - x_1}$$

$$\frac{11 - 6}{10 - 5} = \frac{5}{5} = 1$$

5. (1) (2, 7)
Since you are not given a graph of the equations, probably your best bet is to solve as in Lesson 48. Substitute $2x + 3$ for y in the equation $y = 3x + 1$.

$$\begin{aligned} y &= 3x + 1 \\ 2x + 3 &= 3x + 1 \\ 3 &= x + 1 \\ 2 &= x \end{aligned}$$

Now put 2 in place of x in one of the original equations. If you put 2 in place of x in $y = 3x + 1$, you get

$$\begin{aligned} y &= 3x + 1 \\ &= 3(2) + 1 \\ &= 6 + 1 \\ &= 7 \end{aligned}$$

So $x = 2$ and $y = 7$. Therefore, the point that the lines have in common is (2, 7).

If you substituted $3x + 1$ for y in $y = 2x + 3$, you should have gotten the same final answer.

Answers: "The Real Thing"

43 Numerical Expressions

(page 302)

1. (4) 82
Here there are only addition and subtraction and no parentheses. To simplify the expression, start at the left and add or subtract the numbers, in order, as you work your way from left to right.

$$\begin{aligned} &32 - 28 + 41 - 19 + 56 \\ &4 + 41 - 19 + 56 \\ &45 - 19 + 56 \\ &26 + 56 \\ &82 \end{aligned}$$

2. (2) 44
Do the work inside the parentheses first.

$$93 - (48 + 23) + (47 - 25)$$
$$93 - 71 + 22$$

Now start at the left and work your way from left to right.

$$22 + 22$$
$$44$$

3. (1) 23
Do the work in the parentheses first. The second set of parentheses you can do right away. To do the first set of parentheses, you'll first have to do the exponent inside.

$$6^2 - (25 - 2^3) + (16 \div 4)$$
$$36 - (25 - 8) + 4$$
$$36 - 17 + 4$$
$$19 + 4$$
$$23$$

(pages 305–306)

4. (3) $(3 \times 85) + (2 \times 75)$
Gather the information from the problem and translate it.

3 days with 85 parts each day
(3×85)
plus 2 days with 75 parts each day
(2×75)

You get the expression
$(3 \times 85) + (2 \times 75)$

5. (2) $(2.54 \div 9) - (1.29 \div 6)$
To find the cost of 1 blade, divide the total cost by the number of blades.

Brand A blade $(1.29 \div 6)$
Brand B blade $(2.54 \div 9)$

Subtract to find how much more a Brand B blade costs.
$(2.54 \div 9) - (1.29 \div 6)$

6. (4) $(7 \times 8) + (5 \times 6.25)$
Gather the information from the problem and translate it.

7 hr at $8/hr: (7×8)
plus
5 hr at $6.25/hr: (5×6.25)
$(7 \times 8) + (5 \times 6.25)$

7. (5) $\frac{20}{4}(0.97 - 0.88)$
To find how much Stefan saves on 4 bars, subtract the sale price from the regular price.

$(0.97 - 0.88)$

Since he bought 20 bars, divide 20 by 4 because the bars are sold in packages of 4.

$\frac{20}{4}$

Multiply to find his total savings.

$\frac{20}{4}(0.97 - 0.88)$

8. (1) $\frac{980 - 200}{9}$
The Ryans paid $200 down so subtract $200 from $980 to find how much they have left to pay.

$980 - 200$

They are going to pay the rest in 9 equal payments. Divide to find the amount they will pay each month.

$$\frac{980 - 200}{9}$$

KEEPING TRACK

Perfect score = 8

Your score ☐

44 Algebraic Expressions

(pages 309–310)

1. (5) 8x

To find the perimeter of a rectangle, you add the lengths of all of its sides. If the length is x yards, then there are two sides of length x yards. If the width is $\frac{1}{3}x$ yards, there are two sides that are $\frac{1}{3}x$ yards long. So the perimeter is represented by the algebraic expression

$$x + x + \tfrac{1}{3}x + \tfrac{1}{3}x$$

Simplify and you get

$$2 \times + \tfrac{2}{3}x \text{ or } 2\tfrac{2}{3}x$$

This is the perimeter in *yards.* To change to *feet,* multiply by 3.

$$3 \times 2\tfrac{2}{3}x = 3 \times \tfrac{8}{3}x = 8x$$

2. (1) $0.05x + 0.08y$

On the x apples that he bought for \$0.15 each, he made a profit of \$0.05 each (since he sold them for \$0.20). His profit on these was \$0.05x.

On the y apples that he bought for \$0.12 each, his profit was \$0.08 each, so the expression for his profit on these is $0.08y$. To get the total profit on all the apples, add: $0.05x + 0.08y$.

3. (1) $2n + 8$

First find the perimeter of the Johnsons' lot. The length is m feet and the width is n feet.

$$m + n + m + n$$

Combine like terms to find the perimeter of the Johnson's lot.

$$2m + 2n$$

Next find the perimeter of the Howards' lot. The length is $m - 3$ feet and the width is $2n + 7$ feet.

$$m - 3 + 2n + 7 + m - 3 + 2n + 7$$

Rearrange the terms so that like terms are together.

$$m + m + 2n + 2n - 3 + 7 - 3 + 7$$

Combine like terms to find the perimeter of the Howards' lot.

$$2m + 4n + 8$$

To find how much shorter the perimeter of the Johnsons' lot is than the perimeter of the Howards' lot, subtract the perimeter of the Johnsons' lot from the perimeter of the Howards' lot.

$$(2m + 4n + 8) - (2m + 2n) = 2m + 4n + 8 - 2m - 2n$$

Rearrange the terms so that like terms are together.

$$2m - 2m + 4n - 2n + 8$$

Combine like terms.

$$2n + 8$$

4. (4) $2b + 5c + (1\tfrac{1}{3})d - 3$

First write all the phrases that tell about the pictures Theodore sold.

	first week	second week
photographs	b	$b - 3$
oil paintings	c	$4c$
water colors	d	$\tfrac{1}{3}d$

Write an algebraic expression to add all of these terms.

$$b + c + d + b - 3 + 4c + \tfrac{1}{3}d$$

Rearrange the terms so that like terms are together.

$$b + b + c + 4c + d + \tfrac{1}{3}d - 3$$

Combine like terms to tell how many pictures he sold in all.

$$2b + 5c + 1\tfrac{1}{3}d - 3$$

(pages 311–312)

5. (4) 48

$$2b + 12 = f$$

Substitute 18 for *b:*

$$2(18) + 12 = f$$

Multiply: $36 + 12 = f$

Add: $48 = f$

6. (1) 6

$$\tfrac{1}{3}b = s$$

Substitute 18 for *b:* $\tfrac{1}{3} \times 18 = s$

Multiply: $6 = s$

7. (5) 42

$$(2b + 12) - \tfrac{1}{3}b = r$$

Substitute 18 for *b:*

$$(2 \times 18 + 12) - \tfrac{1}{3} \times 18 = r$$

Multiply: $36 + 12 - 6 = r$

Add: $48 - 6 = r$

Subtract: $42 = r$

8. (3) 23

$$x + 2y + 3z = t$$

Substitute $x = 7$, $y = 2$, and $z = 4$:

$$7 + 2(2) + 3(4) = t$$

Multiply: $7 + 4 + 12 = y$

Add: $23 = y$

9. (5) \$3.65

$$0.25x + 0.10(2y) + 0.05(3z) = c$$

Substitute $x = 3$, $y = 7$, and $z = 10$:

$$0.25(3) + 0.10(2 \times 7) + 0.05(3 \times 10) = c$$

Multiply the numbers inside the parentheses:

$$0.25(3) + 0.10(14) + 0.05(30) = c$$

Multiply: $0.75 + 1.40 + 1.50 = c$

Add: $3.65 = c$

10. (5) \$77.50

$$1.50x\ 2.75y = t$$

Substitute $x = 15$ and $y = 20$:

$$1.50(15) + 2.75(20) = t$$

Multiply: $22.50 + 55.00 = t$

Add: $77.50 = t$

11. (1) \$13.50

First find out how much money Chris collected.

$$1.50x + 2.75y = t$$

Substitute $x = 13$ and $y = 18$:

$$1.50(13) + 2.75(18) = t$$

Multiply: $19.50 + 49.50 = t$

Add: $69.00 = t$

Next find out how much money Kim collected.

$$1.50x + 2.75y = t$$

Substitute $x = 15$ and $y = 12$:

$$1.50(15) + 2.75(12) = t$$

Multiply: $22.50 + 33.00 = t$

Add: $55.50 = t$

Subtract to find how much more money Chris collected than Kim.

$$69.00 - 55.50 = 13.50$$

12. (4) $\frac{(13n + 45)}{9}$
Ellen practiced 3 different lengths of time. Write phrases that tell how long she practiced each time.

n minutes on 2 days: $2n$
$n + 15$ on 3 days:
$3(n + 15) = 3n + 45$
$2n$ on 4 days: $4(2n) = 8n$

Add to find the total amount of time she practiced.

$2n + 3n + 45 + 8n$

Combine like terms.

$13n + 45$

To find the average, divide the total amount of time $(13n + 45)$ by the number of days (9).

$\frac{13n + 45}{9}$

13. (5) $2f + 4g + 1.5h + 5$
First decide how many coins of each kind Bill and Pam have.

	Bill	Pam
nickels	f	$f + 5$
dimes	g	$3g$
dollars	h	$0.5h$

Write an algebraic expression that tells how many coins they have together.

$f + g + h + f + 5 + 3g + 0.5h$

Combine like terms.

$2f + 4g + 1.5h + 5$

KEEPING TRACK
Perfect score = 13
Your score ☐

45 Algebraic Equations

(page 316)

1. (4) 25
Algebraic solution: $x - 7 = 18$
Change − to +: $x + (-7) = 18$
Add 7: $+7 \quad +7$
$x = 25$

Check: $x - 7 = 18$

Substitute 25 for x: $25 - 7 = 18$
Subtract: $18 = 18$

2. (2) 7
Algebraic solution:
$13 = x + 6$
Add −6: $-6 \quad -6$
$7 = x$
Check: $13 = x + 6$
Substitute 7 for x: $13 = 7 + 6$
Add: $13 = 13$

3. (1) 3
Algebraic solution:
$3x - 4 = 5$
Change − to +:
$3x + (-4) = 5$
Add 4: $+4 \quad +4$
$3x = 9$
Multiply by $\frac{1}{3}$: $(\frac{1}{3})3x = (\frac{1}{3})9$
Simplify: $x = 3$
Check: $3x - 4 = 5$
Substitute 3 for x:
$3(3) - 4 = 5$
Multiply: $9 - 4 = 5$
Subtract: $5 = 5$

4. (3) 10
Algebraic solution:
$(\frac{4}{5})x + 7 = 15$
Add −7: $(\frac{4}{5})x + 7 = 15$
$-7 \quad -7$
$(\frac{4}{5})x = 8$
Divide by $\frac{4}{5}$ by multiplying by $\frac{5}{4}$:
$(\frac{5}{4})(\frac{4}{5})\,x = (\frac{5}{4})8$
Simplify: $x = 10$
Check: $(\frac{4}{5})x + 7 = 15$
Substitute 10 for x:
$(\frac{4}{5})10 + 7 = 15$
Multiply: $8 + 7 = 15$
Add: $15 = 15$

5. (5) $x + 7 = 23$
Let x be the number of quarters he has. Translate.

number he has	plus	7	equals	23
↓	↓	↓	↓	↓
x	+	7	=	23

(pages 317–318)

6. (2) $x + 9 = 2x + 4$
Let x be the number of pencils on Becky's desk. Then $x + 9$ is this number increased by 9. She says that $x + 9$ is 4 more than twice the number she has $(2x)$. So $x + 9 = 2x + 4$ translates her statement into an equation.

7. (3) 11
A number (x) is subtracted from 25; $(25 - x)$. This difference is 8 less than twice the number $(2x - 8)$. Therefore, the equation is $25 - x = 2x - 8$.
Algebraic solution:
$25 - x = 2x - 8$
Change − to +:
$25 + (-x) = 2x + (-8)$
Add 8: $25 + (-x) = 2x + (-8)$
$+8 \quad +8$
$33 + -x = 2x$
Add x: $+x \quad +x$
$33 = 3x$
Divide by 3: $\frac{33}{3} = \frac{3x}{3}$
$11 = x$
Check: $25 - x = 2x - 8$
Substitute 11 for x:
$25 - 11 = 2(11) - 8$
Multiply: $25 - 11 = 22 - 8$
Subtract: $14 = 14$

8. (4) 24.736
Algebraic solution:
$4x - 8.275 = 3(x + 5.487)$
Multiply:
$4x - 8.275 = 3x + 16.461$
Change − to +:
$4x + (-8.275) = 3x + 16.461$
Add 8.275:
$+8.275 = \quad + 8.275$
$4x = 3x + 24.736$
Add $-3x$:
$-3x \quad -3x$
$x = 24.736$

Check:
$4x - 8.275 = 3(x + 5.487)$
Substitute 24.736 for x:
$4(24.736) - 8.275 = 3(24.736 + 5.487)$
Add in the parentheses:
$4(24.736) - 8.275 = 3(30.223)$
Multiply:
$98.944 - 8.275 = 90.669$
Subtract:
$90.669 = 90.669$

9. (2) 1.37

Algebraic solution:

$$9.42 - 2x = 3x + 2.57$$

Add -2.57:

$$\underline{-2.57 \qquad\qquad -2.57}$$

$$6.85 - 2x = 3x$$

Change $-$ to $+$:

$$6.85 + (-2x) = 3x$$

Add $2x$: $\quad +2x \quad +2x$

$$6.85 = 5x$$

Divide by 5: $\frac{6.85}{5} = \frac{5x}{5}$

$$1.37 = x$$

Check: $9.42 - 2x = 3x + 2.57$

Substitute 1.37 for x:

$$9.42-2(1.37) = 3(1.37) + 2.57$$

Multiply:

$$9.42 - 2.74 = 4.11 + 2.57$$

Subtract and add:

$$6.68 = 6.68$$

10. (5) $2\frac{1}{2}$

Algebraic solution:

$$(\tfrac{2}{3})x - \tfrac{1}{2} = \tfrac{3}{4} + (\tfrac{1}{6})x$$

Change $-$ to $+$:

$$(\tfrac{2}{3})x + (-\tfrac{1}{2}) = \tfrac{3}{4} + (\tfrac{1}{6})x$$

Add $\frac{1}{2}$: $\quad +\frac{1}{2} \quad +\frac{1}{2}$

$$(\tfrac{2}{3})x = \tfrac{5}{4} + (\tfrac{1}{6})x$$

Add $-(\frac{1}{6})x$: $\quad -(\frac{1}{6})x \quad -(\frac{1}{6})x$

$$(\tfrac{3}{6})x = \tfrac{5}{4}$$

Simplify: $(\frac{1}{2})x = \frac{5}{4}$

Divide by ½ by multiplying by 2:

$$2(\tfrac{1}{2})x = 2(\tfrac{5}{4})$$

Simplify: $x = 2\frac{1}{2}$

Check your answer.

(pages 319–320)

11. (1) $6x = 36$

Replace x by 6 in each equation to see which one has a solution of 6.

(1) $6(6) = 36$
$36 = 36$ **Solution**

(2) $(6) - 4 = 3$
$2 = 3$ **Not a solution.**

(3) $5 + (6) = 13$
$11 = 13$ **Not a solution.**

(4) $\frac{(6)}{4} = 2$
$\frac{3}{2} = 2$ **Not a solution.**

(5) $21 = 3(6)$
$21 = 18$ **Not a solution.**

12. (3) $18 + x = 25$

The equation is:

$$18 + x = 25$$

↓	↓	↓	↓	↓
x	$+$	18	$=$	25

13. (2) 2.11

Algebraic solution:

$$3x - 2.87 = 3.46$$

Change $-$ to $+$:

$$3x + (-2.87) = 3.46$$

Add 2.87:

$$\underline{+ 2.87 \qquad +2.87}$$

$$3x = 6.33$$

Divide by 3; that is, multiply by ⅓:

$$(\tfrac{1}{3})3x = (\tfrac{1}{3})6.33$$

$$x = 2.11$$

Check your answers.

$$3x - 2.87 = 3.46$$

Replace x with 2.11:

$$3(2.11) - 2.87 = 3.46$$

Multiply: $6.33 - 2.87 = 3.46$

Subtract: $3.46 = 3.46$

14. (4) $\frac{11}{24}$

Algebraic solution:

$$\tfrac{3}{4} + 2x = 1\tfrac{2}{3}$$

Change 1⅔ to 5/3, then add $-$¾:

$$\underline{-\tfrac{3}{4} \qquad -\tfrac{3}{4}}$$

$$2x = \tfrac{11}{12}$$

Divide by 2; that is, multiply by ½:

$$(\tfrac{1}{2})2x = (\tfrac{1}{2})(\tfrac{11}{12})$$

$$x = \tfrac{11}{24}$$

Check your answer.

15. (3) $0.22(\frac{75}{5}) = 2x - 0.18$

Translation: Let x = cost of paper. Juan sold all the paper for 0.22(75/5) dollars. This is \$0.18 less than twice the cost. Therefore, the equation is

$$0.22(\tfrac{75}{5}) = 2x - 0.18$$

16. (5) 1.601

Algebraic solution:

$$4.271 - 5x = 2(x - 3.468)$$

Multiply:

$$4.271 - 5x = 2x - 6.936$$

Change $-$ to $+$:

$$4.271 + (-5x) = 2x + (-6.936)$$

Add $5x$:

$$\underline{+5x \qquad +5x}$$

$$4.271 = 7x + (-6.936)$$

Add 6.936:

$$\underline{+6.936 \qquad +6.936}$$

$$11.207 = 7x$$

Divide by 7; that is, multiply by 1/7.

$$(\tfrac{1}{7})11.207 = (\tfrac{1}{7})7x$$

$$1.601 = x$$

Check your answer.

17. (2) $-\frac{1}{2}$

Algebraic solution:

$$6x - \tfrac{2}{3} = 9x + \tfrac{5}{6}$$

Change $-$ to $+$:

$$6x + (-\tfrac{2}{3}) = 9x + \tfrac{5}{6}$$

Add $-6x$: $\quad -6x \quad -6x$

$$-\tfrac{2}{3} = 3x + \tfrac{5}{6}$$

Add $-\frac{5}{6}$: $\quad -\frac{5}{6} \quad -\frac{5}{6}$

$$-\tfrac{9}{6} = 3x$$

Divide by 3; that is, multiply by ⅓.

$$(\tfrac{1}{3})(-\tfrac{9}{6}) = (\tfrac{1}{3})(3x)$$

$$-\tfrac{1}{2} = x$$

Check your answer.

18. (5) \$39.50

Let x be the amount Tina earned.

Translation:

(Ling earned \$46.85) (which is)
46.85 =
(\$32.15 less than)
$-$ 32.15
(twice the amount Tina earned)
$2x$

The equation is:

$$46.85 = 2x - 32.15$$

Algebraic solution:

$$46.85 = 2x - 32.15$$

Change $-$ to $+$:

$$46.85 = 2x + (-32.15)$$

Add 32.15:

$$\underline{+32.15 = +32.15}$$

$$79.00 = 2x$$

Divide by 2; that is, multiply by ½:

$$(\tfrac{1}{2})79.00 = (\tfrac{1}{2})x$$

$$39.50 = x$$

Check: $46.85 = 2x - 32.15$
Replace x by 39.50:
$$46.85 = 2(39.50) - 32.15$$
Simplify:
$$46.85 = 79.00 - 32.15$$
$$46.85 = 46.85$$

19. (4) 48
Let r be Roger's age. His sister's age is 15. Her age is 3 years more than ¼ Roger's age, or $(1/4)r + 3$.

The equation is: $15 = \frac{1}{4}r + 3$

Add -3: $\quad -3 \quad -3$
$$12 = \tfrac{1}{4}r$$
Multiply by 4: $4(12) = 4 \times \frac{1}{4}r$
$$48 = r$$
Check: $15 = \frac{1}{4}r + 3$

Replace r by 48:
$$15 = \tfrac{1}{4} \times 48 + 3$$
$$15 = 12 + 3$$
$$15 = 15$$

KEEPING TRACK
Perfect Score = 19
Your score ☐

46 Inequalities

(pages 322–323)

1. (3) $-2, 2, 10$
Solve $-3x < 12$ by dividing each side by -3. You can multiply by $-1/3$. Remember that when you multiply by a *negative* number, you must reverse the direction of the inequality sign.
$$-\tfrac{1}{3}(-3x) > -\tfrac{1}{3}(12)$$
$$x > -4$$
Look for the choice where *all* the numbers are greater than -4; the correct option is (3).

2. (4) $x \geq 36$
To solve the inequality, first add -5 to both sides.
$$5 + \tfrac{1}{3}x \geq 17$$
$$\underline{-5 \qquad\quad -5}$$
$$\tfrac{1}{3}x \geq 12$$
Divide both sides by ⅓, which means multiply both sides by 3.
$$3(\tfrac{1}{3}x) \geq 3(12)$$
$$x \geq 36$$

3. (2) 7
Put the numbers in the answer choices in place of x in each inequality, starting with the first answer choice (3). Simplify.
$$2x - 11 \leq 10$$
$$2(3) - 11 \leq 10$$
$$6 - 11 \leq 10$$
$$-5 \leq 10 \text{ TRUE}$$
$$16 - 3x \leq 4$$
$$16 - 3(3) \leq 4$$
$$116 - 9 \leq 4$$
$$7 \leq 4 \text{ FALSE}$$
Since $x = 3$ does not make the second inequality true, it is not the answer.

Go to choice (2). Put 7 in place of x in both inequalities.
$$2x - 11 \leq 10$$
$$2(7) - 11 \leq 10$$
$$14 - 11 \leq 10$$
$$3 \leq 10 \text{ TRUE}$$
$$16 - 3x \leq 14$$
$$16 - 3(7) \leq 14$$
$$16 - 21 \leq 14$$
$$-5 \leq 14 \text{ TRUE}$$
So 7 makes both inequalities true. Option (2) is correct.

4. (1) $2x + 3 \leq 15$
Replace x with 5 in each inequality, starting with option (1).
$$2x + 3 \leq 15$$
$$2(5) + 3 \leq 15$$
$$10 + 3 \leq 15$$
$$13 \leq 15 \text{ TRUE}$$
Since there will only be one correct answer choice, you don't need to test any other option.

5. (2)$x > 1\frac{1}{8}$
To solve the inequality, first add -3 to both sides.
$$3 - 2x < \tfrac{3}{4}$$
$$\underline{-3 \qquad\qquad -3}$$
$$-2x < -2\tfrac{1}{4}$$
Next divide both sides by -2. To do this, multiply by $-1/2$. Reverse the direction of the inequality sign.
$$-\tfrac{1}{2}(-2x) > -\tfrac{1}{2}(-2\tfrac{1}{4})$$
$$x > -\tfrac{1}{2}(-\tfrac{9}{4})$$
$$x > \tfrac{9}{8}$$
$$x > 1\tfrac{1}{8}$$

6. (4) $x < 2.45$
First add -2.75 to both sides.
$$10.1 > 3x + 2.75$$
$$\underline{-2.75 \qquad\quad -2.75}$$
$$7.35 > 3x$$
Next divide both sides by 3.
$$\frac{7.35}{3} > x$$
$$2.45 > x$$
Since $2.45 > x$ is equivalent to $x < 2.45$, option (4) is correct.

7. (5) $x \leq -14.378$
First add -5.438 to both sides.
$$5.438 - 4x \geq 62.95$$
$$\underline{-5.438 \qquad\qquad - 5.438}$$
$$-4x \geq 57.512$$
Next divide both sides by -4. Since you are dividing by a negative number, you must change $\geq$ to $\leq$ in the result.
$$\frac{-4x}{-4} \leq \frac{57.512}{-4}$$
$$x \leq -14.378$$

8. (1) $x \geq \frac{2}{3}$
First add $2\frac{5}{6}$ to both sides.
$$x - 2\tfrac{5}{6} \geq \tfrac{1}{2} - 4x$$
$$\underline{+2\tfrac{5}{6} \qquad +2\tfrac{5}{6}}$$
$$x \geq 3\tfrac{1}{3} - 4x$$
Next add $4x$ to both sides.
$$x \geq 3\tfrac{1}{3} - 4x$$
$$\underline{+4x \qquad\quad +4x}$$
$$5x \geq 3\tfrac{1}{3}$$
Finally, divide both sides by 5.
$$\tfrac{5x}{5} \geq 3\tfrac{1}{3} \div 5$$
$$x \geq \tfrac{10}{3} \times \tfrac{1}{5} \geq \tfrac{2}{3}$$

9. (1) 0
You will need to replace x in both of the inequalities with the answer choices until you find the choice that makes both inequalities true. Start with the first choice and replace x with 0.

$x \geq -\frac{1}{2}$ $\qquad$ $x < 5$

$0 \geq -\frac{1}{2}$ TRUE $\qquad$ $0 < 5$ TRUE

Since 0 makes both inequalities true and since there is only one correct answer, you are done.

KEEPING TRACK
Perfect score = 9
Your score ☐

47 Algebraic Equations with Two Unknowns

(pages 324–325)

1. (4) $x = 5, y = 10$
You can solve the equations to find the values of x only and then look for the choice that has the numbers you got. You could also go through the choices one at a time, putting the values for x and y into the equations until you find the values that make both equations true. Suppose you use the first method. To solve the equations, put $2x$ in place of y in $3x + y = 25$.

$$3x + y = 25$$
$$3x + 2x = 25$$
$$5x = 25$$
$$x = 5$$

Put 5 in place of x in the first equation ($y = 2x$).

$$y = 2x$$
$$y = 2(5) = 10$$

So $x = 5$ and $y = 10$. The correct answer is option 4.

2. (2) $x = 2, y = 1$
Solve $y + 1 = x$ and $y = 2x - 3$. Put $2x - 3$ in place of y in the first equation.

$$2x - 3 + 1 = x$$
$$2x - 2 = x$$
$$2x = x + 2$$
$$2x - x = 2$$
$$x = 2$$

Put 2 in place of x in $y = 2x - 3$.

$$y = 2x - 3$$
$$y = 2(2) - 3$$
$$y = 4 - 3 = 1$$

So the values of x and y are $x = 2$ and $y = 1$ —option (2).

3. (1) $x = -3, y = 3$
Solve $y = -x$ and $x + 2y = 3$. First put $-x$ in place of y in $x + 2y = 3$.

$$x + 2y = 3$$
$$x + 2(-x) = 3$$
$$x + (-2x) = 3$$
$$-x = 3$$
$$x = -3$$

Put -3 in place of x in $y = -x$.

$$y = -x$$
$$y = -(-3) = 3$$

The values for x and y are $x = -3, y = 3$—option (1).

(page 326)

4. (4) $0.25
First translate the problem into equations. Let x be the price of a daily paper and let y be the price of a Sunday paper.

April: $25x + 5y = \$12.25$
May: $27x + y = \$7.95$

To solve, rewrite $27x + y = 7.95$ so that you can have a value for y.

$$27x + y = 7.95$$

Subtract $27x$ from each side.

$$y = 7.95 - 27x$$

Use the first equation.

$$25x + 5y = \$12.25$$

Substitute $7.95 - 27x$ for y.

$$25x + 5(7.95 - 27x) = 12.25$$

Multiply.

$$25x + 39.75 - 135x = 12.25$$

Subtract 39.75 from each side.

$$25x - 135x = 12.25 - 39.75$$

Combine like terms.

$$-110x = -27.50$$

Divide each side by -110.

$$x = 0.25$$

The price of a daily paper is $0.25.

5. (1) $14.17
Let x be Sam's hourly wage and let y be Judy's hourly wage.

$$x = y - 0.66$$
$$x + y = 29.00$$

To solve, substitute $y - 0.66$ for x in the second equation.

$$y - 0.66 + y = 29.00$$

Add $0.66 to each side.

$$y + y = 29.00 + 0.66$$

Combine like terms.

$$2y = 29.66$$

Divide each side by 2.

$y = 14.83$ in the first equation.
$x = 14.83 - 0.66 = 14.17$

Sam earned $14.17 per hour.

6. (3) $0.72
Let x be the number of pencils and y the number of notebooks.

$$12x + 5y = \$4.08$$
$$x + 8y = \$5.80$$

Rewrite $x + 8y = \$5.80$ so that you have a value for x.

$$x + 8y = 5.80$$

Subtract $8y$ from each side.

$$x = 5.80 - 8y$$

Use the second equation.

$$12x + 5y = 4.08$$

Substitute $5.80 - 8y$ for x.

$$12\,(5.80 - 8y) + 5y = 4.08$$

Multiply.

$$69.60 - 96y + 5y = 4.08$$

Subtract 69.60 from each side.

$$-96y + 5y = 4.08 - 69.60$$

Combine like terms.

$$-91y = -65.52$$

Divide each side by -91.

$$y = 0.72$$

One notebook costs $0.72.

7. (2) 14
Let x be the length in inches and y be the width in inches. Recall that the formula for finding the perimeter of a rectangle is

$$P = 2\ell + 2w$$
$$x = 2y$$
$$42 = 2x + 2y$$

The first equation gives $2y$ as a value for x. Use the second equation, $42 = 2x + 2y$, and substitute $2y$ for x.

$$42 = 2(2y) + 2y$$
$$42 = 4y + 2y$$
$$42 = 6y$$
$$7 = y$$

Now use the first equation.

$$x = 2y$$

Substitute 7 for y.

$$x = 2(7)$$
$$x = 14$$

The length is 14 inches.

8. (2) 12
Let x be the number of rabbits and y the number of pigeons. Use the fact that rabbits have 4 feet and pigeons have 2 feet.

$$4x + 2y = 78$$
$$x + y = 27$$

Rewrite $x + y = 27$ so that you have a value for y.

$$x + y = 27$$

Subtract x from each side.

$$y = 27 - x$$

Use the first equation.

$$4x + 2y = 78$$

Substitute $27 - x$ for y.

$$4x + 2(27 - x) = 78$$
$$4x + 54 - 2x = 78$$
$$2x = 24, \text{ or } x = 12$$

There are 12 rabbits.

9. (5) 55
Let x be the number of men and y be the number of women.

$$x + y = 92$$
$$x = 2y - 19$$

The second equation gives you a

value for x, so substitute it in the first equation.

$2y - 19 + y = 92$
$2y + y = 92 + 19$
$3y = 111$
$y = 37$

Use the first equation to find the value of x.

$x + y = 92$

Substitute 37 for y.

$x + 37 = 92$
$x = 92 - 37$
$x = 55$

So the number of men is equal to 55.

10. (4) 11
Let x be Wanda's age now and y be Wanda's grandfather's age.

$y + 3 = 6(x - 1)$
$x + y = 68$

Use the second equation to find a value for y.

$x + y = 68$

Subtract x from each side.

$y = 68 - x$

Substitute $68 - x$ for y in the first equation.

$68 - x + 3 = 6(x - 1)$
$68 - x + 3 = 6x - 6$
$68 - x + 3 + 6 = 6x$
$68 + 3 + 6 = 6x + x$
$77 = 7x$, or $7 = x$

Wanda's present age is 11.

KEEPING TRACK
Perfect score = 10
Your score ☐

48 Factoring

(page 330)

1. (1) $2(c + d)$
Find the common factor of $2c$ and $2d$. Write the common factor 2 outside a pair of parentheses.

$2(\quad)$

Extract the factor 2 from each term of $2c + 2d$ and write the result in the parentheses.

$2(c + d)$

2. (5) $3j^2 + 12j = 3j(j + 4)$
Each statement gives the product on one side of the equal sign and the two factors on the other side. To find which statement is true, multiply the factors given in each choice and check your answer.

In (1), $2rs(r + 1) = 2r^2s + 2rs$, not $r^2 + 2rs$.

In (2), $3n(3n + 2) = 9n^2 + 6n$, not $3n + 6n^2$.

In (3), $4(m + 4) = 4m + 16$, not $4m + 4$.

In (4), $4(v + w) = 4v + 4w$, not $2v + 2w$.

In (5), $3j(j + 4) = 3j^2 + 12j$, so (5) is the correct answer.

3. (1) $5k(j + k)$
To find the correct choice, it is enough to find the choice in which $5k$ is a factor of both terms. In (1), $5k$ is a factor of $5jk$ and of $5k^2$. So (1) must be the correct answer.

4. (3) 5
Try dividing 10 by each answer option. Only 5, Option (3) works.

5. (4) r
Notice that if you factor $2r^2 - rs$ (total paid), you get $r(2r - s)$. But $2r - s$ is the number of oranges in the bag. So r must represent the price per orange.

6. (2) $8de(2de - 3)$
To prove, factor out $8de$.

$8de(2de - 3)$

Option (2) is the only one that has $8de$ as a factor.

KEEPING TRACK
Perfect score = 6
Your score ☐

49 Quadratic Equations

(pages 336–337)

1. (4) + 2
You can factor $b^2 - 4$ to get $(b + 2)(b - 2)$.

If $b + 2 = 0$, then $b = -2$
If $b - 2 = 0$, then $b = +2$

The other root is -2.

2. (3) $(x + 3)(x - 2)$
Factor x first and place each x in parenthesis.

$(x)(x)$

The negative sign before the 6 tells you the signs in factored form must be + and −. Which numbers when added equal 1 and when multiplied equal -6? The answer is $+3$ and -2. The factors are $(x + 3)(x - 2)$.

3. (2) -5
Find the factors of $a^2 + a - 20$. You know that $a - 4$ is one factor. Since the final term (-20) is negative and the second term (x) is positive, one sign is negative and one sign is positive.

$(a - 4)(a + \quad)$

What number multiplied by -4 gives -20 and added to -4 gives $+1$? The answer is $+5$.

$(a - 4)(a + 5)$

Now consider the equation $(a - 4)(a + 5) = 0$ If $a - 4 = 0$, then $a = +4$. This is the root you were told. If $a + 5 = 0$, then $a = -5$. So the other root is -5.

4. (5) $+3$
Factor the left side of $y^2 - 6y + 9 = 0$ by writing the parentheses with the unknown in each. Next determine what signs will be in the parentheses. Since the final term (9) is positive, both signs are the same. Since the second term $(-6y)$ has a minus sign, you need minus signs in both pairs of parentheses.

Determine the numbers. When the numbers are multiplied, the answer is $+9$. When they are added, the answer is -6. Since $(-3) \times (-3) = 9$ and $(-3) + (-3) = -6$, the factors are

$(y - 3)(y - 3)$.

Set each factor equal to 0 to find the roots.

If $y - 3 = 0$, then $y = +3$
If $y - 3 = 0$, then $y = +3$

In this case, the roots you get from the factors are the same.

5. (1) -6
Factor the left side of $c^2 + 12c + 36 = 0$ by writing out the parentheses with the unknown in each. You must find two numbers that multiply to give $+36$ and that add to give $+12$. Both must be positive. The positive whole numbers that multiply to give $+36$ are

+1 and +36
+2 and +18
+3 and +12
+4 and +9
+6 and +6

Only the last pair (+6 and +6) add to give +12. So the factors are $(c + 6)$ and $(c + 6)$. These two factors are the same. So $(c + 6)(c + 6) = 0$ only if $c + 6 = 0$. If $c + 6 = 0$, $c = -6$.

6. (3) +5 or +6
Factor the left-hand side of the equation. You must find two numbers that multiply to give +30. They must add to give −11. So both numbers must be negative. The pairs of negative numbers that you can multiply to get +30 are these:

−1 and −30
−2 and −15
−3 and −10
−5 and −6

Only −5 and −6 add to give −11. The factors are $x - 5$ and $x - 6$ for the roots of $(x - 5)(x - 6) = 0$.

Set each factor equal to 0.

If $x - 5 = 0$, then $x = 5$
If $x - 6 = 0$, then $x = 6$

So the roots are +5 and +6.

7. (2) $m^2 + 2m - 8 = 0$
You know that −4 and +2 are the roots of an equation.

If $m = -4$, then $m + 4 = 0$
If $m = +2$, then $m - 2 = 0$

Work back to an equation by multiplying $m + 4$ and $m - 2$:

$$(m + 4)(m - 2) = 0$$
$$m^2 - 2m + 4m - 8 = 0$$
$$m^2 + 2m - 8 = 0$$

8. (4) $a^2 - 3a - 28 = 0$
You know that +7 and −4 are the roots of an equation.

If $a = +7$, then $a - 7 = 0$
If $a = -4$, then $a + 4 = 0$

Use $a - 7$ and $a + 4$ as factors in an equation.

$$(a - 7)(a + 4) = 0$$
$$a^2 + 4a - 7a - 28 = 0$$
$$a^2 - 3a - 28 = 0$$

KEEPING TRACK
Perfect score = 8
Your score ☐

50 Line Graphs and Slopes

(page 340)

1. (2) (1, −1)
Read the point of intersection from the graph as (1, −1). If you wish, you can check by using 1 for x and −1 for y to see if these numbers make both equations true.

$$2x - y = 3$$
$$(2 \times 1) - (-1) = 3$$
$$2 + 1 = 3$$
$$3 = 3 \quad \text{True}$$

$$3x + y = 2$$
$$(3 \times 1) + (-1) = 2$$
$$3 - 1 = 2$$
$$2 = 2 \quad \text{True}$$

2. (4) (−2, −5)
You can try each choice, replacing x by the first number of the ordered pair and y by the second number. Substitute in both equations till you find the pair that makes both equations true. You'll find that (−2, −5) is the only pair that works.

You could also draw graphs and read the point of intersection from the graph.

Yet another possibility is to solve the pair of equations for x and y. Use the value you get for x as the first number of an ordered pair and the value you get for y as the second number. Then find your pair in the list of choices.

3. (1) $y = (-\frac{2}{3})x + 4$
The three points identified on line AB fit only one equation: $y = (-\frac{2}{3})x + 4$. To see this, substitute the values for x and y for each point in $y = (-\frac{2}{3})x + 4$.

Point A is (−3, 6).

$$y = (-\tfrac{2}{3})x + 4$$
$$6 = (-\tfrac{2}{3}) - 3 + 4$$
$$6 = 2 + 4$$
$$6 = 6$$

Point B is (9, −2).

$$y = (-\tfrac{2}{3})x + 4$$
$$-2 = (-\tfrac{2}{3})(9) + 4$$
$$-2 = -6 + 4$$
$$-2 = -2$$

The third point is (3, 2).

$$y = (-\tfrac{2}{3})x + 4$$
$$2 = (-\tfrac{2}{3})(3) + 4$$
$$2 = -2 + 4$$
$$2 = 2$$

(pages 342–343)

4. (3) $\frac{2}{3}$
If you draw a horizontal line through M and a vertical line through N, you get a right triangle. The vertical leg is 6 units and the horizontal leg is 9 units.

$$\frac{\text{length of the vertical leg}}{\text{length of the horizontal leg}} = \frac{6}{9} = \frac{2}{3}$$

You know the slope is *positive* because the line goes *up* as you move from left to right. So the slope is $\frac{2}{3}$.

5. (4) $-\frac{2}{3}$
Graph the points on a coordinate grid.

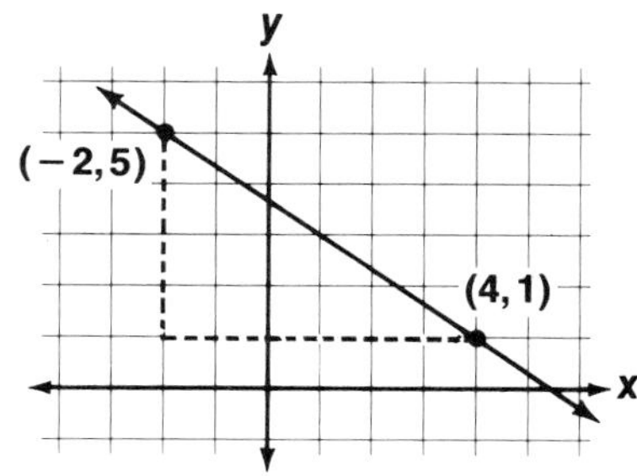

The line formed by these points slopes downward from left to right, which means the slope is a negative number. Construct a right triangle by dropping a line straight down from (−2, 5) and another line across to the left from (4, 1). Count the units in the legs of the right triangle.

Slope =

$$= \frac{\text{length of the vertical leg}}{\text{length of the horizontal leg}} = -\frac{4}{6} = -\frac{2}{3}$$

You could also use the formula

$$\text{slope} = \frac{y_2 - y_1}{x_2 - x_1}.$$

Let (x_2, y_2) be (−2, 5) and let (x_1, y_1) be (4, 1). Then $x_2 = -2$, $y_2 = 5$ and $x_1 = 4$, $y_1 = 1$. Use these values in the formula

$$\text{slope} = \frac{5 - 1}{(-2) - 4} = \frac{4}{-6} = -\frac{2}{3}.$$

6. (3) 0.24
Use the counting method.

$$\frac{\text{vertical leg}}{\text{horizontal leg}} = \frac{6}{25} = .24$$

Since the line rises from left to right, you know that the slope is positive. The slope of the line is option (3), 0.24.

7. (2) (4, −5)
A line that has no slope is a vertical line. All of the x-coordinates on a vertical line are the same. In (4, 6), 4 is the x-coordinate, so any ordered pair with 4 as the x-coordinate is on this line. Only (4, −5) can be a correct choice.

8. (4) (−3, −2)
To get to the point of intersection, start at the origin. Go left 3 units and down 2 units. The ordered pair (−3, −2) is the point where lines d and c intersect.

9. (5) 2.00
Let (x_1, y_1) be (2, −4) and let (x_2, y_2) be (7, 6). So $x_1 = 2$, $y_1 = -4$, and $x_2 = 7$, $y_2 = 6$. Use these values in the slope formula.

$$\text{slope} = \frac{y_2 - y_1}{x_2 - x_1} = \frac{6 - (-4)}{7 - 2} = \frac{10}{5} \text{ or } 2$$

10. (4) 30.0
The grade is the same as the slope. Use a grid to find 15% of 200.

percent	15	?	vertical leg
	100	200	horizontal leg

$15 \times 200 = 3{,}000$
$3{,}000 \div 100 = 30$

The vertical leg is 30 feet.

11. (1) d and k
Lines that are parallel have the same slope.

12. (2) 45°
Take a point such as, say, (4, 4). It lies on the line for $y = x$.

Form a right triangle. The horizontal and vertical legs are equal, so the triangle is isosceles. Therefore, the angles opposite them have the same measure, say a. All the angles of the triangle add up to 180°.

$a + a + 90 = 180$

Solve: $2a + 90 = 180$
$2a = 90$, or $a = 45$

KEEPING TRACK
Perfect score = 12

Your score ☐

Keeping Track

Now enter all your scores from the Keeping Track boxes on the lines below.

Lesson	Perfect Score	Your Score
Numerical Expressions	8	______
Algebraic Expressions	13	______
Algebraic Equations	19	______
Inequalities	9	______
Algebraic Equations with Two Unknowns	10	______
Factoring	6	______
Quadratic Equations	8	______
Line Graphs and Slopes	12	______
TOTAL	85	______

How did you do in *Algebra?* Were you able to get perfect scores in some lessons? Now is the time to review those lessons that gave you trouble. Then you can go on to the Extra Practice that follows.

Extra Practice: Algebra

1. Alex's pay is $4.50 per hour. His deductions are 22% of his earned pay. If he works h hours, which of the following expressions shows his take-home pay in dollars?

   (1) $4.50\,h$ (2) $4.50\,h - .22$
   (3) $4.50(.78)$ (4) $(.78)(4.50)\,h$
   (5) $4.50\,h - .22h$

   (1) (2) (3) (4) (5)

2. If $3y - 2(y - 1) = 4(y + 2)$, $y =$

   (1) 1 (2) -2 (3) -1
   (4) 0 (5) 2

   (1) (2) (3) (4) (5)

3. If $-3x \leq 18$, then

   (1) $x \geq -6$ (2) $x \leq -6$ (3) $x = 6$
   (4) $x \leq -54$ (5) $x \geq -54$

   (1) (2) (3) (4) (5)

4. If $7x - \frac{2}{3}y = 10$ and $y = 3x$, then $y =$

   (1) 2 (2) $3\frac{1}{3}$ (3) $5\frac{1}{7}$
   (4) 6 (5) $7\frac{1}{2}$

   (1) (2) (3) (4) (5)

5. What is the temperature in degrees Celsius when it is 77°F? $°C = \frac{5}{9}(°F - 32)$

   (1) 10.8° (2) 25° (3) 30.6°
   (4) 45° (5) 60.6°

   (1) (2) (3) (4) (5)

Questions 6 and 7 refer to the following information.

A truck and a compact car are 280 miles apart. The truck is moving at 30 miles per hour and the compact car is moving at 40 miles per hour. The vehicles are moving toward each other.

6. After how many hours will they meet?

   (1) 4 (2) 7 (3) 9.3
   (4) 14 (5) 28

   (1) (2) (3) (4) (5)

7. How many miles has the truck traveled when it meets the compact car?

   (1) 120 (2) 140 (3) 160
   (4) 210 (5) 280

   (1) (2) (3) (4) (5)

8. If $2x - y = 4$ and $x + 2y = -3$, find y.

   (1) -2 (2) $3\frac{1}{3}$ (3) $\frac{1}{4}$
   (4) 0 (5) 1

   (1) (2) (3) (4) (5)

9. Joan is 11 years old. At what age will she be twice as old as her brother?

   (1) 8 (2) 11 (3) 18 (4) 19
   (5) Insufficient data is given to solve the problem.

   (1) (2) (3) (4) (5)

10. At a clearance sale, a pair of boots was selling for $10 less than half of the original price. Which of the following shows the sale price of the boots?

    (1) $2x - 10$ (2) $\frac{1}{2}(x - 10)$
    (3) $\frac{x}{2} - 10$ (4) $10 - \frac{x}{2}$ (5) $\frac{10 - x}{2}$

    (1) (2) (3) (4) (5)

11. Solve $2x - 4 > 6$.

    (1) $x > 7$ (2) $x > 5$ (3) $x > 1$
    (4) $x > -1$ (5) $x > -5$

    (1) (2) (3) (4) (5)

12. Evaluate $2x^2 + 4y - 3yz$ when $x = 0$, $y = 3$ and $z = -1$.

    (1) 3 (2) 6 (3) 12 (4) 21 (5) 23

    (1) (2) (3) (4) (5)

13. Solve $6x^2 + 7x = 3$.

    (1) $\frac{1}{3}$ or $\frac{3}{2}$ (2) -3 or 2 (3) -2 or 3
    (4) $\frac{1}{3}$ or $-\frac{3}{2}$ (5) $\frac{2}{3}$ or $\frac{1}{3}$

    (1) (2) (3) (4) (5)

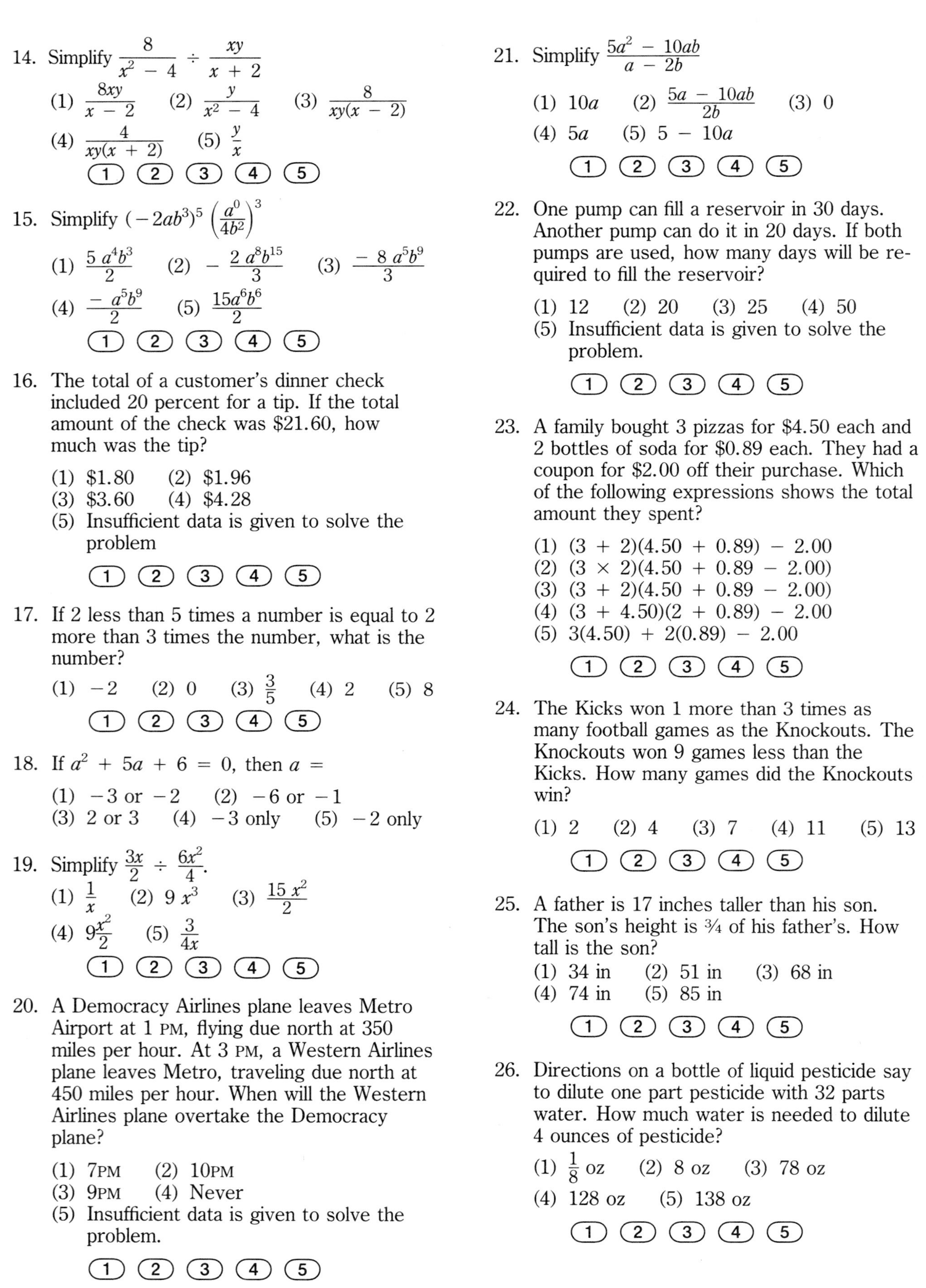

14. Simplify $\dfrac{8}{x^2 - 4} \div \dfrac{xy}{x + 2}$

(1) $\dfrac{8xy}{x - 2}$ (2) $\dfrac{y}{x^2 - 4}$ (3) $\dfrac{8}{xy(x - 2)}$
(4) $\dfrac{4}{xy(x + 2)}$ (5) $\dfrac{y}{x}$

① ② ③ ④ ⑤

15. Simplify $(-2ab^3)^5 \left(\dfrac{a^0}{4b^2}\right)^3$

(1) $\dfrac{5\,a^4b^3}{2}$ (2) $-\dfrac{2\,a^8b^{15}}{3}$ (3) $\dfrac{-8\,a^5b^9}{3}$
(4) $\dfrac{-a^5b^9}{2}$ (5) $\dfrac{15a^6b^6}{2}$

① ② ③ ④ ⑤

16. The total of a customer's dinner check included 20 percent for a tip. If the total amount of the check was $21.60, how much was the tip?

(1) $1.80 (2) $1.96
(3) $3.60 (4) $4.28
(5) Insufficient data is given to solve the problem

① ② ③ ④ ⑤

17. If 2 less than 5 times a number is equal to 2 more than 3 times the number, what is the number?

(1) -2 (2) 0 (3) $\frac{3}{5}$ (4) 2 (5) 8

① ② ③ ④ ⑤

18. If $a^2 + 5a + 6 = 0$, then $a =$

(1) -3 or -2 (2) -6 or -1
(3) 2 or 3 (4) -3 only (5) -2 only

19. Simplify $\frac{3x}{2} \div \frac{6x^2}{4}$.

(1) $\frac{1}{x}$ (2) $9\,x^3$ (3) $\frac{15\,x^2}{2}$
(4) $9\frac{x^2}{2}$ (5) $\frac{3}{4x}$

① ② ③ ④ ⑤

20. A Democracy Airlines plane leaves Metro Airport at 1 PM, flying due north at 350 miles per hour. At 3 PM, a Western Airlines plane leaves Metro, traveling due north at 450 miles per hour. When will the Western Airlines plane overtake the Democracy plane?

(1) 7PM (2) 10PM
(3) 9PM (4) Never
(5) Insufficient data is given to solve the problem.

① ② ③ ④ ⑤

21. Simplify $\dfrac{5a^2 - 10ab}{a - 2b}$

(1) $10a$ (2) $\dfrac{5a - 10ab}{2b}$ (3) 0
(4) $5a$ (5) $5 - 10a$

① ② ③ ④ ⑤

22. One pump can fill a reservoir in 30 days. Another pump can do it in 20 days. If both pumps are used, how many days will be required to fill the reservoir?

(1) 12 (2) 20 (3) 25 (4) 50
(5) Insufficient data is given to solve the problem.

① ② ③ ④ ⑤

23. A family bought 3 pizzas for $4.50 each and 2 bottles of soda for $0.89 each. They had a coupon for $2.00 off their purchase. Which of the following expressions shows the total amount they spent?

(1) $(3 + 2)(4.50 + 0.89) - 2.00$
(2) $(3 \times 2)(4.50 + 0.89 - 2.00)$
(3) $(3 + 2)(4.50 + 0.89 - 2.00)$
(4) $(3 + 4.50)(2 + 0.89) - 2.00$
(5) $3(4.50) + 2(0.89) - 2.00$

① ② ③ ④ ⑤

24. The Kicks won 1 more than 3 times as many football games as the Knockouts. The Knockouts won 9 games less than the Kicks. How many games did the Knockouts win?

(1) 2 (2) 4 (3) 7 (4) 11 (5) 13

① ② ③ ④ ⑤

25. A father is 17 inches taller than his son. The son's height is ¾ of his father's. How tall is the son?

(1) 34 in (2) 51 in (3) 68 in
(4) 74 in (5) 85 in

① ② ③ ④ ⑤

26. Directions on a bottle of liquid pesticide say to dilute one part pesticide with 32 parts water. How much water is needed to dilute 4 ounces of pesticide?

(1) $\frac{1}{8}$ oz (2) 8 oz (3) 78 oz
(4) 128 oz (5) 138 oz

① ② ③ ④ ⑤

Answers: Extra Practice

Arithmetic

(pages 218–219)

1. (2) \$7.74

Add the amounts Carla spent and subtract from the amount she started with.

$$\begin{array}{r} \$29.76 \\ +12.50 \\ \hline \$42.26 \end{array} \qquad \begin{array}{r} \$50.00 \\ -42.26 \\ \hline \$\ 7.74 \end{array}$$

2. (2) 6%

Since he paid \$112 a month for 2 years (or 24 months), the total amount he repaid was

\$112 × 24 = \$2,688.

Since the principal of the loan was \$2,400, the difference \$2,688 − \$2,400 is the amount of interest for 2 years.

$$\begin{array}{r} \$2,688 \\ -2,400 \\ \hline \$\ \ 288 \end{array}$$

Half this amount, \$144, is the amount of interest per year. Now you can find the rate of interest.

$$\text{Rate} = \frac{\text{Interest Amount}}{\text{Principal}} =$$

$$\frac{\$144}{\$2,400} = .06 \text{ or } 6\%$$

3. (5) Not enough information is given.

Since we do not know the distance traveled, we are not able to solve the problem.

4. (5) \$119.54

Add together their expenses.

	\$267.84	meals
	309.52	gas
	71.10	tolls
+	32.00	campsites
	\$680.46	total expenses

Subtract this from the amount they started with.

	\$800.00	starting amount
−	680.46	expenses
	\$119.54	left

5. (3) 2 qt 2 cups

Multiply the number of people times the amount they each drink.

	20	oz each
×	4	people
	80	oz per day

Convert ounces to quarts and cups. (1 cup = 8 oz, 1 qt = 4 cups)

8 oz in 1 c $\quad \begin{array}{l} \underline{10 \text{ c}} \text{ per day} \\ \overline{)80 \text{ oz}} \end{array}$

4 c in 1 qt $\quad \begin{array}{l} \underline{2 \text{ qt } 2 \text{ c}} \text{ per day} \\ \overline{)10 \text{ cups}} \\ \ \ \underline{8} \\ \ \ 2 \end{array}$

6. (2) 15

Since this is a room, we may think of the floor plan this way:

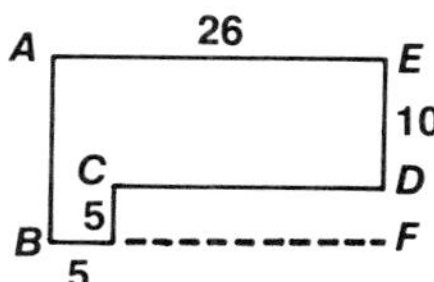

We see that the wall *AB* must be as long as the wall *ED* plus the imaginary wall *DF.* So *AB* = 10 + 5 = 15.

7. (3) 187.5

Volume = length × width × depth. First find the volume of the waterbed in cubic feet (express 10 inches depth as 10/12 or 5/6 foot!).

$$V = \ell \times w \times d$$

$$= \frac{\overset{1}{\cancel{6}}}{1} \times \frac{5}{1} \times \frac{5}{\underset{1}{\cancel{6}}}$$

$$= 25 \text{ cu ft}$$

Then use a grid to find how many gallons of water it needs.

gal	7.5	?	gal
cu ft	1	25	cu ft

7.5 × 25 = 187.5
187.5 ÷ 1 = 187.5 gal.

8. (5) Insufficient data is given to solve the problem.

You do not know how much the agency charges for miles above the 100 free miles.

9. (2) 3

Find the area that can be cleaned with one can. (Multiply length × width.)

10 × 14 = 140 sq ft for 1 can

Find the total area of both rugs.

9 × 12 = 108 sq ft
13 × 15 = 195 sq ft

$$\begin{array}{r} 108 \\ +195 \\ \hline 303 \end{array} \text{ sq ft total area}$$

Then divide by the area for one can.

$$\begin{array}{r} 2\frac{23}{140} \\ 140\overline{)303} \\ \underline{280} \\ 23 \end{array} \text{ or 3 cans are needed}$$

10. (2) \$73,500

First use the grid to find the amount of increase in the price of a house.

amt of increase	?	75	% increase
1976 price	\$42,000	100	

\$42,000 × 75 = \$3,150,000
\$3,150,000 ÷ 100 = \$31,500 increase

Add the increase to the 1976 price to find the 1980 price.

	\$42,000	1976 price
+	31,500	increase
	\$73,500	1980 price

11. (3) 6

You have to change 2 hours 47 minutes to minutes

(2 × 60) + 47 = 120 + 47
= 167 minutes

Then you subtract

$$\begin{array}{r} 173 \\ -167 \\ \hline 6 \end{array}$$

The faster horse finished 6 minutes before the other horse.

12. (2) 67,500 sq yd
First find the area of the rectangle by multiplying length times width: 1,350 × 450 = 607,500 sq ft. Then convert the *square feet* to *square yards* by using a grid. (1 sq yd = 9 sq ft)

sq yd	1	?	sq yd
sq ft	9	607,500	sq ft

1 × 607,500 = 607,500
607,500 ÷ 9 = 67,500

The area is 67,500 sq yd.

13. (2) 20
Use the grid.

no. of games won	17	85	% of games won
total games played	?	100	

17 × 100 = 1,700
1,700 ÷ 85 = 20 games played

14. (5) 96
Divide the total length of tubing by the length of each piece to be cut to find the number of pieces he will have.

$72\text{m} \div \frac{3}{4}\text{m}$

$\frac{\overset{24}{\cancel{72}}}{1} \times \frac{4}{\underset{1}{\cancel{3}}} = 96$ pieces

15. (2) 8
Find how many reams the office uses per year by multiplying the number used per month by 12 months.

12 months
× 4 reams per month
48 reams per year

Then divide by the number of reams per carton to find how many cartons are used.

8 cartons per year
6 rpc)48 reams
48
0

16. (2) $\frac{1}{10}$
Find the amount of increase since the question asks you to compare the *increase* with the old price.

22¢ new price
−20¢ old price
2¢ increase

Then express the increase as a part of the *old* price.

$\frac{2¢ \text{ increase}}{20¢ \text{ old price}} = \frac{1}{10}$

17. (1) $2\frac{1}{6}$ c
Add the amounts of liquid poured into the container. Use a common denominator of 6.

$2\frac{1}{2} = 2\frac{3}{6}$ c orange juice
$1\frac{1}{3} = +\ 1\frac{2}{6}$ c pineapple juice
$3\frac{5}{6}$ c liquid

Subtract from the total amount the container will hold.

$6 = 5\frac{6}{6}$ c total
$-3\frac{5}{6} = 3\frac{5}{6}$ c juice
$2\frac{1}{6}$ c more that the container will hold

18. (1) 916
The area can be divided into 3 pieces, each area found separately and the three areas added together.

Then 26 × 9 = 234
26 × 9 = 234
32 × 14 = 448
916 sq ft

19. (1) $1.11
Find the cost of a weekday call.

$0.34 each additional min
× 4 min
$1.36

$0.52 1st min
+ 1.36 next 4 min
$1.88 for 5 min

Find the cost of a weekend call.

$0.14 each additional min
× 4 min
$0.56

$0.21 1st min
+ 0.56 next 4 min
$0.77 for 5 min

Subtract to find how much is saved by calling on the weekend.

$1.88 weekday call
− 0.77 weekend call
$1.11 saved by calling weekend

20. (3) 180
Find the number of hours less by multiplying 1/7 times 210 kilowatt-hours

$\frac{1}{\underset{1}{\cancel{7}}} \times \frac{\overset{30}{\cancel{210}}}{1} = 30$ hrs less

Subtract this from the number of hours in April to find the number of hours in May.

210 kilowatt-hours in April
− 30 kilowatt-hours less
180 kilowatt-hours in May

Geometry

(pages 297–298)

1. (5) *e*
By definition, a perpendicular crosses another line at right angles. Only line *e* forms a right angle with the *x*-axis.

2. (3) 47°
The sum of the interior angles in any triangle is 180°. ∠*B* is a right angle (90°) shown by the box. Add the measurements of the given angles, then subtract from 180° to find the missing angle.

∠*A* = 43°
+∠*B* = 90°
133°

180°
−133°
47° = ∠*C*

3. (4) 60

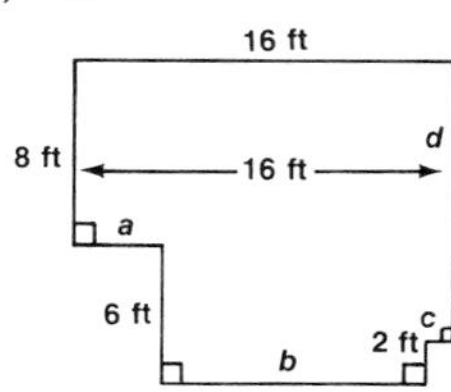

The width of the room is 16 ft, and widths *a* + *b* + *c* will also equal 16 ft. The length of the room is 14 ft (8 ft + 6 ft) and length d + 2 ft is also equal to 14 ft. The baseboard is the measure of the room's perimeter or 16 ft + 16 ft + 14 ft + 14 ft = 60 ft

4. (5) 28
Use the formula $A = \pi r^2$ with $r = 3$ feet and $\frac{22}{7}$ for π.

$A = (\frac{22}{7})(3)^2$
$= (\frac{22}{7})(9) = \frac{198}{7}$
$= 28.3$, or about 28 sq ft.

5. (4) 17.27
Find the volume of the cylinder ($V = \pi r^2 h$).

volume of cylinder
$V = \pi r^2 h$
$V = (3.14)(1)^2(5.5)$
$V = 17.27$ cu in

6. (3) (2, −1)
To find the *x*-coordinate of the midpoint, take the average of the *x*-coordinates of *M* and *N*. Find the *y*-coordinate the same way.

$x = \frac{-2+6}{2} = \frac{4}{2} = 2$

$y = \frac{3-5}{2} = -\frac{2}{2} = -1$

The midpoint is (2, −1).

7. (4) 37.68
Use the formula $C = 2\pi r$, where *r* is the radius. (You may also use $C = \pi d$, where *d* is the diameter. Remember the diameter is twice the radius, $d = 2r$.)

$C = 2(3.14)\,6$
$= 37.68$

8. (5) 120°
If *x* represents the measure of $\angle b$, then $2x$ will be the measure of $\angle a$. $\angle a$ and $\angle b$ are supplementary, so their sum is 180°. $\angle a$ and $\angle h$ are alternate exterior angles and have the same measure.

$\angle a + \angle b = 180°$
$2x + x = 180°$
$3x = 180°$
$x = 60° = \angle b$
$2x = 120° = \angle a = \angle h$

9. (3) (3, 2)
Solve the two equations. Note that *y* can be easily eliminated by adding the two equations. Then solve for *x*.

$x + y = 5$
$+\ 2x - y = 4$
$3x = 9$
$x = 3$

Now substitute for *x* in the first equation and solve for *y*.

$x + y = 5$
$3 + y = 5$
$y = 2$

The point of intersection is (3, 2).

10. (2) 2.5
Similar triangles have corresponding sides in proportion to each other. Use a grid to find the length of the missing side. *BC* corresponds with *CD*, and *AC* corresponds with *EC*. ($BC = 8.5 + 1.5 = 10$)

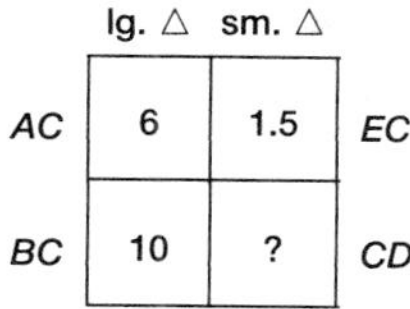

	lg. △	sm. △	
AC	6	1.5	EC
BC	10	?	CD

$10 \times 1.5 = 15$
$15 \div 6 = 2.5$

11. 4. $(-\frac{1}{2}, \frac{1}{2})$

To find the midpoint, find the average of the *x*-coordinates and the average of the *y*-coordinates.

Adding the *x*-coordinates and dividing by 2 you got $(-3) + (2) = -1$, and $-1 \div 2 = -\frac{1}{2}$. For the *y*-coordinates you got $(-1) + (2) = 1$, and $1 \div 2 = \frac{1}{2}$. The ordered pair for this midpoint is $(-\frac{1}{2}, \frac{1}{2})$.

12. (4) 102°
Supplementary angles have a sum of 180°. Subtract the given angle from 180° to find its supplement.

$180° - 78° = 102°$

13. (3) 15

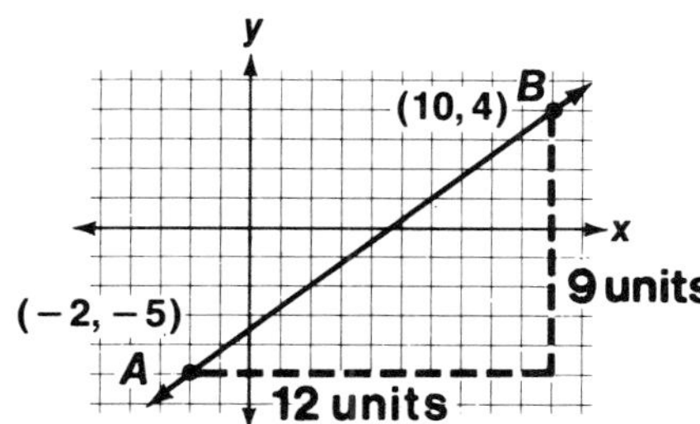

Draw a vertical line down from point *B* and a horizontal line from point *A* to meet it. (Label the new point *C*.) Count the number of units in $\overline{AC}$ and $\overline{BC}$. Use the Pythagorean Theorem to find hypotenuse *AB*.

$a^2 + b^2 = c^2$
$9^2 + 12^2 = c^2$
$81 + 144 = c^2$
$225 = c^2$
$c = 15$

14. (1) 2
To find the perpendicular distance, you draw a perpendicular line from the point of intersection (−3, −2) to the *x*-axis. The point of intersection is 2 units below the *x*-axis. So the length of the perpendicular is 2 units.

15. (3) 5

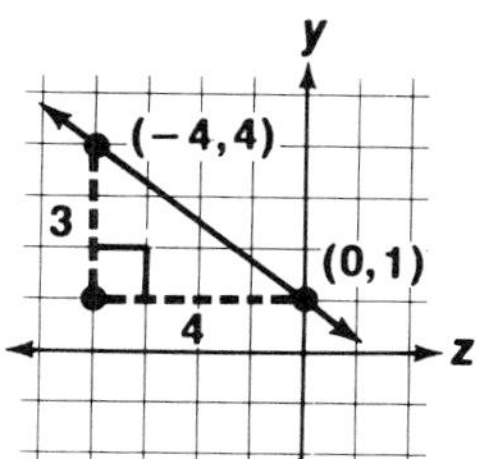

You should have drawn a vertical line through one point and a horizontal line through the other (*T*). Then you should have used the Pythagorean Theorem ($a^2 + b^2 = c^2$), where *a* and *b* are legs of a right triangle and *c* is the hypotenuse.

The vertical distance for the ordered pair (−4, 4) to point *T* is 3 units; the horizontal distance from point *T* to the ordered pair (0, 1) is 4 units. 3 and 4 are the legs of the right triangle. So

$a^2 + b^2 = c^2$
$3^2 + 4^2 = c^2$
$9 + 16 = c^2$
$25 = c^2$
$\sqrt{25} = c$
$5 = c$

The diagonal distance is 5.

16. (2) 24 m
First find the length of the third side, which is the hypoteneuse of a right triangle. Use the Pythagorean formula with $a = 8$ and $b = 6$.

$c^2 = a^2 + b^2$
$c^2 = 8^2 + 6^2$
$c^2 = 64 + 36$
$c^2 = 100$
$c = \sqrt{100}$
$c = 10$ meters

The perimeter is the sum of the three sides.

Perimeter $= 8$ m $+ 6$ m $+ 10$ m
$= 24$ meters

Algebra

(pages 357–358)

1. (4) (.78)(4.50)h
Alex's pay before deductions is 4.50h, where h is the number of hours he works. His deductions are 22% of this, so he takes home 100% − 22% = 78% or .78. So his take-home pay is (.78)(4.50)h.

2. (2) − 2

$$3y - 2(y - 1) = 4(y + 2)$$
$$3y - 2y + 2 = 4y + 8$$
$$y + 2 = 4y + 8$$
$$+2 = 3y + 8$$
$$-6 = 3y$$
$$-2 = y$$

3. (1)$x \geq -6$
First solve for x when $-3x = 18$. Divide both sides by -3.

$$\frac{-3x}{-3} = \frac{18}{-3}$$
$$x = -6$$

Now try a number greater than −6 in the original inequality. Try 1.

$$-3(1) \leq 18$$
$$-3 \leq 18$$

Since the numbers *greater* than −6 (and also −6) make the original inequality true, then $x \geq -6$.

If you try a number less than −6 in the original inequality, it doesn't work.

4. (4) 6
Substitute 3x for y in the first equation, and solve for x. Then multiply x by 3 to solve for y. (y = 3x) Substitute 3x for y.

$$7x - (\tfrac{2}{3})y = 10$$
$$7x - (\tfrac{2}{3})(3x) = 10$$
$$7x - 2x = 10$$
$$7x + -2x = 10$$
$$5x = 10$$
$$x = 2$$

Then find y.

$$y = 3x$$
$$y = 3(2)$$
$$y = 6$$

5. (2) 25°
Substitute 77 for °F in the formula to solve for Celsius.

$$°C = \tfrac{5}{9}(°F - 32)$$
$$°C = \tfrac{5}{9}(77 - 32)$$
$$°C = \frac{5}{\cancel{9}_1}(\cancel{45}^5)$$
$$°C = 25$$

6. (1) 4
Since the truck and the car start at 280 miles apart and move toward each other, the distances they have traveled when they meet will add up to 280 miles. Let t be the number of hours they travel until they meet. The truck's speed is 30 miles per hour, so after t hours, its distance is 30t. The car's speed is 40 miles per hour, so after t hours its distance is 40t.

$$40t + 30t = 280$$
$$70t = 280$$
$$t = 4$$

7. (1) 120
To find the distance the truck traveled, multiply its speed, 30 miles per hour, by the time it travels before meeting the car, 4 hours.

$$30(4) = 120$$

8. (1) −2
Solve the second equation for x.

$$x + 2y = -3$$
$$x = -2y - 3$$

Now substitute for x in the first equation.

$$2(x) - y = 4$$
$$2(-2y - 3) - y = 4$$
$$-4y - 6 - y = 4$$
$$-5y - 6 = 4$$
$$-5y = 10$$
$$y = -2$$

9. (5) Insufficient data is given to solve the problem.
You need the brother's present age, or his age at the time when the sister is twice as old as he is.

10. (3) $\frac{x}{2} - 10$
Express the original price with any letter. Divide it by two. Then subtract 10 from the result.

$\frac{x}{2}$ → half price; − 10 → less $10

11 (2) $x > 5$
First solve for x as if it were an equation. $2x - 4 = 6$. Then try substituting values greater than x and less than x in the original inequality to decide whether x should be greater or less than the value found.

$$2x - 4 = 6$$
$$2x + -4 = 6$$
$$+4 + 4$$
$$2x = 10$$
$$x = 5$$

Try $x < 5$; say $x = 4$

$$2x - 4 > 6$$
$$2(4) - 4 > 6$$
$$8 - 4 > 6$$
$$8 + -4 > 6$$
$$4 > 6$$

NO

Try $x > 5$; say $x = 6$

$$2x - 4 > 6$$
$$2(6) - 4 > 6$$
$$12 - 4 > 6$$
$$12 + -4 > 6$$
$$8 > 6$$

YES

So $x > 5$

12. (4) 21
Substitute the values of each variable into the equation and solve.

$$2x^2 + 4y - 3yz$$
$$2(0)^2 + 4(3) - 3(3)(-1)$$
$$0 + 12 - (-9)$$
$$12 + 9$$
$$21$$

13. (4) $\frac{1}{3}$ or $-\frac{3}{2}$
Remember to get zero on one side before you factor.

$$6x^2 + 7x = 3$$
$$6x^2 + 7x - 3 = 0$$
$$(2x + 3)(3x - 1) = 0$$

Either $2x + 3 = 0$, or $3x - 1 = 0$.

$$2x + 3 = 0$$
$$2x = -3$$
$$x = -\frac{3}{2}$$

or

$$3x - 1 = 0$$
$$3x = 1$$
$$x = \frac{1}{3}$$

14. (3) $\frac{8}{xy(x-2)}$

$$\frac{8}{x^2 - 4} \div \frac{xy}{x + 2} =$$
$$\frac{8}{x^2 - 4} \times \frac{x + 2}{xy} =$$
$$\frac{8}{(x - 2)\cancel{(x + 2)}} \cdot \frac{\cancel{x + 2}}{xy} =$$
$$\frac{8}{(x - 2)xy}$$

15. (4) $\frac{-a^5b^9}{2}$

$$(-2ab^3)^5 \cdot \left(\frac{a^0}{4b^2}\right)^3 =$$
$$-(2ab^3)^5 \left(\frac{1}{4b^2}\right)^3 =$$
$$-2^5a^5(b^3)^5 \times \frac{1}{4^3(b^2)^3} =$$
$$\frac{-32\,a^5b^{15}}{64b^6} =$$
$$\frac{-a^5b^9}{2}$$

16. (3) $3.60
Let x represent the amount of the check before tip, and $.20x$ the tip.

$$x + .20x = \$21.60$$
$$1.20\,x = 21.60$$
$$x = \frac{21.60}{1.20} = 18$$

Then $.20x = .20(18) = \$3.60$.

If you use a grid:

total bill	21.60	120	% food cost plus tip
	?	100	food cost

$$100 \times 21.60 = 2160$$
$$2160 \div 120 = 18$$

Check without tip = $18.
Tip = $21.60 − $18.00 = $3.60

17. (4) 2
Write the words as an equation and solve for the missing number.

(Two less than 5 times a number)
$5x - 2$
(is equal to)
$=$
(2 more than 3 times a number)
$3x + 2$

$$5x - 2 = 3x + 2$$
$$5x + -2 = 3x + 2$$
$$-3x \qquad -3x$$
$$2x + -2 = 2$$
$$+2 = +2$$
$$2x = 4$$
$$x = 2$$

18. (1) −3 or −2
When you substitute (−3) or (−2) for a in the equation, both sides of the equation balance.

$$a^2 + 5a + 6 = 0$$
$$(-3)^2 + 5(-3) + 6 = 0$$
$$9 + (-15) + 6 = 0$$
$$-6 + 6 = 0$$
$$0 = 0$$

$$(-2)^2 + 5(-2) + 6 = 0$$
$$4 + (-10) + 6 = 0$$
$$6 + -6 = 0$$
$$0 = 0$$

19. (1) $\frac{1}{x}$

$$\frac{3x}{2} \div \frac{6x^2}{4} = \frac{3x}{2} \cdot \frac{4}{6x^2}$$
$$= \frac{3x}{2} \cdot \frac{2 \cdot 2}{2 \cdot 3x^2}$$
$$= \frac{\cancel{3x}}{\cancel{2}} \cdot \frac{\cancel{2} \cdot \cancel{2}}{\cancel{2} \cdot \cancel{3x}^2}$$
$$= x \cdot \frac{1}{x^2}$$
$$= \frac{1}{x}$$

20. (2) 10 PM
Note that when the second plane overtakes the first, the planes have flown equal distances. Find the time when this happens. Let t be the number of hours the Democracy plane has flown. (Since it left at 1 PM, we will be able to find the time when it is overtaken.) The number of hours flown by the Western plane then is $t - 2$ because it left at 3 PM, 2 hours later. When overtaken, the Democracy plane has flown 350 t miles. At this time the Western plane has flown 450 $(t - 2)$ miles.

$$350t = 450\,(t - 2)$$
$$350t = 450t - 900$$
$$900 = 100t$$
$$9 = t$$

The Democracy plane has been flying 9 hours, so it is overtaken at 10 PM.

21. (4) $5a$

$$\frac{5a^2 - 10ab}{a - 2b} = \frac{5a(a - 2b)}{a - 2b}$$
$$= 5a$$

22. (1) 12
Let t be the number of days for both pipes working together. Note that the rates of the two pipes add. The rate of the first pipe is 1 reservoir/30 days, and the rate of the second pipe is 1 reservoir/20 days. The part of the job done by the first pipe in the t days is $\frac{1}{30}$ (t) and the part done by the second pipe is $\frac{1}{20}$ (t). To fill the reservoir, these two parts must add up to 1 complete job.

$$\frac{1}{30}\,t + \frac{1}{20}\,t = 1$$

Multiply by 60 (the LCD) to clear fractions.

$$60\left(\tfrac{1}{30}\right)t + 60\left(\tfrac{1}{20}\right)t = 60(1)$$
$$2t + 3t = 60$$
$$5t = 60$$
$$t = 12$$

23. (5) 3(4.50) + 2(0.89) − 2.00
Multiply the number of each item by its price, then add together, and subtract $2.00 from the total.

3(4.50) + 2(.89) − 2.00

24. (2) 4
Let x represent the number of games the Kicks won; let y represent the number of games the Knockouts won. Express the information in terms of x and y.

$x = 3y + 1$

Kicks = one more than 3 times the Knockouts

AND

$y = x - 9$
Knockouts = nine less than Kicks

Substitute $(3y + 1)$ for x in the second equation.

$$y = x \quad - 9$$
$$y = (3y + 1) - 9$$

Solve for y.

$$y = 3y + 1 + (-9)$$
$$y = 3y + (-8)$$

Add $-3y$ to both sides.

$$\begin{array}{rcrr} y & = & 3y + (-8) & \\ -3y & & -3y & 8 \\ \hline -2y & = & & -8 \end{array}$$

Divide each side by -2.

$$\frac{-2y}{-2} = \frac{-8}{-2}$$
$$y = 4$$

The Knockouts won 4 games.

25. (2) 51
Let x stand for the father's height and y for the son's height. Express the given information in terms of the x and y. Substitute $\frac{3}{4}x$ for y in the first equation.

$x = (y + 17)$
father is 17 inches taller than son

AND

$y = \frac{3}{4}x$
son is $\frac{3}{4}$ of father's height

$$x = y + 17$$
$$1x = \tfrac{3}{4}x + 17$$

$$\begin{array}{lrr} \text{Add } -\tfrac{3}{4}x: & -\tfrac{3}{4}x & -\tfrac{3}{4}x \\ \hline & \tfrac{1}{4}x = & 17 \end{array}$$

Multiply by 4: $4(\tfrac{1}{4}x) = 4(17)$

$x = 68$ in = father's height

$y = \frac{3}{\cancel{4}_{1}}(\cancel{68}^{17})$

$y = 51$ in = son's height

26. (4) 128 oz
Use the grid to solve the proportion.

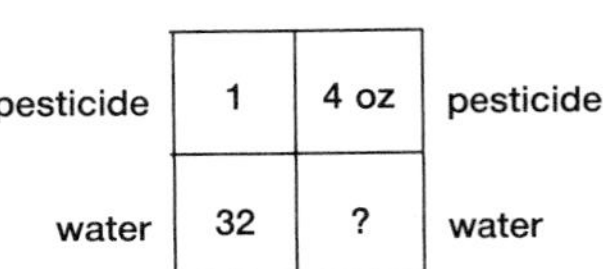

$4 \times 32 = 128$
$128 \div 1 = 128$ oz

The Posttests

You are now ready to take the final step in your study for passing the GED Mathematics Test. Both Posttests in this book resemble the actual GED Mathematics Test in the number and kinds of questions asked. When taking each Posttest, take no more than the allotted time for one entire test. Remember to guess whenever you don't know an answer. To get the best possible score, don't leave a question unanswered.

Take Posttest A. Figure your score using the Answer Key on page 374. You may want or need to take Posttest B based on your score.

If possible, try to make arrangements to take the actual GED Mathematics Test as soon as you can after achieving passing scores on the Posttests in this book. Your test-taking skills will be sharp, and you will feel more confident.

MATHEMATICS POSTTEST A

Directions

The Mathematics Posttest consists of 56 problems. Whenever possible, you should arrive at your own answer to a question before looking at the choices; otherwise, you may be misled by wrong answers that look possible. You may refer to the formula page on the inside back cover of this book.

You should take approximately 1½ hours to complete this test. There is no penalty for guessing. Try to answer as many questions as you can. Work rapidly but carefully, without spending too much time on any one question. If a question is too difficult for you, skip it and come back to it later.

For each answer, mark one answer space. See how the following example is done.

EXAMPLE

A secretary bought a desk calendar for \$5.95 and a pencil container for \$1.89. How much change did she get from a \$10 bill?

(1) \$2.16
(2) \$2.61
(3) \$7.84
(4) \$8.74
(5) \$8.84

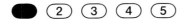

The amount is \$2.16; therefore, answer space (1) has been marked.

Answers to the questions are in the Answer Key on page 374. Explanations for the answers are on pages 371–374.

Directions: Choose the one best answer for each problem.

1. An employee's gross pay is $364. Her total deductions are 6% more than last year. Her employer deducts $67 for taxes, $25 for social security, and $18 for health insurance. What is her net pay after deductions?

   (1) $110 (2) $244 (3) $254
   (4) $272 (5) $474

   (1) (2) (3) (4) (5)

2. How many 2-pound bags of flour can be made from 13 barrels of flour?

   (1) 8 (2) 33 (3) 40 (4) 75
   (5) Not enough information is given.

   (1) (2) (3) (4) (5)

3. Susan bought a skirt for $9.00, a blouse for $12.00, and shoes for $20.00. The sales tax is $5\frac{1}{2}$%. Which of the following expressions is a step in finding the amount Susan must pay?

   (1) 5.5(9.00) (2) $\frac{20.00}{.055}$
   (3) $100 - 5\frac{1}{2}$ (4) $\frac{9.00}{.055}$
   (5) 9.00 + 12.00 + 20.00

   (1) (2) (3) (4) (5)

4. A bridge spans 800 feet of United States territory and 750 feet of Canadian territory. What fraction of the bridge is over Canadian territory?

   (1) $\frac{1}{750}$ (2) $\frac{15}{31}$ (3) $\frac{1}{2}$ (4) $\frac{16}{31}$
   (5) Not enough information is given.

   (1) (2) (3) (4) (5)

5. How many liters are in 5 gallons of gasoline if 3.8 liters equal 1 gallon?

   (1) 0.76 (2) 1.32 (3) 1.90
   (4) 8.8 (5) 19

   (1) (2) (3) (4) (5)

6. Stock for a mining company was up $1\frac{1}{4}$ on Monday, down $\frac{5}{8}$ on Tuesday, down $\frac{1}{2}$ on Wednesday, and up $\frac{3}{8}$ on Thursday. What was the change from Monday to Thursday?

   (1) down $\frac{1}{2}$ (2) down $\frac{1}{4}$ (3) down $\frac{1}{8}$
   (4) up $\frac{1}{2}$ (5) up $2\frac{3}{4}$

   (1) (2) (3) (4) (5)

Items 7–9 refer to the table below.

Parcel Post Rate Schedule

Pounds up to	Zones: Local	1 and 2	3	4	5
2	1.52	1.55	1.61	1.70	1.83
3	1.58	1.63	1.73	1.86	2.06
4	1.65	1.71	1.84	2.02	2.29
5	1.71	1.79	1.96	2.18	2.52
6	1.78	1.87	2.07	2.33	2.74
7	1.84	1.95	2.18	2.49	2.89
8	1.91	2.03	2.30	2.64	3.06
9	1.97	2.11	2.41	2.75	3.25
10	2.04	2.19	2.52	2.87	3.46

7. How much does it cost to send a package weighing 4 pounds 10 ounces to a local address?

   (1) $1.65 (2) $1.71 (3) $2.02
   (4) $2.18 (5) $2.52

   (1) (2) (3) (4) (5)

8. How much would it cost to mail a $4\frac{1}{2}$ pound package to zone 1 and an 11-ounce package to zone 4?

   (1) $1.79 (2) $3.34 (3) $3.41
   (4) $3.49 (5) $3.88

   (1) (2) (3) (4) (5)

9. A customer sent 2 packages, one weighing 3 pounds 7 ounces and the other weighing 5 pounds 4 ounces to zone 3. How much could the customer save by combining the contents of the two packages into one?

   (1) $1.50 (2) $1.84 (3) $2.07
   (4) $2.41 (5) $3.91

   (1) (2) (3) (4) (5)

10. A customer could receive 12 issues of a magazine for $15.00 or 20 issues for $22.00. The magazine produces 10 issues each year. What would the savings per issue be on 20 issues for $22.00?

(1) $0.15 (2) $0.35 (3) $0.91
(4) $1.10 (5) $1.25

① ② ③ ④ ⑤

Items 11 and 12 refer to the following graph.

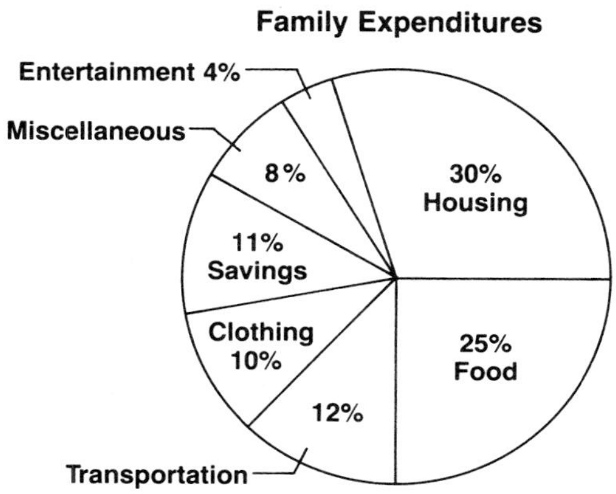

Family Expenditures

Entertainment 4%
Miscellaneous 8%
30% Housing
11% Savings
Clothing 10%
25% Food
12%
Transportation

11. If the Ling family's monthly income is $2,800, how much is spent on transportation?

(1) $336.00 (2) $236.00 (3) $233.33
(4) $33.60 (5) $12.00

① ② ③ ④ ⑤

12. Of the amount budgeted for housing, 20% is spent on heat. How much do the Lings pay for heat each month?

(1) $20 (2) $128 (3) $168
(4) $560 (5) $840

① ② ③ ④ ⑤

13. A customer bought $3\frac{1}{2}$ pounds of cheese for $6.30. What was the price of 1 pound?

(1) $0.56 (2) $0.90 (3) $1.80
(4) $2.21 (5) $2.80

① ② ③ ④ ⑤

14. Find the amount of simple interest paid on a loan of $1,700 at 12 percent annual rate for 9 months.

(1) $153.00 (2) $173.00 (3) $183.60
(4) $1,275.00 (5) $1,836.00

① ② ③ ④ ⑤

15. Last year the Franklins paid $375 per month in rent. This year they pay $420 per month. By what percent did their rent increase?

(1) 833.3% (2) 112% (3) 89.3%
(4) 12% (5) 10.7%

① ② ③ ④ ⑤

16. A customer wanted to buy a tool set that was on sale for 35% off. If the sale price was $146.25, what was the set's original price?

(1) $51.19 (2) $5.06 (3) $225.00
(4) $241.25 (5) $417.86

① ② ③ ④ ⑤

17. Six pounds of plums, when dried, yield $1\frac{1}{2}$ pounds of prunes. How many pounds of plums yield 5 pounds of prunes?

(1) 4 (2) $7\frac{1}{2}$ (3) 9
(4) 10 (5) 20

① ② ③ ④ ⑤

18. The directions on the package say to use 1 part ground coffee beans to 12 parts water. How many cups of ground coffee would be needed to make 9 cups of coffee?

(1) $\frac{1}{12}$ (2) $\frac{3}{4}$ (3) $1\frac{1}{3}$ (4) 3 (5) 108

① ② ③ ④ ⑤

19. How much simple interest is charged on a loan of $1,500 at an annual rate of 18% for 6 months?

(1) $45 (2) $135 (3) $270
(4) $1,620 (5) $13,500

① ② ③ ④ ⑤

20. When arranged from least to greatest, which number is fourth?

0.0205 0.002 0.023 0.021 0.02

(1) 0.0205 (2) 0.002 (3) 0.023
(4) 0.021 (5) 0.02

① ② ③ ④ ⑤

21. A child grew 6.2 centimeters one year, 5.9 the next year, and 5.5 the next. Approximately what was her average rate of growth in centimeters over the 3 years?

(1) 4.0 (2) 5.5 (3) 5.8
(4) 5.9 (5) 17.6

① ② ③ ④ ⑤

22. The odometer on the family car read 42,527 last Monday. This Monday it read 51,809. What was the average number of miles per day the family drove the car?

(1) 1,183 (2) 1,326 (3) 1,327
(4) 1,340 (5) 13,477

1 2 3 4 5

23. Another way to write .030962 is

(1) 3.0962×10^3 (2) 3.0962×10^{-2}
(3) 3.0962×10^1 (4) 309.62×10^{-3}
(5) $.30962 \times 10^2$

1 2 3 4 5

24. Kelly is 3 times as old as her brother and weighs 26 pounds more than he does. Their ages combined equal 24. How old is Kelly?

(1) 6 (2) 7 (3) 8 (4) $10\frac{1}{2}$ (5) 18

1 2 3 4 5

Item 25 is based on the following figure.

25. Which letter best represents the point $\frac{-3}{2}$?

(1) A (2) B (3) C (4) D (5) E

1 2 3 4 5

26. Six people put together a submarine sandwich using a 5-foot long loaf of bread. If they divided the sandwich evenly among themselves, what was the size in inches of each piece?

(1) $1\frac{1}{5}$ (2) $8\frac{1}{3}$ (3) 10
(4) 12 (5) $14\frac{2}{5}$

1 2 3 4 5

27. A customer ordered items that weighed 9 ounces, 1 pound 12 ounces, and 2 pounds 3 ounces. The items cost, respectively, \$0.85, \$1.20, and \$2.00. If the freight charge is 20 cents per pound and the customer pays the freight, what is the total shipping cost?

(1) \$0.70 (4) \$0.90 (3) \$0.85
(2) \$0.79 (5) 1.05

1 2 3 4 5

28. According to the directions, 2 ounces of concentrated orange juice should be diluted with 3 ounces of water. How many ounces of concentrated orange juice should be used to make 1 quart of diluted orange juice?

(1) 12.8 (2) 13.2 (3) 21.3
(4) 48 (5) 80

1 2 3 4 5

29. Find the perimeter of triangle *ABC* in the figure above.

(1) 6 (2) 12 (3) 15
(4) 16 (5) 24

1 2 3 4 5

30. An artist wants to put a 2-inch border around her painting. The painting is 14 inches by 18 inches. The border material is 1/8 inch thick. What will be the perimeter in inches of the painting with the border?

(1) 36 (2) 64 (3) 72
(4) 80 (5) 288

1 2 3 4 5

31. Give the circumference in inches of a compact disc that has a radius of 6 inches.

(1) 9.14 (2) 15.14 (3) 18.84
(4) 37.68 (5) 113.04

1 2 3 4 5

32. A gallon of paint covers 400 square feet. How many gallons of paint are needed to cover 2 walls measuring 22 feet wide and 1 foot thick, and 2 walls 18 feet wide and 1 foot thick?

(1) $\frac{3}{10}$ (2) $\frac{9}{10}$ (3) 1 (4) $1\frac{1}{10}$
(5) Not enough information is given.

1 2 3 4 5

33. How many square yards of carpeting are needed to cover a floor 15 feet long and 12 feet wide?

(1) 9 (2) 18 (3) 20 (4) 60 (5) 180

1 2 3 4 5

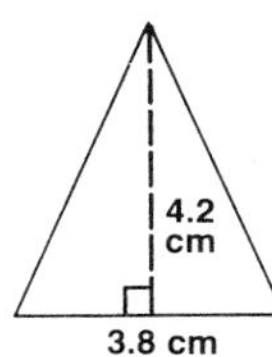

34. How much fabric will it take to make the sail in the figure? The area, in square centimeters, of the sail in the figure above is

(1) 17.64 (2) 15.96 (3) 7.98
(4) 7.93 (5) 4.0

(1) (2) (3) (4) (5)

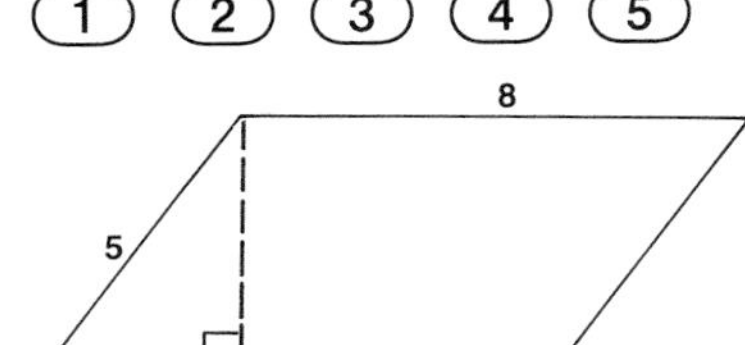

35. Find the area of the parking lot in the figure above.

(1) 16 (2) 20 (3) 32 (4) 40 (5) 55

(1) (2) (3) (4) (5)

36. A sink is 18 inches long and 7 inches deep. How many gallons of water can it hold?

(1) 0.203 (2) 1.2 (3) 1.71 (4) 2.42
(5) Not enough information is given.

(1) (2) (3) (4) (5)

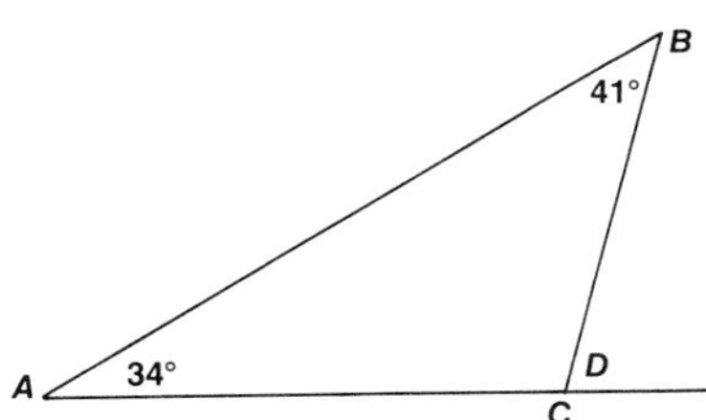

37. In triangle ABC above, what is the measurement of angle D?

(1) 15° (2) 53° (3) 75° (4) 95°
(5) Not enough information is given.

(1) (2) (3) (4) (5)

38. Find the length of side DE.

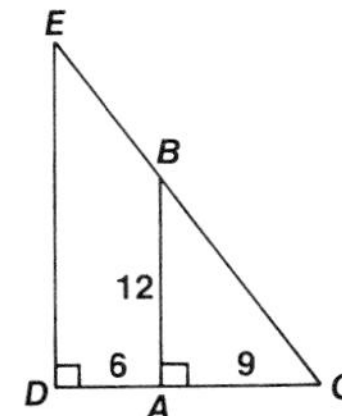

(1) $7\frac{1}{2}$ (2) 9 (3) 12 (4) 15 (5) 20

(1) (2) (3) (4) (5)

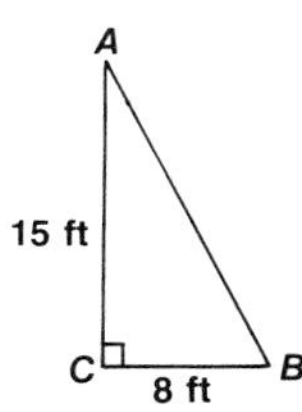

39. Find the length in feet of the ladder leaning against a wall as shown above.

(1) 17 (2) 23 (3) 43 (4) 60 (5) 289

(1) (2) (3) (4) (5)

40. Line QS is parallel to the y-axis. Find the perpendicular distance from point P to line QS.

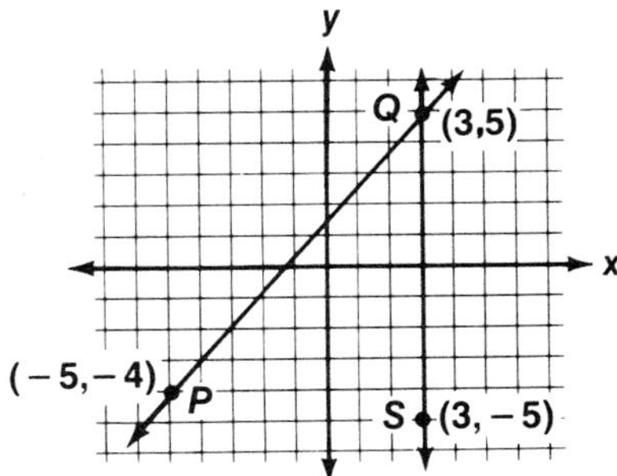

(1) 2 (2) 7 (3) 8 (4) 9 (5) 10

(1) (2) (3) (4) (5)

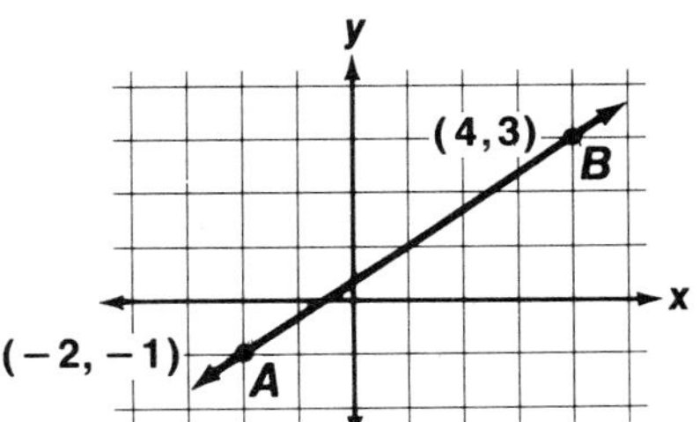

41. What is the slope of line AB?

(1) -1 (2) $\frac{-2}{3}$ (3) $\frac{2}{3}$ (4) 1 (5) $1\frac{1}{2}$

(1) (2) (3) (4) (5)

42. Tickets to a play cost $5.50 for adults and $3.75 for children. If a family of 2 adults and 3 children bought tickets, which of the following is equal to how much they spent?

(1) (2 + 3) (5.50 + 3.75)
(2) 2(5.50) × 3(3.75)
(3) 2 + 3(5.50 + 3.75)
(4) (2 + 5.50) (3 + 3.75)
(5) 2(5.50) + 3(3.75)

(1) (2) (3) (4) (5)

43. Find the value of $5^2 - 3^3 + 1^4$.

(1) -1 (2) 1 (3) 2 (4) 3 (5) 5

(1) (2) (3) (4) (5)

44. The Bittnei family bought a piano for \$1,250. They paid \$800 in cash and the rest in equal installments over 2 years. Which expression shows their monthly payment in dollars?

(1) $12(800 + 1{,}250)$ (2) $\frac{1{,}250 + 800}{12}$
(3) $\frac{1{,}250 - 800}{24}$ (4) $800 + \frac{1{,}250}{12}$
(5) $\frac{800 + 1{,}250}{24}$

(1) (2) (3) (4) (5)

45. Three less than the square root of the difference of two numbers is written

(1) $a + b - c$ (2) $\sqrt{a - b} - 3$
(3) $a - b - 3$ (4) $3 - a - b$
(5) $\sqrt{a - b - 3}$

(1) (2) (3) (4) (5)

46. Seven less than four times a number is 29. What is the number?

(1) -9 (2) 9 (3) 18
(4) 26 (5) 32

(1) (2) (3) (4) (5)

47. If $\frac{5(3x + 7)}{2} = 55$, then $x =$

(1) 5 (2) 6 (3) $6\frac{2}{3}$
(4) 7 (5) 25

(1) (2) (3) (4) (5)

48. Three consecutive even numbers add up to 42. Find the middle number.

(1) 12 (2) 14 (3) 16
(4) 40 (5) 42

(1) (2) (3) (4) (5)

49. Solve $K = \frac{ab - 5}{2}$ for b.

(1) $\frac{2K + 5}{a}$ (2) $\frac{2K}{-5a}$
(3) $\frac{2K - 5}{a}$ (4) $\frac{K - 10}{a}$
(5) $K - 2a - 10$

(1) (2) (3) (4) (5)

50. Which of the following ordered pairs is the intersection of the lines $y = 2x + 1$ and $y = x - 3$?

(1) $(-7, -4)$ (2) $(-7, 4)$ (3) $(4, 7)$
(4) $(-4, 7)$ (5) $(-4, -7)$

(1) (2) (3) (4) (5)

51. If $2x + 4y + 7 = 31$ and $x = 2y$, then $x =$

(1) 2 (2) 3 (3) 4 (4) 6 (5) 8

(1) (2) (3) (4) (5)

52. Solve $x - 10 < -6 - x$.

(1) $x > 2$ (2) $x > -2$
(3) $x < \frac{7}{2}$ (4) $x < \frac{-1}{2}$
(5) $x < 2$

(1) (2) (3) (4) (5)

53. Solve $2x^2 + 5x = 12$ for x
(1) 3 (2) 4 (3) 3 or 4
(4) $\frac{3}{2}$ or -4 (5) $\frac{-3}{2}$ or 4

(1) (2) (3) (4) (5)

54. Solve $2x^2 + 5x = 3$.

(1) 3 or $\frac{1}{2}$ (2) -3 or $\frac{1}{2}$
(3) 0 or $\frac{1}{2}$ (4) $-\frac{1}{2}$ or 2
(5) 3 or 1

(1) (2) (3) (4) (5)

55. 902.64 written in scientific notation is

(1) 9.0264×10^{-4}
(2) 9.0264×10^{-2}
(3) 9.0264×10^{2}
(4) 9.0264×10^{3}
(5) 9.0264×10^{4}

(1) (2) (3) (4) (5)

56. In the figure, line G is parallel to line J. Find the measure of angle M.

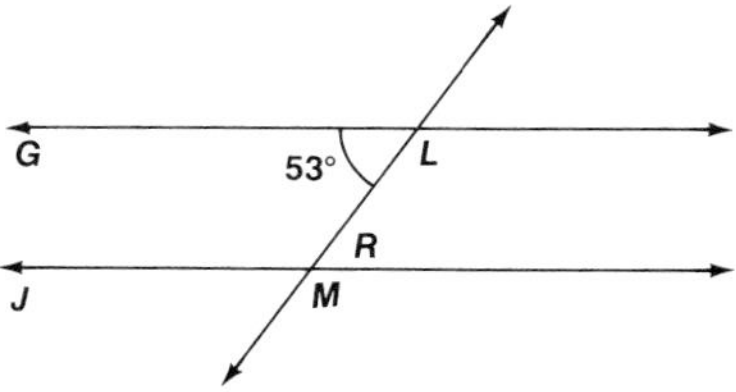

(1) 37° (2) 53° (3) 127°
(4) 133° (5) 143°

(1) (2) (3) (4) (5)

Answer Explanations: Posttest A

1. (3) $254
Add together all the deductions. Then subtract the total amount of deductions from the gross pay to find the net pay.

$67	taxes
25	Social Security
+ 18	health ins.
$110	total deductions

$364	gross
− 110	deductions
$250	net (take-home) pay

2. (5) Not enough information is given.
You do not know the weight of the flour in the barrels.

3. (5) 9.00 + 12.00 + 20.00
Since the tax is on all the items, add up the amounts first. Then multiply by the rate, changed from percent to decimal.
(5½% = .05½)

4. (2) $\frac{15}{31}$
Find the total span of the bridge, then express the Canadian portion as a part of the whole span and reduce.

800 ft	American
+ 750 ft	Canadian
1,550 ft	total

$$\frac{750}{1,550} = \frac{15}{31}$$

5. (5) 19
Use a grid. Put known information on one side.

gal	1	5	gal
L	3.8	?	L

5 × 3.8 = 19
19 ÷ 1 = 19

6. (4) up $\frac{1}{2}$
Add the stocks that went up; subtract the ones that went down. (Make sure you find a common denominator first!)

Or, think of the stocks going up as positive, those going down as negative. Add them all together.

$$\left(+1\frac{1}{4}\right) + \left(-\frac{5}{8}\right) + \left(\frac{-1}{2}\right) + \left(\frac{+3}{8}\right) =$$

$+\frac{1}{2}$, or up $\frac{1}{2}$

7. (2) $1.71
Find the package size in the first column. It comes between 4 and 5. Read the higher weight (5). Then read across to the price under the local zone, $1.71.

8. (4) $3.49
Find the cost of the two packages and add. The 4½ pound package to zone 1 will cost $1.79. The 11-ounce package to zone 4 is $1.70. Together they will cost $3.49.

9. (1) $1.50
Find the cost of the two individual packages to zone 3. Then find the combined weight of those packages (to be the one large package) and the cost of sending that to zone 3. Find the difference in cost by subtracting.

3 lb. 7 oz. →	$1.84		
+5 lb. 4 oz. →	+ 2.07		
8 lb. 11 oz.	$3.91	2 sm.	
	−$2.41	1 lg.	
	$1.50	saved	

10. (1) $0.15
Find the cost per issue for each subscription rate, then subtract to find the amount of the difference.

$1.25 per issue: 12)15.00 — 12, 3 0, 2 4, 60, 60

$1.10 per issue: 20)22.00 — 20, 2 0, 2 0, 00, 00

$1.25	
− 1.10	
$0.15	savings with 20 issues

11. (1) $336.00
Transportation is 12% of the budget. Find 12% of $2,800.

amt. transp.	?	12	% trans. of budget
total budget	2,800	100	

12 × 2,800 = 33,600
33,600 ÷ 100 = $336

12. (3) $168
First find the amount budgeted for housing. Heating cost will be 20% of the housing amount.

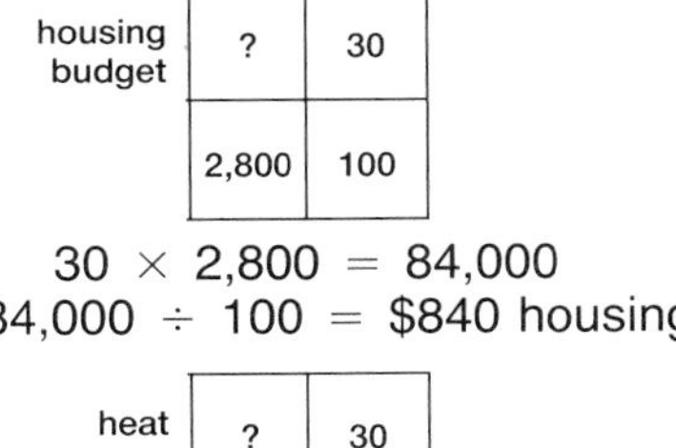

housing budget	?	30
	2,800	100

30 × 2,800 = 84,000
84,000 ÷ 100 = $840 housing

heat budget	?	30
	840	100

20 × 840 = 16,800
16,800 ÷ 100 = $168 heat

13. (3) $1.80
Divide the total cost by the number of pounds to find the cost per pound.

$$\frac{6.30}{1} \div 3\frac{1}{2} =$$

$$\frac{\overset{0.90}{\cancel{6.30}}}{1} \times \frac{2}{\underset{1}{\cancel{7}}} = \$1.80 \text{ per lb}$$

14. (1) $153.00
Use the formula Interest = Principle × Rate × Time. To find the interest, multiply the amount borrowed (principle) times the percent (rate) times the part of a year (time).

9 mo = $\frac{9}{12}$ yr = $\frac{3}{4}$ yr

I = $1700 × 12% × $\frac{9}{12}$ yr

$$= \frac{\overset{17}{\cancel{1700}}}{1} \times \frac{\overset{1}{\cancel{12}}}{\underset{1}{\cancel{100}}} \times \frac{9}{\underset{1}{\cancel{12}}}$$

= $153.00 interest

15. (4) 12%
Find the dollar amount that their rent increased per month. Then use the grid:

$420	
− 375	
$ 45	increase

amt. of increase	$45	?	% increase
original amount	$375	100	

$45 \times 100 = 4{,}500$
$4{,}500 \div 375 = 12\%$ increase

16. (3) $225.00
When 35% is taken off a price, you will pay the remaining 65%.

(100% − 35% = 65%)
original price — off — sale price

sale price	146.25	65	% corresponding to sale price
original price	?	100	

$146.25 \times 100 = 14{,}625$
$14{,}625 \div 65 = \$225$ original price

17. (5) 20
Use the grid for this proportion problem as follows:

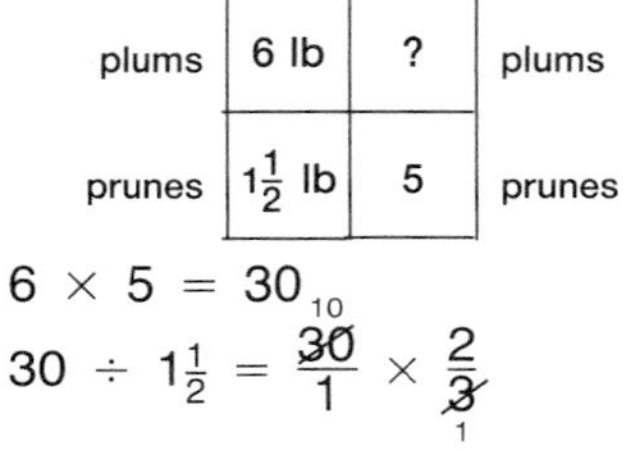

plums
6 lb
?
plums
prunes
$1\frac{1}{2}$ lb
5
prunes

$6 \times 5 = 30$

$30 \div 1\frac{1}{2} = \frac{\cancel{30}^{10}}{1} \times \frac{2}{\cancel{3}_{1}}$

$= 20$ lbs plums

18. (2) $\frac{3}{4}$
Use a grid for this problem.

ground coffee	1	?	ground coffee
water	12	9 cups	water

$1 \times 9 = 9$
$9 \div 12 = \frac{9}{12} = \frac{3}{4}$ cup

19. (2) $135
Interest =
Principal (amount borrowed) × Rate (percent) × Time (how many years?)
↓ ↓ ↓

$\frac{\cancel{1500}^{15}}{1} \times \frac{\cancel{18}^{9}}{\cancel{100}_{1}} \times \frac{1}{\cancel{2}_{1}}$ year

$1,500 × 18% × 6 months.
Since the rate is annual, the time (months) must also be expressed as part of a year.

$15 \times 9 \times 1 = 135$

20. (4) 0.021
Remember that the place value gets *smaller* as you go to the *right* of the decimal point. In order from least to greatest, the numbers are .002, .02, .0205, .021, .023.

You may also do this problem by writing the decimals as fractions. Then convert to the same denominator to easily compare.

$.002 = \frac{2}{1{,}000} = \frac{20}{10{,}000}$

$.02 = \frac{2}{100} = \frac{200}{10{,}000}$

$.0205 = \frac{205}{10{,}000}$

$.021 = \frac{21}{1{,}000} = \frac{210}{10{,}000}$

$.023 = \frac{23}{1{,}000} = \frac{230}{10{,}000}$

21. (4) 5.9
Find the total amount of growth, then divide by the number of years to find the average growth per year.

```
  6.2 cm        5.86    round to
  5.9 cm     3)17.60    tenths = 5.9
+ 5.5 cm       15
 17.6           2 6
                2 4
                  20
```

22. (2) 1,326
Find the number of miles traveled in the week.

```
  51,809 mi
− 42,527 mi
   9,282 mi
```

Find the number of miles per day.

$\frac{9282}{7} = 1{,}326$

23. (2) 3.0962×10^{-2}
Move the decimal point to the right of the 3. Since you have moved the point 2 places to the *right*, the exponent is −2.

24. (5) 18
Let Kelly's brother's age be x.
Then Kelly $= 3x$
and $x + 3x = 24$
↑ brother ↑ Kelly ↑ combined ages

Combine terms: $4x = 24$
Divide by 4: $x = 6$ ← Kelly's brother
Kelly $= 3x = 3(6) = 18$ years

25. (2) B
Point B lies halfway between −1 and −2, or at −1½ or −3/2.
(Point *A* is −3, point *C* is −½, points *D* and *E* are positive.)

26. (3) 10
First convert feet to inches, then divide the length by the number of people.

```
  12 in           10  in for each person
×  5            6)60
  60 in long
```

27. (4) $0.90
First find the total weight in pounds, then multiply by 20 cents per pound.

```
       9 oz
1 lb  12 oz
2 lb   3 oz
3 lb  24 oz = 4.5 lb
```

(4.5 lb × 0.20/lb = $0.90)

28. (1) 12.8
Since 2 oz concentrated juice is diluted with 3 ounces water, this makes 5 ounces dilute juice. The ratio of concentrated juice to dilute juice is 2 to 5. The problem asks for the amount of concentrated juice needed to make 1 *quart.* (1 qt = 32 oz.) Set up a proportion.

$\frac{\text{conc. juice}}{\text{dilute mix}} = \frac{2}{5} = \frac{x}{32}$

$x = \frac{32}{5}(2) = 12.8$

29. (4) 16
Note that the two base angles are equal, so this is an isosceles triangle and the sides opposite the base angles are equal. The perimeter is 6 + 5 + 5 = 16.

30. (4) 80

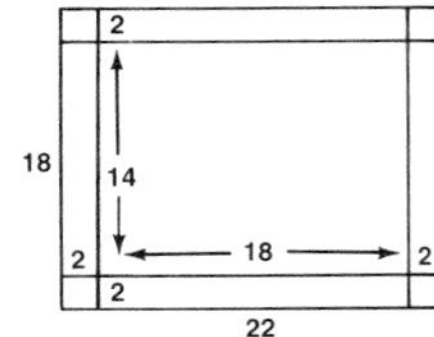

A sketch of the picture with the border around it shows that 2 inches must be added to each end of the length and to each end of the width. The resulting length will be 18 + 2 + 2, or 22 inches, and width is 14 + 2 + 2, or 18 inches. The perimeter will be 18 + 18 + 22 + 22, or 80 inches.

31. (4) 37.68
Circumference = π × diameter,

and diameter = 2 × radius. Find the diameter, then multiply by π (3.14).

$6 \times 2 = 12$ in diameter
$C = 3.14(12) = 37.68$ in

32. (5) Not enough information is given.
The heights of the walls are not given, so you cannot find the area to be painted.

33. (3) 20
Find the area of the floor by multiplying length by width.

$15 \times 12 = 180$ sq ft

Then convert square feet to square yards. (9 sq ft = 1 sq yd)

$180 \div 9 = 20$ sq yds

34. (3) 7.98
Area of a triangle is (½) *bh.*

$A = \frac{1}{2}(3.8)(4.2)$

$A = \frac{1}{2}(15.96) = 7.98 \text{ cm}^2$

35. (3) 32
Use the Pythagorean Theorem to find the altitude of the triangle.

$h = \sqrt{5^2 - 3^2} = \sqrt{25 - 9} = \sqrt{16} = 4$

Now use the formula for area of a parallelogram.

$A = bh$
$= 4(8) = 32$

36. (5) Not enough information is given.
You do not know the third dimension of the sink, so you cannot find its volume.

37. (3) 75°
The three angles of a triangle add up to 180°. Add the two known angles, then subtract from 180° to find the missing angle measurement.

$\angle A = 34° \quad \angle B = 41°$

$34° + 41° = 75° \qquad 180° - 75° = 105°$

$\angle C = 105°$

Then $\angle D = 180 - 105 = 75°$

38. (5) 20
Triangles *EDC* and *BAC* are similar because their corresponding angles are equal. They share angle *C* and each has a right angle (*D* and *A*).

$\frac{DE}{BA} = \frac{DC}{AC}$ and $\frac{DE}{12} = \frac{15}{9}$

$DE = \frac{\overset{5}{\cancel{15}}}{\underset{\cancel{3}1}{\cancel{9}}}(\overset{4}{\cancel{12}}) = 20$

39. (1) 17
Since this is a right triangle, find the length of the hypotenuse using the Pythagorean Theorem:

$a^2 + b^2 = c^2$ (*a* and *b* are the lengths of the legs; *c* is the hypotenuse).

Square each leg: $(15)^2 + (8)^2 = c^2$
$225 + 64 = c^2$
$289 = c^2$

Take the square root: $\sqrt{289} = c$
$17 = c$

40. (3) 8
Draw a perpendicular line from *P* to *QS*. Then count the units from *P* to the point of intersection. It is 5 units to the *y*-axis and then 3 more units to the right.

41. (3) $\frac{2}{3}$
First notice that the line goes up to the right. The slope will be a positive number. Next, make a right triangle, making AB the hypotenuse. Count the number of units in the vertical leg and the units in the horizontal leg. Express the vertical units over the horizontal as a fraction and reduce.

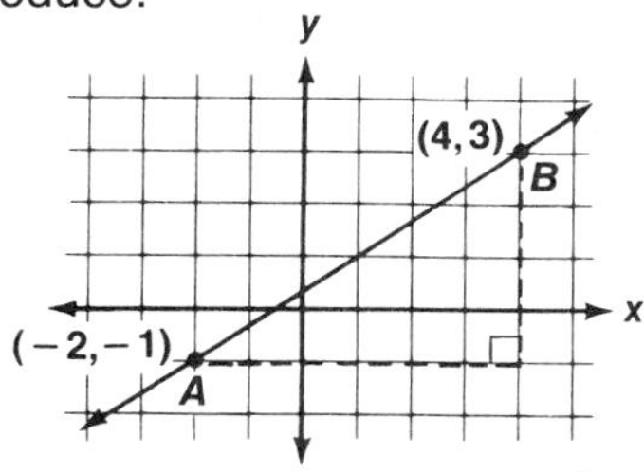

$\frac{4 \text{ vertical units}}{6 \text{ horizontal units}} = \frac{2}{3}$

42. (5) 2(5.50) + 3(3.75)
Find the total cost of adult tickets, then the total cost of children's tickets, then add those two amounts:

2 adults at $5.50 each
2(5.50)
PLUS 3 children at $3.75 each
3($3.75)
OR 2(5.50) + 3(3.75)

43. (1) −1
First evaluate each term. Then combine terms, being careful to look at positives and negatives.

$5^2 \rightarrow 5 \times 5 \rightarrow 25$
$3^3 \rightarrow 3 \times 3 \times 3 \rightarrow 27$
$1^4 \rightarrow 1 \times 1 \times 1 \times 1 \rightarrow 1$
$5^2 - 3^3 + 1^4 =$
$25 - 27 + 1 = -1$

44. (3) $\frac{1{,}250 - 800}{24}$
After they pay $800, the Bittners have to spread $1,250 − $800 over 2 years. The question asks for the *monthly* payment, so use 24 months for 2 years.

45. (2) $\sqrt{a - b} - 3$
Three less than the square root means the −3 follows the radical sign. *The square root of the difference of 2 numbers* means the square root is applied *after* the difference is found.

46. (2) 9
Let the missing number be represented by *x*.

4*x*	−7	= 29
4 times a number	seven less than	is

$4x - 7 = 29$

Change − to +: $4x + -7 = 29$
Add 7: $+7 = +7$
Divide by 4: $4x = 36$

$\frac{4x}{4} = \frac{36}{4}$
$x = 9$

47. (1) 5
By substituting the answer choices for *x* in the equation, you will find the one answer which will make both sides of the equation equal.

First try 5 in place of *x*:

$\frac{5[3(5) + 7]}{2} = 55$

$\frac{5[15 + 7]}{2} = 55$

$\frac{5(22)}{2} = 55$

$\frac{110}{2} = 55$

$55 = 55$

If you went on to try 6, 6⅔, 7 and 25 in place of *x,* you would find that both sides of the equation were not equal.

48. (2) 14
Consecutive even numbers are separated by 2 on the number line. Let the first be x. Then the next one is $x + 2$, and the last one is $(x + 2) + 2$, or $x + 4$,

$$x + (x+ 2) + [(x+ 2)+ 2] = 42.$$
$$3x+ 6 = 42$$
$$3x = 36$$
$$x = 12.$$

So $x + 2 = 14$, and $x + 4 = 16$.

49. (3) $\frac{2K - 5}{a}$
Use the same steps to get b alone on one side, just as if you had numbers instead of letters. Start with undoing division by 2, since that was the last operation performed.

$$K = \frac{ab - 5}{2}$$
$$2K = 2b - 5$$
$$2K - 5 = ab$$
$$\frac{2K - 5}{a} = b$$

50. (5) $(-4, -7)$
The point of intersection is the ordered pair which fits both equations. Substitute each ordered pair into both equations. $(-4, -7)$ will make both equations true.

$$y = 2x + 1$$
$$-7 = 2(-4) + 1$$
$$-7 = (-8) + 1$$
$$-7 = -7 \quad \text{TRUE}$$
$$y = x - 3$$
$$-7 = -4 - 3$$
$$-7 = -4 + (-3)$$
$$-7 = -7 \quad \text{TRUE}$$

51. (4) 6
Substitute $2y$ back into the first equation for x, then solve for y. The question asks for the value of x. Once you have found y, substitute its value in the second equation to find x.

$$2x + 4y + 7 = 31$$

Substitute $2y$ for x.

$$2(2y) + 4y + 7 = 31$$
$$4y + 4y + 7 = 31$$

Combine terms: $8y + 7 = 31$
Add -7: $8y = 24$
Divide by 8: $y = 3$
Then find x. $x = 2y$

$$x = 2(3)$$
$$x = 6$$

52. (5) $x < 2$

$$x - 10 < -6 - x.$$
$$x < 4 - x$$
$$2x < 4$$
$$x < 2$$

53. (4) $\frac{3}{2}$ or -4
Always try factoring first. If a quadratic is factorable, that's the easiest way to solve it. Remember to get zero on one side before you factor.

$$2x^2 + 5x = 12$$
$$2x^2 + 5x - 12 = 0$$
$$(2x - 3)(x + 4) = 0$$
$$2x - 3 = 0 \text{ or } x + 4 = 0$$
$$x = \tfrac{3}{2} \text{ or } x = -4$$

54. (2) -3 or $\frac{1}{2}$
To factor, first get zero on one side.

$$2x^2 + 5x - 3 = 0$$

Then $(2x - 1)(x + 3) = 0$
Now either $2x - 1 = 0$

$$2x = 1$$
$$x = \tfrac{1}{2}$$

or $x + 3 = 0$

$$x = -3$$

55. (3) 9.0264×10^2
The decimal point was moved two places to the left.

56. (3) 127°
Angle R and the 53° angle are alternate interior angles, so $\angle R = 53°$. Since $\angle M$ and $\angle R$ are supplementary angles, $\angle M = 180° - 53 = 127°$.

Finding Your Score

When you finish Posttest A, be sure to record your score on the Progress Chart on the inside back cover of the book.

How did you do on Posttest A? If you didn't pass or just barely passed, there are several things you can study to improve your test score. For instance, read the answer explanations. If you had trouble with a particular topic (arithmetic, geometry, or algebra), you might want to reread the opening pages for that section. You also may review the sections on graphs or using grids in the appropriate lessons. After looking again at those pages, take Posttest B and compare your scores on the Progress Chart.

If you did well on Posttest A, you're probably ready to take the GED Mathematics Test. However, you'll no doubt feel more comfortable with the test and more confident of your ability if you do Posttest B as further practice. If you can't take the GED Mathematics Test soon after you study this book, you can use Posttest B as a refresher just before exam time.

Answer Key

1. (3)	**9.** (1)	**17.** (5)	**25.** (2)	**33.** (3)	**41.** (3)	**49.** (3)
2. (5)	**10.** (1)	**18.** (2)	**26.** (3)	**34.** (3)	**42.** (5)	**50.** (5)
3. (5)	**11.** (1)	**19.** (2)	**27.** (4)	**35.** (3)	**43.** (1)	**51.** (4)
4. (2)	**12.** (3)	**20.** (4)	**28.** (1)	**36.** (5)	**44.** (3)	**52.** (5)
5. (5)	**13.** (3)	**21.** (4)	**29.** (4)	**37.** (3)	**45.** (2)	**53.** (4)
6. (4)	**14.** (1)	**22.** (2)	**30.** (4)	**38.** (5)	**46.** (2)	**54.** (2)
7. (2)	**15.** (4)	**23.** (2)	**31.** (4)	**39.** (1)	**47.** (1)	**55.** (3)
8. (4)	**16.** (3)	**24.** (5)	**32.** (5)	**40.** (3)	**48.** (2)	**56.** (3)

MATHEMATICS POSTTEST B

Directions

The Mathematics Posttest consists of 56 problems. Whenever possible, you should arrive at your own answer to a question before looking at the choices; otherwise, you may be misled by wrong answers that look possible.

You should take approximately 1½ hours to complete this test. There is no penalty for guessing. Try to answer as many questions as you can. Work rapidly but carefully, without spending too much time on any one question. If a question is too difficult for you, omit it and come back to it later.

For each answer, mark one answer space. See how the following example is done.

EXAMPLE

A secretary bought a desk calendar for \$5.95 and a pencil container for \$1.89. How much change did she get from a ten-dollar bill?

(1) \$2.16
(2) \$2.61
(3) \$7.84
(4) \$8.74
(5) \$8.84

The amount is \$2.16; therefore, answer space (1) has been marked.

Answers to the questions are in the Answer Key on page 384. Explanations for the answers are on pages 381–384.

Directions: Choose the one best answer for each problem.

1. An employee's gross pay was \$418. Her employer deducted \$68 in taxes and \$21 for insurance. What was her net pay after deductions?

   (1) \$319 (2) \$329 (3) \$371
   (4) \$471 (5) \$507

   (1) (2) (3) (4) (5)

2. A new box of laundry detergent has $\frac{1}{4}$ more detergent in it than the regular box. If the regular size box has 3 pounds of detergent, how much is in the new box?

   (1) $\frac{3}{4}$ lb (2) $2\frac{3}{4}$ lb (3) $3\frac{1}{4}$ lb
   (4) $3\frac{3}{4}$ lb (5) 12 lb

   (1) (2) (3) (4) (5)

3. Admission to the movies is \$3.75 for adults and \$2.50 for children. How much would it cost for 2 adults and 3 children to see the movie?

   (1) \$12.00 (2) \$12.50 (3) \$15.00
   (4) \$16.25 (5) \$31.25

   (1) (2) (3) (4) (5)

4. A customer bought three items costing \$4.29, \$1.98 and \$0.64. How much change did he receive from a twenty-dollar bill?

   (1) \$1.67 (2) \$3.09 (3) \$6.91
   (4) \$12.19 (5) \$13.09

   (1) (2) (3) (4) (5)

5. In 1955, the average wage was \$1.85 per hour. In 1980, the average wage was about \$7.40. What fraction of the 1980 wage was the 1955 wage?

   (1) $\frac{4}{1}$ (2) $\frac{3}{4}$ (3) $\frac{1}{3}$ (4) $\frac{1}{4}$ (5) $\frac{1}{5}$

   (1) (2) (3) (4) (5)

6. There were 12 yards of material on a bolt which was sold at 30% off. The clerk cut off one piece $2\frac{7}{8}$ yards long and another piece $1\frac{3}{4}$ yards long. How many yards of material were left on the bolt?

   (1) $9\frac{1}{8}$ (2) $8\frac{3}{8}$ (3) $7\frac{5}{8}$ (4) $7\frac{3}{8}$ (5) $4\frac{5}{8}$

   (1) (2) (3) (4) (5)

7. A customer bought $3\frac{1}{2}$ pounds of 10% fat-free cheese for \$6.30. What was the price of two pounds of cheese?

   (1) \$0.56 (2) \$0.90 (3) \$1.80
   (4) \$2.21 (5) \$2.80

   (1) (2) (3) (4) (5)

8. In Company A, 37% of the employees are women. If there are 5,400 employees, how many men are employed by Company A?

   (1) 14,595 (2) 9,195 (3) 3,412
   (4) 3,402 (5) 1,998

   (1) (2) (3) (4) (5)

9. During one football game, 85% of the stadium was full. If 15,300 people attended the game, how many seats are in the stadium?

   (1) 18,000 (2) 18,305 (3) 33,300
   (4) 113,005 (5) 180,000

   (1) (2) (3) (4) (5)

10. A retail chain spends \$563,000 annually in advertising. If 70% of its advertisements are television commercials, approximately how much does it spend on non-television advertising?

    (1) \$169,000 (2) \$219,000 (3) \$393,900
    (4) \$463,000 (5) \$496,500

    (1) (2) (3) (4) (5)

11. The price of gasoline dropped from \$1.22 a gallon to \$0.99 a gallon. By about what percent did the price of gasoline drop?

    (1) 2% (2) 12% (3) 19%
    (4) 23% (5) 81%

    (1) (2) (3) (4) (5)

12. A baseball player's batting average is the ratio of the number of hits to the number of times at bat. If a player has a batting average of .275, and has been at bat 80 times, how many hits has he made?

(1) 21 (2) 22 (3) 29 (4) 35 (5) 320

1 2 3 4 5

13. A recipe calls for 2 cups flour and $1\frac{1}{2}$ cups cocoa. How much flour would be needed if the baker had only 1 cup cocoa?

(1) 3 c (2) $1\frac{1}{2}$ c (3) $1\frac{1}{3}$ c

(4) 1 c (5) $\frac{3}{4}$ c

1 2 3 4 5

14. In a large company, blue-eyed employees outnumber other employees 5 to 1. About how many other employees would you expect to find in a department of 100?

(1) 16 (2) 17 (3) 20 (4) 84 (5) 85

1 2 3 4 5

15. Find the amount of simple interest paid on a loan of $1,700 at 12 percent annual rate.

(1) $153.00 (2) $173.00
(3) $183.60 (4) $1,275.00
(5) Not enough information is given.

1 2 3 4 5

16. A student has test scores of 88, 85, 91, and 84. What was her average test score?

(1) 62 (2) 82 (3) 87 (4) 88 (5) 348

1 2 3 4 5

17. Another way to write 2,590 is

(1) 2.59×10^{-3} (2) 2.59×10^{4}
(3) 2.59×10^{3} (4) $.259 \times 10^{-3}$
(5) $.259 \times 10^{3}$

1 2 3 4 5

18. Written in the usual form, 3.59×10^{-2} is

(1) 0.0359 (2) 3.059 (3) 359
(4) 3,590 (5) 35,900

1 2 3 4 5

19. Which number is smallest?

(1) 4° (2) 4^{-3} (3) $(-6)^{5}$
(4) -2 (5) 4^{3}

1 2 3 4 5

20. In order from smallest to largest, which number is third? $1\frac{1}{4}$, $\frac{-3}{2}$, $\frac{1}{12}$, $\frac{1}{11}$, $\frac{7}{2}$

(1) 1.25 (2) $\frac{-3}{2}$ (3) $\frac{1}{12}$

(4) $\frac{1}{11}$ (5) $\frac{7}{2}$

1 2 3 4 5

21. A survey shows that teenagers watch television an average of 22 hours 31 minutes per week. They prefer 30-minute programs. According to the survey, how long do they watch television, on an average, per day?

(1) 4 hr 31 min (2) 3 hr 19 min
(3) 3 hr 13 min (4) 3 hr 5 min
(5) 2 hr 13 min

1 2 3 4 5

22. A plane leaves Boston, Massachusetts, at 2:35 PM Eastern Standard Time and arrives in Denver, Colorado, at 9:10 PM. Mountain Standard Time. Mountain Standard Time is 2 hours earlier than Eastern Standard Time (12:00 noon EST = 10:00 AM MST). How long was the flight from Boston to Denver?

(1) 4 hr 35 min (2) 6 hr 25 min
(3) 6 hr 35 min (4) 8 hr 25 min
(5) 8 hr 35 min

1 2 3 4 5

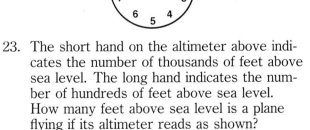

23. The short hand on the altimeter above indicates the number of thousands of feet above sea level. The long hand indicates the number of hundreds of feet above sea level. How many feet above sea level is a plane flying if its altimeter reads as shown?

(1) 2,000 (2) 2,800 (3) 8,000
(4) 8,200 (5) 10,000

1 2 3 4 5

Items 24 and 25 refer to the following graph.

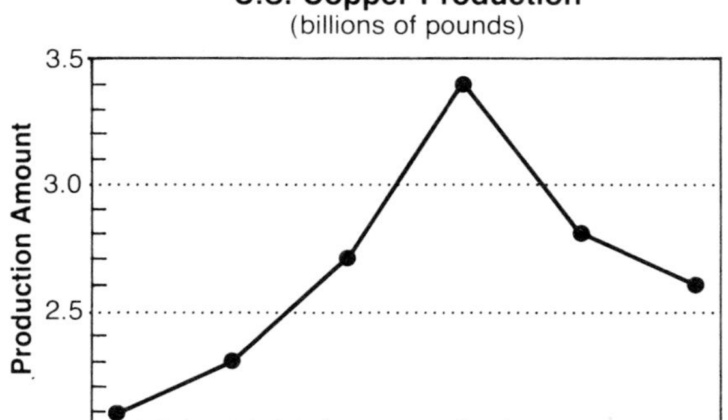

24. During which five-year period did copper production increase the most?

(1) 1955–1960 (2) 1960–1965
(3) 1965–1970 (4) 1970–1975
(5) 1975–1980

① ② ③ ④ ⑤

25. About how many more pounds of copper were produced in 1970 than in 1955?

(1) 1,350,000,000 (2) 1,000,000,035
(3) 135,000,000 (4) 13,500,000
(5) 1,350,000

① ② ③ ④ ⑤

Items 26 and 27 refer to the following table.

Shipping Charges

Order Total	via Surface	via Air
Up to $50	add $3.50	add $6.00
$50.01–$100	add $4.25	add $7.00
$100.01–$200	add $4.75	add $8.00
$200.01–$300	add $5.25	add $9.00
Over $300	add $5.75	add $10.00

26. According to the table, how much would the shipping charge be if a customer's order totaled $47.50 and the package was shipped by air?

(1) $3.50 (2) $14.25 (3) $4.75
(4) $5.75 (5) $6.00

① ② ③ ④ ⑤

27. How much would a customer save by having one $200 order sent by air rather than two $100 orders sent by surface?

(1) $0.50 (2) $2.75 (3) $3.25
(4) $3.75 (5) $4.50

① ② ③ ④ ⑤

28. How many inches of framing are needed to go around a picture that is $14\frac{1}{2}$ inches wide and $6\frac{3}{4}$ inches long?

(1) $42\frac{1}{2}$ (2) $41\frac{1}{2}$ (3) $32\frac{1}{2}$
(4) $21\frac{1}{4}$ (5) $20\frac{1}{4}$

① ② ③ ④ ⑤

29. What is the area in square centimeters of a label on a can that has a radius of 3.5 centimeters? ($A = \frac{22}{7}dh$, where d = diameter and h = height.)

(1) 220 (2) 200 (3) 110 (4) $31\frac{3}{7}$
(5) Not enough information is given.

① ② ③ ④ ⑤

30. Find the area of triangle *ABC* in the figure below.

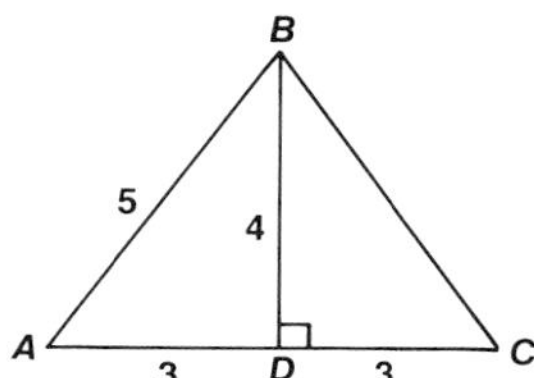

(1) 3 (2) 6 (3) 12
(4) 15 (5) 22

① ② ③ ④ ⑤

31. How many square yards of padding are needed to go under 2 carpets, each measuring 9 feet wide, 12 feet long, and 2 inches thick?

(1) 216 (2) 72 (3) 36
(4) 24 (5) 18

① ② ③ ④ ⑤

32. The inside measurements of a refrigerator are 3 feet long, 2 feet wide, and 30 inches deep. How many cubic feet of space does the refrigerator have?

(1) $7\frac{1}{2}$ (2) 15 (3) 180
(4) 1,080 (5) 25,920

① ② ③ ④ ⑤

Items 33 and 34 refer to the figure Line ℓ is parallel to line m.

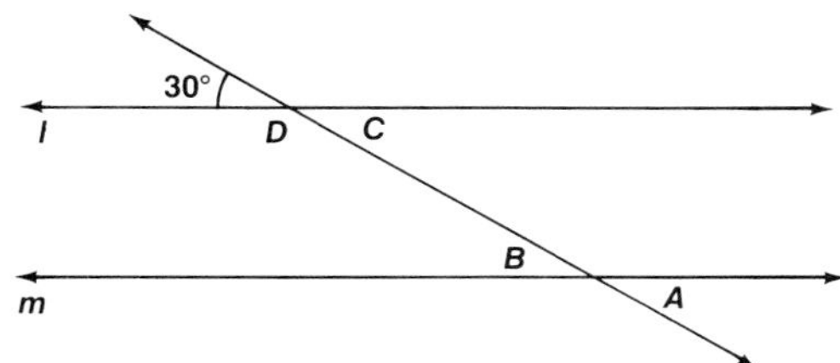

33. What is the measure of angle A?

(1) 30° (2) 60° (3) 90°
(4) 150° (5) 160°

① ② ③ ④ ⑤

34. What is the measure of angle B?

(1) 30° (2) 60° (3) 90°
(4) 150° (5) 160°

① ② ③ ④ ⑤

35. What is the length in yards of the other leg of a right triangle that has a hypotenuse of 26 yards and a leg of 24 yards?

(1) 5 (2) 8 (3) 10
(4) 50 (5) 100

① ② ③ ④ ⑤

36. If a pizza has an area of 254.34 square inches, what is its radius in inches?

(1) 3.14 (2) 9 (3) 18
(4) 27 (5) 81

① ② ③ ④ ⑤

37. In the figure below, what is the measure of angle C?

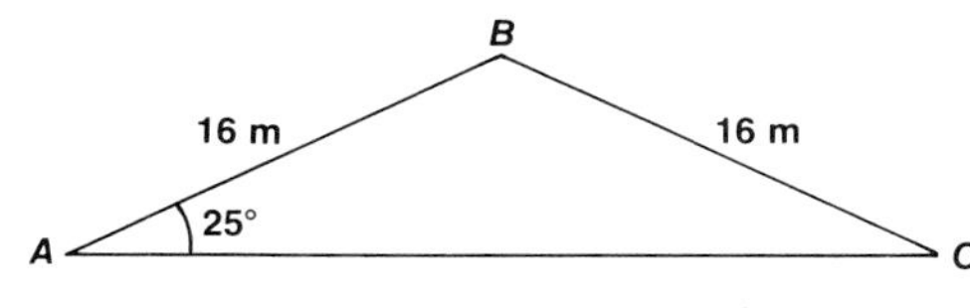

(1) 25° (2) $65\frac{1}{2}°$ (3) $77\frac{1}{2}°$ (4) 155°
(5) Not enough information is given.

① ② ③ ④ ⑤

38. Find the length of line EF in the figure below.

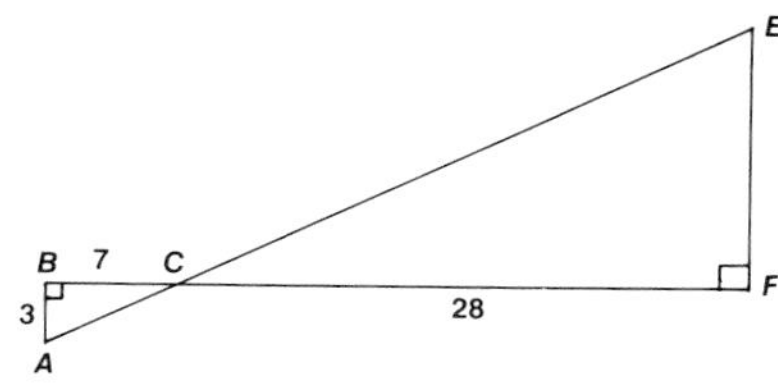

(1) $\sqrt{58}$ (2) 12 (3) 2.5
(4) 7 (5) 12.9

① ② ③ ④ ⑤

39. Fred's shadow is 12 feet long. The shadow of a utility pole is 30 feet long. Fred is 5 feet tall. How tall in feet is the pole?

(1) 12.5 (2) 13 (3) 36
(4) 72 (5) $\sqrt{30}$

① ② ③ ④ ⑤

40. A bass string on an upright piano is strung diagonally on the frame. If the piano frame is 32 inches high and 60 inches long, approximately how long in inches is the string?

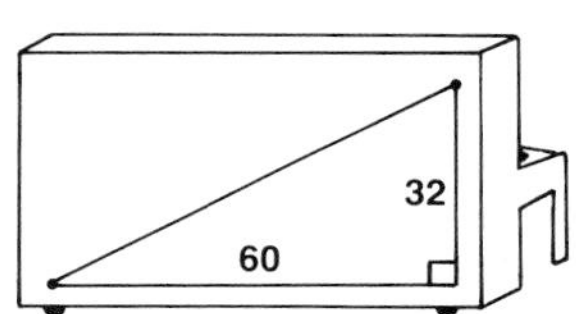

(1) 14 (2) 51 (3) 62
(4) 68 (5) 92

① ② ③ ④ ⑤

41. Simplify $3^4 - 2$.

(1) 10 (2) 25 (3) 62
(4) 79 (5) 81

① ② ③ ④ ⑤

42. Find the value of $a^2(b - c)$ when $a = -4$, $b = 5$, and $c = 2$.

(1) 48 (2) 54 (3) 18
(4) -24 (5) -48

① ② ③ ④ ⑤

43. Find $x + 4(y - z)$ when $x = 3$, $y = 2$, and $z = 3$.

(1) -23 (2) -7 (3) -1
(4) 0 (5) 1

① ② ③ ④ ⑤

44. A painter charges \$9.50 per hour for labor plus the cost of materials. If it took him $4\frac{1}{2}$ hours to paint a room, and the materials cost \$37.80, which of the following shows the amount he charged his customer?

(1) 4.5 (9.50) + 37.80
(2) 4.5 (9.50 + 37.80)
(3) 9.50 (4.5 + 37.80)
(4) 4.5 (37.80) (9.50)
(5) (4.5 + 9.50) (37.80)

(1) (2) (3) (4) (5)

45. A laborer earned \$90 per day working in a mine on Monday, Tuesday, and Wednesday. On Thursday and Friday, he earned \$70 per day resurfacing a road. What was his average wage per day in dollars?

(1) $90 + 70$ (2) $\frac{90 + 70}{2}$
(3) $\frac{90 + 70}{5}$ (4) $3(90) + 2(70)$
(5) $\frac{3(90) + 2(70)}{5}$

(1) (2) (3) (4) (5)

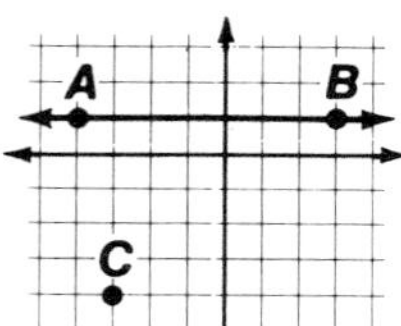

46. What is the perpendicular distance from C to line AB in the figure above?

(1) 3 (2) 4 (3) 5 (4) 6 (5) 8

(1) (2) (3) (4) (5)

47. If $8a - 9 = 47$, then a equals

(1) 56 (2) 9 (3) 7 (4) -7 (5) -9

(1) (2) (3) (4) (5)

48. Mr. Sanchez makes \$100 less than twice the amount Mrs. Sanchez makes per week. Together they spend \$48 on transportation. How much does Mrs. Sanchez make per week?

(1) \$113.34 (2) \$146.67
(3) \$180.00 (4) \$220.00
(5) Not enough information is given.

(1) (2) (3) (4) (5)

49. Solve $xy + 5x - 6y = 10w$ for x.

(1) $\frac{10 + 6y - 5x}{y}$ (2) $\frac{10w - 6y}{5 - y}$
(3) $\frac{10w + xy + 5x}{-6}$ (4) $\frac{10w + y}{y}$
(5) $\frac{10w + 6y}{5 + y}$

(1) (2) (3) (4) (5)

50. The sum of 5 consecutive numbers is 55. Find the first number.

(1) 5 (2) 9 (3) 10 (4) 12
(5) Not enough information is given.

(1) (2) (3) (4) (5)

51. If $2x + 4y + 7 = 31$ and $y = \frac{x}{2}$, then $x =$

(1) -6 (2) -4 (3) 4 (4) 6 (5) 8

(1) (2) (3) (4) (5)

52. Solve $\frac{1}{3}x + 5 < 4 - 9x$.

(1) $x < -\frac{1}{7}$ (2) $x < -\frac{3}{28}$ (3) $< x \frac{3}{28}$
(4) $x < \frac{19}{20}$ (5) $x > \frac{19}{20}$

(1) (2) (3) (4) (5)

53. To solve $8x^2 + \frac{2x - 5}{3} = 9$, first

(1) take the square root
(2) multiply by 3 (3) divide by 2
(4) divide by 8 (5) factor

(1) (2) (3) (4) (5)

54. If $d^2 + 3d - 4 = 0$, then d equals

(1) -4 and 1 (2) 4 and -1
(3) -4 only (4) -1 only (5) 1 only

(1) (2) (3) (4) (5)

55. Solve $3x^2 - 2x = 1$.

(1) 3 or -3 (2) 3 or -1
(3) -1 or 1 (4) $-\frac{1}{3}$ or 1
(5) $\frac{1}{3}$

(1) (2) (3) (4) (5)

56. Solve $3x^2 + 5x = 12$.

(1) -3 (2) -3 or 1 (3) $\frac{4}{3}$ or $\frac{-4}{3}$
(4) $\frac{4}{3}$ (5) -3 or $\frac{4}{3}$

(1) (2) (3) (4) (5)

Answer Explanations: Posttest B

1. (2) $329
Add together the amounts to be deducted from the paycheck.

	$68	taxes
+	21	insurance
	$89	total deductions

Then subtract this amount from the gross pay.

	$\overset{3\,10\,18}{\$418}$
−	89
	$329

2. (4) $3\frac{3}{4}$ lb
First multiply ¼ by 3 lb to find how much more the new box has than the old. Then add this to the old box amount to find how much the new box has.

$\frac{1}{4} \times \frac{3}{1} = \frac{3}{4}$ lb more than the old

3 lb	+	$\frac{3}{4}$ lb	=	$3\frac{3}{4}$ lb
old box		more		new box

3. (3) $15.00
Multiply the cost per adult times the number of adults. Multiply the cost per child times the number of children. Then add these costs together to find the total.

	$3.75	per adult
×	2	adults
	$7.50	

	$2.50	per child
×	3	children
	$7.50	

	$7.50	adults
+	7.50	children
	$15.00	total

4. (5) $13.09
Add the costs of the items, then subtract from $20.00

	$\overset{1\;2}{\$4.29}$	
	1.98	
+	0.64	
	$6.91	total cost

	$\overset{1\,19\,191}{20.00}$	
−	6.91	
	$13.09	change

5. (4) $\frac{1}{4}$
The 1955 wage compared to the 1980 wage written as a fraction would be:

$$\frac{\text{1955 wage}}{\text{1980 wage}}$$

Put in the values and reduce.

$$\frac{\text{1955 wage}}{\text{1980 wage}} = \frac{1.85}{7.40} = \frac{1.85 \times 100}{7.40 \times 100} = \frac{185}{740} = \frac{1}{4}$$

6. (4) $7\frac{3}{8}$
Add together the amounts that were cut off the bolt. Then subtract this amount from the number of yards that was on the bolt to find how much was left. (Be sure to give fractions a common denominator.)

$$2\tfrac{7}{8} \rightarrow 2\tfrac{7}{8}$$
$$+\ 1\tfrac{3}{4} \rightarrow 1\tfrac{6}{8}$$
$$3\tfrac{13}{8} = 4\tfrac{5}{8} \text{ yds cut off}$$

$$\overset{1}{12}\tfrac{8}{8}$$
$$-\ 4\tfrac{5}{8}$$
$$7\tfrac{3}{8} \text{ yds left}$$

7. (3) $1.80
Divide the total cost by the number of pounds to find the cost per pound.

$$\frac{6.30}{1} \div 3\frac{1}{2} =$$

$$\frac{\overset{0.90}{\cancel{6.30}}}{1} \times \frac{2}{\underset{1}{\cancel{7}}} = \$1.80 \text{ per lb}$$

OR

```
        $ 1.80 per lb
  3.5.)6.3.00
        3 5
        2 80
        2 80
        0 0
```

8. (4) 3,402
Use the grid to find the number of women employed by Company A. Then subtract this amount from the total number of employees to find the number of men.

no. of women	?	37	% women
total employees	5,400	100	

37 × 5,400 = 199,800
199,800 ÷ 100 = 1,998 women

	5,400	total
−	1,998	
	3,402	men

9. (1) 18,000
Use the grid to work this percent problem.

no attended	15,300	85	% attended
total seats	?	100	

15,300 × 100 = 1,530,000
1,530,000 ÷ 85 = 18,000 seats in the stadium

10. (1) $169,000
Since 70% is spent on television commercials, 30% is spent on non-television advertising.
.30 × $563,000 = $169,000

11. (3) 19%
First subtract to find the amount of decrease in price. Then use the grid to solve the percentage.

	$1.22	old price
−	0.99	new price
	$0.23	decrease

amt. of decrease	0.23	?	% decrease
original	1.22	100	

0.23 × 100 = 23
23 ÷ 1.22 =
Round 18.85 to 19%.

12. (2) 22
Use a grid for this proportion. (First change the batting average decimal to a fraction.)

$.275 = \frac{275}{1000}$

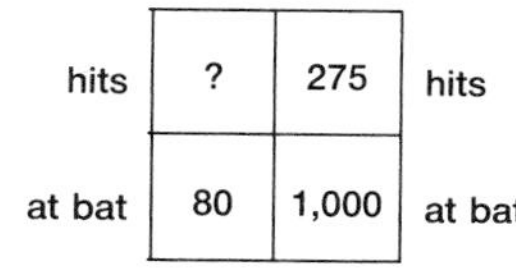

hits
?
275
hits
at bat
80
1,000
at bat

275 × 80 = 22,000
22,000 ÷ 1,000 = 22 hits

13. (3) $1\frac{1}{3}$ c
Use a grid for this proportion.

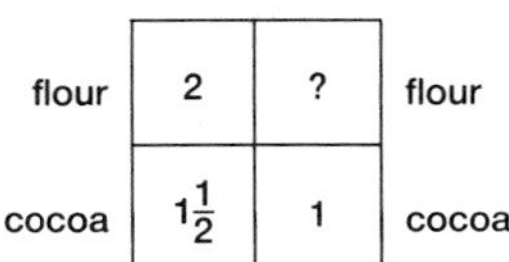

$2 \times 1 = 2$

$2 \div 1\frac{1}{2} = 2 \div \frac{3}{2} =$

$2 \times \frac{2}{3} =$

$\frac{4}{3}$, or $1\frac{1}{3}$ cups flour

14. (2) 17
Since blue-eyed employees outnumber the other employees 5 to 1, for every 6 employees there are 5 blue-eyed people and 1 other person. That is, one person in 6 is not blue-eyed. So in a group of 100, you would expect 1/6 of the people not to have blue eyes.

$\frac{1}{6} \times 100 = 16.6$, or about 17 people.

15. (5) Not enough information is given.
The problem does not give the period of the loan.

16. (3) 87
Add together the 4 test scores. Then divide this total by 4 to find the average score.

```
   1
   88
   85
   91
 + 84
  348  total points

   87  average score
 4)348
   32  tests
    28
    28
     0
```

17. (3) 2.59×10^3
Place the decimal point after the 2. You have moved it 3 places to the left, so the exponent is 3.

18. (1) .0359

$3.59 \times 10^{-2} = .0359$

Remember the exponent, − 2, tells you to move the decimal point 2 places to the left.

19. (3) -6^5
Think of a number line. Which of these numbers will be negative when simplified? The smallest number will be a negative one.

$4^0 = 1$

$4^{-3} = \frac{1}{4^3}$ (Small, but positive)

$(-6)^5 =$

$(-6)(-6)(-6)(-6)(-6) =$

$(-6^5)\ (-2)\ 4^3$

The negative ones are −2 and -6^5. The latter is certainly the smaller. We do not need to calculate the exact value of -6^5 because -6^5 is plainly to the left of −2 on the number line.

20. (4) $\frac{1}{11}$
You may convert to a common denominator to compare the numbers, but the common denominator will be very large in this case. Think about the numbers. The negative number of course will be smallest. Of the others, two are less than 1, namely 1/12 and 1/11. Twelfths are smaller than elevenths, so we have so far −3/2, 1/12, 1/11. Now convert the last two numbers to a common form:

$1.25 = \frac{5}{4} = 1\frac{1}{4}$

$\frac{7}{2} = 3\frac{1}{2}$

So the numbers from smallest to largest are

$-\frac{3}{2}, \frac{1}{12}, \frac{1}{11}, 1\frac{1}{4}, \frac{7}{2}$

21. (3) 3 hr 13 min
Divide the amount of time per week by 7 to find the average amount of time per day.

```
   3 hrs 13   min per day
 7)22 hrs 31  min
   21
   (1 hr = 60 min)
           91 min
            7
           21
           21
            0
```

22. (5) 8 hr 35 min
First put the departure and arrival times into the same time frame by converting Denver time (Mountain Standard) to Boston time (Eastern Standard). When it is 10:00 AM MST, it is 12:00 noon EST. Therefore, when it is 9:10 MST, it is 11:10 EST. Now subtract the departure time from the arrival time to find the length of the flight.

```
     (60 + 10)
     0  7 0
  1  1:10  EST arrived
  −  2:35  EST departed
     8:35
```

Flight time is 8 hr 35 min

23. (2) 2,800
The short hand is at the 2 and represents thousands. The long hand is at the 8 and represents hundreds. Add these together.

2,000 + 800 = 2,800 ft above sea level.

24. (3) 1965–1970
The line graph shows an increase (goes up) from 1955 through 1970. After 1970 it decreases (goes down). Production increased the most between 1965 and 1970. The line is the sharpest or longest upward line between those times. OR you can also compare the amounts of productions (subtract) between any two given years in which there is an increase. 1965–1970 shows the greatest difference or increase.

25. (1) 1,350,000,000
Find each year along the bottom of the graph. Move up from each year, then directly to the left column to find the actual amount of production. Notice the production is given in billion pounds. Then subtract the amounts to find the difference.

1970 = 3.45 billion
1955 = 2.10 billion

```
   3,450,000,000
 − 2,100,000,000
   1,350,000,000 lb
```

26. (5) $6.00
Since $47.50 is less than $50, you would be charged from the row labeled "Up to $50." Go across to the "VIA AIR" column. The charge would be $6.00.

27. (1) $0.50
Find the two charges. Then subtract to find the savings.

$200 via Air → $8.00
$100 via Surface → $4.25

```
2 pkg × $4.25 =     $8.50
1 pkg × $8.00 =   −  8.00
        savings of  $0.50
```

28. (1) $42\frac{1}{2}$

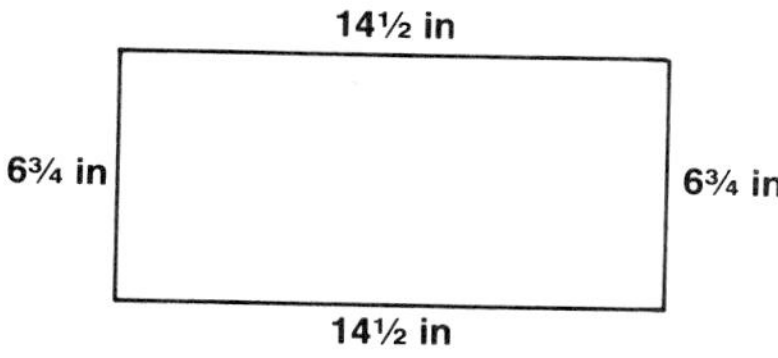

The length of framing needed will be the total of each side of the picture. Add these lengths together. (Be sure to give fractions a common denominator.)

$$\begin{aligned} 14\tfrac{1}{2} &= 14\tfrac{2}{4} \\ 14\tfrac{1}{2} &= 14\tfrac{2}{4} \\ 6\tfrac{3}{4} &= 6\tfrac{3}{4} \\ +\ 6\tfrac{3}{4} &= 6\tfrac{3}{4} \\ \hline & 40\tfrac{10}{4} = 42\tfrac{2}{4} = 42\tfrac{1}{2} \text{ in} \end{aligned}$$

29. (5) Not enough information is given.
You need the height of the can, it is one dimension of the label.

30. (3) 12
Use the formula for the area of a triangle.

Area $\triangle ABC = \frac{1}{2}(4)(6) = 12$

31. (4) 24
Since the answer is in square yards, change all measures to yards.
9 ft = 3 yds
12 ft = 4 yds
Then find the area by multiplying length times width.
3 × 4 = 12 sq yds
Multiply that by 2 since there are 2 carpets.
12 × 2 = 24 sq yds.

32. (2) 15
The final measurement is in cubic feet; change all measurements to feet. Then multiply length times width times depth to find the volume.
3 ft × 2 ft × 30 in
↓
3 ft × 2 ft × $2\frac{1}{2}$ ft

$$\frac{3}{1} \times \frac{\overset{1}{\cancel{2}}}{1} \times \frac{5}{\underset{1}{\cancel{2}}} = 15 \text{ cubic feet}$$

33. (1) 30°
Angle A and the 30° angle are alternate exterior angles, so they are equal.

34. (1) 30°
Angle B and the 30° angle are corresponding angles, so they are equal.

35. (3) 10
Use the Pythagorean Theorem to find the length of the leg.

$$\begin{aligned} c^2 &= a^2 + b^2 \\ 26^2 &= 24^2 + b^2 \\ 676 &= 576 + b^2 \\ 100 &= b^2 \\ \sqrt{100} &= b \\ 10 \text{ yd} &= b \end{aligned}$$

36. (2) 9
Use the formula $A = \pi r^2$

$$\begin{aligned} \frac{254.34}{3.14} &= \frac{3.14}{3.14} \times r^2. \\ 81 &= r^2 \\ \sqrt{81} &= r \\ 9 \text{ in} &= r \end{aligned}$$

37. (1) 25°
Note two sides are of equal length. The angles opposite these sides must be equal.
$\angle A = \angle C$
So $\angle C = 25°$

38. (2) 12
Since triangles ABC and CEF are both right triangles and share angle C, they are similar. Set up a proportion.

$$\frac{EF}{3} = \frac{28}{7}$$
$$EF = \frac{28}{7}(3) = 4\,(3) = 12$$

39. (1) 12.5
These right triangles are similar. Set up a proportion.

$$\frac{5}{12} = \frac{x}{30}$$
$$x = \frac{5}{\underset{2}{\cancel{12}}}(\overset{5}{\cancel{30}})$$
$$x = \frac{5}{2}(5)$$
$$x = \frac{25}{2}$$
$$x = \frac{25}{2} = 12.5 \text{ ft}$$

40. (4) 68
Use the Pythagorean Theorem to find the length of the string (hypotenuse).

$$\begin{aligned} a^2 + b^2 &= c^2 \\ (32)^2 + (60)^2 &= c^2 \\ (32 \times 32) + (60 \times 60) &= c^2 \\ 1{,}024 + 3{,}600 &= c^2 \\ 4{,}624 &= c^2 \\ \sqrt{4{,}624} = 68 &= c \end{aligned}$$

41. (4) 79
3^4 means 3 multiplied by itself 4 times. Carry out the multiplication, then change subtraction to addition and add.

$$\begin{aligned} &3^4 - 2 \\ &3^4 + -2 \\ (3 \times 3 \times &3 \times 3) - 2 \\ &81 - 2 = 79 \end{aligned}$$

42. (1) 48
Simply substitute numbers for the letters and solve.

$$\begin{aligned} a^2\,(b - c) &= \\ (-4)^2\,(5 - 2) &= \\ (16)\,(3) &= 48 \end{aligned}$$

43. (2) −1
Substitute the given values for x, y, and z into the equation.

$$\begin{aligned} &x + 4(y - z) \\ &(3) + 4(2 - 3) \\ &(3) + 4(2 + -3) \\ &(3) + 4(-1) \\ &(3) + (-4) = -1 \end{aligned}$$

44. (1) 4.5(9.50) + 37.80
First multiply the labor charge per hour by the number of hours, then add the cost of all the materials.
(4.5 hours × 9.50 per hour) + 37.80 for materials

OR

4.5(9.50) + 37.80

45. (5) $\frac{3(90) + 2(70)}{5}$

The laborer earns \$90 a day for 3 days, or 3(90) dollars, plus \$70 a day for 2 days, or 2(70). His total earnings are 3(90) + 2(70). Since he worked 5 days a week, his average wage per day is

$$\frac{3(90) + 2(70)}{5}$$

46. (3) 5
Count units along the vertical grid line that passes through C. The number of units from C to line AB is 5.

47. (3) 7
Substitute the choices given for the value of *a* in the equation to find the one which will fit or make the equation true.

$8a - 9 = 47$

Substitute 7: $8(7) - 9 = 47$
$56 - 9 = 47$
$56 + -9 = 47$
$47 = 47$

TRUE, so 7 is the value of a.

Or, you could solve the equation for *a*.

$8a - 9 = 47$
$8a + -9 = 47$
$8a + -9 = 47$
Add 9: $+9 = +9$
$8a = 56$

Divide by 8: $\frac{8a}{8} = \frac{56}{8}$
$a = 7$

48. (5) Not enough information is given.
The amount Mr. Sanchez earns is needed.

49. (5) $\frac{10w + 6y}{5 + y}$
$xy + 5x - 6y = 10w$
To solve for *x*, gather all the *x* terms alone on one side of the equation, then get *x* itself alone.
$xy + 5x = 10w + 6y$
$x(y + 5) = 10w + 6y$
$x = \frac{10w + 6y}{y + 5}$

50. (2) 9
Let the first number be *x*. Then the next ones, in order, are $x + 1, x + 2, x + 3$, and $x + 4$.
$x + (x + 1) + (x + 2) +$
$(x + 3) + (x + 4) = 55$
$5x + 10 = 55$
$5x = 45$
$x = 9$

The numbers are 9,10,11,12,13.

51. (4) 6
Substitute the value of *y* in the second equation into the first, and then solve for *x*.

If $x = 2y, y = \frac{x}{2}$

Substitute: $2x + 4(\frac{x}{2}) + 7 = 31$

$2x + 2x + 7 = 31$
Combine: $4x = 24$
Divide: $x = 6$

52. (2) $x < -\frac{3}{28}$
$\frac{1}{3}x + 5 < 4 - 9x$
First clear fractions.
$3(\frac{1}{3})x + 3(5) < 3(4) - 3(9x)$
$x + 15 < 12 - 27x$
$28x + 15 < 12$
$28x < -3$
$x < -\frac{3}{28}$

53. (2) Multiply by 3
Remember the long fraction bar is a grouping symbol. You must clear the 3 from the denominator to solve for *x*, which is in the numerator.

54. (1) −4 and 1
Substitute the options given for *d* into the equation. −4 and 1 fit, or make the equation true.
$d^2 + 3d - 4 = 0$
$(-4)^2 + 3(-4) - 4 = 0$
$16 + (-12) - 4 = 0$
$16 + (-12) + (-4) = 0$
$16 + -16 = 0$
$0 = 0$ TRUE

$d^2 + 3d - 4 = 0$
$(1)^2 + 3(1) - 4 = 0$
$1 + 3 - 4 = 0$
$1 + 3 + (-4) = 0$
$4 + 4 = 0$
$0 = 0$ TRUE

55. (4) $-\frac{1}{3}$ or 1
Always try factoring first for a quadratic equation. To factor, you must have zero alone on one side.
$3x^2 - 2x = 1$
$3x^2 - 2x - 1 = 0$
$(3x + 1)(x - 1) = 0$
$3x + 1 = 0$ OR $x - 1 = 0$
$3x = -1,$
$x = -\frac{1}{3}$ OR $x = 1$

56. (5) -3 or $\frac{4}{3}$
Before you can factor, you must get zero alone on one side.
$3x^2 + 5x = 12$
$3x^2 + 5x - 12 = 0$
$(3x - 4)(x + 3) = 0$

Then either
$3x - 4 = 0$ OR $x + 3 = 0$
$3x = 4$ $x = -3$
$x = \frac{4}{3}$

Answer Key

1.	(2)	**29.**	(5)
2.	(4)	**30.**	(3)
3.	(3)	**31.**	(4)
4.	(5)	**32.**	(2)
5.	(4)	**33.**	(1)
6.	(4)	**34.**	(1)
7.	(3)	**35.**	(3)
8.	(4)	**36.**	(2)
9.	(1)	**37.**	(1)
10.	(1)	**38.**	(2)
11.	(3)	**39.**	(1)
12.	(2)	**40.**	(4)
13.	(3)	**41.**	(4)
14.	(2)	**42.**	(1)
15.	(5)	**43.**	(2)
16.	(3)	**44.**	(1)
17.	(3)	**45.**	(5)
18.	(1)	**46.**	(3)
19.	(3)	**47.**	(3)
20.	(4)	**48.**	(5)
21.	(3)	**49.**	(5)
22.	(5)	**50.**	(2)
23.	(2)	**51.**	(4)
24.	(3)	**52.**	(2)
25.	(1)	**53.**	(2)
26.	(5)	**54.**	(1)
27.	(1)	**55.**	(4)
28.	(1)	**56.**	(5)

Index

When a word and page number are in **bold** type, the word is defined in a Coming to Terms on that page.